Drug-like Properties: Concepts, Structure Design and Methods

Drug-like Properties: Concepts, Structure Design and Methods: *from ADME to Toxicity Optimization*

Edward H. Kerns

and

Li Di

AMSTERDAM • BOSTON • HEIDELBERG • LONDON
NEW YORK • OXFORD • PARIS • SAN DIEGO
SAN FRANCISCO • SINGAPORE • SYDNEY • TOKYO

Academic Press is an imprint of Elsevier

Academic Press is an imprint of Elsevier
30 Corporate Drive, Suite 400, Burlington, MA 01803, USA
525 B Street, Suite 1900, San Diego, California 92101-4495, USA
84 Theobald's Road, London WC1X 8RR, UK

This book is printed on acid-free paper. ♾

Library of Congress Cataloging-in-Publication Data

Kerns, Edward.
  Drug-like properties: concepts, structure design, and methods: from ADME to toxicity optimization/Edward Kerns and Li Di.—1st ed.
    p. ; cm.
  Includes bibliographical references and index.
  ISBN-13: 978-0-12-369520-8 (alk. paper)   1. Pharmaceutical chemistry.
    2. Drugs—Structure-activity relationships.   3. Drug development.
    4. Drugs—Design.   I. Di, Li.   II. Title.
  [DNLM: 1. Drug Design.   2. Drug Evaluation, Preclinical.   3. Drug Toxicity.
    4. Pharmaceutical Preparations—metabolism.   5. Pharmacokinetics.
    6. Structure-Activity Relationship. QV 744 K395d 2008]
RS420.K47 2008
615'.19—dc22

                                                                                2007035586

British Library Cataloguing-in-Publication Data
A catalogue record for this book is available from the British Library.

ISBN: 978-0-1236-9520-8

For information on all Academic Press publications
visit our Web site at www.books.elsevier.com

Printed in the United States of America
   08   09   10   10 9 8 7 6 5 4 3 2

Contents

Preface

Drug research is a fulfilling career, because new drugs can improve human health, quality of life, and life span. For scientists dedicated to drug research, it can also be a supremely challenging mission, owing to the numerous attributes that must be simultaneously optimized to arrive at an efficacious drug-like compound. ADME/Tox (absorption, distribution, metabolism, elimination, toxicity) is one of these challenges. Of the thousands of novel compounds that a drug discovery project team invents and that bind to the therapeutic target, typically only a fraction of these have sufficient ADME/Tox properties to become a drug product. This book is devoted to providing you, the drug research scientist or student, with an introduction to ADME/Tox property concepts, structure design, and methodology to help you succeed with these challenges.

Chemists will be aided by the case studies, structure-property relationships, and structure modification strategies in this book. These assist in diagnosing the substructures of a lead structure that are not drug-like and suggest ideas for ADME/Tox structure design. Overviews of property methods provide the background needed to accurately interpret and apply the data for informed decisions. For ADME/Tox scientists, insights on property assays assist with selecting methods and generating data that impacts projects.

Biologists/pharmacologists will benefit from an increased understanding of ADME/Tox concepts. This is especially important, because in recent years the application of property data has expanded from optimizing *in vivo* pharmacokinetics and safety to biological assays. Low solubility, chemical instability, and low permeability can greatly affect bioassay data. Equipped with this understanding, biologists are better able to optimize bioassays and include property affects in data interpretation.

Accordingly, understanding ADME/Tox is important for all drug researchers, owing to its increasing importance in advancing high quality candidates to clinical studies and the processes of drug discovery. ADME/Tox properties are a crucial aspect of clinical candidate quality. If the properties are weak, the candidate will have a high risk of failure or be less desirable as a drug product. ADME/Tox has become integrated in the drug discovery process and is a tremendous asset in guiding selection and optimization of precious leads. This book is a tool and resource for scientists engaged in, or preparing for, the selection and optimization process. The authors wish you success in creating the pharmaceuticals of the future that will benefit all people.

In preparing this book, the authors had the support and counsel of many drug research colleagues. The leadership of Magid Abou-Gharbia, Guy T. Carter, and Oliver J. McConnell of Wyeth Research, Chemical and Screening Sciences are greatly appreciated. The careful manuscript review and feedback by Christopher P. Miller was highly beneficial. The thoughtful comments of several anonymous reviewers are greatly appreciated. LD thanks

Prof. Donald M. Small, Prof. Bruce M. Foxman, and Prof. Ruisheng Li for guidance. EK thanks Prof. David M. Forkey, William L. Budde, and Charles M. Combs for mentorship. We thank Prof. Ronald T. Borchardt and Christopher A. Lipinski for their friendship, collaboration, and leadership in the ADME/Tox and medicinal chemistry fields. The enthusiastic feedback of students in the American Chemical Society short course on Drug-like Properties was highly valuable. The collaborative adventure of understanding drug-like properties in drug discovery was shared with numerous Wyeth Research colleagues in Pharmaceutical Profiling and Medicinal Chemistry and their respectful, innovative collaboration is greatly appreciated.

Dedication

Ed Kerns dedicates this book to:
William, Virginia, Nancy, Chrissy, and Patrick: for your love and support.

Li Di dedicates this book to:
My parents: I am infinitely in debt to you.
My sisters, Ning and Qing: for being my best friends.
My children, Kevin and Sophia: I am very proud of you.

Part 1
Introductory Concepts

Chapter 1

Introduction

Overview

> ▶ *Drug-like properties confer good ADME/Tox characteristics to a compound.*

> ▶ *Medicinal chemists control properties through structure modification.*

> ▶ *Biologists use properties to optimize bioassays and interpret biological experiments.*

Drug discovery is an exceedingly complex and demanding enterprise. Discovery scientists must pursue multiple lines of investigation involving diverse disciplines, often with conflicting goals, and integrate the data to achieve a balanced clinical candidate. In recent years there has been considerable discussion of the importance of optimizing the absorption, distribution, metabolism, excretion, and toxicity (ADME/Tox) properties (e.g., physicochemical, metabolic, toxicity) of compounds in addition to pharmacology (e.g., efficacy, selectivity) to increase drug discovery success. Christopher A. Lipinski[1] has commented:

> Drug-like is defined as those compounds that have sufficiently acceptable ADME properties and sufficiently acceptable toxicity properties to survive through the completion of human Phase I clinical trials.

Drug properties have always been a prominent component of the *development* phase, after *discovery*, during which detailed studies are performed on formulation, stability, pharmacokinetics (PK), metabolism, and toxicity. However, in recent years it is has become imperative to integrate drug properties into drug discovery research. Ronald T. Borchardt[2] has commented:

> . . . drug-like properties are . . . intrinsic properties of the molecules and it is the responsibility of the medicinal chemists to optimize not only the pharmacological properties but also the drug-like properties of these molecules.

This integration enables optimization of drug discovery leads for ADME/Tox during drug discovery. Comprehensive, simultaneous optimization of in vivo pharmacology, pharmacokinetics, and safety has been an important advance.

Another important advance has been the recognition that properties have a major effect on the performance of drug research biological experiments. Low solubility, permeability, or stability in assay media alters the biological data used to develop structure–activity relationships (SARs), a key aspect of drug discovery. Biologists use property data to optimize bioassays, dosing vehicles, and in vivo routes of administration. Thus, drug-like properties have become important for discovery biological research.

The various drug properties, terminology, and assays can be overwhelming to drug discovery scientists and students without sufficient introduction. Some texts on drug properties are daunting because they are written from the perspective of experts in pharmaceutics or metabolism and contain detail and math that are not useful for discovery scientists. This book is a practical guide for medicinal chemists, biologists, managers, and students. It provides background material and real-world, practical examples for practicing discovery scientists who need to make sense of the data and arrive at informed decisions.

This book provides tools for working with drug-like properties. First, the interactions of drug molecules with the in vivo barriers they encounter after oral administration are described, in order to understand why properties limit drug exposure to the therapeutic target. Next, each key drug property is explored (Figure 1.1) in terms of

1. Fundamentals of each property

2. Effects of each property on ADME/Tox and biological experiments

3. Structure–property relationship (SPR) case studies, to see how structure affects properties

4. Structure modification strategies, to guide property optimization

5. Property method descriptions, for accurate measurement and application of data

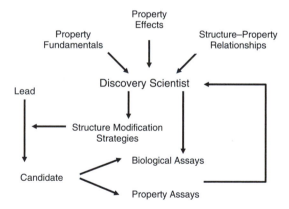

Figure 1.1 ▶ This book provides discovery scientists and students with a practical understanding of property fundamentals, effects, structure–property relationships, and structure modification strategies that can be applied to improving leads and bioassays. Quality property assays and data are critical for making informed decisions.

Knowledge of these properties equips discovery scientists for increased effectiveness in lead selection, optimization, and enhancement of discovery biology.

Property-related concepts are described with a minimum of math and with emphasis on practical application. Specific property applications in diagnosis of poor pharmacokinetics, design of prodrugs, formulation for in vivo dosing, and strategies for applying property data are discussed.

Drug discovery has diverse elements that must be delicately integrated and balanced. Drug-like properties are important characteristics of quality clinical candidates.

◻ Problems

(Answers can be found in Appendix I.)

1. Define the term *drug-like*.

2. What are two major lead optimization areas in drug discovery?

3. How can understanding compound properties assist discovery biologists?

4. Compound properties can affect which of the following?: (a) pharmacokinetics, (b) bioavailability, (c) IC_{50}, (d) safety.

◻ References

1. Lipinski, C. A. (2000). Drug-like properties and the causes of poor solubility and poor permeability. *Journal of Pharmacological and Toxicological Methods, 44*, 235–249.

2. Borchardt, R. T. (2004). Scientific, educational and communication issues associated with integrating and applying drug-like properties in drug discovery. In R. Borchardt, E. Kerns, C. Lipinski, D. Thakker, & B. Wang (Eds.), *Pharmaceutical profiling in drug discovery for lead selection*, pp. 451–466. Arlington, VA: AAPS Press.

Advantages of Good Drug-like Properties

Overview

▶ *Structural properties determine in vivo pharmacokinetics and toxicity.*

▶ *Inefficient research, attrition, and costs are reduced if compounds have good properties.*

▶ *ADME/Tox property assessment and optimization are important aspects of drug discovery.*

▶ *Optimal clinical candidates have a balance of activity and properties.*

2.1 Drug-like Properties Are an Integral Part of Drug Discovery

Drug discovery is continuously advancing as new fundamental knowledge, methods, technologies, and strategies are introduced. These new capabilities result in changes in the discovery process. For example:

▶ Pharmacology screening has changed from direct testing in living systems to in vitro high-throughput screening.

▶ Initial leads (hits) for optimization have changed from natural products and natural ligands to large libraries of diverse structures.

▶ Compound design has been enhanced from structure–activity relationships by the addition of x-ray crystallography and NMR binding studies and computational modeling.

▶ Lead optimization chemistry has been enhanced from one-at-a-time synthesis by the addition of parallel synthesis.

▶ Traditional sequential experiments have been enhanced with parallel experiments, such as microtiter plate formats.

Drug discovery is constantly reevaluating itself in order to advance in speed, efficiency, and quality and thus remain successful.

Drug-like property optimization is another area of drug discovery advancement. It offers significant opportunities for enhancing discovery success. This book focuses on the fundamental knowledge, methods, and strategies of absorption, distribution, metabolism, excretion, and toxicity (ADME/Tox) and how structures can be optimized. As background for this information, this chapter describes how optimization of ADME/Tox has progressed.

The term *drug-like* captures the concept that certain properties of compounds are most advantageous in their becoming successful drug products. The term became commonly used following the pivotal work of Lipinski and colleagues.[1] Their work examined the structural properties that affect the physicochemical properties of solubility and permeability and their effect on drug absorption. Since that article, the term *drug-like properties* has expanded and been linked with all properties that affect ADME/Tox.

2.1.1 Many Properties Are of Interest in Discovery

Drug-like properties are an integral element of drug discovery projects. Properties of interest to discovery scientists include the following:

- ▶ Structural properties
    - ▶ Hydrogen bonding
    - ▶ Lipophilicity
    - ▶ Molecular weight
    - ▶ pK_a
    - ▶ Polar surface area
    - ▶ Shape
    - ▶ Reactivity
- ▶ Physicochemical properties
    - ▶ Solubility
    - ▶ Permeability
    - ▶ Chemical stability
- ▶ Biochemical properties
    - ▶ Metabolism (phases I and II)
    - ▶ Protein and tissue binding
    - ▶ Transport (uptake, efflux)
- ▶ Pharmacokinetics (PK) and toxicity
    - ▶ Clearance
    - ▶ Half-life
    - ▶ Bioavailability
    - ▶ Drug–drug interaction
    - ▶ LD_{50}

The structure determines the compound's properties (Figure 2.1). When these structural properties interact with the physical environment, they cause physicochemical properties (e.g., solubility). When these structural properties interact with proteins, they cause biochemical properties (e.g., metabolism). At the highest level, when the physicochemical and biochemical properties interact with living systems they cause PK and toxicity. Medicinal chemists control the PK and toxicity properties of the compound by modifying the structure.

Figure 2.1 ▶ Compound structure determines the fundamental properties that determine physicochemical and biochemical properties, which ultimately determine pharmacokinetics and toxicity.

2.1.2   Introduction to the Drug Discovery and Development Process

Before exploring how properties affect drug candidates, it is useful to briefly review the process of drug discovery and development. New drug candidates are found during the discovery stage (Figure 2.2). They then enter clinical development and, if approved by the Food and Drug Administration (FDA), become drug products that are used in patient therapy. The major activities in each stage are listed in Figure 2.2. This book focuses on the discovery stage. However, the later stages impose stringent drug-like requirements on the properties of candidates. Thus, it is necessary to anticipate these requirements during drug discovery and promote to development only those compounds that have the highest chances of success.

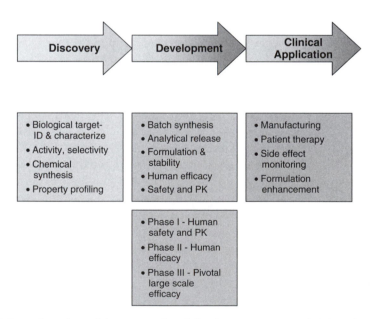

Figure 2.2 ▶ Overview of drug research and development stages and their major activities.

Drug discovery is diagramed in greater detail in Figure 2.3. In general, successive stages involve increasing depth of study and more stringent advancement criteria. The discovery screening process initially casts a broad net, to explore diverse pharmacophore structural

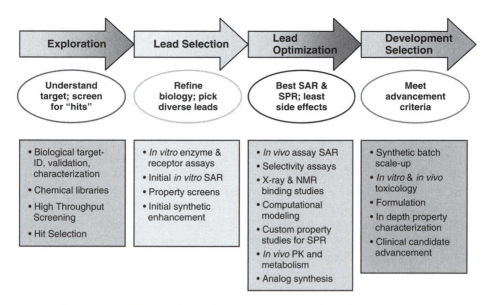

Figure 2.3 ▶ Stages of drug discovery, primary goals, and major activities.

space. It then narrows these possibilities to select a few lead scaffolds (templates). These are structurally modified to explore SARs, the cornerstone of modern drug discovery, during the lead optimization stage. Finally, candidates for development are subjected to in-depth studies to qualify or disqualify them for development.

2.1.3 Development Attrition is Reduced by Improving Drug Properties

Much of the early history of drug discovery focused on finding active compounds. Issues such as PK, toxicity, solubility, and stability were addressed during the development phase. In 1988 a pivotal paper on the reasons for failure of drugs in development revealed a startling problem.[2] Approximately 39% of drugs were failing in development because of poor biopharmaceutical properties (PK and bioavailability). With the high cost of development, this failure represented a major economic loss for the companies. Furthermore, years of work on discovery and development were lost, and the introduction of a new drug product was delayed.

This great need for enhancement was actively addressed by adding resources to assess biopharmaceutical properties during late discovery. Sorting out the compounds with acceptable properties at this stage did not require the rigorous methods applied during development. Thus, for this task, methods used during development were adapted to use fewer resources and to operate at higher throughput. Criteria were relaxed to reflect the reduced accuracy and precision of the revised methods and the lower level of detail needed for decisions at this stage. The assessment of PK was implemented in the late-discovery/predevelopment stage. This testing succeeded in keeping poor candidates from progressing into development and reduced development attrition.

2.1.4 Poor Drug Properties Also Cause Discovery Inefficiencies

Once late-discovery biopharmaceutical assessment was in place and an attrition burden was lifted from development, another discovery need was revealed. Candidates that were failing in late discovery because of poor properties still caused a great burden on drug discovery. Failure late in discovery meant that the project to discover a new drug had lost valuable time and resources on the failed candidate and had to start over. This recognition led to the implementation of property assessment even earlier in discovery so that such losses would be reduced. In pharmaceutical companies, implementation has been accomplished by different approaches. In one approach, higher-throughput animal PK capabilities are added earlier in discovery in order to screen more compounds in vivo for PK. This strategy measures the key PK properties that can predict in vivo candidate ADME success. In a second strategy, higher-throughput in vitro property assays are used. These assays measure fundamental physicochemical and biochemical properties, such as solubility, permeability, and metabolic stability, which determine higher-level properties, such as PK. In vitro studies require fewer resources and animals per compound than do PK studies, so more compounds can be assessed using in vitro assays. Also, physicochemical and biochemical properties, determined with in vitro methods, can be more useful to medicinal chemists in deciding how to modify structures to improve properties.[3-5] Medicinal chemists can correlate physicochemical and biochemical properties with structural features more closely than with PK properties. Physicochemical and biochemical methods typically measure a single property (e.g., passive diffusion permeability). On the other hand, PK properties result from multiple variables operating in a dynamic manner, and they do not indicate which discrete structural modifications to make. Most pharmaceutical companies use a combination of these two strategies during discovery.

As a result of these enhancements of discovery, the property-induced failure of compounds in development declined dramatically from 39% in 1988 to 10% in 2000.[6] Figure 2.4 shows that pharmaceutical companies have been successful in improving the biopharmaceutical properties of development candidates. The 2002 study also suggests that other property issues (toxicity, formulation) still are challenges.

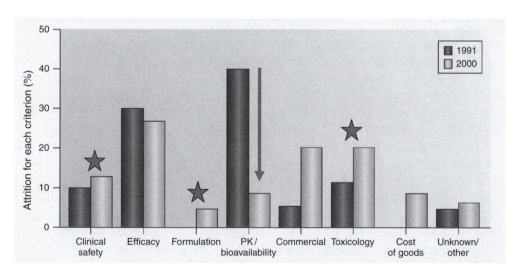

Figure 2.4 ▶ Between 1991 and 2000, development attrition due to pharmacokinetics (PK) and bioavailability was greatly reduced. Toxicology, clinical safety, and formulation remain significant drug-like property issues. (Reprinted with permission from [6].)

2.1.5 Marginal Drug Properties Cause Inefficiencies During Development

Although the rate of outright candidate failure in development has decreased due to early termination of candidates with inadequate properties, candidates with marginal properties still progress into development. Even though they might not fail in development, they impose significant inefficiencies on development by increasing development costs and prolonging development time lines.

For example, compounds with poor solubility and stability usually require a longer development time line and more resources, owing to more difficult formulation development, stability testing, and dissolution studies. Sophisticated formulations can improve the dissolution rate and reduce active compound degradation. It is tempting for discovery scientists to shift the burden to development pharmaceutics scientists who will fix the problems by using sophisticated formulations. Although this may be an acceptable choice for new first-in-class therapies, for other drug products it can impose a burden on development resources and delay the introduction of a new drug product.

For a new drug that would produce hundreds of millions of dollars of sales in its first year, $5 to $10 million of sales are potentially lost for each week of delay in discovery or development. Furthermore, if a patent has been filed, each week of delay could result in 1 less week of patent exclusivity during the time of highest sales for the inventing company. Thus, real economic considerations drive enhancement of compound quality.

Another result of poor-to-mediocre properties is that the patient may take on a greater burden. For example, if the drug is poorly absorbed, the dose must be increased to reach therapeutic levels. The dosing regimen might need to be shifted from oral to intravenous, which is not acceptable for dosing among the wide patient population. If the drug has a short half-life in vivo because of metabolic instability, then the drug must be dosed more frequently. Patients are less likely to consistently self-administer drugs that require higher and more frequent daily doses. Once-per-day dosing of a solid dosage form by mouth is preferable. Pharmaceutical companies and academic laboratories have a strong commitment and mission to enhance the quality and length of patient life; thus, patient burdens, needs, and benefits are a primary focus.

In most cases, it is more advantageous to try to improve drug-like properties (e.g., solubility, stability, and permeability) during discovery. This is best accomplished by modifying the chemical structure. Modifications usually are performed at sites in the molecule that are shown by SAR to not be critical for therapeutic target binding. In some cases, the structural requirements for ligand binding to the target do not permit structural modifications to improve properties. Under these conditions, discovery scientists and managers must decide if the drug candidate still has viability as a drug product. Attention to properties and a workflow that includes property optimization during discovery allow for the best chance of discovering a candidate that combines all of the qualities of a successful drug product.

Some people have commented retrospectively that if properties had been assessed in the past, then some of our current drug products with poor properties would not have become available for clinical therapy. It is true that some current drugs have poor properties and may not have been submitted to the FDA under current criteria. However, it is widely recognized that early property assessment and optimization provide the opportunity for earlier correction of property limitations. If the current property awareness and assessment had been available at the time of discovery of those drugs with poor properties, then better structural analogs having comparable potency without the property limitations may have been discovered. In this way, even better drugs with reduced patient burden and costs may become available sooner.

2.1.6   Poor Properties Can Cause Poor Discovery Research

In addition to development problems, poor properties can cause problems during drug discovery. Once property data became available during discovery, their value to discovery in ways other than PK began to be recognized. We now know that when discovery project teams encounter unexplained problems, some of the problems are due to poor properties.[3,7,8]

Following are examples of how poor drug properties can reduce the quality of drug discovery biological research:

▶ Low or inconsistent bioactivity responses for in vitro bioassays can be due to precipitation, owing to low solubility of the compound in the bioassay medium or in dilutions prior to the assay.

▶ Low activity in bioassays may be due to chemical instability of the compound in the test matrix.

▶ An unexpectedly high drop in activity can result when transitioning from enzyme or receptor activity assays to cell-based assays. This can be due to poor permeability of the compounds through the cell membrane, which must be penetrated for the compound to reach intracellular targets.

▶ Compounds may be unstable or insoluble in the DMSO solutions that are stored in microtiter plates and experience freeze–thaw cycles, or they may be exposed to various physicochemical conditions in the laboratory.

▶ Poor efficacy of a central nervous system (CNS) drug in vivo may be due to poor penetration of the blood–brain barrier.

▶ Poor efficacy in vivo may be due to low concentrations in the plasma and target tissue because of poor PK, low bioavailability, or instability in the blood.

These effects of poor properties may go unrecognized if discovery scientists are unaware of them and are not vigilant in checking for the effects or ensuring that discovery experiments are designed and interpreted to account for the properties. Poor properties can limit exposure of the compound to the target protein in the discovery biological experiment. This property effect may be misinterpreted as an actual SAR, and a valuable pharmacophore may be overlooked. If the potential effects of poor properties on bioassays are taken into consideration, then the active pharmacophore may be rescued by testing under more appropriate conditions to obtain accurate biological data. Structural modification then can improve the deficient property.

2.2 Changing Emphasis on Properties in Discovery

In the past, the focus on binding to the active site of the target protein has been a strong priority in discovery for medicinal chemistry. Exploration of SAR by synthesis of analogs having systematic modifications of the core structural scaffold has allowed optimization of binding by orders of magnitude. It is so important for discovery project teams to arrive at a potent candidate that other considerations, such as drug-like properties, might be given comparatively less attention.

There is a strong emphasis on driving potency from the micromolar IC_{50} range of high-throughput screening (HTS) hits to the low nanomolar range of good clinical candidates. However, if the focus is solely on activity, the research team can arrive at a candidate with properties that are worse than the original HTS hit. For example, the candidate may be too polar to penetrate the blood–brain barrier and reach the intended CNS target, it may be unstable and rapidly cleared by first-pass metabolism, or it may be too insoluble to be absorbed from the intestine. These findings may hopelessly misdirect a discovery program.

Once nanomolar activity is obtained, it is hard to go back and fix properties by structural modifications because it may be necessary to modify the substructures that were added in order to enhance binding affinity.

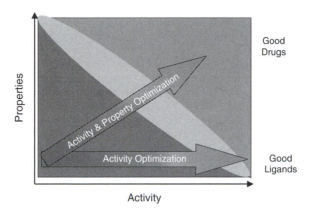

Figure 2.5 ▶ Changing strategy for drug candidates, from a focus on activity to balanced attention to activity and properties. (Reprinted with permission from [5].)

This situation is demonstrated in Figure 2.5. A primary focus on activity can yield compounds that are very effective as ligands for the target protein, but the properties may be inadequate for the compounds to become successful drugs. For example, increased lipophilicity can enhance target protein binding; however, it also can reduce solubility and metabolic stability. Balanced attention to both activity and properties (holistic approach) yields candidates that can become good drugs. The balanced approach suggested by Figure 2.6 now is common in drug discovery. Good activity and drug-like properties are complementary, and both are necessary for a good drug product. The most active or selective compound may not make the best drug product because of property limitations that cause poor PK or safety. A less potent compound with better properties may produce a better in vivo therapeutic response and be a better drug product for patients. As in the sport decathlon, the candidate is tested by many events/challenges, and it is the combined performance that determines success, not being the best in individual events.

Figure 2.6 ▶ Pharmaceuticals balance activity and properties. (Reprinted with permission from [11].)

The multitude of challenges faced by discovery scientists has been variously characterized. One useful image is to characterize them as a series of hurdles that a compound must pass.[9] Another useful analogy is juggling (Figure 2.7). A diverse ensemble of crucial elements must be simultaneously monitored and kept in balance in order to achieve success. Neglecting one element can cause the whole ensemble to crash.

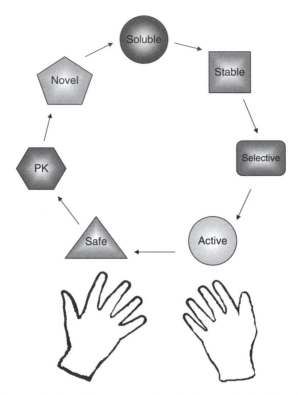

Figure 2.7 ► Success in drug discovery requires simultaneously juggling diverse variables.

2.3 Property Profiling in Discovery

In the past, it was not possible to systematically enhance drug properties during drug discovery because the data were not available. Methods were not in place to provide discovery scientists with the necessary information. If methods were available, they were too slow to provide data on a time schedule that was useful for medicinal chemists. Few resources were devoted to this activity, so data were not available for many compounds. The data were not sufficiently comprehensive for all of the major properties that had to be addressed. If data were available, the data quality either was low or unknown, and discovery scientists did not trust the data for making crucial decisions.

With the increased emphasis on drug-like properties in discovery came an infusion of property knowledge from development colleagues (pharmaceutics, metabolism, toxicology, PK, process, analytical). Experts in these disciplines assisted discovery scientists with understanding and measuring properties. However, discovery applications require distinctly different perspectives and strategies than does development because of the differences in goals and activities of the stages.[9]

Current methods for property prediction and measurement are discussed in Chapters 22 to 37. Those chapters provide information on the various tools available for property assessment. They also provide insight on how data are produced by drug-like property profiling colleagues for project teams. Knowledge of methods and the data they produce leads to better interpretation and application of property data by discovery scientists. The method information also allows chemists to decide on which methods to implement in their organization.

2.4 Drug-like Property Optimization in Discovery

This book provides resource material for medicinal chemists, discovery biologists, managers, and students who are interested in integrating pharmaceutical properties into their selection and optimization of leads and candidates.

A new strategy introduced into discovery is structure–property relationships (SPRs). This is complementary to SAR. The structures of compounds are correlated to their property performance. SPR allows medicinal chemists to understand how structural modifications improve properties for their scaffold. Thus, the established strategy of structure-based design is supplemented with the new strategy of "property-based design" by van de Waterbeemd et al.,[10] the study and modification of structure to achieve property improvement.

There are many reasons for a drug discovery project team to strive toward selecting leads with good drug-like properties and optimizing properties for their compound series during drug discovery. Property optimization can be approached in balance with activity and selectivity optimization. The advantages of good drug-like properties include the following:

▶ Better planning, execution, and interpretation of discovery experiments

▶ Reduced discovery time lag from not having to fix property-based problems at a later time

▶ Faster and more economical pharmaceutical development

▶ Candidates with lower risk and higher future value

▶ Longer patent life

▶ Higher patient acceptance and compliance

Problems

(Answers can be found in Appendix I.)

1. How do medicinal chemists control compound properties?

2. In addition to structure, what determines physicochemical (e.g., solubility) and biochemical properties?

3. How can drug-like properties be used in each stage of drug discovery (Figure 2.3)?

4. What assays are available to discovery scientists for property assessment and optimization?

5. How do poor properties affect development, clinical application, and product lifetime?

6. How do drug properties affect discovery biological experiments?

7. Define and describe SPR.

8. Which of the following are advantages of optimizing drug-like properties?: (a) better-quality drug product, (b) lower risk of failure, (c) faster and less expensive development, (d) lower cost of goods, (e) more reliable discovery biological data, (f) easier synthesis.

References

1. Lipinski, C. A., Lombardo, F., Dominy, B. W., & Feeney, P. J. (1997). Experimental and computational approaches to estimate solubility and permeability in drug discovery and development settings. *Advanced Drug Delivery Reviews, 23*, 3–25.

2. Prentis, R. A., Lis, Y., & Walker, S. R. (1988). Pharmaceutical innovation by the seven UK-owned pharmaceutical companies (1964–1985). *British Journal of Clinical Pharmacology, 25*, 387–396.

3. Di, L., & Kerns, E. H. (2003). Profiling drug-like properties in discovery research. *Current Opinion in Chemical Biology, 7*, 402–408.

4. Curatolo, W. (1998). Physical chemical properties of oral drug candidates in the discovery and exploratory development settings. *Pharmaceutical Science & Technology Today, 1*, 387–393.

5. Kerns, E. H., & Di, L. (2003). Pharmaceutical profiling in drug discovery. *Drug Discovery Today, 8*, 316–323.

6. Kola, I., & Landis, J. (2004). Opinion: Can the pharmaceutical industry reduce attrition rates? *Nature Reviews Drug Discovery, 3*, 711–716.

7. Lipinski, C. A. (2000). Drug-like properties and the causes of poor solubility and poor permeability. *Journal of Pharmacological and Toxicological Methods, 44*, 235–249.

8. Lipinski, C. A. (2001). Avoiding investment in doomed drugs. *Current Drug Discovery, 2001*, 17–19.

9. Venkatesh, S., & Lipper, R. A. (2000). Role of the development scientist in compound lead selection and optimization. *Journal of Pharmaceutical Sciences, 89*, 145–154.

10. van de Waterbeemd, H., Smith, D. A., Beaumont, K., & Walker, D. K. (2001). Property-based design: Optimization of drug absorption and pharmacokinetics. *Journal of Medicinal Chemistry, 44*, 1313–1333.

11. Kerns, E. H., & Di, L. (2002). Multivariate pharmaceutical profiling for drug discovery. *Current Topics in Medicinal Chemistry, 2*, 87–98.

Chapter 3

Barriers to Drug Exposure in Living Systems

Overview

▶ *Physiological barriers reduce the amount of dosed compound that reaches the target.*

▶ *Barriers include membranes, pH, metabolic enzymes, and transporters.*

▶ *Good properties enable good absorption, distribution, low metabolism, reasonable elimination, and low toxicity.*

Drugs encounter many barriers in living systems from the time the dosage form is administered until the time the drug molecules reach the therapeutic target.

The behavior of a drug molecule at each barrier is a direct result of the drug's chemical structure. The combined performance at all the barriers in the body (i.e., the portion that passes the barriers) determines the concentration of drug (exposure) at the therapeutic target. Along with inherent activity at the therapeutic target, this determines the drug's in vivo efficacy. Pharmacokinetics (PK) is often used as a surrogate for exposure. PK parameters are used to determine the dosage form and dosing regimen that are used in clinical practice.

Performance at a particular barrier (e.g., intestinal epithelium) may be due primarily to one or several drug properties (e.g., permeability, efflux transport). For a particular compound, one or two property deficiencies (e.g., metabolic stability, solubility) can greatly limit its PK performance. In discovery, we have the opportunity to make structural modifications that greatly improve the performance of the limiting properties, that is, to create a better drug product.

3.1 Introduction to Barriers

When a drug molecule encounters a barrier, the amount of drug that reaches the other side is diminished. The barrier concept is illustrated in Figure 3.1. Penetration of drug molecules to the therapeutic target is slowed and attenuated by the barrier. The behavior of the molecules at each of the barriers determines the rate at which the molecules progress to the target as well as their target exposure concentration at a particular time after dosing. If we are successful in optimizing the performance of a discovery lead series at these barriers, then we may achieve exposure of drug to the target that is consistent with sustainable efficacy.

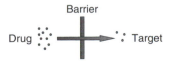

Figure 3.1 ▶ Model for drug barriers in living systems. Drug delivery to the therapeutic target is attenuated by barriers in the organism. (Reprinted with permission from [4].)

Barriers include a diverse ensemble of physicochemical and biochemical processes that drug molecules encounter. Examples of barriers include cell membranes, metabolic enzymes, solution pH, efflux transporters, and binding proteins.

It is important to remember that efficacy is driven by the inherent activity of the molecule at the target (e.g., IC_{50}) and by exposure (e.g., concentration and duration). In a discovery project, the lead structures are constrained by the substructures and orientations that are needed for binding to the target. If these constraints cause a compound to perform poorly at one or more of the barriers in a living system, then the compound might not achieve sufficient exposure at a safe dose and will not become a drug product.

This is the ying–yang relationship of successful drugs (see Figure 2.6). The process of drug discovery balances a relentless search for molecules that have structural features that produce:

1. Strong target binding using structure-based design and the structure–activity relationship (SAR)

2. High performance at in vivo barriers, using property-based design[1] and the structure–property relationship (SPR)

In the same way that design of structural features using SAR is known as *structure-based design*, the design of structural features using SPR has become known as *property-based design*.[1] How a medicinal chemist goes about balancing these often disparate processes is a matter of experience and strategy.

It is useful to explore the barriers to exposure in living organisms from the standpoint of the drug molecule. Its passage through the various organ systems and the physicochemical and biochemical environment it encounters is a fascinating journey. The molecule's physicochemical properties determine its behavior in solutions and at membrane barriers. Its binding and reactivity at particular enzymes determine its behavior at metabolic barriers. Its binding to various transporters and plasma proteins, as well as nonspecific binding to macromolecules throughout the body, affects its absorption, distribution, and excretion behaviors. The reactivity of the drug and its metabolites affects toxicity.

For the purposes of this introduction to barriers, it is useful to consider barriers in a linear sequence as the drug molecule moves toward the therapeutic target. In living systems, different molecules encounter different barriers at the same time, so the process is very dynamic and interactive. The various barriers also dynamically affect each other. For example, in the intestinal epithelium, efflux transporters remove some compounds from the cell, allowing metabolizing enzymes in the cell to operate in a concentration range of greater efficiency. As another example, a high rate of metabolism limits brain exposure, even if the compound has good blood–brain barrier permeability.

3.2 Drug Dosing

A common goal of pharmaceutical researchers is to develop a drug dosage form that is a low-dose tablet with a dosing regimen of oral administration once per day. Administration by mouth is termed *oral* and is abbreviated PO (*per os*). A drug product of this type has reasonable manufacturing and storage costs and high patient compliance. If a compound has limited performance at one or more in vivo barriers, it may have poor PK performance and will require adjustments to this approach, such as:

▶ More frequent dosing (if the half-life is short)

▶ Higher doses (if the bioavailability is low)

▶ Administration by a different route (if absorption is low)

▶ A different vehicle or formulation (if solubility is low)

If oral dosing does not produce sufficient exposure, another route of administration, such as intravenous, is necessary (Table 3.1; see Section 41.1). Nonoral routes are also used during discovery before properties are optimized for good absorption after oral dosing.

Formulations can be developed to increase the absorption of molecules. This is accomplished by increasing the solubility or dissolution rate of the drug product (see Chapter 41).

TABLE 3.1 ▶ Dosing Routes

Administration	Description	Abbreviation
Oral	Swallowed by mouth	PO
Intravenous	Injected directly into the bloodstream as a bolus (rapidly) or by infusion (continuously)	IV
Subcutaneous	Injected under the skin	SC
Transdermal	Applied as a patch or other device and transported through the skin	
Topical	Applied as a solution or suspension onto the skin	
Intramuscular	Injected into the muscle	IM
Epidural	Injected into the epidural space just inside the bone of the lower vertebrae	
Suppository	Placed in the rectum	
Intranasal	Sprayed into the nose	
Buccal	Tablet is held inside the mouth between cheek and gum until dissolved	
Sublingual	Tablet is held underneath the tongue until dissolved	
Intraperitoneal	Injected within the peritoneal (abdominal) cavity	IP

3.3 Barriers in the Mouth and Stomach

In oral dosing, the compound first encounters the mouth. A portion of the drug can be absorbed in the mouth if it stays in the mouth for some time. Buccal and sublingual dosing involve keeping the drug in the mouth. The drug is absorbed through the membranes of the mouth into blood capillaries.

The drug tablet is ingested via the esophagus and arrives at the initial portion of the gastrointestinal (GI) tract, the stomach (Figure 3.2). The drug tablet is broken into smaller particles by the aqueous environment and by movements of the stomach. Here the drug tablet encounters several physicochemical and stability barriers. The first barrier is its *dissolution rate* (see Sections 7.1.2 and 7.5). Another barrier is its solubility at the low pH of the stomach. Molecules must be in solution in order to diffuse to the membrane surface for absorption. Factors that affect solubility are discussed in Chapter 7. As solubility increases, the concentration of compound in solution and at the membrane surface increase. This is favorable for increased absorption. *Permeability* through the membrane is the next barrier for drug molecules in reaching systemic circulation. Higher permeability results in higher drug absorption.

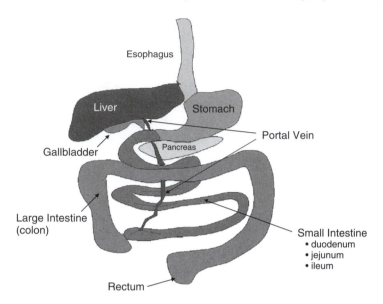

Figure 3.2 ▶ Diagram of the gastrointestinal tract. (see Plate 1)

For most drugs, absorption from the stomach is limited. This is because the stomach surface area is relatively low (approximately $1\,m^2$), and blood flow around the stomach (perfusion) is low. In addition, drug material does not stay in the stomach very long. The gastric emptying time varies from 0.5 to 1 hour in the fasted state to several hours after a heavy meal.

Another barrier to drug absorption from the stomach is the acidic solution. In the fasted state, stomach pH is between 1 and 2 and in the fed state is between pH 3 and 7. Compounds that have *chemical instability* at acid pH might be decomposed by hydrolysis reactions. Molecules continue to be exposed to acidic pH and potential acidic decomposition in the upper portions of the small intestine (see Section 3.4). Compounds also encounter *hydrolytic enzymes*. These enzymes naturally catalyze the breakdown of polymeric macromolecules to monomers as nutrients. Some drug molecules can bind to these enzymes and be catalyzed for hydrolysis. Solution stability in enzymatic and varying pH solutions is discussed in Chapter 13.

3.4   Gastrointestinal Tract Barriers

The stomach contents empty into the *duodenum*, the first region of the small intestine. Following regions are termed the *jejunum* and *ileum*, in sequence. The pH in the small intestine is higher than in the stomach, varying from pH 4.4 in the duodenum in the fasted state to pH 7.4–8 at the end of the ileum. The pH values of the intestinal regions are listed in Table 3.2. This progression of pH creates a pH gradient from the stomach through the small intestine. The transit time is the amount of time available for drugs to be absorbed in that region.

TABLE 3.2 ▶ **pH Values and Transit Times of Gastrointestinal Tract Regions**

GI tract region	Average pH, fasted	Average pH, fed	Transit time (h)
Stomach	1.4–2.1	3–7	0.5–1
Duodenum	4.4–6.6	5.2–6.2	
Jejunum	4.4–6.6	5.2–6.2	2–4
Ileum	6.8–8	6.8–8	

In the small intestine, the solubility of the drug continues to be an absorption barrier. Solubility varies throughout the length of the intestine. The solubility of a compound is a function of the pH and the pK_a of the molecule (see Section 7.1.3). Basic compounds are mostly in the charged cationic (protonated) state throughout the stomach and upper intestine, where the pH is acidic. This favors good solubility because the charged form is more soluble than the neutral form. Acids are neutral in the stomach and upper intestine. Therefore, an acid's solubility in acidic regions is limited. As the pH increases throughout the intestine, the relative amount of the anionic form of the acid increases, resulting in higher solubilities. These behaviors are typical of the solubility differences among compounds in different regions of the intestine. The fundamentals and effects of pK_a on solubility are discussed in Chapter 6.

The material coming from the stomach is mixed with bile in the intestine. Bile enters from the gallbladder. Bile acids enhance the solubility of lipophilic drug molecules. Bile acids work like detergents to enhance solubility. Their natural function is to solubilize food lipids to enhance absorption. Lipophilic drug molecules gain similar benefits in solubility. Bile acids form micelles to which lipophilic molecules adsorb, enhancing their diffusion to the intestinal membrane for absorption.

Pancreatic fluid also is added in the duodenum. It enters from the pancreas and contains hydrolytic enzymes. Major enzymes in pancreatic fluid include amylases, lipases, and proteases. They can catalyze the hydrolysis of some drug molecules that contain hydrolyzable functional groups (see Chapter 13).

Membrane permeability is a major absorption barrier in the intestine. As with solubility, permeability varies with the pH of the intestinal region and the compound's pK_a. A neutral molecule has much greater permeability than does its charged (ionic) form. Conversely, a neutral molecule is less soluble than is its charged form. Thus, permeability and solubility vary inversely with pH.

In the small intestine, drug molecules encounter an anatomy that greatly enhances the absorption of nutrients. The inner surface area of the intestinal lumen is enhanced approximately 400 times by three morphological features. Along the length of the intestinal lumen are folds, which are up and down undulations along the inner surface. Villi add to the surface area by projecting 1 mm into the intestinal lumen (Figure 3.3). A layer of epithelial

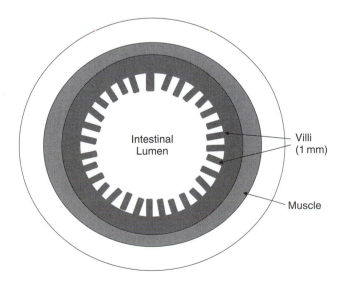

Figure 3.3 ▶ Diagram of the cross-section of the small intestine.

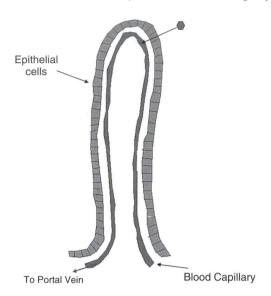

Figure 3.4 ▶ Diagram of a gastrointestinal villus.

cells cover the surface of the villi (Figure 3.4) and forms the primary permeation barrier to drug molecules. A compound must pass through this cellular membrane to reach the blood capillary and subsequent systemic circulation. Another morphological feature that enhances surface area is the microvilli on the luminal side of the epithelial cells (Figure 3.5). The microvilli extend 1 μm into the lumen and are called the *brush border*.

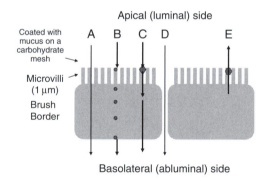

Figure 3.5 ▶ Permeation mechanisms through the gastrointestinal epithelial cells: (A) passive diffusion, (B) endocytosis, (C) uptake transport, (D) paracellular transport, (E) efflux transport.

3.4.1 Permeation of the Gastrointestinal Cellular Membrane

Compounds permeate through cellular membrane barriers by several different mechanisms (Figure 3.5). In passive diffusion, molecules diffuse through the lipid bilayer membranes and cytoplasm. Some types of transporters in the membrane perform active uptake of compounds that are ligands. Other types of transporters perform efflux of compounds from the cells (see Chapter 9). P-glycoprotein (Pgp) is a well-known efflux transporter. Paracellular permeation between the cells is available to smaller, more polar compounds. In endocytosis, molecules are engulfed by membrane and move through the cell in vesicles. Permeation mechanisms are discussed in greater detail in Chapter 8.

3.4.2 Passive Diffusion at the Molecular Level

Passive diffusion is generally the predominant permeation mechanism for most drugs. The cellular lipid bilayer membrane is shown in Figure 3.6. It consists of phospholipid molecules that self-assemble as a bilayer, with the aliphatic portion on the inside, away from the water molecules, and the polar phosphate and head groups oriented toward the water molecules. Passive diffusion involves movement of drug molecules through the bilayer as follows. The hydrating water molecules around the drug molecule are shed, and hydrogen bonds are broken. The molecule then passes through the region of polar head groups of the phospholipid molecules. It encounters the tightly packed lipid chains around the glycerol backbone and moves to the more disordered lipophilic region of the lipid aliphatic chains in the middle region of the bilayer. Molecules with a larger molecular size (i.e., higher molecular weight [MW]) do not pass through the tightly packed region as readily as smaller molecules. Molecules with higher lipophilicity typically are more permeable than less lipophilic molecules through the highly nonpolar central core of the lipid bilayer membrane. Molecules then move through the side chains and polar head groups of the other leaflet of the bilayer and are rehydrated by water molecules and form hydrogen bonds again. The highest energy barriers for passive diffusion through the bilayer appear to be the tightly packed, highly ordered regions of the phospholipids side chains in the region near the glycerol backbone[2] on both sides of the bilayer.

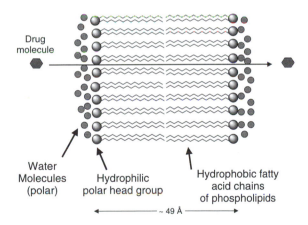

Drug molecule

Water Molecules (polar)

Hydrophilic polar head group

Hydrophobic fatty acid chains of phospholipids

~ 49 Å

Figure 3.6 ▶ Passive diffusion of drug molecule through lipid bilayer membrane.

The chemical structures of representative phospholipid molecules are shown in Figure 3.7. One of the alcohol groups of the glycerol backbone is attached to a phosphate group, which in turn is attached to a head group. Examples of head groups of common phospholipids are shown in Figure 3.7. Phosphatidylcholine is a common phospholipid found in many membranes. The head groups impart a charge and polarity to the outside of the membrane. Membranes also contain other components, such as cholesterol and transmembrane proteins (e.g., channels, transporters, receptors). The membranes in a specific tissue are composed of a specific mixture of phospholipids and other components, which may be different from other tissues. Thus, a compound might have different passive diffusion membrane permeability in different tissues (e.g., GI tract vs blood–brain barrier).

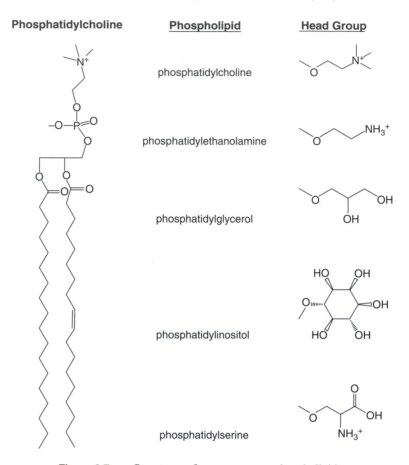

Figure 3.7 ▶ Structures of some common phospholipids.

3.4.3 Metabolism in the Intestine

Drug molecules can be metabolized in the intestine. Cytochrome P450 3A4 isozyme (CYP3A4) is a major metabolic enzyme in intestinal epithelial cells. This isozyme metabolizes diverse compound structures. Intestinal metabolism is considered part of "first-pass metabolism," which is the initial metabolism of drugs before they reach systemic circulation.

CYP3A4 has similar substrate specificity to Pgp. The two seem to work in concert; Pgp reduces the intracellular concentration of drug in the epithelial cells, which allows CYP3A4 to catalyze drug oxidation in an efficient manner, without being saturated.[1]

3.4.4 Enzymatic Hydrolysis in the Intestine

The natural function of the GI system is the digestion and absorption of nutrients to sustain the living system. Food contains macromolecules that are made up of the monomers that are needed to produce energy and build specific proteins, carbohydrates, lipids and nucleic acids for that organism. Breakdown of the macromolecules is accomplished by a variety of GI enzymes from the pancreas, stomach, and saliva. Protein digestion to amino acids is accomplished by peptidases, such as pepsin, which is secreted into the stomach, and trypsin and chymotrypsin, which are secreted by the pancreas into the small intestine. Fat digestion to fatty acids is performed by esterases, such as lipase, which is secreted by the pancreas into the small intestine. Ribonuclease and deoxyribonuclease digest RNA and DNA, respectively. Phosphatases and phosphodiesterases are other common enzymes.

GI enzymes can also catalyze drug hydrolysis. Drugs that contain derivatives of carboxylic acids, such as esters, amides, and carbamates, are especially susceptible. Enzymes are found in the intestinal lumen and are present in high concentration at the brush border. Thus, drugs can be hydrolyzed before they reach the bilayer membrane. Solution stability is discussed in Chapter 13.

Prodrugs are designed to take advantage of hydrolysis. Compounds that have a desirable pharmacological effect, but lack sufficient solubility for absorption, have been modified to add a substructure that increases solubility in the intestine (e.g., phosphate). The increased solubility allows the modified compound to diffuse to the cell surface. The hydrolytic enzyme cleaves off this substructure, just before it reaches the bilayer membrane. The active drug is released and permeates through the epithelial cells to reach systemic circulation. Prodrugs are the subject of Chapter 39.

The barriers to drug absorption in the GI tract, discussed in this chapter, are summarized in Figure 3.8 and Table 3.3. The dynamic balance of permeation mechanisms (passive, active uptake, efflux), pH and enzyme-induced hydrolysis, CYP metabolism, solubility, and dissolution rate affect the net rate of absorption in the intestine. Medicinal chemists should consider all of these mechanisms when trying to diagnose the causes of poor PK (see Chapter 38).

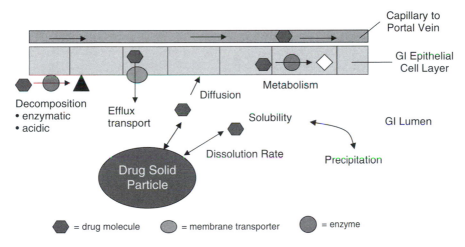

Figure 3.8 ▶ Composite diagram of barriers to drug absorption in the gastrointestinal tract.

TABLE 3.3 ▶ Drug Barriers in the Gastrointestinal Tract

Property	Description
Dissolution rate	Rate of transfer of compound from the surface of the particle to aqueous solution
Solubility	Maximum concentration that can be reached under the present conditions
Permeability	Movement from an aqueous solution through a lipid membrane to the aqueous solution on the other side
Chemical instability	Reaction of compound as a result of an environmental condition (e.g., pH, water)
Hydrolyzing enzymes	Naturally occurring enzymes that catalyze hydrolysis of food molecules and can catalyze hydrolysis of some drugs

3.4.5 Absorption Enhancement in the Intestine

A number of factors can enhance absorption of molecules in the intestine (Figure 3.9). One major factor that can be manipulated by discovery scientists is increasing the dissolution rate. This can be accomplished by reducing the particle size, which increases surface area. Particle size is reduced by grinding the solid material or by using techniques that produce nanoparticles. With greater surface area, more of the compound is solubilized in the same time. Salt form also can be manipulated to increase dissolution rate. Several possible counter ions are often screened to select a salt form having a higher dissolution rate. Formulation also is manipulated to enhance dissolution rate. Embedding the compound in excipients that break apart in an aqueous environment can rapidly disperse the compound material in the stomach, thus increasing the dissolution rate (see Chapter 41).

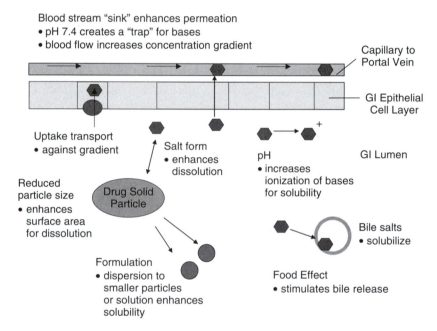

Figure 3.9 ▶ Composite diagram of factors that enhance drug absorption in the gastrointestinal tract through enhancement of permeation, dissolution, and solubility.

Another factor that contributes to absorption is solubility. This can be enhanced for a discovery lead compound by structural modifications that introduce a solubilizing functional group (see Chapter 7). Solubility is enhanced by bile salts (e.g., taurocholate, glycocholate) that are released by the gallbladder into the duodenum during stomach emptying and form micelles. Their natural function is the solubilization of lipophilic food components, such as lipids (e.g., triacylglycerols). In the same manner, bile salts serve to solubilize lipophilic drug molecules. This produces a greater concentration of lipophilic drugs in luminal solution. Bile salts carry the lipophilic molecules to the surface of the epithelium for enhanced absorption. Food intake stimulates the release of bile salts.

The pH of the gastric lumen can enhance solubility. For example, the solubility of basic compounds is enhanced at lower pHs values, and the solubility of acidic compounds is enhanced at higher pH values.

Uptake transporters enhance the absorption of some drugs. The natural function of uptake transporters is to enhance nutrient absorption. If the molecule has affinity for a transporter, its absorption might be enhanced (see Chapter 9).

Absorption is facilitated by removal of drug molecules from the intestine by blood flow. Capillary vessels sweep drug molecules into the portal vein and quickly away from the intestine. This creates a concentration gradient that increases passive diffusion in the absorptive direction.

3.5 Barriers in the Bloodstream

In the bloodstream, three barriers affect drugs: enzymatic hydrolysis, plasma protein binding, and red blood cell binding. Each of these barriers can reduce the free unchanged drug in systemic circulation, reducing penetration into the tissues.

3.5.1 Plasma Enzyme Hydrolysis

A large number of enzymes are present in blood for natural functions. They include cholinesterase, aldolase, lipase, dehydropeptidase, alkaline and acid phosphatase, glucuronidase, dehydrogenase, and phenol sulfatase. The substrate specificity and relative amount of these enzymes vary with species, disease state, gender, age, and race. A drug molecule may also be a substrate for an enzyme in the blood. The most common reaction is hydrolysis. Stability in plasma is the subject of Chapter 12.

3.5.2 Plasma Protein Binding

Approximately 6% to 8% of plasma is protein, and a large percentage of this serves as carrier protein for naturally occurring compounds. Drugs often reversibly bind to these proteins as well, which reduces the concentration of free drug in solution. The affinity of binding determines the ratio of bound and unbound ("free") drug in solution. There is high capacity for drug binding in plasma, and it is normally not saturated unless the drug concentration is very high. Protein binding results in a constant fraction of bound and free drug over a wide total drug concentration range. The concentrations of plasma proteins can vary with disease state and age. The three types of binding proteins in plasma are albumin, α_1-acid glycoprotein, and lipoproteins.

Human serum albumin has at least six binding sites with molecular binding specificity. Two sites bind fatty acids and another binds bilirubin. Two sites bind acidic drugs. Warfarin and phenylbutazone bind at one site and diazepam and ibuprofen at another.[3] Other drugs can also bind to sites on human serum albumin.

Basic drugs can bind to α_1-acid glycoprotein. This protein has one binding site. Examples of drugs that bind to α_1-acid glycoprotein include disopyramide and lignocaine. This protein can be saturated at higher drug concentrations.

Lipoproteins occur as particles and serve the natural function of transporting cholesterol and triacylgylcerols (hydrophobic lipids). They consist of nonpolar lipids, surrounded by more polar lipids and protein. Binding of drug molecules involves nonspecific lipophilic interactions.

Protein binding has several effects on drug disposition, which can have complex and counteracting effects. For example:

► Only unbound "free" drug permeates the membranes of the capillary blood vessels to enter the tissues. The drug must reach a therapeutic concentration in the tissues to produce the desired pharmacologic action. The free drug in the tissues reaches equilibrium with the free drug in the plasma. Thus, high binding limits free drug concentration in tissue.

▶ Only unbound drug permeates into the liver and kidney for clearance. Thus, high binding reduces clearance of a drug from the body and increases PK half-life.

▶ The kinetics and affinity of binding are important. For some drugs, binding to the plasma protein has a fast "off rate," and free drug depletion from plasma into tissues will result in rapid re-equilibration of the concentration of free drug in the plasma. If a compound is a kidney active secretion transporter substrate (see Section 3.7), its affinity for the transporter likely is higher than its affinity for plasma protein. If re-equilibration is fast from plasma-protein bound to free drug in plasma, re-equilibration can occur within the kidney, thus making formerly bound drug available for extraction in the kidney during the same pass. The result is a high renal extraction rate. Plasma protein binding is discussed in Chapter 14.

3.5.3 Red Blood Cell Binding

Drug molecules can bind to red blood cells. This is primarily a lipophilic interaction with the cell membrane. Drug discovery projects often check for red cell binding of lead compounds.

3.6 Barriers in the Liver

The liver presents two major barriers to drugs: metabolism and biliary extraction. The liver is one of the two major organs of drug clearance from the body.

Within the liver, the portal vein, which carries drug molecules from the intestine, branches into successively smaller vessels. The narrowest are called *venous sinusoid*. A portion of the molecules in solution permeate through the walls of the vessels into the hepatocytes, which form narrow sheets that are highly vascularized by the venous sinusoids (Figure 3.10). A small duct is created where the hepatocytes meet, called the *bile canaliculus*. Within hepatocytes, compounds encounter an array of metabolic enzymes that can modify the structure. Drug molecules and metabolites permeate with bile into the bile canaliculus, from which the fluid moves into the hepatic ductile and into the gallbladder. Bile is released from the gallbladder into the small intestine, resulting in excretion of a significant amount of drug and metabolites in the feces. The fraction of a drug excreted by this route depends on the properties of the compound.

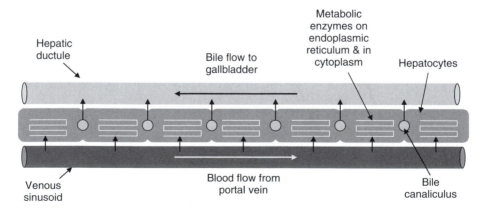

Figure 3.10 ▶   Diagram of hepatic clearance by metabolism in hepatocytes and extraction into bile.

3.6.1 Metabolism

Two types of metabolic reactions occur in the liver. The first is phase I, which causes changes to the drug molecule (e.g., hydroxylation), which are primarily oxidative. The second type of reaction is phase II, which adds polar groups to these oxidized positions or other substructures of the molecule. Metabolism serves the natural function of making xenobiotic compounds more polar so that they have higher partitioning into the aqueous bile and urine for excretion from the body. A high rate of metabolism results in rapid clearance, low exposure, and low bioavailability. Metabolism is a major route of drug clearance and is discussed in Chapter 11.

3.6.2 Biliary Excretion

A portion of the metabolite and unmetabolized drug molecules move into the bile by passive diffusion and active transport into the bile canaliculus. Transporters such as Pgp are present on the bile canaliculus membrane to actively transport some compounds into the bile. Transporters are discussed in Chapter 9. The distribution of various metabolites and the drug molecules from hepatocytes into the bile and blood depends on their properties (e.g., transporter affinity, passive diffusion, metabolic stability). After the bile is secreted into the intestine, some metabolite and drug molecules are reabsorbed from the intestinal lumen back into systemic circulation in a process called *enterohepatic circulation*. In addition to biliary excretion, some of the metabolite molecules and unmetabolized drug molecules move from the hepatocytes back into systemic circulation. This occurs by passive diffusion and active transport into the venous sinusoid. They eventually are extracted by the kidney.

3.7 Barriers in the Kidney

The liver and kidney are the major organs of elimination for most compounds. In the kidney, molecules of dosed compounds and metabolites permeate from the bloodstream into the urine for excretion. The permeation mechanisms are the same as found in other tissues of the body, primarily passive diffusion, paracellular, active uptake, and efflux.

The primary unit of renal (kidney) elimination is the nephron (Figure 3.11). The kidney contains thousands of nephrons. The first stage of renal elimination is termed *filtration*. Approximately 10% of total renal blood flow passes through the glomerulus, a complex mesh of blood capillaries. These present a high surface area to the Bowman's capsule, which is connected to the urinary system. The membranes of the Bowman's capsule have loose junctions, which allow a high rate of paracellular permeation of water, drug molecules, and other blood components but normally not proteins or cells. Molecules can also permeate passively.

In the proximal tubule, some drug molecules can be actively secreted from the bloodstream by transporters. For example, penicillins and glucuronides are transported by organic anion transporters, morphine and procaine are transported by organic cation transporters, and digoxin is transported by Pgp.[3] Much of the water (99%) and some of the drug molecules are reabsorbed by passive diffusion.

Reabsorption of molecules back into the bloodstream also occurs by passive and active transport mechanisms. The behavior of a particular drug is dependent on the same physico-chemical properties (e.g., pK_a, lipophilicity) and transporter affinity as permeation in other tissues. The physiological factors affecting transport include blood flow, surface area, pH of the fluids, and transporter expression. Generally, metabolites in the blood are more readily eliminated by the kidney than their parent molecules because increasing polarity enhances extraction by the nephron. Urine flows through the ureter to the bladder from which it is excreted.

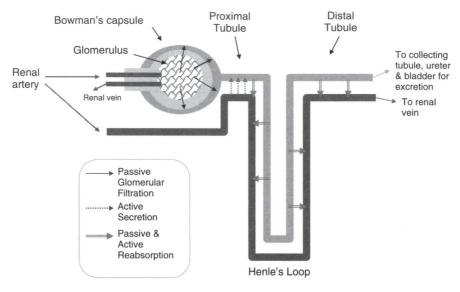

Figure 3.11 ▶   Diagram of drug extraction in the kidney nephrons.

3.8   Blood–Tissue Barriers

Some organs have barriers that reduce the penetration of drugs into the organ tissue. Such barriers exist at the placenta, testes, and brain. The barrier consists of a cellular layer that has properties that attenuate permeation. These barriers can include tight intercellular junctions that reduce paracellular permeation, efflux transporters that actively remove compound molecules from inside the cells or membrane, and a different lipid composition with different passive diffusion characteristics than the GI tract. A major issue in discovering drugs for neurological disorders is permeation of the blood–brain barrier, which is discussed in Chapter 10.

3.9   Tissue Distribution

The bloodstream carries compound molecules throughout all the tissues of the body. Distribution of drug into nontarget tissues effectively keeps it away from the target disease tissue. Compounds may depot in some tissues. For example, lipophilic compounds tend to accumulate in adipose tissues. Acidic compounds accumulate in muscle, which has a pH of approximately 6. The pH values of various physiological fluids and organs are listed in Table 3.4.

TABLE 3.4 ▶   **pH Values of Physiological Fluids and Organs**

Physiological fluid	pH
Blood	7.4
Stomach	1–3
Small intestine	5.5–7
Saliva	6.4
Cerebrospinal fluid	7.4
Muscle	6
Urine	5.8

Blood flow to an organ affects the time needed for the organ tissue drug concentration to equilibrate with the blood. High cardiac output (blood flow) to heart, lungs, kidney, and brain allows rapid equilibration of drugs with those organs. Cardiac output is lower to skin, bone, and fat, resulting in slower equilibration in these tissues.

3.10 Consequences of Chirality on Barriers and Properties

Chirality can have a significant effect on the behavior of compounds in vivo. It affects many properties owing to the different interaction of enantiomers with proteins in vivo. This affects the compound's PK and pharmacodynamics. Examples of properties affected by chirality and the causes (in parentheses) are as follows:

► Solubility (crystal forms of enantiomers are different)

► Efflux and uptake transport (binding to transporter)

► Metabolism (binding, orientation of molecule's positions to the reactive moiety)

► Plasma protein binding (binding)

► Toxicity, such as CYP inhibition, hERG blocking (binding)

An example is shown in Table 3.5. These drugs have differences in renal clearance owing to chirality. These differences likely are caused by differences in active transport in the nephrons (secretion or reabsorption) or by plasma protein binding. Additional discussion on the effects of chirality is found in chapters on specific properties.

TABLE 3.5 ► **Stereoselectivity of Renal Clearance**

Drug	Renal clearance enantiomeric ratio[a]
Quinidine	4.0
Disopyramide	1.8
Terbutaline	1.8
Chloroquine	1.6
Pindolol	1.2
Metoprolol	1.2

[a] ratio of renal clearances of the two enantiomers

3.11 Overview of In Vivo Barriers

The effects of individual barriers on in vivo delivery of compound to the therapeutic target are discussed in greater detail in following chapters. Poor delivery of the compound to the target results in reduced in vivo efficacy for a given dose. In vivo barriers are summarized in Figure 3.12. It is important for discovery project teams to assess how in vivo barriers affect their lead compounds. This is accomplished by assaying the compounds in vitro for key properties that predict performance at these barriers and making structural modifications to improve these properties.

There is often a tradeoff between structural features that enhance therapeutic target binding and structural features that enhance delivery to the target through optimal performance

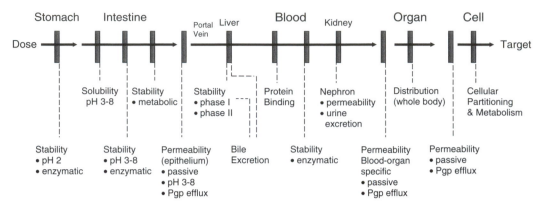

Figure 3.12 ▶ Overview of in vivo barriers to drug delivery to the target. (Reprinted with permission from [4].)

at in vivo barriers. If the sole focus of a drug discovery program is activity optimization, poor properties can result, leading to:

▶ Low absorption, owing to low solubility or permeability

▶ High clearance, owing to metabolism

▶ Clearance by hydrolysis in the GI tract or blood

▶ Efflux that opposes uptake in many membranes and enhances extraction in the liver and kidney

▶ High protein binding that limits free drug at the target

▶ Poor penetration of a blood–organ barrier at the target organ

▶ High volume of distribution due to lipophilicity

Each of these factors usually can be improved by medicinal chemists through structural modification.

Problems

(Answers can be found in Appendix I at the end of the book.)

1. List two factors that affect drug efficacy in vivo.

2. What is the preferred drug dosage form and regimen?

3. List some physicochemical and metabolic property limitations that reduce drug exposure in vivo.

4. What is the relationship of solubility to absorption?

5. What is the relationship of permeability to absorption?

6. What factors make drugs have lower absorption in the stomach than in the small intestine?

7. Is the pH higher or lower in the fasted state than in the fed state?

8. A greater portion of molecules of a basic compound is neutral in the: (a) upper intestine or (b) lower intestine? A greater portion of molecules of an acidic compound is ionized in the: (a) upper intestine or (b) lower intestine?

9. What is mixed with stomach contents as it enters the intestine, and what are the effects on drugs?

10. Charged versus neutral molecules are: (a) more permeable, (b) less permeable, (c) more soluble, or (d) less soluble?

11. Passive diffusion across lipid bilayer membranes is generally higher for molecules with: (a) lower lipophilicity or (b) higher lipophilicity?

12. List three barriers in the bloodstream.

13. For most drugs, the organs primarily involved in elimination are: (a) stomach, (b) large intestine, (c) portal vein, (d) small intestine, (e) liver, (f) kidney?

14. List two clearance mechanisms in the liver.

15. What barrier limits drug penetration to brain tissue?

16. Why are metabolites that are circulating in the blood generally more readily extracted by the kidney than the drug from which they were formed?

17. For most drugs, absorption occurs primarily in the: (a) stomach, (b) large intestine, (c) portal vein, (d) small intestine, (e) liver, (f) kidney.

18. Total absorption from the intestinal lumen into the bloodstream can be affected by which of the following properties of the compound?: (a) solubility, (b) permeability, (c) pK_a, (d) Pgp efflux, (e) metabolic stability, (f) molecular size, (g) enzymatic hydrolysis, (h) blood–brain barrier permeation.

19. Which of the following are effects of high plasma protein binding in vivo?: (a) reduced distribution to tissues, (b) increased metabolism, (c) reduced clearance.

20. Which of the following can be improved by structural modification of the lead?: (a) phase I metabolism, (b) efflux, (c) enzymatic decomposition, (d) solubility, (e) passive diffusion permeability.

References

1. van de Waterbeemd, H., Smith, D. A., Beaumont, K., & Walker, D. K. (2001). Property-based design: Optimization of drug absorption and pharmacokinetics. *Journal of Medicinal Chemistry*, *44*, 1313–1333.

2. Xiang, T.-X., & Anderson, B. D. (2002). A computer simulation of functional group contributions to free energy in water and a DPPC lipid bilayer. *Biophysical Journal*, *82*, 2052–2066.

3. Birkett, D. (2002). *Pharmacokinetics made easy*. Sydney: McGraw-Hill.

4. Kerns, E. H., & Di, L. (2003). Pharmaceutical profiling in drug discovery. *Drug Discovery Today*, *8*, 316–323.

Part 2
Physicochemical Properties

Chapter 4

Rules for Rapid Property Profiling from Structure

Overview

▶ *Lipinski and Veber rules are guidelines for structural properties of drug-like compounds.*

▶ *Rules are effective and efficient means of rapidly assessing structural properties.*

The fastest method for evaluating the drug-like properties of a compound is to apply "rules." Rules are a set of guidelines for the structural properties of compounds that have a higher probability of being well absorbed after oral dosing. The values for the properties associated with rules are quickly counted from examination of the structure or calculated using software that is widely available. These guidelines are not absolute, nor are they intended to form strict cutoff values for which property values are drug-like and which are not drug-like. Nevertheless, they can be quite effective and efficient.

4.1 Lipinski Rules

Although medicinal chemists and pharmaceutical scientists had used structural properties in various ways for many years, rules became more prominent and defined in the field with the report by Lipinski et al.[1] of the "rule of 5," or what has become known as the "Lipinski rules." These rules are a set of property values that were derived from classifying the key physicochemical properties of drug-like compounds. The rules were used at Pfizer for a few years prior to their publication and since then have become widely used. The impact of these rules in the field has been very high. This acceptance can be attributed to many factors:

▶ The rules are easy, fast, and have no cost to use.

▶ The "5" mnemonic makes the rules easy to remember.

▶ The rules are intuitively evident to medicinal chemists.

▶ The rules are a widely used standard benchmark.

▶ The rules are based on solid research, documentation, and rationale.

▶ The rules work effectively.

It is important to keep in mind the intended purpose of the rule of 5. The article[1] states: poor absorption or permeation are more likely when:

▶ >5 H-bond donors (expressed as the sum of all OHs and NHs)

▶ MW > 500

▶ logP > 5 (or MlogP > 4.15)

▶ >10 H-bond acceptors (expressed as the sum of all Ns and Os)

▶ Substrates for biological transporters are exceptions to the rule

Examples of counting hydrogen bond donors and acceptors are given in Table 4.1. For example, an R-OH counts as both one H-bond donor (HBD) and one H-bond acceptor (HBA). Although "violation" of one rule may not result in poor absorption, the likelihood of poor absorption increases with the number of rules broken and the extent to which they are exceeded.

TABLE 4.1 ▶ Examples of Counting Hydrogen Bonds for Lipinski Rules

Functional group	H-bond donor	H-bond acceptors
Hydroxyl	1 (OH)	1 (O)
Carboxylic acid	1 (OH)	2 (2 Os)
–C(O)-N-R$_2$	0	2 (N, O)
Primary amine	2 (NH$_2$)	1 (N)
Secondary amine	1 (NH)	1 (N)
Aldehyde	0	1 (O)
Ester	0	2 (O)
Ether	0	1 (O)
Nitrile	0	1 (N)
Pyridine	0	1 (N)

The rules were derived by examining the structural properties of compounds that, by examination of the United States Adopted Names (USAN) Directory, had survived phase I clinical trials and had moved on to phase II studies. Phase I studies involve human dosing to determine the toxicity and pharmacokinetics. The fact that the compounds had moved on to phase II studies indicates that the compounds had sufficient absorption in humans for pharmaceutical companies to continue investment in their development. The structural properties of this set of 2,200 compounds were examined and clear trends were observed, which became the rules. The rules were set at the 90th percentile of the compound set; thus, 90% of the compounds that had sufficient absorption after oral dosing had molecular property values within the Lipinski rules. Compounds that approach or exceed these values have a higher risk of poor absorption after oral dosing.

The rules are based on a strong physicochemical rationale. Hydrogen bonds increase solubility in water and must be broken in order for the compound to permeate into and through the lipid bilayer membrane. Thus, an increasing number of hydrogen bonds reduces partitioning from the aqueous phase into the lipid bilayer membrane for permeation by passive diffusion. Molecular weight (MW) is related to the size of the molecule. As molecular size increases, a larger cavity must be formed in water in order to solubilize the compound, and solubility decreases. Increasing MW reduces the compound concentration at the surface of the intestinal epithelium, thus reducing absorption. Increasing size also impedes passive diffusion through the tightly packed aliphatic side chains of the bilayer membrane. Increasing Log P also decreases aqueous solubility, which reduces absorption. Finally, membrane transporters can either enhance or reduce compound absorption by either active uptake transport or efflux, respectively. Thus, transporters can have a strong impact on increasing or decreasing absorption.

Lipinski et al. discussed important implications of these rules in light of current drug discovery strategies. The discovery lead optimization stage often increases target binding by adding hydrogen bonds and lipophilicity. Thus, activity optimization can reduce the drug-like properties of a compound series. Combinatorial chemistry and parallel array synthesis tend to be more facile with more lipophilic groups; thus, analog series often increase in lipophilicity. In biology, high-throughput screening (HTS) tends to favor more lipophilic compounds than screening strategies in previous decades because compounds are first dissolved in DMSO and not in aqueous media, as in the past. Therefore, to obtain favorable biological data from modern in vitro biology techniques, a compound need not have significant aqueous solubility. Compounds previously were tested initially by dissolution in aqueous media for in vivo testing and, thus, were required to have aqueous solubility in order to be successful in biological testing. The use of screening libraries having good drug-like properties was recommended. This concept has been extended in recent years with the concept of "lead-like" compounds (see Chapter 20).

4.2 Veber Rules

Additional rules were proposed by Veber et al.[2] They studied structural properties that increase oral bioavailability in rats. They concluded that molecular flexibility, polar surface area (PSA), and hydrogen bond count are important determinants of oral bioavailability. Rotatable bonds can be counted manually or using software. PSA is calculated using software and is closely related to hydrogen bonding.

Veber rules for good oral bioavailability in rats are as follows:

► $\leq$10 rotatable bonds

► $\leq$140 Å^2 PSA, or $\leq$12 total hydrogen bonds (acceptors plus donors)

4.3 Other Rules

Other researchers have proposed other rule sets. Pardridge[3] proposed rules for blood–brain barrier permeability. These rules are used for predicting compounds that have a greater likelihood of permeating the blood–brain barrier (see Chapter 10). Oprea et al.[4–7] proposed the "rule of 3" for lead-like compounds. These rules are used to guide the selection of leads for the lead optimization stage of discovery (see Chapter 20).

4.4 Application of Rules for Compound Assessment

Rules are typically used for the following purposes:

► Anticipating the drug-like properties of potential compounds when planning synthesis

► Using the drug-like properties of "hits" from HTS as one of the selection criteria

► Evaluating the drug-like properties of compounds being considered for purchase from a compound vendor

An example of the counting and calculation of rules the compound doxorubicin is shown in Figure 4.1. Doxorubicin has a very low oral bioavailability, as would be anticipated from the structural properties covered by the rules.

Figure 4.2 shows an example of how rules could help anticipate absorption in a drug discovery project.[8] Structural modifications of the compound on the left were made to optimize

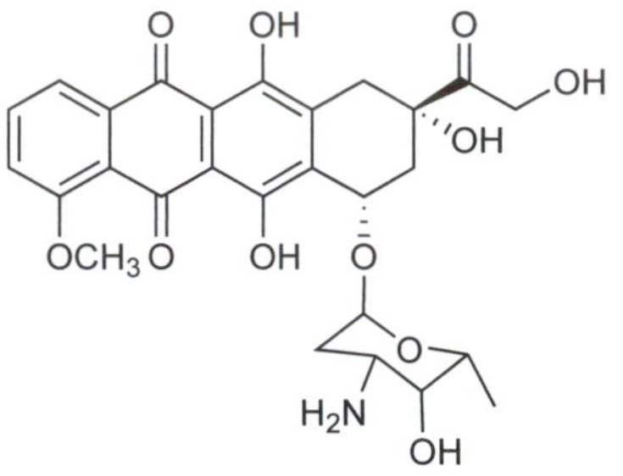

Lipinski Rules

- H-bond donors = 7
- MW = 543
- ClogP = −1.7
- H-bond acceptors = 12

Veber Rules

- Rotatable bonds = 11
- PSA = 206
- Total H-bonds = 19

Figure 4.1 ▶ Example of counting and calculations for the Lipinski and Veber rules for doxorubicin, which has an oral bioavailability of approximately 5%. Guidelines are exceeded for all rules except ClogP.

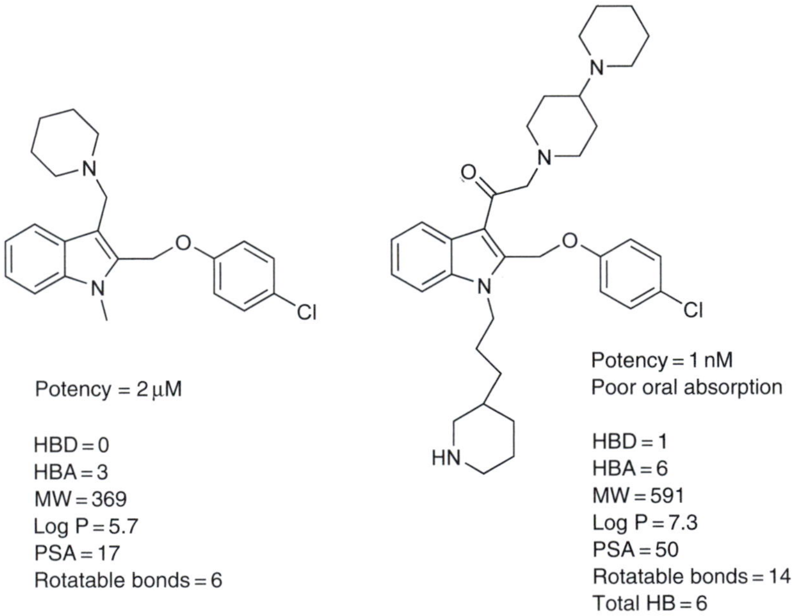

Potency = 2 μM

HBD = 0
HBA = 3
MW = 369
Log P = 5.7
PSA = 17
Rotatable bonds = 6

Potency = 1 nM
Poor oral absorption

HBD = 1
HBA = 6
MW = 591
Log P = 7.3
PSA = 50
Rotatable bonds = 14
Total HB = 6

Figure 4.2 ▶ Structural optimization for activity in this neuropeptide Y Y1 antagonist discovery project modified the lead on the **left** to the compound on the **right**. Although a 2,000-fold increase in potency was achieved, the resulting compound had poor absorption properties after oral dosing, as anticipated from the structural rules.

activity, resulting in the compound on the right. Unfortunately, the binding-optimized compound had poor absorption after oral dosing, as could be predicted from the structural properties prior to synthesis.

Exceeding the rules often reduces absorption after oral dosing. Poor absorption properties result in low bioavailability or the need for dosing via an alternate route, either of which limits the potential scope of the drug product.

Problems

(Answers can be found in Appendix I at the end of the book.)

1. Low absorption is more likely for a compound that has which of the following?: (a) 7 H-bond donors, (b) 2 H-bond donors, (c) MW 350, (d) MW 580, (e) ClogP 7.2, (f) ClogP 2.7, (g) 5 H-bond acceptors, (h) 13 H-bond acceptors, (i) high permeability by uptake transporter, (j) PSA 155, (k) PSA 35.

2. Why are H-bonds important in absorption?

3. Why is high Log P unfavorable in absorption?

4. Rules are best used for: (a) strict guidelines for compound rejection, (b) assessing compounds for which no in vitro property data are available, (c) sole basis for selecting compounds for in vivo studies, (d) anticipating the metabolism of compounds?

5. Count the number of hydrogen bond donors (HBD) and hydrogen bond acceptors (HBA) in the following using the Lipinski rule: (a) $-COOCH_3$, (b) $R_1-NH-C(O)-R_2$.

6. Determine the rule of 5 values for the following compounds. Indicate which structural properties exceed the rules and are a problem.

Structure	#HBD	#HBA	MW	cLogP	PSA	Problem
1 Buspirone			385	1.7	7.0	
2			418	−3.3	143	
3 Paclitaxel			852	4.5	209	
4 Cephalexin			347	0.5	138	

Structure	#HBD	#HBA	MW	cLogP	PSA	Problem

5

Cefuroxime

| | | | 424 | −1.5 | 199 | |

6

Olsalazine

| | | | 302 | 3.2 | 141 | |

7. Which of the following characteristics puts a compound at risk for poor absorption?: (a) MW 527, (b) 5 H-bond acceptors, (c) ClogP 6.1, (d) is a substrate for an uptake transporter, (e) 7-H-bond donors, (f) PSA 152.

8. Which of the following is an effect of H-bonding?: (a) H-bonds increase lipid solubility, (b) H-bonds increase water solubility, (c) H-bonds decrease water solubility, (d) H-bonds must be broken for the molecule to partition into the bilayer membrane.

9. Which of the following is a positive effect of a lower MW?: (a) water solubility increases, (b) acid decomposition decreases, (c) passive diffusion increases.

References

1. Lipinski, C. A., Lombardo, F., Dominy, B. W., & Feeney, P. J. (1997). Experimental and computational approaches to estimate solubility and permeability in drug discovery and development settings. *Advanced Drug Delivery Reviews, 23*, 3–25.

2. Veber, D. F., Johnson, S. R., Cheng, H.- Y., Smith, B. R., Ward, K. W., & Kopple, K. D. (2002). Molecular properties that influence the oral bioavailability of drug candidates. *Journal of Medicinal Chemistry, 45*, 2615–2623.

3. Pardridge, W. M. (1995). Transport of small molecules through the blood-brain barrier: Biology and methodology. *Advanced Drug Delivery Reviews, 15*, 5–36.

4. Oprea, T. I., Davis, A. M., Teague, S. J., & Leeson, P. D. (2001). Is there a difference between leads and drugs? A historical perspective. *Journal of Chemical Information and Computer Sciences, 41*, 1308–1315.

5. Oprea, T. I. (2002). Chemical space navigation in lead discovery. *Current Opinion in Chemical Biology, 6*, 384–389.

6. Oprea, T. I. (2002). Current trends in lead discovery: Are we looking for the appropriate properties? *Journal of Computer-Aided Molecular Design, 16*, 325–334.

7. Hann, M. M., & Oprea, T. I. (2004). Pursuing the leadlikeness concept in pharmaceutical research. *Current Opinion in Chemical Biology, 8*, 255–263.

8. Smith, D. A. (2002). Ernst Schering Research Foundation Workshop, *37*, 203–212.

Chapter 5

Lipophilicity

Overview

▶ *The lipophilicity of a compound is commonly estimated using Log P from octanol/water partitioning.*

▶ *Lipophilicity is a major determinant of many ADME/Tox properties.*

Lipophilicity is a property that has a major effect on absorption, distribution, metabolism, excretion, and toxicity (ADME/Tox) properties as well as pharmacological activity. Lipophilicity has been studied and applied as an important drug property for decades. It can be quickly measured or calculated. Lipophilicity has been correlated to many other properties, such as solubility, permeability, metabolism, toxicity, protein binding, and distribution.

5.1 Lipophilicity Fundamentals

Lipophilicity is the tendency of a compound to partition into a nonpolar lipid matrix versus an aqueous matrix. It is an important determinant of most other drug properties. Lipophilicity is readily calculated, thanks to the work of Hansch and Leo.[1] It is a rapid and effective tool for initial compound property assessment, as indicated by its inclusion in the "rule of 5."

One traditional approach for assessing lipophilicity is to partition the compound between immiscible nonpolar and polar liquid phases. Traditionally, octanol has been widely used as the nonpolar phase and aqueous buffer as the polar phase. The partitioning values that are measured are termed Log P and Log D. It is important to recognize that these terms are different.

Log P: Log of the partition coefficient of the compound between an organic phase (e.g., octanol) and an aqueous phase (e.g., buffer) at a pH where all of the compound molecules are in the neutral form.

$$\text{Log P} = \log([\text{Compound}_{\text{organic}}]/[\text{Compound}_{\text{aqueous}}]).$$

Log D: Log of the distribution coefficient of the compound between an organic phase (e.g., octanol) and an aqueous phase (e.g., buffer) at a specified pH (x). A portion of the compound molecules may be in the ionic form and a portion may be in the neutral form.

$$\text{Log D}_{\text{pHx}} = \log([\text{Compound}_{\text{organic}}]/[\text{Compound}_{\text{aqueous}}]).$$

Log P depends on the partitioning of the neutral molecules between the two matrices. Log D depends on the partitioning of the neutral portion of the molecule population plus the partitioning of the ionized portion of the molecule population. Ions have greater affinity for the polar aqueous phase than for the nonpolar organic phase. The fraction of the molecule population that is ionized depends on the pH of the aqueous solution, the pK_a of the compound, and whether the compound is an acid or base. The effects of pK_a and pH are discussed in Chapter 6. For acids, the neutral/anion ratio of molecules in solution decreases with increasing pH; therefore, log D decreases with increasing pH. Conversely, for bases, the neutral/cation ratio of molecules in solution increases with increasing pH; therefore, log D increases with increasing pH.

Abraham et al.[2,3] have shown that Log P is affected by several fundamental structural properties of the compound:

▶ Molecular volume

▶ Dipolarity

▶ Hydrogen bond acidity

▶ Hydrogen bond basicity

Molecular volume is related to molecular weight and affects the size of the cavity that must be formed in the solvent to solubilize the molecule. Dipolarity affects the polar alignment of the molecule with the solvent. Hydrogen bond acidity is related to hydrogen bond donation, and hydrogen bond basicity is related to hydrogen bond acceptance. They affect hydrogen bonding with the solvent. In-depth study of these effects by Abraham et al. resulted in linear free energy equations from which partitioning behavior can be predicted. Researchers frequently apply this approach to evaluating the predictability of methods for partition-based properties and enhancing the predictability of assay methods by adding calculations in order to better model particular properties.[4]

Remember that lipophilicity changes with the conditions of the phases, including the following:

▶ Partitioning solvents/phases

▶ pH

▶ Ionic strength

▶ Buffer

▶ Co-solutes or co-solvents

For example, partitioning between octanol and water is different than between cyclohexane and water. This is due to the differences in the molecular properties of the phases, which lead to different interactions of the solvent and solute molecules. pH affects the degree of ionization, as discussed previously. Increasing ionic strength results in increasing polarity of the aqueous phase. The buffer also affects the polarity, molecular interactions and formation of in situ salts (as counter ions) with drug molecules. Co-solvents, such as dimethylsulfoxide (DMSO), can interact with solutes and change their partitioning behavior, even at low percentage compared to the phases. Thus, the effect of solution conditions should be considered when predicting the effect of lipophilicity and reporting assay data.

5.2 Lipophilicity Effects

Lipophilicity has been correlated to various models of drug properties affecting ADME/Tox.[5,6] They include permeability, absorption, distribution, plasma protein binding, metabolism, elimination, and toxicity. Lombardo et al.[5,6] correlated pharmacokinetic volume of distribution (V_d) to lipophilicity.

A general guide for optimal gastrointestinal absorption by passive diffusion permeability after oral dosing is to have a moderate Log P (range 0–3), as suggested in Figure 5.1. In this range, a good balance of permeability and solubility exists. Compounds with a lower Log P are more polar and have poorer lipid bilayer permeability. Compounds with a higher Log P are more nonpolar and have poor aqueous solubility.

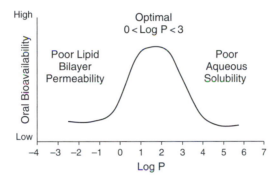

Figure 5.1 ▶ Hypothetical example of how Log P can affect oral bioavailability for a compound series. Absorption by passive diffusion permeation after oral dosing is generally considered optimal for compounds having a moderate Log P and decreases for compounds having higher and lower Log P values.

Table 5.1 is useful for estimating the impact of log $D_{7.4}$ on drug-like properties in discovery[7]:

▶ Log $D_{7.4}$ < 1: There is good solubility but low absorption and brain penetration, owing to low passive diffusion permeability. These compounds tend to have high clearance by the kidney, owing to their polarity. These compounds may exhibit paracellular permeation if the molecular weight is low.

▶ 1 < Log $D_{7.4}$ < 3: This is an ideal range. These compounds generally have good intestinal absorption, owing to a good balance of solubility and passive diffusion permeability. Metabolism is minimized, owing to lower binding to metabolic enzymes.

▶ 3 < Log $D_{7.4}$ < 5: These compounds have good permeability but absorption is lower, owing to lower solubility. Metabolism is increased in this range, owing to increased binding to metabolic enzymes.

▶ Log $D_{7.4}$ > 5: Compounds in this range tend to have low absorption and bioavailability, owing to low solubility. Metabolic clearance is high because of high affinity for metabolic enzymes. V_d and half-life (see Chapter 19) are high because compounds partition into and stay in tissues.

TABLE 5.1 ▶ Impact of Log D$_{7.4}$ on Drug-like Properties[7]

Log D$_{7.4}$	Common Impact on Drug-like Properties	Common Impact *In Vivo*
< 1	Solubility high Permeability low by passive transcellular diffusion Permeability possible via paracellular if MW < 200 Metabolism low	Volume of distribution low Oral absorption and BBB penetration unfavorable Renal clearance may be high
1 to 3	Solubility moderate Permeability moderate Metabolism low	Balanced volume of distribution Oral absorption and BBB penetration favorable
3 to 5	Solubility low Permeability high Metabolism moderate to high	Oral bioavailability moderate to low Oral absorption variable
> 5	Solubility low Permeability high Metabolism high	High volume of distribution (especially amines) Oral absorption unfavorable and variable

5.3 Lipophilicity Case Studies and Structure Modification

Lipophilicity is an underlying structural property that affects higher-level physicochemical and biochemical properties. It often is an effective guide for modifying the structure of a lead series to improve a property. The effects of lipophilicity on specific properties and structure modification strategies are discussed in most of the subsequent chapters on properties.

ΔLog P has been used to predict permeation of the blood–brain barrier (BBB).[8] ΔLog P is the Log P from partitioning between octanol and aqueous phases minus the Log P from partitioning between cyclohexane and aqueous phases. The difference is attributed to the contribution of hydrogen bonding to Log P$_{ow}$ (octanol/water) compared to Log P$_{cw}$ (cyclohexane/water). As ΔLog P increases, BBB permeability generally decreases. Its correlation to BBB permeation has been interpreted in terms of the negative effect of hydrogen bonding on BBB permeability (see Chapter 10 and Section 28.2.1.4).

Lipophilicity also has been correlated to activity. One example is shown in Figure 5.2 [9] for a series of 11 compounds with anticonvulsant activity. Log P correlated with –Log ED with an R^2 of 0.83. Thus, activity increased (ED decreased) as Log P increased.

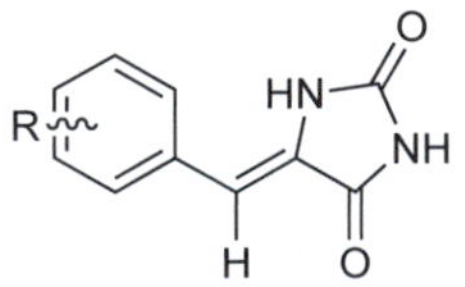

$$-\text{Log ED} = -1.247 + 0.795\ \text{LUMO} + 0.150\ \text{Log P}$$
$$n = 11,\ r^2 = 0.834,\ r^2_{cv} = 0.793,\ \text{SE} = 0.063$$

Figure 5.2 ▶ Correlation between anticonvulsant activity and Log P.[9] For this series, activity increased with Log P.

Problems

(Answers can be found in Appendix I at the end of the book.)

1. What is the major difference between Log P and Log D?

2. What factors affect Log P?

3. What is the most favorable Log $D_{7.4}$ range for drugs?

4. Why is a low Log P unfavorable for absorption? Why is a high Log P unfavorable for absorption?

5. Which is measured for the neutral form of a compound?: (a) Log D, (b) Log P.

6. At a Log $D_{7.4}$ of 2, which of the following can be predicted?: (a) high intestinal absorption, (b) low solubility, (c) high permeability, (d) high metabolism, (e) high central nervous system penetration.

7. At a Log $D_{7.4}$ greater than 5, which of the following can be predicted?: (a) high intestinal absorption, (b) low solubility, (c) high metabolism, (d) low bioavailability.

References

1. Hansch, C., Leo, A., & Hoekman, D. (1995). *Exploring QSAR. Fundamentals and applications in chemistry and biology, volume 1. Hydrophobic, electronic and steric constants, volume 2*. New York: Oxford University Press.

2. Abraham, M. H., Chadha, H. S., Leitao, R. A. E., Mitchell, R. C., Lambert, W. J., Kaliszan, R., et al. (1997). Determination of solute lipophilicity, as log P(octanol) and log P(alkane) using poly(styrene-divinylbenzene) and immobilized artificial membrane stationary phases in reversed-phase high-performance liquid chromatography. *Journal of Chromatography. A, 766*, 35–47.

3. Abraham, M. H., Gola, J. M. R., Kumarsingh, R., Cometto-Muniz, J. E., & Cain, W. S. (2000). Connection between chromatographic data and biological data. *Journal of Chromatography. B, Biomedical Sciences and Applications, 745*, 103–115.

4. Valko, K., Du, C. M., Bevan, C. D., Reynolds, D., & Abraham, M. H. (2001). High throughput lipophilicity determination: Comparison with measured and calculated log P/log D values. In B. Testa, H. van de Waterbeemd, G. Folkers, & R. Guy (Eds.), *Pharmacokinetic optimization in drug research: Biological, physiological, and computational strategies*. Zurich: Verlag Helvetica Chimica Acta, pp. 127–182.

5. Hansch, C., Leo, A., Mekapati, S. B., & Kurup, A. (2004). QSAR and ADME. *Bioorganic & Medicinal Chemistry, 12*, 3391–3400.

6. Lombardo, F., Obach, R. S., Shalaeva, M. Y., & Gao, F. (2002). Prediction of volume of distribution values in humans for neutral and basic drugs using physicochemical measurements and plasma protein binding data. *Journal of Medicinal Chemistry, 45*, 2867–2876.

7. Comer, J. E. A. (2003). High throughput measurement of logD and pKa. In P. Artursson, H. Lennernas, & H. van de Waterbeemd (Eds.), *Methods and principles in medicinal chemistry 18* (pp. 21–45). Weinheim: Wiley-VCH.

8. Young, R. C., Mitchell, R. C., Brown, T. H., Ganellin, C. R., Griffiths, R., Jones, M., et al. (1988). Development of a new physicochemical model for brain penetration and its application to the design of centrally acting H2 receptor histamine antagonists. *Journal of Medicinal Chemistry, 31*, 656–671.

9. Thenmozhiyal, J. C., Wong, P. T.-H., & Chui, W.-K. (2004). Anticonvulsant activity of phenylmethylene-hydantoins: A structure-activity relationship study. *Journal of Medicinal Chemistry, 47*, 1527–1535.

pK_a

Overview

▶ *The ionizability of a compound is indicated by pK_a.*

▶ *Ionizability is a major determinant of solubility and permeability.*

▶ *When pH = pK_a, the concentrations of ionized and neutral molecules in solution are equal.*

▶ *Basicity of bases increases as pK_a increases; acidity of acids increases at pK_a decreases.*

The great majority of drugs contain ionizable groups (Figure 6.1). Most are basic, and some are acidic. Only 5% are not ionizable. pK_a indicates a compound's ionizability. It is a function of the acidity or basicity of group(s) in the molecule. Medicinal chemists can modify the acidic or basic substructures on the scaffold in order to obtain the desired pK_a, which affects solubility and permeability.

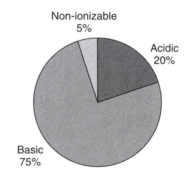

Figure 6.1 ▶ Most drugs are ionizable.

▢ 6.1 pK_a Fundamentals

pK_a is the negative log of the ionization constant K_a. It is common to use pK_a for both acids and bases.

For acids:

$$HA = H^+ + A^-$$

$$pK_a = -\log([H^+] \bullet [A^-]/[HA]).$$

For bases:

$$HB^+ = H^+ + B$$

$$pK_a = -\log([H^+] \bullet [B]/[HB^+]).$$

Useful aspects of the behavior of acids and bases can be derived from the above relationships.

For acids:

▶ As pH decreases, there is a greater concentration of neutral acid molecules (HA) and a lower concentration of anionic acid molecules (A^-) in solution.

▶ Acids with a lower pK_a are stronger (greater tendency to form A^-).

For bases:

▶ As pH decreases, there is a lower concentration of neutral base molecules (B) and a higher concentration of cationic base molecules (HB^+) in solution.

▶ Bases with lower pK_a are weaker (lower tendency to form HB^+).

The Henderson-Hasselbach equation is a useful relationship for discovery.
For acids:

$$pH = pK_a + \log([A^-]/[HA]) \quad \text{or} \quad [HA]/[A^-] = 10^{(pK_a - pH)}.$$

For bases:

$$pH = pK_a + \log([B]/[HB^+]) \quad \text{or} \quad [BH^+]/[B] = 10^{(pK_a - pH)}.$$

These relationships provide a means of calculating the concentration of ionic and neutral species at any pH, if pK_a is known. Moreover, it is useful to note that when pH is the same as pK_a, then there is an equal concentration of ionic and neutral species in solution. This relationship as well as the change in concentration of each species is shown in Figure 6.2.

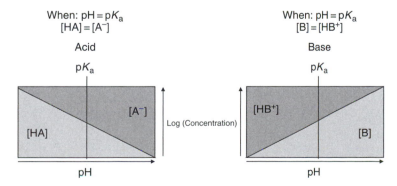

Figure 6.2 ▶ Concentration of neutral and ionic species of acids and bases at pHs above and below their pK_a.

6.2   pK_a Effects

Ionized molecules are more soluble in aqueous media than neutral molecules because they are more polar. Solubility is determined by both the intrinsic solubility of the neutral molecule and the solubility of the ionized species, which is much greater.

Conversely, ionized molecules are less permeable than neutral molecules. The neutral molecules are much more lipophilic than the ionized molecules and are considered to be the dominant form that permeates by passive diffusion.

Because pK_a determines the degree of ionization, it has a major effect on solubility and permeability. These, in turn, determine intestinal absorption after oral dosing. The effects of ionization suggest a relationship frequently encountered by medicinal chemists: highly permeable compounds often have low solubility and vice versa. Thus, there is a tradeoff between solubility and permeability because of the opposite effects of ionization on these properties.

An example of this effect is shown in Figure 6.3. An acidic compound with a pK_a of 5 exhibits decreasing permeability as the pH of the solution increases. Conversely, the solubility increases at increasing solution pH.

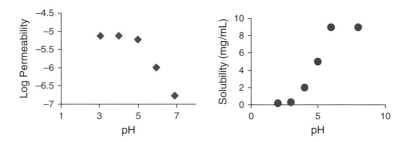

Figure 6.3 ▶ Permeability and solubility profiles for an acidic compound with a pK_a of 5. Permeability and solubility are pH dependent for ionizable compounds. The properties exhibit opposite effects with pH because of the effects of ionization.

pK_a also affects the activity of a structural series. Presumably, this is due to changes in interactions at the active site of the target protein.

6.3   pK_a Case Studies

Examples of the pK_a values of a number of substructures that commonly appear in drug molecules are listed in Table 6.1. A more extensive listing is available.[1] The pK_a values of some drugs are listed in Table 6.2.

An example of the effect of pK_a and molecular size on the activity of a structural series is shown in Figure 6.4.[2] The piperidine with a pK_a around 10 was modified with aliphatic groups that have little effect on pK_a. Their increasing size appears to be responsible for some loss of activity. However, when the aromatic ring was added, the basicity of the amine greatly decreased and caused the activity to be much lower than would be expected from the size of the aromatic ring compared to changes in size from other added moieties.

An example of the effect of pK_a on activity is shown in Figure 6.5.[3] IC$_{50}$ decreased as pK_a decreased (increasing acidity).

Basic drugs tend to penetrate the blood–brain barrier more effectively than acids do. An example is shown in Figure 6.6, where the basic trifluoroperizine (pK_a 7.8) permeates the blood–brain barrier, whereas the acidic indomethacin (pK_a 4.2) does not.[4]

TABLE 6.1 ► Examples of Acidic and Basic Substructures and Respective pK_a

Acids	pK_a
CF_3COOH	0.23
CCl_3COOH	0.9
CCl_2HCOOH	1.3
$CClH_2COOH$	2.9
HCOOH	3.8
C_6H_5COOH	4.2
Succinic acid	4.2, 5.6
H_3COOH	4.8
Thiophenol	6.5
p-Nitrophenol	7.2
m-Nitrophenol	9.3
C_6H_5OH	10.0

Bases	pK_a
Guanidine	13.6
Acetamide	12.4
Pyrrolidine	11.3
Piperidine	11.1
Methyl amine	10.6
Piperazine	9.8, 5.3
Trimethyl amine	9.8
Glycine	9.8
Morpholine	8.4
Imidazole	6.8
Pyridine	5.2
Quinoline	4.9
Aniline	4.9
Triazole	2.5
Purine	2.4
Pyrimidine	1.2
Diphenylamine	0.8

TABLE 6.2 ► Example Drugs and pK_a

Acids	pK_a
Penicillin V	2.7
Salicylic acid	3.0, 13.8
Acetylsalicylic acid	3.5
Diclofenac	4.1
Sulfathiazole	7.1
Phenobarbital	7.4, 11.8
Phenytoin	8.3
Acetaminophen	9.9
Caffeine	14

Bases	pK_a
Caffeine	0.6
Quinidine	4.1, 8.0

Continued

TABLE 6.2 ▶ *Continued*

Bases	pK_a
Tolbutamide	5.3
Cocaine	8.4
Ephedrine	9.4
Imipramine	9.5
Atropine	9.7

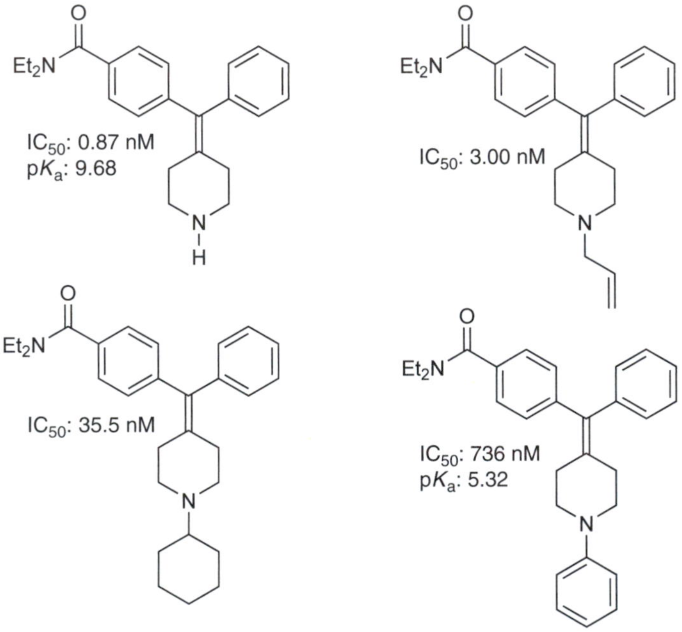

Et2N
IC50: 0.87 nM
pKa: 9.68
IC50: 3.00 nM
IC50: 35.5 nM
IC50: 736 nM
pKa: 5.32

Figure 6.4 ▶ Effect of pK_a and size on activity.[2]

H2N
R1
R2
R3
Log IC50 (uM)
pKa
100
10
1
5
6
7
8
9
10

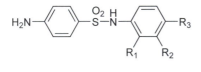

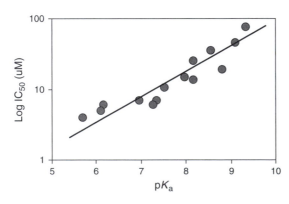

Compounds	IC_{50} (uM)	pK_a
4-OCH_3	75	9.34
H	45	9.10
4-Cl	35	8.56
4-I	25	8.17
2-Cl, 4-OCH_3	19	8.81
3-CF_3	15	7.98
2-Cl	14	8.18
4-$COCH_3$	11	7.52
4-CN	7	7.36
4-NO_2	7	6.97
2-OCH_3, 4-NO_2	6	7.27
2-Cl, 4-NO_2	6	6.17
2-NO_2, 4-CF_3	5	6.10
2-Br, 4-NO_2	4	5.70

Figure 6.5 ▶ Effect of pK_a on activity of a structural series.[3]

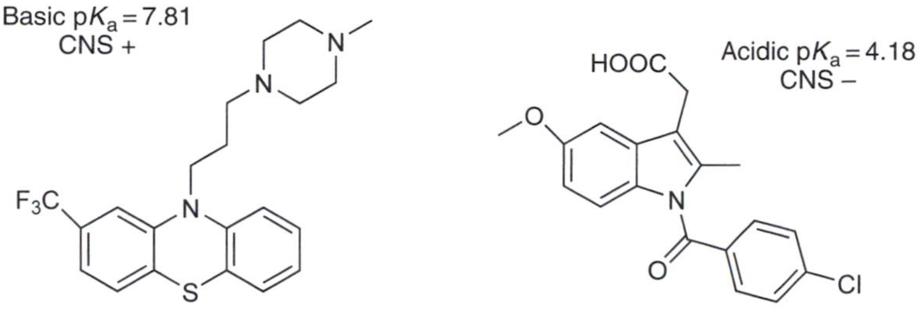

Basic pK_a = 7.81
CNS +

Trifluoroperazine

Acidic pK_a = 4.18
CNS −

Indomethacin

Figure 6.6 ▶ Basic drugs tend to permeate the blood–brain barrier, whereas acidic drug generally do not.[4]

The effect of pK_a on water solubility for a set of bile acids is shown in Figure 6.7.[5] As the acidity of the bile acids increased, water solubility increased. Figure 6.8 shows two SRC kinase inhibitors.[6] Increasing basicity increased solubility due to ionization.

Wohnsland and Faller[7] reported that the parallel artificial membrane permeability assay (PAMPA) permeability of diclofenac (acidic pK_a 5.6) was higher at lower pH than at higher pH, whereas desipramine (basic pK_a 6.5) had the opposite behavior. Avdeef[8] showed acid

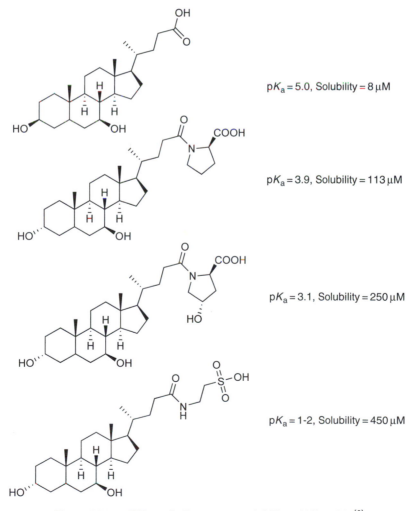

pK_a = 5.0, Solubility = 8 μM

pK_a = 3.9, Solubility = 113 μM

pK_a = 3.1, Solubility = 250 μM

pK_a = 1-2, Solubility = 450 μM

Figure 6.7 ▶ Effect of pK_a on water solubility of bile acids.[5]

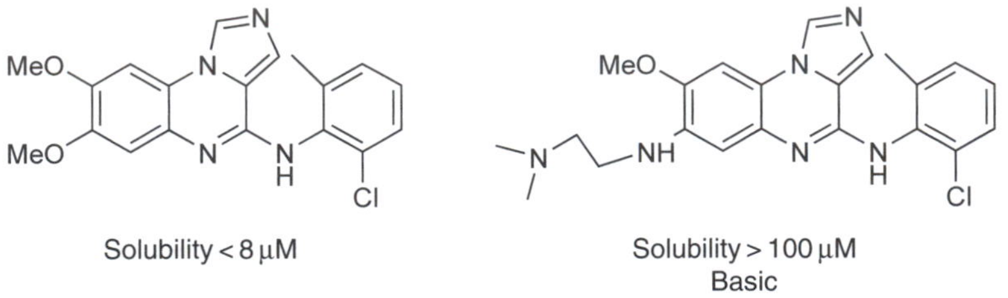

Solubility < 8 µM

Solubility > 100 µM
Basic

Figure 6.8 ▶ Effect of pK_a on solubility for SRC kinase inhibitors.[6]

(ketoprofen, pK_a 3.98), base (verapamil, pK_a 9.07), and ampholyte (piroxicam, pK_a 5.07, 2.33) had unique permeability profiles.

6.4 Structure Modification Strategies for pK_a

When synthetic modifications are planned for the purpose of improving the water solubility or permeability of a structural series, a wide selection of substructures can be used. It is important to remember that structural modifications that increase solubility will also decrease permeability.

By modifying the substructures of a molecule to introduce groups with differing pK_a values, medicinal chemists can modify the solubility and permeability of the compound. Examples of pK_a modifications to enhance solubility and permeability can be found in Chapters 7 and 8, respectively. Electron donating and withdrawing groups can be added or removed to add electron density at the acid or base, depending on the desired effect. For example, the strength of an acid can be increased by adding α-halogen(s) or other electron withdrawing groups (e.g., carboxy, cyano, nitro). Addition of aliphatic groups has little effect. Removing electron-withdrawing groups can decrease acid strength.

The strength of a base is decreased by attachment an aromatic group (e.g., aniline). The lone pair is delocalized into the aromatic ring. It is also decreased slightly with addition of an aliphatic group. The basicity of an aniline can be increased with attachment of a methoxy group, which donates electrons, or it can be decreased with the attachment of a nitro group, which withdraws electrons.

Problems

(Answers can be found in Appendix I at the end of the book.)

1. For acids, as pH decreases, are there: (a) more anions, (b) more neutral molecules, (c) higher solubility, (d) lower solubility, (e) higher permeability, (f) lower permeability?

2. For bases, as pH decreases, are there: (a) more cations, (b) more neutral molecules, (c) higher solubility, (d) lower solubility, (e) higher permeability, (f) lower permeability?

3. At pH 6.8, a basic compound of pK_a 9.5 is mostly in what form?: (a) ionized, (b) neutral.

4. For benzoic acid (pK_a 4.2), estimate the degree of ionization in the fasted state for the stomach, duodenum, and blood. For $HA = H^+ + A^-$, use the relationship: $[HA]/[A^-] = 10^{(pK_a - pH)}$

Location	pH	$[HA]/[A^-] = 10^{(pK_a - pH)}$	Ionization
Stomach	1.5		
Duodenum	5.5		
Blood	7.4		

5. For piperazine (pK_a 9.8), estimate the degree of ionization in the fasted state for the stomach, duodenum, and blood. For $BH^+ = H^+ + B$, use the relationship:
$[BH^+]/[B] = 10^{(pK_a - pH)}$

Location	pH	$[BH^+]/[B] = 10^{(pK_a - pH)}$	Ionization
Stomach	1.5		
Duodenum	5.5		
Blood	7.4		

6. If the pH is 2 units above the pK_a of an acid, the predominant species is: (a) neutral, (b) anion. If the pH is 2 units below the pK_a of a base, the predominant species is: (c) neutral, (d) cation.

References

1. Martin, A. (1993). *Physical pharmacy* (4th ed). Philadelphia: Lea & Febiger.

2. Wei, Z.- Y., Brown, W., Takasaki, B., Plobeck, N., Delorme, D., Zhou, F., et al. (2000). N,N-Diethyl-4-(phenylpiperidin-4-ylidenemethyl)benzamide: A novel, exceptionally selective, potent d-opioid receptor agonist with oral bioavailability and its analogues. *Journal of Medicinal Chemistry, 43*, 3895–3905.

3. Miller, G. H., Doukas, P. H., & Seydel, J.K. (1972). Sulfonamide structure-activity relation in a cell-free system. Correlation of inhibition of folate synthesis with antibacterial activity and physicochemical parameters. *Journal of Medicinal Chemistry, 15*, 700–706.

4. Clark, D. E. (2003). In silico prediction of blood-brain barrier permeation. *Drug Discovery Today, 8*, 927–933.

5. Roda, A., Cerre, C., Manetta, A. C., Cainelli, G., Umani-Ronchi, A., & Panunzio, M. (1996). Synthesis and physicochemical, biological, and pharmacological properties of new bile acids amidated with cyclic amino acids. *Journal of Medicinal Chemistry, 39*, 2270–2276.

6. Chen, P., Doweyko, A. M., Norris, D., Gu, H. H., Spergel, S. H., Das, J., et al. (2004). Imidazoquinoxaline Src-family kinase p56[Lck] inhibitors: SAR, QSAR, and the discovery of (S)-N-(2-Chloro-6-methylphenyl)-2-(3-methyl-1-piperazinyl)imidazo- [1,5-a]pyrido[3,2-e]pyrazin-6-amine (BMS-279700) as a potent and orally active inhibitor with excellent in vivo antiinflammatory activity. *Journal of Medicinal Chemistry, 47*, 4517–4529.

7. Wohnsland, F., & Faller, B. (2001). High-throughput permeability pH profile and high-throughput alkane/water log P with artificial membranes. *Journal of Medicinal Chemistry, 44*, 923–930.

8. Avdeef, A. (2001). Physicochemical profiling (solubility, permeability and charge state). *Current Topics in Medicinal Chemistry, 1*, 277–351.

Chapter 7

Solubility

Overview

> ► *Solubility is the maximum dissolved concentration under given solution conditions.*

> ► *Solubility is a determinant of intestinal absorption and oral bioavailability.*

> ► *Solubility is increased by adding ionizable groups or reducing Log P and MW.*

> ► *Salt forms increase dissolution rate.*

Solubility is one of the most important properties in drug discovery. Insoluble compounds can plague discovery. Many negative effects can occur for low-solubility compounds, including the following:

> ► Poor absorption and bioavailability after oral dosing

> ► Insufficient solubility for IV dosing

> ► Artificially low activity values from bioassays

> ► Erratic assay results (biological and property methods)

> ► Development challenges (expensive formulations and increased development time)

> ► Burden shifted to patient (frequent high-dose administrations)

These are major hurdles and deserve serious consideration by discovery project teams. Chapter 40 is devoted to solving solubility problems in biological assays. Lipinski et al.[1] caution that solubility is a much larger issue for drug discovery than is permeability.

Solubility problems can intensify during discovery because the molecular characteristics needed for strong binding to the target protein can be deleterious to solubility. For example, lipophilic structures may seem to be the best available leads for a discovery project, or lipophilic groups are often added during optimization to enhance target binding. Unfortunately, these examples, while helping to meet discovery's primary goal of finding active compounds, can reduce solubility and have a negative impact on other discovery and development goals.

Under short time lines and high expectations, discovery scientists can make the mistake of placing too high of a reliance on advanced formulation and delivery technologies to improve the pharmacokinetics and bioavailability of insoluble compounds. This reliance can lead a discovery project team toward a clinical candidate that cannot achieve sufficiently high absorbed doses to produce an effective therapeutic treatment. It is wise to solve solubility insufficiencies during discovery with structural modifications.

▗ 7.1 Solubility Fundamentals

7.1.1 Solubility Varies with Structure and Physical Conditions

Solubility is the maximum concentration that a compound reaches in a solvent matrix at equilibrium with solid compound. It is important to remember that there is no single solubility value for a compound. Solubility is determined by many factors:

- ▶ Compound structure

- ▶ Physical state of compound that is introduced into solution

    - ▶ Solid: Amorphous, crystalline, polymorphic form

    - ▶ Liquid: Predissolved in solvent (e.g., dimethylsulfoxide [DMSO])

- ▶ Composition and physical conditions of solvent(s)

    - ▶ Types of solvents

    - ▶ Amount (%) of co-solvents

    - ▶ Solution components (e.g., salts, ions, proteins, lipids, surfactants)

    - ▶ pH

    - ▶ Temperature

- ▶ Methods of measurement

    - ▶ Equilibration time

    - ▶ Separation techniques (e.g., filter, centrifuge)

    - ▶ Detection (e.g., ultraviolet, mass spectrometry, turbidity)

For example, the solubility of a compound can be very different in pH 7.4 buffer, simulated intestinal fluid, blood, and biological assay media containing 1% DMSO. Therefore, it is necessary to specify the conditions of the compound and solution for proper use of the solubility data. In drug discovery, various solubility experiments are performed to estimate the effect of solubility in different systems to better mimic the actual in vitro and in vivo conditions.

7.1.2 Dissolution Rate

Dissolution rate is the speed at which a compound dissolves from a solid form into a solvent. Solid forms vary among the neutral form (free acid or base), salt form, or formulated dosage forms. Modification of the solid form to achieve different rates of dissolution is discussed in Sections 7.5 and 7.6. Such modifications allow control of the rate of absorption and pharmacokinetics of a compound.

7.1.3 Structural Properties Affect Solubility

Solubility is affected by physicochemical properties, which can be estimated using in vitro assays or software calculations. Each of these is determined by underlying structural properties.

▶ Lipophilicity: Determined by van der Waals, dipolar, hydrogen bonds, ionic interactions

▶ Size: Molecular weight, shape

▶ pK_a: Determined by functional group ionizability

▶ Crystal lattice energy: Determined by crystal stacking, melting point

Medicinal chemists have the ability to change solubility by modifying the structure, which affects these physicochemical properties. Structural modification of polarity, hydrogen bonding, molecular size, ionizability, and crystal stacking are discussed in a later section of this chapter. The crystalline forms can be modified by crystallization conditions and salt form.

A demonstration of the effect of lipophilicity and crystal lattice energy on solubility is the empirically derived general solubility equation of Yalkowsky and Banerjee[2] for estimating the aqueous solubility of a compound based on measurable or calculable properties:

$$\text{Log S} = 0.8 - \text{Log P}_{ow} - 0.01(\text{MP} - 25),$$

here S is solubility, Log P_{ow} is the octanol/water partition coefficient (a measure of lipophilicity), and MP is the melting point (a measure of crystal lattice strength). This relationship shows that solubility decreases 10-fold as:

▶ Log P increases by 1 unit

▶ Melting point increases by 100°C

pK_a and solution pH are important because the charged form of a drug compound is more soluble than the neutral form. At a particular pH, there is a distribution of molecules between the neutral and ionized state. Therefore, the solubility of a compound at a particular pH is the sum of the "intrinsic solubility" (solubility of the neutral compound) of the neutral species portion of molecules in solution, plus the solubility of the charged species portion of molecules in solution. This has implications for the solubility of compounds in various physiological fluids and solutions that have different pHs in drug discovery.

This phenomenon also can be described mathematically, as follows. Drugs ionize according to the reactions:

$$HA + H_2O = H_3O^+ + A^- \quad \text{(Acid)}$$
$$B + H_2O = OH^- + HB^+ \quad \text{(Base)}.$$

At equilibrium, the solubility of a mono-acid or mono-base can be described as follows:

$$S = [HA] + [A^-] \quad \text{(Acid)}$$
$$S = [B] + [HB^+] \quad \text{(Base)},$$

where S = solubility.

A mathematical derivation of the Henderson-Hasselbalch equation provides insights for solubility:

$$S = S_0(1 + 10^{(pH - pKa)}) \quad \text{(Acid)}$$

$$S = S_0(1 + 10^{(pKa - pH)}) \quad \text{(Base)},$$

where S_0 = intrinsic solubility (solubility of the neutral compound). Solubility changes linearly with S_0 and exponentially with the difference between pH and pK_a. Examples of the effects of intrinsic solubility and pK_a of acids are listed in Table 7.1.[3] Barbital and amobarbital have the same pK_a (7.9, weak acid), but barbital has higher intrinsic solubility than amobarbital (7.0 mg/mL vs. 1.2 mg/mL). Therefore, barbital has higher total solubility than amobarbital at all pH values, including pH 9 (95 mg/mL vs. 15 mg/mL). Naproxen and phenytoin have similar intrinsic solubility but different pK_a values. This results in dramatically different solubility at pH 9. Naproxen is much more acidic (pK_a 4.6) than phenytoin (pK_a 8.3) and therefore is much more soluble because solubility increases exponentially with the difference in pH and pK_a. This example demonstrates that introducing an ionization center is an effective structure modification for increasing solubility.

TABLE 7.1 ► Solubility at a Given pH is a Function of the Intrinsic Solubility of the Neutral Portion of Molecules and Solubility of the Ionized Portion of Molecules[3]

	pK_a	Intrinsic Solubility (mg/mL)	Solubility @ pH 9 (mg/mL)
Barbital	7.9	7.0	95
Amobarbital	7.9	1.2	15
Naproxen	4.6	0.016	430
Phenytoin	8.3	0.02	0.12

Barbital   Amobarbital   Naproxen   Phenytoin

One mistake that discovery scientists should avoid is confusing the pH solubility curve for a pK_a titration curve. The pK_a titration curve (Figure 7.1) plots the change in ionization with pH. The sharp rise occurs in the region of pK_a, and the inflection point occurs where pH equals pK_a. Scientists may think that the sharp rise in solubility should correspond to where pH equals pK_a. In fact, the pH region of the sharp rise in solubility is dependent on the intrinsic solubility (S_0), as shown in Figure 7.2. This figure plots the pH solubility profiles of four acidic compounds with the same pK_a of 4.5 but with different intrinsic solubilities (0.1, 1, 10, and 100 mg/mL). The inflection point does not correspond to the pK_a of the compound. The pK_a of the compound is at the pH when total solubility is two times the intrinsic solubility because solubility and pK_a follow a log-linear correlation. A plot of solubility on a log scale versus pH on a linear scale (Figure 7.2, inset) shows that the turning point pH is the pK_a of the compound.

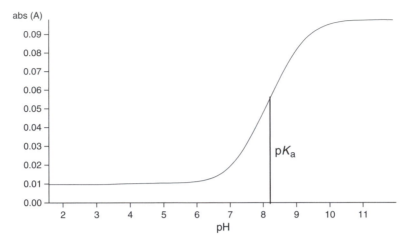

Figure 7.1 ► pK_a titration curve for a compound with pK_a of 8.2.

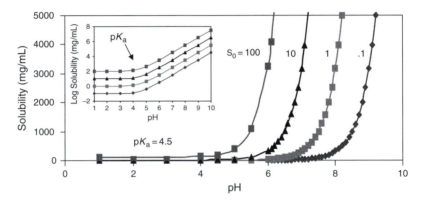

Figure 7.2 ► pH solubility profiles for compounds all having pK_a = 4.5 but different S_0 values. When pH = pK_a, then total solubility equals 2S_0.

7.1.4 Kinetic and Thermodynamic Solubility

It is important to distinguish between "kinetic" and "thermodynamic" solubility. Discovery scientists should know how solubility was measured to properly interpret the data or to apply the data to project decision-making.

Kinetic solubility has two distinguishing characteristics: (a) the compound initially is fully dissolved in an organic solvent (e.g., DMSO) then added to the aqueous buffer, and (b) equilibrium is not reached between dissolved compound and solid compound. If the compound precipitates from solution, it can be in a "metastable" crystalline form, such as amorphous or mixture of crystal forms. The kinetic solubility conditions mimic the conditions of discovery biological and property assays, in which compounds are predissolved in DMSO prior to addition into an in vitro assay solution, exposure times vary, equilibrium usually is not established, and assay concentrations are comparable (μM). Kinetic solubility data are useful in drug discovery to:

► Alert teams to potential absorption or bioassay liabilities

► Diagnose erratic bioassay results

► Develop structure–solubility relationships

▶ Select compounds for NMR-binding and x-ray co-crystallization experiments

▶ Develop generic formulations for animal dosing

Thermodynamic solubility is distinguished by (a) the addition of aqueous solvent directly to solid crystalline material and (b) establishment of equilibrium between the dissolved and solid material. Excess solid is used, so solid crystalline material is always present in the original solid form to which the aqueous solution was added. Long mixing time is applied to ensure equilibrium (e.g., 24–72 hours). Higher target concentrations (mg/mL) are used to guide formulation development. The thermodynamic solubility value varies with the crystal form of the solid (amorphous, crystalline, different polymorphs, hydrates, and solvates). High-energy crystal forms (less stable forms) tend to have higher solubility than low-energy forms. For example, amorphous material has higher solubility than crystalline material. Bioassays in drug discovery do not start with solid material (DMSO stock solution is used), nor do they generally reach equilibrium or have solid present, so the relevance of thermodynamic solubility to bioassays in drug discovery is limited. Using thermodynamic solubility data in early drug discovery can be counterproductive because (a) it can vary among different synthetic batches of the same compound due to different crystal forms, and (b) it is not relevant to amorphous solids typically made in discovery. Thermodynamic solubility is most useful in late discovery and early development, where a large batch has been synthesized and its crystal form has been characterized to:

▶ Guide formulation development

▶ Diagnose in vivo results

▶ Plan development strategy

▶ File regulatory submissions

Kinetic solubility in general tends to be higher than thermodynamic solubility. This is because the presence of DMSO helps enhance solubility, metastable crystal forms with higher crystal packing energy, and supersaturation of the solution occurs due to a nonequilibrium state.

Discovery and development scientists have very different viewpoints on solubility.[4,5] The goal of discovery scientists is to dissolve the compound into solution in any way possible to enable biological assays and demonstrate proof of concept. Amorphous and metastable crystal forms are okay, and discovery scientists love to use DMSO. Kinetic solubility is consistent with this viewpoint. In development, the goal is to develop a human dosage form and perform the detailed technical studies required for regulatory approval. Solubilization options are constrained, and unrealistically solubilized systems can be misleading. Crystal forms are well characterized, and development scientists never use DMSO. In development, all that matters is thermodynamic solubility.

7.1.4.1 Consequences of Chirality on Solubility

Chirality affects solubility because of the crystal form. The two enantiomers crystallize in different forms from each other and from the racemate. The Wallach rule states that racemate crystals are more stable and dense than their chiral counterparts. For example, *S*-ketoprofen has a melting point of 72°C, whereas *RS*-ketoprofen has a melting point of 94°C. The increased stability of the crystal results in a higher melting point. This affects aqueous thermodynamic solubility values, which are 2.3 mg/mL for *S*-ketoprofen and 1.4 mg/mL for *RS*-ketoprofen. Kinetic solubility values are not affected by crystal stability.

7.2 Effects of Solubility

7.2.1 Low Solubility Limits Absorption and Causes Low Oral Bioavailability

In order for a compound to be absorbed in the intestine after oral dosing, the solid dosage form or suspension must disintegrate, dissolve, and diffuse to the surface of the intestinal epithelium to be absorbed into systemic circulation. As the concentration of compound increases when it dissolves, more drug molecules are present at the surface of the epithelial cells and a greater amount of drug per unit time per surface area (flux) is absorbed. This is the reason why solubility is so important for absorption. Insoluble compounds tend to have incomplete absorption and, therefore, low oral bioavailability.

Optimization of oral bioavailability is a goal of the discovery stage. Oral bioavailability incorporates both the extent of intestinal absorption, which is affected by solubility and permeability, and the presystemic metabolism (Figure 7.3). Insoluble compounds, such as the compound YH439 (Figure 7.4) have low oral bioavailability.[6] YH439 has low oral bioavailability (0.9%–4.0% in rat) due to poor aqueous solubility. When the compound was formulated in a mixed-micelle formulation, the oral bioavailability increased to 21%. Formulation helped solubilization of the compound and improved absorption and oral bioavailability. When the compound was tested in a dose-escalating toxicity study, area under the curve AUC_{0-t} remained unchanged when the dose increased from 100 mg/kg ($AUC = 32 \, \mu g \, min/mL$) to 500 mg/kg ($AUC = 37 \, \mu g \, min/mL$), indicating solubility limits its absorption.[6] Low solubility hampered toxicity study at higher doses.

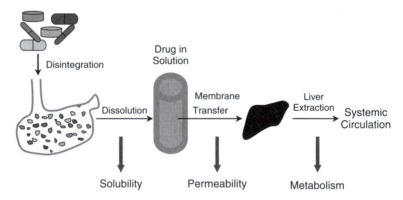

Figure 7.3 ▶ Solubility, permeability, and metabolic stability affect oral absorption and bioavailability following oral dosing.

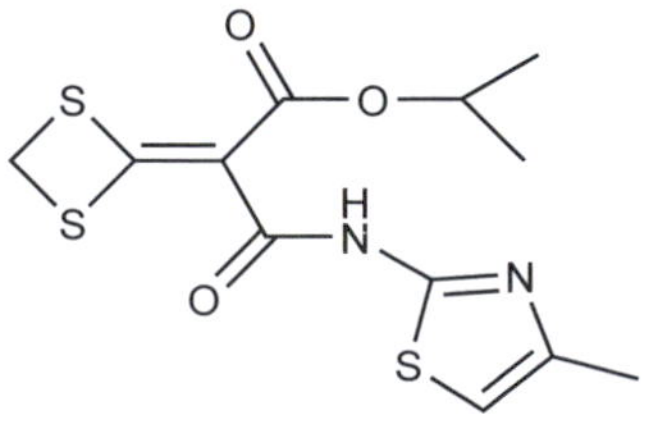

Figure 7.4 ▶ Structure of YH439.[6] Insoluble compounds have low bioavailability after oral dosing.

Another example of how solubility affects oral bioavailability is shown in Figure 7.5 for two protease inhibitors.[7] The early lead compound L-685,434 had good potency in vitro in both enzyme and cell-based assays but was completely inactive in vivo after oral dosing due to poor solubility. The compound was modified by introducing ionizable centers into the

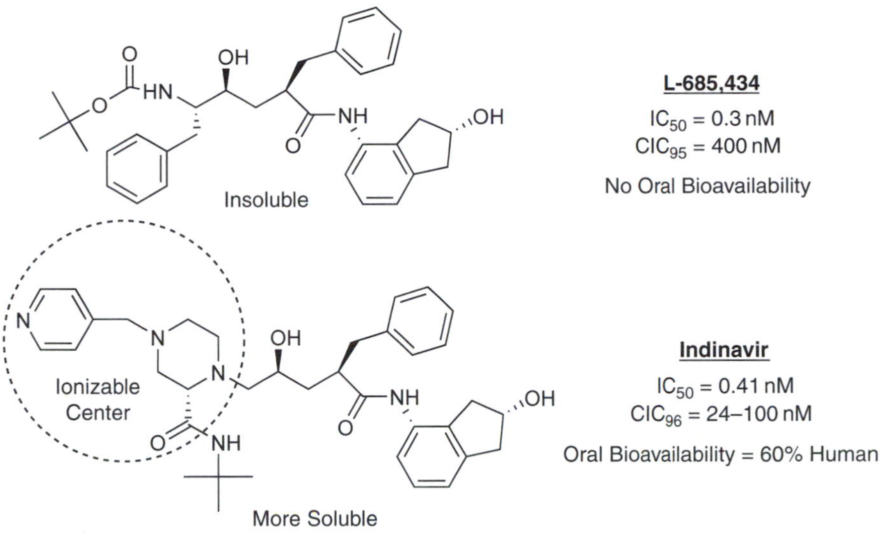

Figure 7.5 ► Effect of solubility on oral bioavailability. (Reprinted with permission from [39].)

molecule to increase solubility. The modified compound (indinavir) is much more soluble, maintains good potency, and has oral bioavailability of 60% in human.

7.2.2 Good Solubility is Essential for IV Formulation

To successfully develop an IV formulation, compounds must have sufficient solubility in the vehicle to deliver the expected dose in a restricted volume. Insoluble compounds can be a challenge to formulate, and the process is not always successful. For example, it was reported that high-dose rat and monkey IV studies could not proceed for LY299501 (Figure 7.6) because the solubility of the compound was too low for the targeted IV dose of 30 mg/kg.[8] In many cases, a high organic solvent content must be used to solubilize insoluble compounds for IV administration in laboratory animals. This can cause homolysis (nonisotonic), dissolution of tissues, artificially higher oral absorption, and brain penetration due to damage of membrane integrity. This can generate erroneous data and mislead project teams. Compounds with low aqueous solubility can precipitate at the site of injection and result in nonlinear pharmacokinetics, such as second peak phenomena, which has been observed for lidocaine, disopyramide, and YH439 as a result of redissolution of the precipitated material.[6]

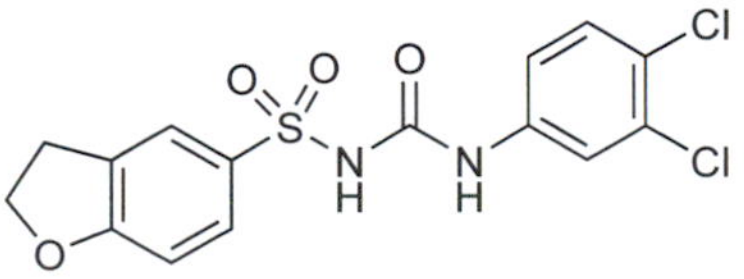

Figure 7.6 ► Drug discovery compound LY295501 for oncology.

7.2.3 Acceptance Criteria and Classifications for Solubility

One question commonly asked by medicinal chemist is: "What is the minimum solubility required for a compound?". This really depends on permeability and dose of a compound.

The maximum absorbable dose (MAD) is the maximum amount of a drug that can be absorbed at a certain dose. MAD is defined as follows[9,10]:

$$MAD = S^*K_a^*SIWV^*SITT,$$

where S = solubility (mg/mL, pH 6.5), K_a = intestinal absorption rate constant (min^{-1}; permeability in rat intestinal perfusion experiment, quantitatively similar to human K_a), SIWV = small intestine water volume (~250 mL), and SITT = small intestine transit time (min; ~270 min).

Thus, solubility and permeability are two major factors in achieving maximum absorption. Table 7.2 shows the minimum acceptable solubility of a drug for humans at a given dose and permeability in order to achieve maximum absorption. This is graphed in Figure 7.7. The more potent (i.e., dose producing the pharmacological effect if fully absorbed) and the more permeable the compound, the lower the solubility required to achieve complete absorption. On the contrary, the less potent and the less permeable the compound is, the higher the solubility required to achieve complete absorption. For example, if the compound has low permeability (0.003 min^{-1}) and low potency (10 mg/kg), a solubility of 3.46 mg/mL would be required to achieve maximum absorption of the dose. In drug discovery, solubility less than 0.1 mg/mL is quite common. Formulation often can help to increase the solubility of insoluble compounds. Therefore, maximum absorption of a compound with minimal solubility depends on high permeability and potency.

TABLE 7.2 ► Minimum Acceptable (Target) Solubility for Human Dosing at a Given Dose and Permeability to Achieve Maximum Absorption[9]

Human dose (mg) (MAD)	7	7	70	70	700	700
Human dose (mg/kg)	0.1	0.1	1	1	10	10
Permeability (K_a, min^{-1})	0.003 (low)	0.03 (high)	0.003 (low)	0.03 (high)	0.003 (low)	0.03 (high)
Minimum acceptable solubility (mg/mL)	0.035	0.0035	0.35	0.035	3.5	0.35

MAD, maximum absorbable dose.

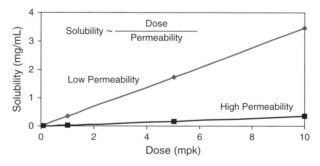

Figure 7.7 ► Relationship of solubility, permeability, and maximum absorbable dose. High-permeable compounds require lower solubility than low-permeability compounds to achieve maximum oral absorption. (Reprinted with permission from [38].)

Lipinski[11] has developed a useful graphical representation for the correlation of solubility, permeability, and dose (Figure 7.8). For example, if the compound has average permeability (shown as "avg K_a") and average potency (shown as "1.0" mg/kg dose if fully

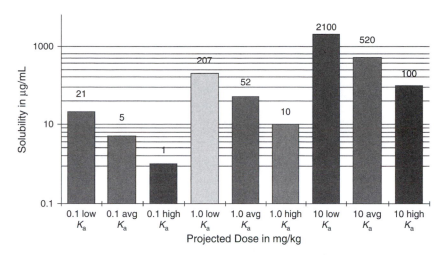

Figure 7.8 ▶ Graph for estimating the solubility (µg/mL) of discovery compounds needed to reach targeted dose levels (mg/kg), depending on the permeability (low, avg, high K_a). (Reprinted with permission from [11].)

absorbed), the compound will need to have a minimum solubility of 52 µg/mL to be completely absorbed. If the compound is not very potent, with a dose of 10 mg/kg and average permeability, then the solubility must be 10 times higher (520 µg/mL). Such estimates provide useful guidelines for optimizing solubility during discovery. In rating the solubility of compounds for discovery project teams, the following solubility classification ranges are suggested for medicinal chemists:

▶ <10 µg/mL   Low solubility

▶ 10–60 µg/mL   Moderate solubility

▶ >60 µg/mL   High solubility

These classification ranges are intended to provide general guidelines on potential solubility issues for *human oral absorption*. However, these criteria usually are too low for *animal dosing* in solution formulation. Table 7.3 gives estimates of the solubility needed for dosing a rat with a solution in discovery in vivo studies. The solubility requirement typically is much higher than 60 µg/mL, which is considered "high" in drug discovery for predicting human absorption.

TABLE 7.3 ▶ Target Solubility of Dosing Solution for Dosing a 250-g Rat at Ideal Dosing Volumes

	Target solubility (mg/mL)	
Dose (mg/kg)	PO	IV
1	0.1–0.2	0.2–1
5	0.5–1	1–5
10	1–2	2–10
Ideal volume (mL/kg)	5–10	1–5

Different solubility classification systems are used in different stages of drug discovery and development. They were developed to provide general guidelines on how to use solubility information to guide compound selection and advancement.

7.2.3.1 Biopharmaceutics Classification System

The solubility classifications used in drug discovery are quite different than those used in drug development. The Biopharmaceutics Classification System (BCS) is widely used in drug development. It divides compounds into four classes based on solubility and permeability (Figure 7.9).[12]

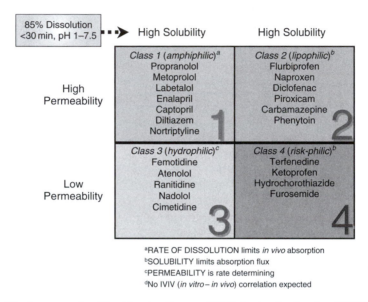

Figure 7.9 ▶ Biopharmaceutics Classification System (BCS) for in vitro/in vivo (IVIV) correlation.[12]

As development colleagues become involved in projects during the late discovery and predevelopment stages, it is often useful for discovery scientists to understand the BCS classifications. This facilitates experiments that address development issues, in order to promote the optimum candidate to development and streamline the transition to development.

▶ Class I is high solubility and high permeability. This is an ideal class for oral absorption.

▶ Class II is low solubility and high permeability. Formulation typically is used to enhance solubility of compounds in this class.

▶ Class III is high solubility but low permeability. Prodrug strategies typically are used for these compounds.

▶ Class IV is low solubility and low permeability. Development of this class of compounds can be risky and costly. No in vitro/in vivo correlations are expected.

The purpose of the BCS is to indicate the similarities and differences between compounds with regard to solubility and permeability. Compounds in the same BCS class tend to behave in a similar manner with regard to absorption and can follow the same regulatory approval process with regard to in vitro/in vivo correlation experiments. The Food and Drug Administration (FDA) can grant a waiver of bioavailability/bioequivalence studies for immediate-release orally administered formulations if the compound is class I and does not have a narrow therapeutic index. This saves a lot of resources and time for the development of new drugs, formulations, and generics.

The "high" solubility classification here is much more strict than the "high" solubility used in drug discovery ($>60\,\mu g/mL$). High solubility in BCS is defined as (a) 85% dissolution of the dose within 30 minutes at all pH values from 1 to 7.5 and (b) dose/solubility (D/S) $\leq 250\,mL$ (e.g., theophylline).[12,13]

For discovery scientists, the BCS emphasizes the critical roles and balance of solubility, permeability, and fully absorbed dose (i.e., potency), which challenge the development of a successful drug product. Beyond the role of solubility, permeability, and potency, discovery scientists must consider additional properties not addressed by the BCS (e.g., metabolic, plasma and solution stability, plasma protein binding, renal and biliary clearance, transporters). Wu and Benet[14] suggested a Biopharmaceutics Drug Disposition Classification System (BDDCS), which includes considerations of routes of elimination.

7.2.4 Molecular Properties for Solubility and Permeability Often are Opposed

Structural properties that determine solubility and permeability are shown in Figure 7.10.[15] All the physicochemical properties are intercorrelated. Changing one property can affect several others. The figure shows how structural features that enhance solubility often reduce permeability. For example, increasing charge, ionization, or hydrogen bonding capacity will increase solubility but will decrease permeability. Increasing lipophilicity and size to some extent will increase permeability but will decrease solubility. Medicinal chemists must balance the different structural features to find a balance between solubility and permeability for the clinical candidate in order to achieve optimal absorption.

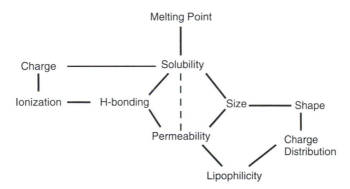

Figure 7.10 ▶ Effects of structural properties on solubility and permeability. (Reprinted with permission from [15].)

Permeability tends to vary over a more narrow range than does solubility.[9] The difference between a high-permeability and a low-permeability compound can be 50-fold (0.001–$0.05\,min^{-1}$). The difference between a high-solubility and low-solubility compound can be one million–fold ($0.1\,\mu g/mL$–$100\,mg/mL$). Therefore, if a structural modification improves solubility by 1,000 fold while it reduces permeability by 10-fold, then there will still be a 100-fold improvement in absorption.

7.3 Effects of Physiology on Solubility and Absorption

The gastrointestinal (GI) tract is a dynamic environment with changing conditions that affect compound solubility. Differences exist between species, and it is important to know what these differences are in order to extend the results of animal species dosing experiments to humans.

7.3.1 Physiology of the Gastrointestinal Tract

Some of the important characteristics of the GI tract are shown in Figure 7.11.[16] The GI tract has a pH gradient throughout its length, from acidic pH at the stomach to acidic–neutral pH in the small intestine to basic pH in colon. The wide pH range, long transit time, and high surface area of the small intestine allow for much higher absorption of drugs than in the stomach and colon. Acidic and basic drugs have different solubilities throughout the GI tract. Bases are more soluble in the stomach and upper small intestine due to ionization at acidic pHs. Acids are more soluble in later sections of the small intestine because the region is more basic.

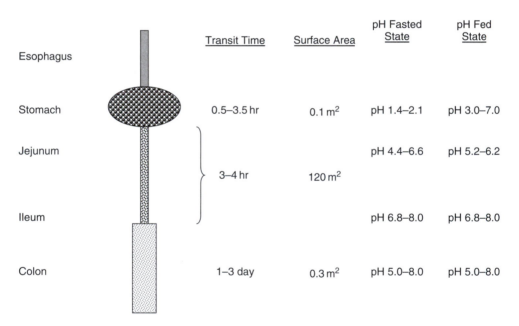

Figure 7.11 ▶ Physiology and biophysics of the gastrointestinal tract. (Reprinted with permission from [40].)

7.3.2 Species Differences in Gastrointestinal Tract

Gastric emptying time varies among species (Table 7.4).[17] If a drug is primarily absorbed in the intestine, it should reach plasma concentrations faster in the rat than in the human as a result of the earlier entry into the gastric lumen. Thus, rapid emptying results in earlier absorption.

The GI tract pH can be different among species. Rats and humans are good acid secretors. However, cats and dogs secrete less acid. Therefore, if the solubility of the discovery compound is pH dependent, differences in solubility between species will result in differences in absorption. An example is shown in Figure 7.12.[18] The compound L-735,524, an HIV

TABLE 7.4 ► **Species Differences In Gastric Emptying Time**

Species	Gastric-emptying time (min)
Rat	~10
Rabbit	30
Dog	40–50
Human	60

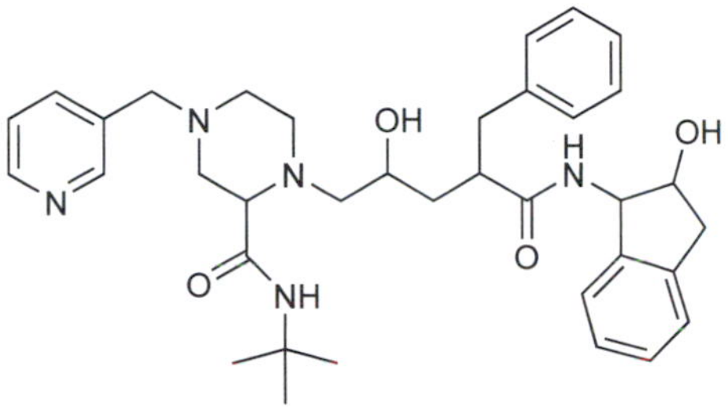

Solubility is pH dependent
- 60 mg/mL at pH 3.5
- <0.03 mg/mL at pH 5

Vehicle	Species	Oral %F, 10 mpk
0.5% Methylcellulose, pH 6.5 Suspension	Rat	16%
	Dog	16%
0.05 M Citric Acid, pH 2.5 Solution	Rat	23%
	Dog	72%

Figure 7.12 ► Species dependence of solubility and oral absorption for L-735,524. Dog is a poor acid secretor and has a gastrointestinal tract pH of 7. First-pass metabolism in dog is less than in rat.[18]

protease inhibitor analog of indinavir, shows steep changes in solubility with pH. It is much more soluble at low pH because of the presence of three basic amines as ionization centers. First-pass metabolism was higher for rat than dog. However, when the compound was dosed in a methylcellulose suspension formulation at pH 6.5 (owing to low solubility), the two species had the same oral bioavailability (16%). Rat is a good acid secretor but dog is not; thus, the compound is more soluble in rat stomach than in dog stomach. Even though the metabolism is faster in rat, the higher solubility in rat, due to lower acidity, results in the same oral bioavailability as in dog. When the compound is formulated in acidic buffer (citric acid), the compound is soluble and the oral bioavailability in dog increases to 72%, whereas the oral bioavailability in rat remains about the same as when a suspension formulation is used. This suggests that when solubility is not limited, first-pass metabolism is the dominant factor affecting oral bioavailability.

7.3.3 Food Effect

It is commonly thought that a high-fat diet will increase the solubility of lipophilic compounds and, therefore, enhance absorption. Actually, food can affect oral bioavailability in many different ways.[19,20] Food can either increase or decrease oral bioavailability by delaying gastric emptying (delays absorption), slowing input into the intestine (delays absorption), stimulating bile salt secretions (increases solubility of lipophilic compounds), altering the pH of the GI fluid (changes solubility), increasing blood flow (improves "sink condition" that enhances absorption and faster metabolism), and increasing competition for metabolic enzymes (slows metabolism). Different buffers have been developed to simulate gastric fluid conditions in fasted and fed states (Table 7.5).[16] It has been found that solubility measured in gastric fluid gave better prediction for oral bioavailability than solubility measured in aqueous buffer alone, when permeability is considered.[21]

TABLE 7.5 ▶   Buffer Composition that Simulates Fasted and Fed States[16]

	Simulated fasted state	Simulated fed state
Sodium taurocholate	5 mM	15 mM
Lecithin	1.5 mM (0.1%)	4 mM (0.3%)
pH	6.8	6.0

7.4   Structure Modification Strategies to Improve Solubility

Through the years, many cutting-edge technologies have been developed to formulate insoluble compounds (Figure 7.13).[22] These are discussed in Chapter 41. In drug discovery, medicinal chemists would like to solve drug delivery problems "with covalent bonds,"[9] to improve solubility through structure modification. The strategies are listed in Table 7.6.

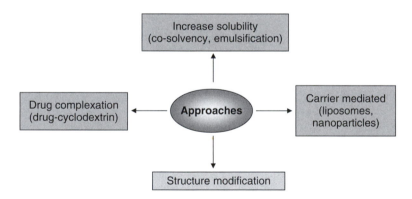

Figure 7.13 ▶   Approaches to improve solubility. The first choice is to modify the structure.

TABLE 7.6 ▶   Structure Modifications Strategies for Solubility Improvement

Structure modification	Section
Add ionizable group	7.4.1
Reduce Log P	7.4.2
Add hydrogen bonding	7.4.3
Add polar group	7.4.4
Reduce molecular weight	7.4.5
Out-of-plane substitution to reduce crystal packing	7.4.6
Construct a prodrug	7.4.7

7.4.1 Add Ionizable Groups

Addition of ionizable groups is commonly used for enhancing solubility. It is one of the most effective structural modifications to increase solubility. Typically, a basic *amine* or a *carboxylic acid* is introduced to the structure. Compounds with an ionizable functional group will be charged in pH buffers and have increased solubility.

One example is solubility enhancement of artemisinin (Figure 7.14), which is an antimalarial agent.[23] The sodium salt of a *carboxylic acid* analog achieved higher solubility, but, in this case, the compound was unstable. Ultimately, the *amine* analogs attained better solubility and stability and were active after oral dosing.[23]

For the series shown in Figure 7.15, the analogs with ether-containing side chains had good potency against a panel of tumor cell lines but had low solubility. Upon incorporation

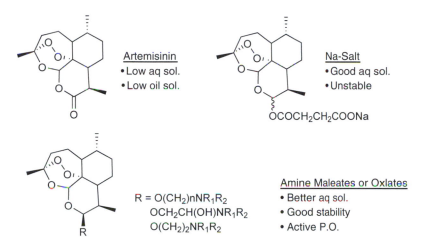

Figure 7.14 ▶ Introduction of a side chain with a carboxylic acid or amine enhances the solubility of artemisinin.

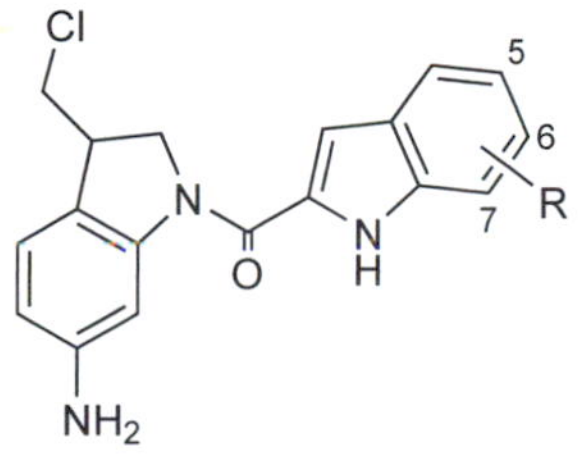

R	Solubility (μM)	IC50 (μM)			
		AA8	UV4	EMT6	SKOV3
5,6,7-triOMe	32	0.35	0.055	0.27	0.63
5-OMe	23	0.31	0.047	0.23	0.67
5-O(CH$_2$)$_2$NMe$_2$	700	0.16	0.044	0.12	0.26
5-OMe, 6-O(CH$_2$)$_2$NMe$_2$	>1200	0.22	0.039	0.11	0.15
5-OMe, 7-O(CH$_2$)$_2$NMe$_2$	47	0.14	0.029	0.09	0.16

Figure 7.15 ▶ Improved solubility of antitumor agents without loss of activity. (Reprinted with permission from [38].)

of a basic *amine*-containing side chain at position 5 or 6, solubility was greatly enhanced.[24] The last compound differed only at the position of substitution (position 7 vs. 5 or 6) but had lower solubility. This could be due to difference in crystal packing.

The compounds shown in Figure 7.16 were tested for solubility in simulated gastric fluid (pH 1.2) and phosphate buffer (pH 7.4).[25] Solubility was much higher for the compound that had a basic amine, especially at acidic pH.

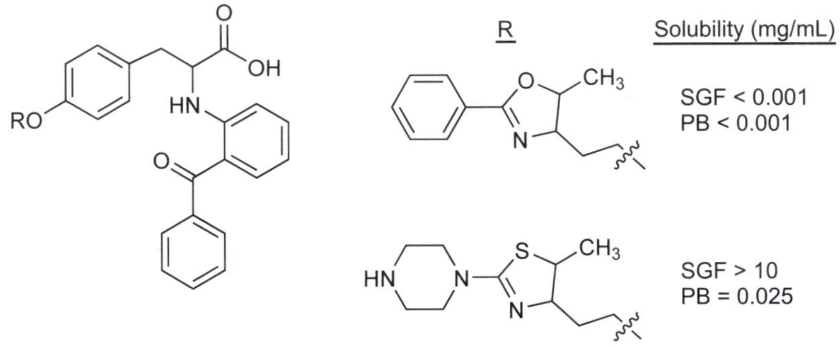

Figure 7.16 ► Solubility improves with increasing basicity. PB, phosphate buffer (pH 7.4); SGF, simulated gastric fluid (pH 1.2).

The most active compound in vitro is not necessarily the most active compound in vivo. The compounds shown in Figure 7.17 are an example.[26] The first compound has an IC$_{50}$ of 0.004 nM; however, it is not active in vivo because of its low solubility. The second compound is 5-fold less active in vitro, but it is active in vivo because of its higher solubility. Adding a basic nitrogen to the molecule as an ionization center enhances solubility. A successful drug possesses a balance of potency and drug-like properties.

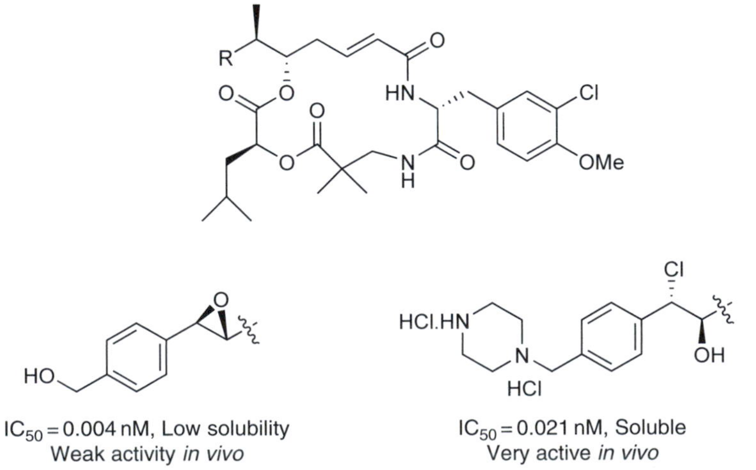

Figure 7.17 ► Series compounds that are more active in vitro and have low solubility may not be as active in vivo as series analogs that have lower in vitro activity but are soluble and, thus, better absorbed. (Reprinted with permission from [38].)

7.4.2 Reduce Log P

Several protease inhibitors are shown in Figure 7.18.[17] Reducing Log P increased solubility and led to higher systemic exposure, as indicated by enhanced maximum concentrations in the blood, C_{max}. Reducing Log P and increasing solubility enhances the in vivo exposure.

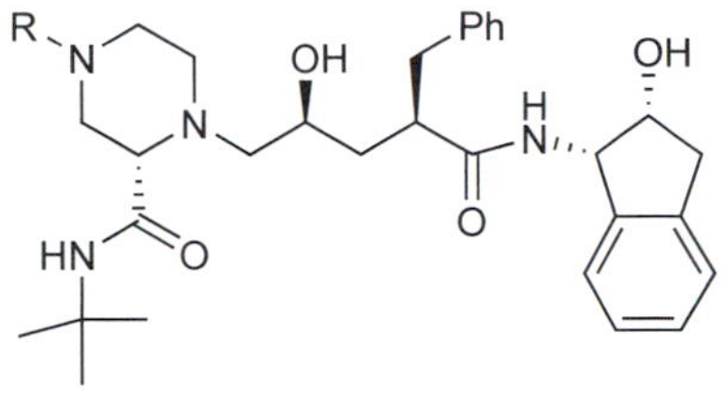

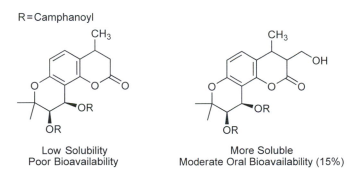

#	R	Cmax (uM)	Solubility (mg/mL) at pH 7.4	Log P
1	benzyloxycarbonyl	<0.10	<0.001	4.67
2	8-quinolinylsulfonyl	<0.10	<0.001	3.7
3	2,4-difluorophenylmethyl	0.73	0.0012	3.69
4	3-pyridylmethyl	11.4	0.07	2.92

Figure 7.18 ▶ For a series of protease inhibitors, absorption increased (as indicated by C_{max}) as solubility increased. Compound 4 in the chart was developed into the commercial drug indinavir.

7.4.3 Add Hydrogen Bonding

Introducing hydrogen bond donors and acceptors, such as OH and NH_2, can enhance aqueous solubility. Two anti-AIDS agents are shown in Figure 7.19.[27] The first compound has poor aqueous solubility and poor oral bioavailability, which limited its further development. Introducing a hydroxyl group into the molecule increased solubility and oral bioavailability.

Figure 7.19 ▶ Effects of H-bonds on solubility for anti-AIDS agents.

An example of antifungal agents is shown in Figure 7.20. Although nystatin, a polyene macrolide, is an effective antifungal agent, its use in medical practice is problematic because of its low solubility and significant human toxicity.[28] Structural modification by introducing hydroxyl groups at positions C31 and C33 increased the solubility by more than 2,000-fold. This dramatic increase in solubility is due, in part, to disruption of aggregate formation.[28]

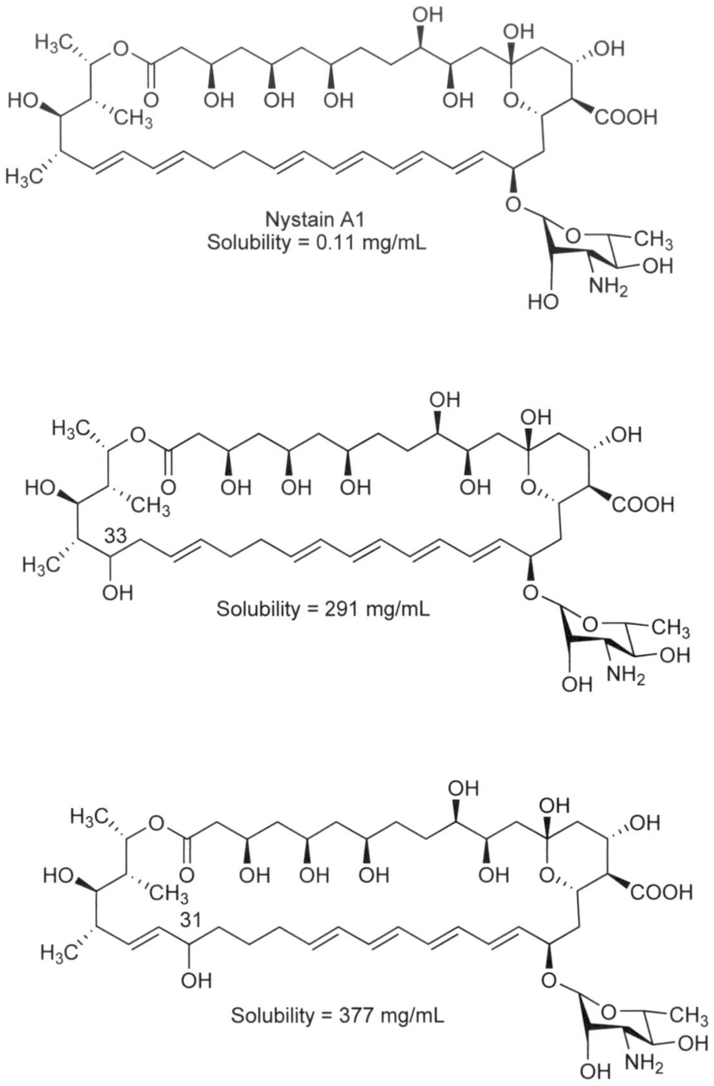

Figure 7.20 ▶ Addition of hydrogen bond increased aqueous solubility.

7.4.4 Add Polar Group

Water solubility usually increases with the addition of a polar group. Figure 7.21 shows a series of epoxide hydrolase inhibitors.[29] Solubility increased with the introduction of the ester group (more polar) and carboxylic acid group (more polar and ionizable).

7.4.5 Reduce Molecular Weight

Reduction in molecular weight is another useful approach for increasing solubility. An example of CDK2 inhibitors is shown in Figure 7.22.[30] The lower molecular weight increased solubility and metabolic stability, while maintaining in vitro activity. In vivo potency was improved because of the increased solubility and stability.

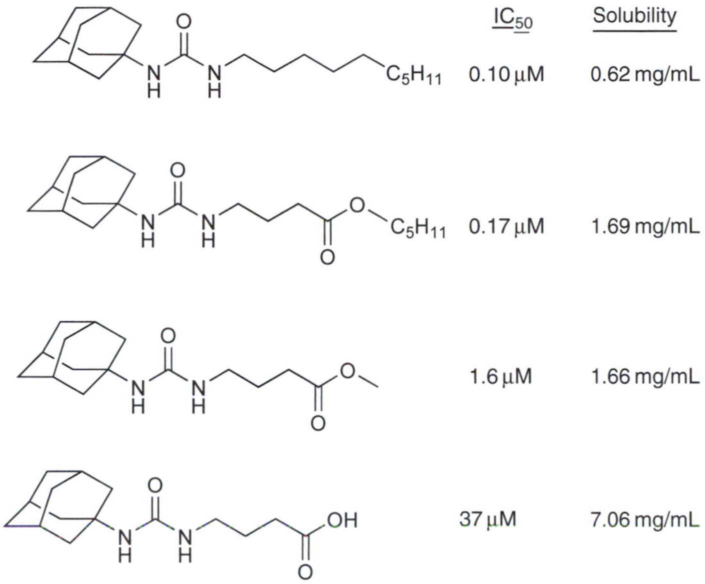

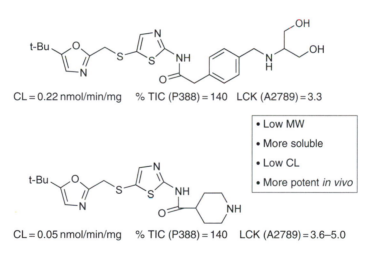

	IC$_{50}$	Solubility
	0.10 µM	0.62 mg/mL
	0.17 µM	1.69 mg/mL
	1.6 µM	1.66 mg/mL
	37 µM	7.06 mg/mL

Figure 7.21 ▶ Water solubility increased with addition of polar and ionizable groups in these epoxide hydrolase inhibitors.

CL = 0.22 nmol/min/mg   % TIC (P388) = 140   LCK (A2789) = 3.3

- Low MW
- More soluble
- Low CL
- More potent *in vivo*

CL = 0.05 nmol/min/mg   % TIC (P388) = 140   LCK (A2789) = 3.6–5.0

Figure 7.22 ▶ Reduction in molecular weight for these CDK2 inhibitors resulted in increased solubility, improved metabolic stability, and increased in vivo potency.

7.4.6 Out-of-Plane Substitution

Out-of-plane substitutions are illustrated in Figure 7.23.[31] Addition of the ethyl group shifts the planarity of the molecule, resulting in a disruption of the crystal packing to form a higher-energy crystal that is more soluble.

Figure 7.24 shows two AMPA/Gly$_N$ receptor antagonists. Although PNQX is very potent in both in vitro and in vivo models, the major disadvantage is its poor solubility (8.6 µg/mL at pH 7.4), which leads to the potential of crystallization in the kidney. Introducing out-of-plane substitutions enhanced solubility to 150 µg/mL.[32]

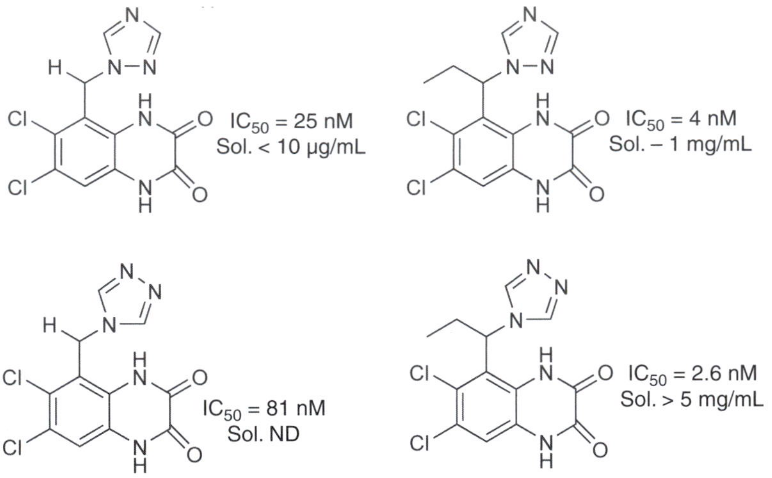

Figure 7.23 ▶ Addition of the ethyl group causes an out-of-plane conformation, which disrupts the crystal packing and increases the solubility. (Reprinted with permission from [38].)

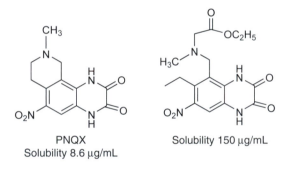

PNQX
Solubility 8.6 µg/mL

Solubility 150 µg/mL

Figure 7.24 ▶ Introducing out-of-plane substitutions increases solubility.

7.4.7   Construct a Prodrug

Charged or polar groups can be added to make prodrugs with increased aqueous solubility. Figure 7.25 shows fosphenytoin, which is a prodrug of phenytoin.[33] The phosphate group greatly increases the solubility, making it much easier to formulate for clinical dosing. Enzymatic hydrolysis in the intestine releases phenytoin for absorption. Prodrugs are discussed in Chapter 39.

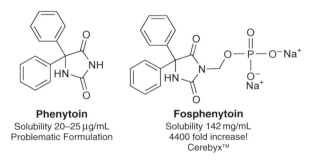

Phenytoin
Solubility 20–25 µg/mL
Problematic Formulation

Fosphenytoin
Solubility 142 mg/mL
4400 fold increase!
Cerebyx™

Figure 7.25 ▶ Fosphenytoin was prepared as a prodrug of phenytoin to increase solubility.

7.5 Strategies for Improving Dissolution Rate

Solubility is *how much* of a compound can dissolve in solution. Dissolution rate is *how fast* a compound can dissolve into solution. Increasing the dissolution rate will make a drug dissolve faster so that it can be absorbed within the GI transit time, even though solubility remains the same. Several approaches for increasing dissolution rate are listed in Table 7.7. These strategies are especially worthwhile for discovery animal dosing experiments to study in vivo efficacy and pharmacokinetics. Formulation strategies for discovery are discussed in Chapter 41.

TABLE 7.7 ▶ Strategies for Increasing Dissolution Rate

Goal	Change	Section
Increase surface area of solid	Reduce particle size	7.5.1
Predissolve in solution	Oral solution	7.5.2
Improve wetting of solid	Formulate with surfactants	7.5.3
	Prepare a salt form	7.5.4

7.5.1 Reduce Particle Size

Milling the solid material to a smaller particle size increases the surface area so that more molecules will be exposed to solvent at the same time. This results in an increased dissolution rate and increased oral absorption. Newer technologies allow the preparation of "nanoparticles" for even higher surface area. Figure 7.26 shows the effect of particle size reduction on oral exposure of MK-0869 in Beagle dogs. Exposure increases with decreasing particle size.

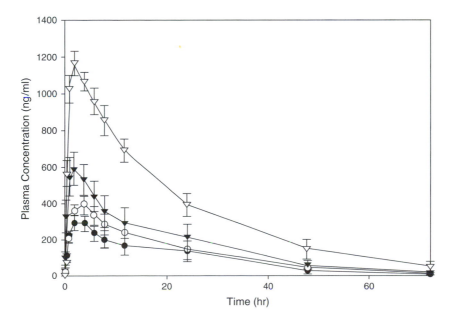

Figure 7.26 ▶ Effects of particle sizes from different milling processes on oral absorption of MK-0869 in Beagle dogs. Oral absorption of MK-0869 increased in Beagle dogs with decreasing particle size. (Reprinted with permission from [41].)

7.5.2 Prepare an Oral Solution

Solid material can be dosed orally as a solid dosage form, suspension (dispersed in a solution), or solution. If a compound is dosed as a solution rather than a suspension or solid dosage form, it has a better chance of being absorbed because it no longer must dissolve throughout the GI tract. Especially in drug discovery, a compound that can be dissolved in solution is the preferred formulation because it has a better chance of being absorbed and demonstrating proof of concept.

7.5.3 Formulate with Surfactants

Surfactants improve the wetting of the solid and increase the rate of disintegration of the solid material to finer particles. This increases dissolution rate and absorption.

7.5.4 Prepare a Salt Form

A salt of an acid or a base typically has a higher dissolution rate than the corresponding free acid or base. This produces increased absorption. The salt form does not change the intrinsic solubility of a free acid or a free base but does increase the overall solubility through ionization. Salt forms are discussed in greater detail in Section 7.6.

7.6 Salt Form

Salt forms typically are selected to modify physicochemical properties (e.g., dissolution rate, crystallinity, hygroscopicity, etc.) and mechanical properties (hardness, elasticity, etc.), leading to increased bioavailability, stability, and manufacturability.[34,35]

Examples of commercial drugs and their salt forms are listed in Table 7.8.[36] The solubility of the salt in pure water is much greater than the intrinsic solubility of the corresponding free acid or base. (Its solubility in buffer is affected by the solution pH, see Section 7.6.2.)

TABLE 7.8 ▶ Example Salts of Commercial Drugs[36]

Name	Solubility in water[a] (mg/mL)
Codeine	8.3
Sulfate	33
Phosphate	44
Atropine	1.1
Sulfate	2,600
Pseudoephedrine	0.02
Hydrochloride	2,000
Cetirizine	0.03
Dihydrochloride	300

[a] Final pH of water after salt dissolves differs with the salt.

7.6.1 Solubility of Salts

Three equilibria govern the relationship between a free base (or acid) and its corresponding salt (Figure 7.27). First is the equilibrium between the salt in the solid state and the salt

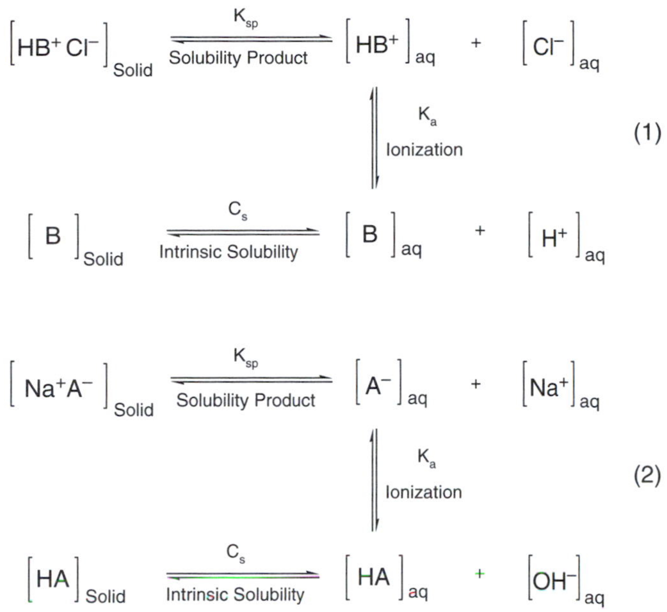

$$\left[HB^{+}Cl^{-} \right]_{Solid} \xrightleftharpoons[\text{Solubility Product}]{K_{sp}} \left[HB^{+} \right]_{aq} + \left[Cl^{-} \right]_{aq}$$

$$\updownarrow \quad \begin{matrix} K_{a} \\ \text{Ionization} \end{matrix} \qquad (1)$$

$$\left[B \right]_{Solid} \xrightleftharpoons[\text{Intrinsic Solubility}]{C_{s}} \left[B \right]_{aq} + \left[H^{+} \right]_{aq}$$

$$\left[Na^{+}A^{-} \right]_{Solid} \xrightleftharpoons[\text{Solubility Product}]{K_{sp}} \left[A^{-} \right]_{aq} + \left[Na^{+} \right]_{aq}$$

$$\updownarrow \quad \begin{matrix} K_{a} \\ \text{Ionization} \end{matrix} \qquad (2)$$

$$\left[HA \right]_{Solid} \xrightleftharpoons[\text{Intrinsic Solubility}]{C_{s}} \left[HA \right]_{aq} + \left[OH^{-} \right]_{aq}$$

Figure 7.27 ▶ Equilibrium of (**1**) free base and its salts and (**2**) free acid and its salts.

in solution (K_{sp}, solubility product constant). Second is the equilibrium between the free base (or acid) in solid state and free base (or acid) in solution (C_{s}, intrinsic solubility of base or acid). Third is the equilibrium between the free base (or acid) in solution and the corresponding salt in solution (K_{a}, ionization constant).

The solubility of a salt form at different pH values follows the curves that are generalized in Figure 7.28. The solubility at lower pH values for the salt of a base (Figure 7.28A) is determined by the K_{sp} of the salt. Different salts have different maximum solubilities. Before reaching maximum solubility, the solubility is determined by the pH, ionization constant K_{a}, and intrinsic solubility of the free base. At higher pH values, the solubility is the intrinsic solubility of the free base. The concentration behavior of the salt of an acid works in the opposite manner with regard to pH.

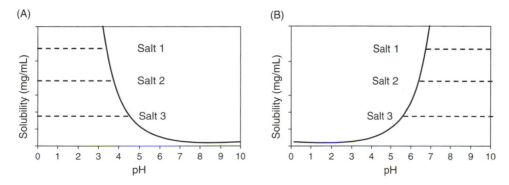

Figure 7.28 ▶ Solubility of hypothetical salts with pH. **A,** Three salts of a basic compound. **B,** Three salts of an acidic compound.

The solubility of a salt is not initially governed by the pH but by the solubility product of the salt (K_{sp}, solubility of the salt):

$$\text{Acid: } K_{sp} = [\text{Counterion}^+][\text{Anion}^-]$$

$$\text{e.g.: } NaA = Na^+ + A^- \qquad K_{sp} = [Na^+][A^-]$$

$$\text{Base: } K_{sp} = [\text{Counterion}^-][\text{Cation}^+]$$

$$\text{e.g.: } BHCl = BH^+ + Cl^- \qquad K_{sp} = [Cl^-][BH^+].$$

The high initial solubility of a salt puts a high concentration of the compound into solution for absorption. Once the compound is in solution, its concentration is governed by the ionization constant K_a and by the pH of the solution. Under these conditions, the compound may precipitate in the GI system, as governed by the solubility of free base or acid.

When the salt of a base (Figure 7.28A) is orally administered, its initial solubility is determined by its K_{sp}, which is independent of the pH. Once it is dissolved, pH and intrinsic solubility of the free base play a role. The acidic pH in the stomach favors the ionized form of the free base, keeping the base in solution, but neutral and basic pH values in the intestine and colon favor the free base, which can shift the equilibrium and lead to precipitation.

When the salt of an acid (Figure 7.28B) is orally administered, its initial solubility is also determined by its K_{sp}, which is independent of pH. Once it is dissolved in the acidic pH of the stomach, the free acid is favored, and its solubility is determined by its intrinsic solubility, which is low compared to the anion, so much of the material precipitates.

7.6.2 Effect of Salt Form on Absorption and Oral Bioavailability

In an aqueous medium at a specific pH with sufficient buffer capacity, a compound will have the same solubility regardless of whether it presents as a salt form or a free acid or base. Salts increase absorption by increasing dissolution rate. When the salt first encounters the aqueous phase, the high dissolution rate rapidly puts a lot of the compound in solution, which enhances absorption. Salts can stay in solution in a supersaturated state and do not precipitate immediately, even if the pH in the GI tract favors free acid (or base) formation and the free acid (or base) has low solubility. Supersaturation leaves a wider time window for compound to be absorbed. Furthermore, if the compound precipitates as a free acid or free base as a result of pH changes, it tends to precipitate as amorphous material and fine particles, which have higher solubility and a higher chance of being absorbed than crystalline material with a large particle size. As a result, the salt has higher absorption than the free acid (or free base) has on its own.

An example of this effect of salt form on increasing absorption is shown in Figure 7.29.[37] The absorption of the free acid p-amino-salicylic acid (PAS) was incomplete. Only 77% of the dose was absorbed because of low solubility and dissolution rate. However, absorption of the salts (Na, K, and Ca) was complete and had faster absorption, as indicated by shorter pharmacokinetic T_{max} (time of maximum concentration) and higher C_{max} (highest concentration). The salt form of PAS has higher oral exposure than the free acid.

For the protease inhibitor indinavir (Figure 7.5), the intrinsic solubility of the free base has very low solubility, and the solubility showed steep dependence on pH. HIV/AIDS patients tend to lack hydrochloric acid in their stomachs. Dosing of the free base caused unpredictable variable blood levels and resulted in rapid development of drug resistance by

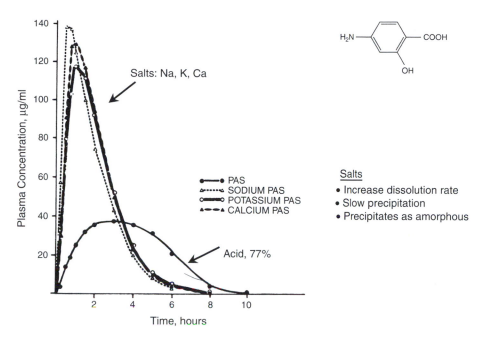

Figure 7.29 ▶ Salt forms have a higher initial solubility and absorption rate than the free acid and precipitate as a fine amorphous solid, resulting in higher initial dissolution and absorption rates. (Reprinted with permission from [37].)

viruses. A sulfate salt of indinavir was developed, which produced more consistent exposure. Indinavir is marketed as the sulfate ethanolate Crixivan.

7.6.3 Salt Selection

The counter ion for a salt should have a pK_a that differs from the drug by 2–3. For human studies, FDA-approved counter ions should be used. If not, enough toxicological data supporting selection of the counter ion must be provided. Approximately 70% of the counter ions used in commercial drugs are anions and 30% are cations. The 10 most commonly used anions and cations for salt formation are shown in Table 7.9.[35,38] The most common anion is Cl^- and the most common cation is Na^+.

TABLE 7.9 ▶ Commonly Used Counter Anions and Cations for Salt Formation[38]

Counter anions	Percent
Chloride	48
Sulfate	5.8
Bromide	5.2
Mesylate	3.2
Maleate	3.1
Citrate	2.8
Tartrate	2.7
Phosphate	2.5
Acetate	2.1
Iodide	1.2

Continued

TABLE 7.9 ► *Continued*

Counter cations	Percent
Sodium	58
Calcium	12
Potassium	9.8
Magnesium	4.5
Meglumine	2.4
Ammonium	2.0
Aluminum	1.4
Zinc	1.1
Piperazine	0.90
Tromethamine	0.90

The salt form of the drug product is also selected for its optimal physicochemical properties, such as crystallinity, morphology, hygroscopicity, stability, and powder properties.

7.6.4 Precautions for Using Salt Forms

The solubility of an HCl salt in the stomach can be limited by the "common ion effect." The high concentration of Cl^- (0.1–0.15 M) in the stomach can limit salt form dissolution, according to the K_{sp} (Figure 7.27). If this is the case, other salts such as sulfate or phosphate can be used.

If a compound is a very weak acid ($pK_a > 6$) or a very weak base ($pK_a < 5$) with very low intrinsic solubility, conversion of a small amount of the salt to the free acid or free base can cause precipitation and lead to various issues. For example, for phenytoin, a very weak acid with very low solubility ($\sim 20\,\mu g/mL$), the IV formulation of the sodium salt can precipitate because of conversion of the salt to the free acid. Precipitation during IV dosing can cause problems.

When the salt form particles enter the stomach or intestine, conversion to free acid, free base, or hydrate on the surface of the particle can cause formation of an insoluble film or coating that prevents further dissolution of the salt. In this case, the salt form will not enhance the dissolution rate.

☐ Problems

(Answers can be found in Appendix I at the end of the book.)

1. A basic amine (pK_a 9) is dissolved in DMSO and tested in phosphate buffer saline (PBS, pH 7.4) for biological activity. An HCl salt of this compound is prepared and tested under the same condition. Will the IC_{50} be the same, lower, or higher?

2. A free acid (pK_a 4) and its sodium salt are tested for solubility. Will they have the same solubility in water? Why? Will they have the same solubility in pH 7.4 potassium phosphate buffer?

3. What approaches can be used to increase solubility? What is the most effective chemical modification to increase solubility? What approaches can be taken to increase dissolution rate?

4. Compound A was dosed in rat as an oral suspension at 100 mg/kg, 200 mg/kg, and 300 mg/kg. C_{max} and AUC of all three doses were the same. What is the potential cause?

5. An acidic compound has intrinsic solubility of 2 µg/mL and pK_a of 4.4. What is the approximate solubility of the compound at pH 7.4?

6. Why does the solubility of subsequent analog compounds in a lead series tend to be lower during lead optimization?

7. List components and characteristics of the aqueous solution matrix that affect solubility.

8. List structural properties that affect solubility.

9. What is the difference between solubility and dissolution rate?

10. Why is thermodynamic solubility not as important as kinetic solubility in early drug discovery?

11. What solubility should the following compounds have for complete human absorption when orally dosed?: (a) dose of 1 mg/kg and high permeability, (b) dose of 10 mg/kg and high permeability, (c) dose of 10 mg/kg and average permeability.

12. Structural modifications to improve solubility often decrease what other property?

13. What usually is the most successful structure modification to improve solubility?

14. Making a salt improves the: (a) intrinsic solubility, (b) dissolution rate.

15. For the following lead structure, what structural modifications could you make to improve solubility?

16. Low solubility can cause which of the following?: (a) low oral bioavailability, (b) low metabolism, (c) low permeability, (d) increased burden on patients, (e) less expensive drug product formulation.

17. Which of the following are true about kinetic solubility measurements?: (a) compound is first dissolved in DMSO then added to aqueous buffer, (b) can be used to develop structure–solubility relationships, (c) is affected by solution pH or components, (d) can be used to recognize solubility limitations and guide structure modifications to improve solubility, (e) better for high throughput analysis than equilibrium solubility.

18. The minimum acceptable solubility to produce in vivo efficacy in humans is predictable using which of the following?: (a) target dose, (b) toxicity, (c) hERG blocking concentration, (d) permeability, (e) intestinal transit time.

19. Which of the following can a salt form change after oral dosing, as compared to the free acid or base?: (a) T_{max}, (b) C_{max}, (c) AUC, (d) oral bioavailability, (e) efficacy.

☐ References

1. Lipinski, C. A., Lombardo, F., Dominy, B. W., & Feeney, P.J. (1997). Experimental and computational approaches to estimate solubility and permeability in drug discovery and development settings. *Advanced Drug Delivery Reviews, 23,* 3–25.

2. Yalkowsky, S., & Banerjee, S. (1992). *Aqueous solubility: Methods of estimation for organic compounds.* New York, NY: Marcel Dekker.

3. Lee, Y.- C., Zocharski, P. D., & Samas, B. (2003). An intravenous formulation decision tree for discovery compound formulation development. *International Journal of Pharmaceutics, 253,* 111–119.

4. Venkatesh, S., & Lipper, R. A. (2000). Role of the development scientist in compound lead selection and optimization. *Journal of Pharmaceutical Sciences, 89,* 145–154.

5. Lipper, R. A. (1999). How can we optimize selection of drug development candidates from many compounds at the discovery stage? *Modern Drug Discovery, 2,* 55–60.

6. Yoon, W. H., Yoo, J. K., Lee, J. W., Shim, C.- K., & Lee, M. G. (1998). Species differences in pharmacokinetics of a hepatoprotective agent, YH439, and its metabolites, M4, M5, and M7, after intravenous and oral administration to rats, rabbits, and dogs. *Drug Metabolism and Disposition, 26,* 152–163.

7. van de Waterbeemd, H., Smith, D. A., Beaumont, K., & Walker, D. K. (2001). Property-based design: Optimization of drug absorption and pharmacokinetics. *Journal of Medicinal Chemistry, 44,* 1313–1333.

8. Ehlhardt, W. J., Woodland, J. M., Toth, J. E., Ray, J. E., & Martin, D. L. (1997). Disposition and metabolism of the sulfonylurea oncolytic agent LY295501 in mouse, rat, and monkey. *Drug Metabolism and Disposition, 25,* 701–708.

9. Curatolo, W. (1998). Physical chemical properties of oral drug candidates in the discovery and exploratory development settings. *Pharmaceutical Science & Technology Today, 1,* 387–393.

10. Johnson, K. C., & Swindell, A. C. (1996). Guidance in the setting of drug particle size specifications to minimize variability in absorption. *Pharmaceutical Research, 13,* 1795–1798.

11. Lipinski, C. A. (2000). Drug-like properties and the causes of poor solubility and poor permeability. *Journal of Pharmacological and Toxicological Methods, 44,* 235–249.

12. Retrieved from http://www.fda.gov/cder/guidance/2062dft.pdf. Guidance for Industry Waiver of in vivo bioavailability and bioequivalence studies for immediate release solid oral dosage forms containing certain active moieties/active ingredients based on a biopharmaceutics classification system, October 27, 2007.

13. Lindenberg, M., Kopp, S., & Dressman, J. B. (2004). Classification of orally administered drugs on the World Health Organization Model list of Essential Medicines according to the biopharmaceutics classification system. *European Journal of Pharmaceutics and Biopharmaceutics, 58,* 265–278.

14. Wu, C.- Y., & Benet, L. Z. (2005). Predicting drug disposition via application of BCS: Transport/absorption/elimination interplay and development of a biopharmaceutics drug disposition classification system. *Pharmaceutical Research, 22,* 11–23.

15. van de Waterbeemd, H. (1998). The fundamental variables of the biopharmaceutical classification system (BCS): A commentary classification. *European Journal of Pharmaceutical Sciences, 7,* 1–3.

16. Dressman, J. B., Amidon, G. L., Reppas, C., & Shah, V. P. (1998). Dissolution testing as a prognostic tool for oral drug absorption: Immediate release dosage forms. *Pharmaceutical Research, 15,* 11–22.

17. Lin, J. H., & Lu, A. Y. H. (1997). Role of pharmacokinetics and metabolism in drug discovery and development. *Pharmacological Reviews, 49,* 403–449.

18. Lin, J. (1995). Species similarities and differences in pharmacokinetics. *Drug Metabolism and Disposition, 23,* 1008–1021.

19. Zimmerman, J., Ferron, G., Lim, H., & Parker, V. (1999). The effect of a high-fat meal on the oral bioavailability of the immunosuppressant sirolimus (rapamycin). *Journal of Clinical Pharmacology, 39,* 1155–1161.

20. Davis, S. S. (1991). Physiological factors in drug absorption. *Annals of the New York Academy of Sciences, 618,* 140–149.

21. Aungst, B. J., Nguyen, N. H., Taylor, N. J., & Bindra, D. S. (2002). Formulation and food effects on the oral absorption of a poorly water soluble, highly permeable antiretroviral agent. *Journal of Pharmaceutical Sciences, 91*, 1390–1395.

22. Singla, A. K., Garg, A., & Aggarwal, D. (2002). Paclitaxel and its formulations. *International Journal of Pharmaceutics, 235*, 179–192.

23. Li, Y., Zhu, Y.- M., Jiang, H.- J., et al. (2000). Synthesis and antimalarial activity of artemisinin derivatives containing an amino group. *Journal of Medicinal Chemistry, 43*, 1635–1640.

24. Milbank, J. B. J., Tercel, M., Atwell, G. J., Wilson, W. R., Hogg, A., & Denny, W. A. (1999). Synthesis of 1-substituted 3-(chloromethyl)-6-aminoindoline (6-amino-seco-CI) DNA minor groove alkylating agents and structure-activity relationships for their cytotoxicity. *Journal of Medicinal Chemistry, 42*, 649–658.

25. Smith, D. A. (2002). *Ernst Schering Research Foundation Workshop, 32*, 203–212.

26. Al-awar, R. S., Ray, J. E., Schultz, R. M., et al. (2003). A convergent approach to cryptophycin 52 analogues: Synthesis and biological evaluation of a novel series of fragment A epoxides and chlorohydrins. *Journal of Medicinal Chemistry, 46*, 2985–3007.

27. Xie, L., Yu, D., Wild, C., et al. (2004). Anti-AIDS agents. 52. *Journal of Medicinal Chemistry, 47*, 756–760.

28. Borgos, S. E. F., Tsan, P., Sletta, H., Ellingsen, T. E., Lancelin, J.- M., & Zotchev, S. B. (2006). Probing the structure-function relationship of polyene macrolides: Engineered biosynthesis of soluble nystatin analogues. *Journal of Medicinal Chemistry, 49*, 2431–2439.

29. Kim, I.- H., Morisseau, C., Watanabe, T., & Hammock, B. D. (2004). Design, synthesis, and biological activity of 1,3-disubstituted ureas as potent inhibitors of the soluble epoxide hydrolase of increased water solubility. *Journal of Medicinal Chemistry, 47*, 2110–2122.

30. Misra, R. N., Xiao, H.- Y., Kim, K. S., et al. (2004). N-(cycloalkylamino)acyl-2-aminothiazole inhibitors of cyclin-dependent kinase 2. N-[5-[[[5-(1,1-dimethylethyl)-2-oxazolyl]methyl]thio]-2-thiazolyl]- 4- piperidinecarboxamide (BMS-387032), a highly efficacious and selective antitumor agent. *Journal of Medicinal Chemistry, 47*, 1719–1728.

31. Fray, M. J., Bull, D. J., Carr, C. L., Gautier, E. C. L., Mowbray, C. E., & Stobie, A. (2001). Structure-activity relationships of 1,4-dihydro-(1H,4H)-quinoxaline-2,3-diones as N-methyl-D-aspartate (glycine site) receptor antagonists. 1. Heterocyclic substituted 5-alkyl derivatives. *Journal of Medicinal Chemistry, 44*, 1951–1962.

32. Nikam, S. S., Cordon, J. J., Ortwine, D. F., et al. (1999). Design and synthesis of novel quinoxaline-2,3-dione AMPA/GlyN receptor antagonists: Amino acid derivatives. *Journal of Medicinal Chemistry, 42*, 2266–2271.

33. Stella, V. J. (1996). A case for prodrugs: Fosphenytoin. *Advanced Drug Delivery Reviews, 19*, 311–330.

34. Garrido, G., Rafols, C., & Bosch, E. (2006). Acidity constants in methanol/water mixtures of polycarboxylic acids used in drug salt preparations: Potentiometric determination of aqueous pKa values of quetiapine formulated as hemifumarate. *European Journal of Pharmaceutical Sciences, 28*, 118–127.

35. Stahl, P. H., & Wermuth, C. G. (Eds.). (2002). *Handbook of pharmaceutical salts: Properties, selection, and use*. Zurich: Weley-VCH.

36. Garad, S. D. (2004). How to improve the bioavailability of poorly soluble drugs. *American Pharmaceutical Review, 7*, 80–93.

37. Wan, S. H., Pentikainen, P. J., & Azarnoff, D. L. (1974). Bioavailability of aminosalicylic acid and its various salts in humans. III. Absorption from tablets. *Journal of Pharmaceutical Sciences, 63*, 708–711.

38. Bighley, L. D., Berge, S. M., & Monkhouse, D. C. (1995). Salt forms of drugs and absorption. In J. Swarbrick & J. C. Boylan (Eds.), *Encyclopedia of pharmaceutical technology* (Vol. 13) (pp. 453–499). New York: Marcel Dekker.

39. Di, L., & Kerns, E. H. (2006). Application of physicochemical data to support lead optimization by discovery teams. In R. T. Borchardt, E. H. Kerns, M. J. Hageman, D. R. Thakker, & J. L. Stevens (Eds.), *Optimizing the drug-like properties of leads in drug discovery*. New York: Springer, AAPS Press.

40. Macheras, P., Reppas, C., & Dressman, J. B. (1995). *Biopharmaceutics of orally administered drugs*. London: Ellis Harwood.

41. Wu, Y., Loper, A., Landis, E., et al. (2004). The role of biopharmaceutics in the development of a clinical nanoparticle formulation of MK-0869: A Beagle dog model predicts improved bioavailability and diminished food effect on absorption in human. *International Journal of Pharmaceutics, 285*, 135–146.

Permeability

Overview

▶ *Permeability is the velocity of molecule passage through a membrane barrier.*

▶ *Permeability is a determinant of intestinal absorption and oral bioavailability.*

▶ *Optimizing passive diffusion is productive because it is the predominant mechanism for absorption of most commercial drugs.*

▶ *Permeability is increased by removing ionizable groups, increasing* Log P, *and decreasing size and polarity.*

Permeability is the velocity of drug passage through a biological membrane barrier. This is a necessary process for absorption in the intestine, passage through blood–organ barriers, penetration into cells containing the therapeutic target, and elimination by the liver and kidney. Permeability also is important in cell-based biological assays in discovery, where the compound must permeate through the cell membrane to reach an intracellular therapeutic target. Prediction of in vitro permeability can enhance a wide range of drug discovery investigations, help with understanding cell-based bioassays, and assist prediction and interpretation of in vivo pharmacokinetics results.

8.1 Permeability Fundamentals

Drug molecules encounter several different membrane barriers in living systems. They include gastrointestinal (GI) epithelial cells, blood capillary wall, hepatocyte membrane, glomerulus, restrictive organ barriers (e.g., blood–brain barrier [BBB]), and the target cell membrane.

Different membranes can have different permeabilities for a compound. These differences are caused by differences in the membrane lipid mixture (passive diffusion), membrane transporter expression (active transport), or tightness of junctions between cells (paracellular).

Different mechanisms of membrane permeation were introduced in Chapter 3. They are passive diffusion, active uptake, endocytosis, efflux, and paracellular (Figure 8.1). Each of these permeation mechanisms is discussed in the following sections.

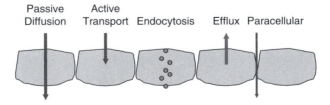

Figure 8.1 ▶ Major permeability mechanisms.[12]

8.1.1 Passive Diffusion Permeability

The most important permeability mechanism for drug discovery is *passive diffusion*. A compound moves by brownian motion from the aqueous phase through the cellular lipid bilayer membrane to the aqueous phase on the other side. The compound must first pass through the luminal (apical) lipid bilayer membrane, then pass through the cytoplasm and exit the cell through the abluminal (basolateral) membrane. Alternatively, a molecule may pass into the membrane, move laterally through the membrane, and exit elsewhere. Passive diffusion is driven by a concentration gradient, with the net movement of molecules from the area of higher concentration to the area of lower concentration.

During intestinal absorption, drug molecules move from the relatively high concentration of the gastric lumen, through the intestinal membrane to the capillary blood vessels, which have a relatively low drug concentration. It has been estimated that 95% of commercial drugs are predominantly absorbed in the GI tract by passive diffusion.[1,2] Even though the compound may be a transporter substrate, transporters can become saturated in the GI tract at the concentrations produced by oral administration.

It is important to remember that permeability is much higher for more lipophilic molecules than for polar molecules. This is primarily because molecules must pass through the highly nonpolar lipid bilayer membrane. Neutral molecules are much more permeable than their charged forms (anionic or cationic). (Ions may permeate, to some extent, by ion pairing to form a neutral species.)

For this reason, pH and pK_a play important roles in passive diffusion. This is illustrated in Figure 8.2 using passive diffusion permeability data from the parallel artificial membrane

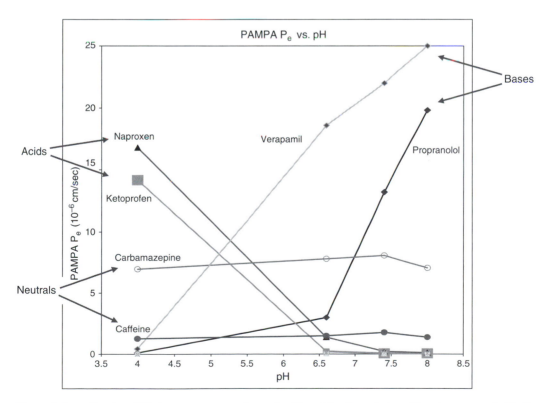

Figure 8.2 ▶ Passive diffusion across a membrane is affected by the solution pH and compound pK_a. In this PAMPA permeability experiment (see Chapter 26), acidic, basic, and neutral compounds have different permeability at different pH values. (Reprinted with permission from [13].)

permeability assay (PAMPA; see Chapter 26), with the same pH on both sides of the membrane. This diagram illustrates the effect of pH on the permeability of acids, bases, and neutrals at different pH values. The passive diffusion of acids is much higher at low pH, where a large percentage of the acid molecules in solution is neutral but drops with increasing pH as the percentage of neutral molecules drops and the percentage of anions increases. Conversely, the passive diffusion of bases is low at low pH, where a large percentage of the basic molecules in solution is protonated (cations) and increases with increasing pH as the percentage of neutral molecules increases. The ionization behaviors of acids and bases and their effects on permeability are illustrated in Figure 8.3. The passive diffusion permeability of neutral molecules is unaffected by pH.

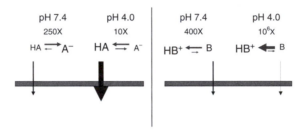

Figure 8.3 ▶ Effects of pH on passive diffusion of acids and bases across a lipid membrane. Permeability is highly favored for the neutral form. Thus, passive diffusion is greatest for bases at higher pH values and for acids at lower pH values. In this example, the acid has $pK_a = 5$ and the base has $pK_a = 10$. The fold ratios of the higher population species are shown.

pH and pK_a also affect passive diffusion across a lipid membrane that has a different pH on each side, as occurs in the GI. For an acid (e.g., $pK_a = 5$), passive diffusion is enhanced in the direction of the higher pH because of the ionization equilibrium. On the other hand, for a base (e.g., $pK_a = 10$), passive diffusion is enhanced in the direction of the lower pH. The reason for this behavior is shown in Figure 8.4. The acid anions (A^-) are "trapped" on the side of the higher pH. The base cations (BH^+) are "trapped" on the side of the lower pH. In the living system, this is not so obvious, because the bloodstream traps the drug molecules and moves them away (often termed *sink effect*). This effect, however, is apparent for in vitro permeability experiments when there is a difference of pH on either side of the membrane. For example, basic compounds may appear to be effluxed in a Caco-2 experiment, but this is really the "secretory" permeability of bases toward the lower pH.

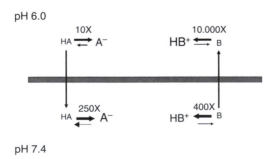

Figure 8.4 ▶ pH differences across a lipid membrane, such as in the GI tract, affect passive diffusion. For an acid (e.g., $pK_a = 5$), passive diffusion is enhanced in the direction of the higher pH because of the ionization equilibrium. On the other hand, for a base (e.g., $pK_a = 10$), passive diffusion is enhanced in the direction of the lower pH. The fold ratio of the higher population species is shown.

8.1.2 Endocytosis Permeability

Another route of permeability is *endocytosis*. Compounds may be engulfed by the membrane, pass through the cell within the vesicle, and be released on the other side. This has been of only minor interest for small molecule drug discovery.

8.1.3 Active Uptake Permeability

Molecules may be permeable by *active uptake transport*, in which a compound binds to a transmembrane protein and moves through the membrane. Active transport requires the expenditure of energy, commonly two ATPs for each molecule transported. Active transport often occurs against the concentration gradient. There must be affinity of the drug for the transporter. Although transporters serve a vital function for the permeability of natural ligands, such as nutrients, they also can be responsible for the permeability of some drug molecules. Transporters are discussed in Chapter 9.

8.1.4 Paracellular Permeability

If molecules are small and polar, they might pass by *paracellular* permeability between the epithelial cells through "pores" or channels that are approximately 8 Å in size. Cells in the GI tract or other organs, such as the glomerulus in the kidney, are sometimes termed *leaky* because of the somewhat loose junctions between the cells that allow molecules to slip between. In other tissues, such as the BBB, the junctions are very tight and there is no appreciable paracellular permeability. In the intestine, this route of permeation is observed for less than 5% of drug compounds. The pores represent less than 0.3% of the total membrane surface, so this route of absorption has limited capacity. Generally, paracellular permeability in the GI tract is available primarily to compounds that have a molecular weight less than 180 Da and are polar.

8.1.5 Efflux Permeability

Another major mechanism of permeability is *efflux*, the active transport of compounds from inside the cell or membrane back into the lumenal space. P-glycoprotein (Pgp) and breast cancer resistance protein (BCRP) are well-known efflux transporters. The net effect of efflux transport is the reduction of drug concentration within the cell or permeation across the membrane. Pgp is a member of the ABC (ATP binding cassette) family of transporters, which utilize the energy from cleavage of two molecules of ATP to ADP and inorganic phosphate for the transport of each drug molecule. Efflux transport is also found in other membranes, such as the BBB, where it serves a protective function by opposing the exposure of brain tissue to some xenobiotic compounds. In liver hepatocytes, efflux enhances removal of drugs and metabolites from within the hepatocytes to the bile canaliculus for elimination from the body. In the nephron, Pgp is one of the transporters involved in active secretion into the proximal tubule. The expression level of transporters, such as Pgp, varies along the length of the small intestine. It has been reported that compounds can move out of systemic circulation and be secreted into the intestinal lumen by efflux transporters.[3] Pgp is discussed further in Chapter 9.

8.1.6 Combined Permeability

The permeability of a compound is the composite of permeability from all of the mechanisms available to it. The term *absorptive transport* is often used to denote compound flux from the GI lumen toward the bloodstream. As shown in Figure 8.5, absorptive transport is the result of passive diffusion, which is driven by the concentration gradient and pH effects, active transport, which is driven by affinity for the transporter, and paracellular permeability, which is driven by size, polarity, and concentration gradient. Conversely, the term *secretory transport* is often used to denote compound flux in the direction of the GI lumen. Secretory transport is the result of passive diffusion and efflux.

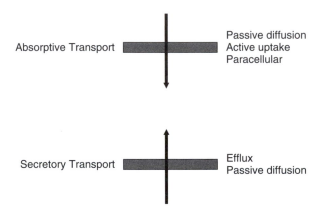

Figure 8.5 ▶ Composite permeability for a particular compound is the result of dynamic interaction of local conditions and how they affect the various permeability mechanisms. Conditions include concentration gradient, pH gradient, transporter affinity, molecular size, and polarity.

The flux of a compound is affected by concentration. For example, after oral administration, the concentration of a compound in the GI tract may be so high that (a) transporters are saturated, and (b) the high concentration gradient drives passive diffusion. Thus, passive diffusion becomes the major route of compound absorption. By contrast, at the BBB, the circulating drug concentration is much lower and usually is not near the transporter saturating concentration. Under these conditions, active transport mechanisms, either uptake or efflux, can have a much greater effect on total permeability if the compound is a substrate for a transporter.

The total permeability of a compound differs with the tissue. Factors that affect permeability are the local concentration gradient, pH polarity across membranes, expression of transporters, transporter K_m, the side of the cell the transporters are present (i.e., apical or basolateral), affinity of the compound for various transporters, and the size of the pores between cells that form the membrane barrier. There is a dynamic modulation of permeability mechanisms as a result of the conditions.

8.2 Permeability Effects

Permeability affects many factors that determine pharmacology in both living systems and in vitro discovery experiments. Two of these factors are bioavailability and cell-based biological activity assays.

8.2.1 Effect of Permeability on Bioavailability

Drug absorption in the GI tract after oral administration depends heavily on permeability. Compounds with low permeability typically have low bioavailability. For example, Table 8.1 shows the case of a potent, highly charged acidic compound. The compound has a bioavailability of less than 1%. This is indicated by the PAMPA passive diffusion permeability method (see Chapter 26), with a low permeability value (0.1×10^{-6} cm/s). When a prodrug

TABLE 8.1 ▶ Example of Effect of Permeability on Oral Bioavailability of an Acidic Compound

	Compound	Prodrug
PAMPA ($P_e \times 10^{-6}$ cm/s)	0.1	7.0
Oral bioavailability	<1%	18%

A compound with good potency ($K_i = 7$nM) had low permeability and bioavailability. Its prodrug had good permeability and bioavailability.
PAMPA, parallel artificial membrane permeability assay.

was made, the compound had much higher PAMPA permeability value (7×10^{-6} cm/s) and a much higher bioavailability of 18%. In this case, passive diffusion is limited for the highly charged acidic compound ($pK_a = 4.5$).

8.2.2 Effect of Permeability on Cell-Based Activity Assays

Permeability can limit the activity of compounds in cell-based assays. For intracellular therapeutic targets, the compound must penetrate the cellular membrane to show activity. Therefore, as discovery project teams progress from non–cell-based assays (e.g., enzymes, receptors) to cell-based assays, the activity can be severely reduced for some compounds. If this is due to the unrecognized cell membrane barrier rather than intrinsic activity, the project team might drop the compound from further consideration. A more successful approach is to recognize when permeability may be limiting cell-based activity. Analogs with improved permeability can be synthesized and tested for activity. It would be unfortunate to discard an active lead that could be improved by synthetic modification if the property (here permeability) causing the poor bioactivity can be identified and the structure modified.

An example of the effect of permeability on cell-based assays is shown in Table 8.2. Both good enzyme activity and permeability were required in order to produce good bioactivity in the cell-based assay. In some cases, compounds that are only moderately potent in the enzyme assay can be the most potent of the group of analogs in the cell-based assay if permeability limits intracellular exposure of the compounds that are more active in the enzyme assay (Table 8.3).

TABLE 8.2 ▶ Example of Effect of Permeability on Cell-based Assay Bioactivity

Compound	In vitro K_i (μM)	PAMPA ($P_e \times 10^{-6}$ cm/s)	Cell-based IC_{50} (μM)
A	0.007	4.9	10.5
B	0.02	1.0	22.1
C	0.01	0.02	Inactive
D	0.05	0.1	Inactive
E	3.5	14.3	Inactive
F	17	6.6	Inactive
G	4.3	0.01	Inactive

White, gray, and dark gray cells categorize data for high-, moderate-, and low-activity ranges, respectively. Cell-based activity requires both good enzyme activity and permeability.
PAMPA, parallel artificial membrane permeability assay.

TABLE 8.3 ▶ Example of How Two Compound Series Can Demonstrate Different Activities in Cell-Based Assays and Enzyme Assays

	Compound series I	Compound series II
Enzyme activity assay	High potency	Moderate potency
PAMPA permeability	Low	High
P-glycoprotein efflux permeability	Yes	No
Cell-based activity assay	Inactive	Active

PAMPA, parallel artificial membrane permeability assay.

8.3 Permeability Structure Modification Strategies

The best way to improve permeability is structural modification. Formulations are not effective in fixing permeability. Thus, it is important to assess permeability early and to build permeability improvement into the synthetic plan from the beginning. This could rescue a chemical series that has great potential and improve drug exposure in animal pharmacology and pharmacokinetic studies.

Several strategies for improving permeability are listed in Table 8.4. These strategies are based on a few fundamental concepts: reduce ionizability, increase lipophilicity, reduce polarity, or reduce hydrogen bond donors or acceptors.

TABLE 8.4 ▶ Strategies for Improving Permeability by Structural Modification

Structure modification strategy	Section
Ionizable group to non-ionizable group	8.3.1
Add lipophilicity	8.3.2
Isosteric replacement of polar groups	8.3.3
Esterify carboxylic acid	8.3.4
Reduce hydrogen bonding and polarity	8.3.5
Reduce size	8.3.6
Add nonpolar side chain	8.3.7
Prodrug	8.3.8

8.3.1 Ionizable Group to Non-ionizable Group

The effect of permeability on absorption after oral administration is shown by the example in Figure 8.6.[4] For the compound where R is CO_2H, in vitro Caco-2 permeability is low and *in vivo* oral bioavailability (%F) is low (4%). When R is the less polar and non-ionizable CH_2OH group, in vitro Caco-2 permeability is 30-fold higher and *in vivo* oral bioavailability is much higher (66%).

8.3.2 Add Lipophilicity

In another example of bioavailability,[5] Figure 8.7 shows that when R is CH_2NHCH_3, in vitro Caco-2 permeability is moderate, which is consistent with the moderate 24% in vivo oral bioavailability. When R is the more lipophilic $CH_2N(CH_3)_2$, in vitro permeability is much higher, resulting in a high 84% in vivo oral bioavailability.

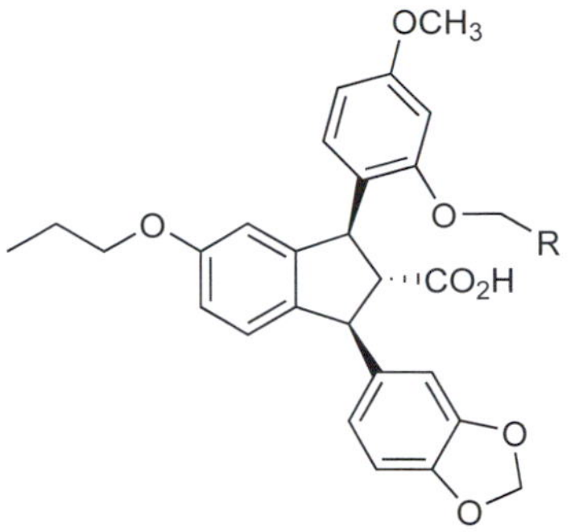

R	ETA, Ki (nM)	Caco-2 (cm/h)	% F (rat)
CO_2H	0.43	0.0075	4
CH_2OH	1.1	0.2045	66

Figure 8.6 ► Effect of permeability on oral absorption.

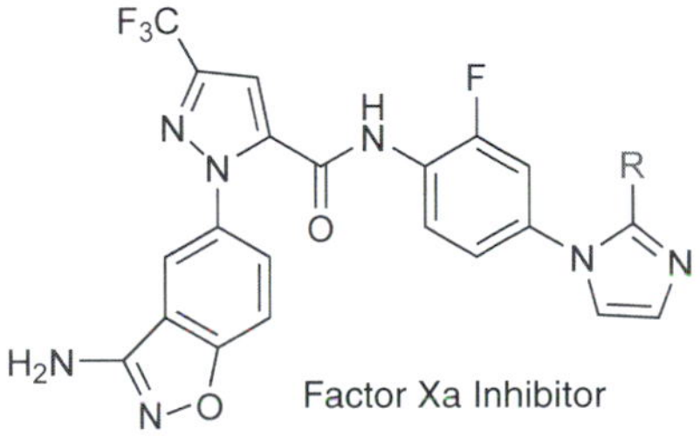

Factor Xa Inhibitor

R	FXa K_i (nM)	Caco-2 P_{app} ($\times 10^{-6}$ cm/s)	CL (L/h/Kg)	$T_{1/2}$ (h)	Vdss (L/Kg)	F (%)
CH_2NHMe	0.12	0.2	1.1	3.7	4.6	24
CH_2NMe_2	0.19	5.6	1.1	3.4	5.3	84

Figure 8.7 ► Substitution at R of $CH_2N(CH_3)_2$ for CH_2NHCH_3 in these factor Xa inhibitors increases Caco-2 permeability and bioavailability. (Reprinted with permission from [14].)

8.3.3 Isosteric Replacement of Polar Groups

When a carboxylic acid was replaced with an isosteric tetrazole moiety,[6] Caco-2 permeability increased (Figure 8.8). The tetrazole had the same PTP1B enzymatic activity ($K_i = 2 \, \mu M$). The carboxylic acid did not have cellular activity in vitro; however, the tetrazole exhibited positive cellular activity.

8.3.4 Esterify Carboxylic Acid

The PTP1B lead in Figure 8.9 was a dicarboxylic acid and was potent and selective in an in vitro enzyme assay.[7] Its activity in a cell-based model was low, which was consistent with the low permeability in the in vitro MDCK cell monolayer permeability assay (see Chapter 26). Synthesis of the diethyl ester prodrug greatly improved permeability and activity in the cell-based assay.

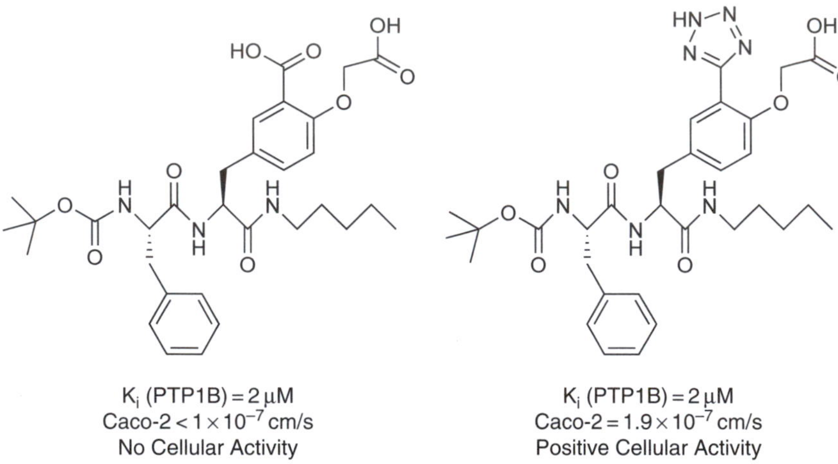

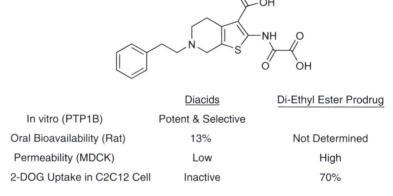

K_i (PTP1B) = 2 μM
Caco-2 < 1 × 10⁻⁷ cm/s
No Cellular Activity

K_i (PTP1B) = 2 μM
Caco-2 = 1.9 × 10⁻⁷ cm/s
Positive Cellular Activity

Figure 8.8 ▶ Replacement of a carboxylic acid with the bioisosteric tetrazole maintained activity and increased permeability, resulting in cell-based assay activity. (Reprinted with permission from [14].)

	Diacids	Di-Ethyl Ester Prodrug
In vitro (PTP1B)	Potent & Selective	
Oral Bioavailability (Rat)	13%	Not Determined
Permeability (MDCK)	Low	High
2-DOG Uptake in C2C12 Cell	Inactive	70%

Figure 8.9 ▶ Effects of permeability on cell-based assay activity for PTP1B lead. (Reprinted with permission from [14].)

8.3.5 Reduce Hydrogen Bonding and Polarity

The deleterious effects of hydrogen bonding and polarity on passive diffusion permeability are shown in the series in Figure 8.10. As Cl is modified to F, polarity increases and passive diffusion permeability decreases. As CH_3 is modified to OCH_3, a hydrogen bond acceptor is added and permeability decreases.

8.3.6 Reduce Size

Figures 8.11[8] and 8.12[9] show examples of permeability structure–property relationships. If we examine the cases where one R group is held the same and the other R group is varied, the permeability effects of size and polarity are observed. Increasing size (e.g., methyl, ethyl, butyl, phenyl) reduced Caco-2 permeation or percent of the dose that was absorbed. Increasing polarity (e.g., CH_3 to CF_3, or CH_2CH_3 to CF_2CF_3) also reduced permeability. Because all of the compounds have similar activity, the structural analog series allows prioritization of compounds based on their properties that will enhance bioavailability and penetration of permeation barriers on the way to the therapeutic target.

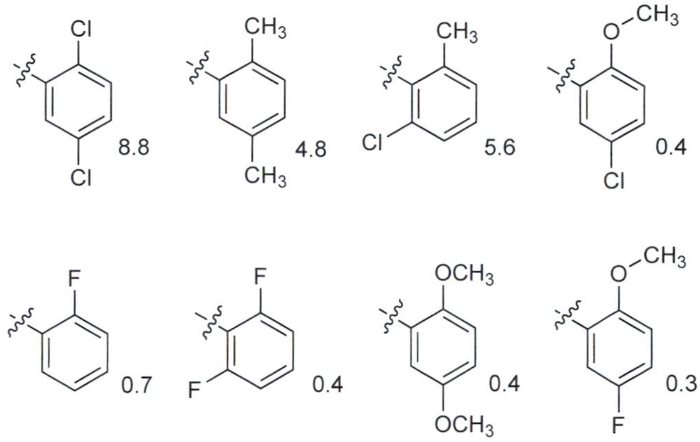

Figure 8.10 ▶ Example of the reduction in permeability by passive diffusion as polarity or hydrogen bonding increases. Permeability values are for PAMPA in units of 10^{-6} cm/s. (Reprinted with permission from [14].)

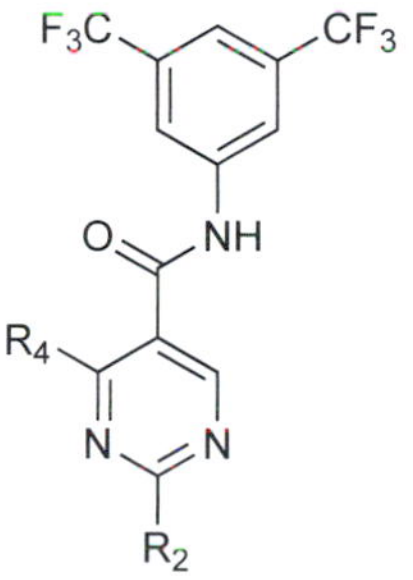

R4	R2	Caco-2 Permeability ($\times 10^{-7}$ cm/s, n = 3, mean $\pm$ SD)
CF_3	Cl	11 ± 4
H	Cl	61 ± 7
CH_3	Cl	62 ± 6
CH_2CH_3	Cl	58 ± 9
$CH_2CH_2CH_3$	Cl	31 ± 9
CF_2CF_3	Cl	9 ± 9
Cl	Cl	31 ± 6
Ph	Cl	9 ± 7
CF_3	F	19 ± 6

Figure 8.11 ▶ Effects of substitutions on permeability.

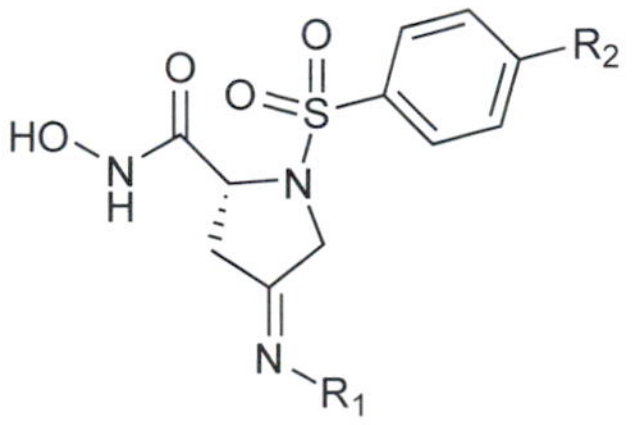

R1	R2	% Dose Absorbed (rat ileum)
OH	OMe	29–35
OH	O^{n}Bu	2–5
OMe	O-4-Pyr	50–68
O^{t}Bu	O-4-Pyr	10–18
OPh	O-4-Pyr	not detected
OMe	OMe	78–81
OMe	OEt	23–42
OMe	O^{n}Bu	28–36
OMe	OPh	15–18

Figure 8.12 ▶ Effects of substitutions on permeability.

8.3.7 Add Nonpolar Side Chain

Modification of a cyclic peptide to increase permeability is shown in Figure 8.13. By adding the nonpolar side chain, the lipophilicity was increased, resulting in an improvement of permeability.[10]

A series of phenylalanine dipeptides (Figure 8.14) was modified with increasingly lipophilic side chains.[11] This modification resulted in increasing Caco-2 permeability.

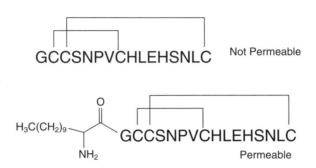

Figure 8.13 ► Adding the nonpolar side chain to this cyclic peptide increased the lipophilicity and resulted in improved permeability.

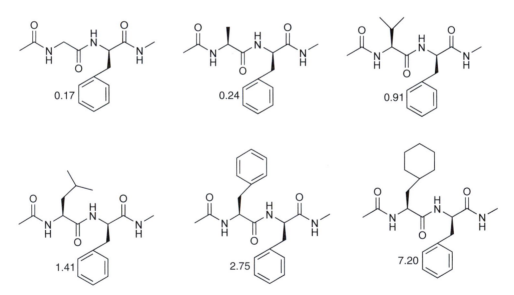

Figure 8.14 ► For this series of phenylalanine dipeptides, Caco-2 permeability (apical to basolateral, units of 10^{-6} cm/s) improved with increasing lipophilicity. (Reprinted with permission from [14].)

8.3.8 Prodrug

Prodrugs have been used to increase permeability. Figure 8.15 shows several prodrugs that have been made for permeability purposes.

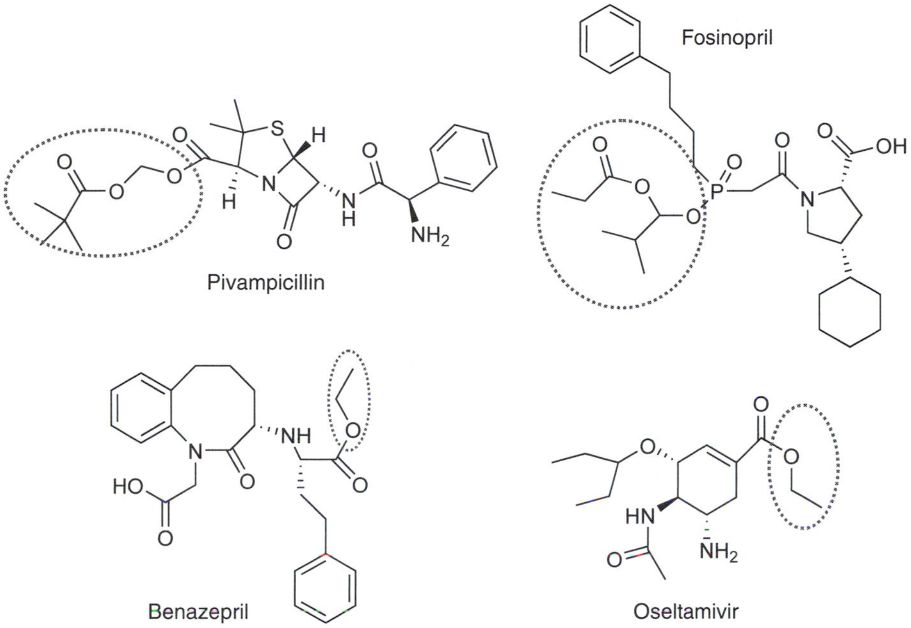

Figure 8.15 ▶ Prodrugs with impróved passive diffusion permeability. The pro-moiety is circled.

▢ Problems

(Answers can be found in Appendix I at the end of the book.)

1. What is the predominant permeability mechanism for absorption of most commercial drugs?

2. What are the structural properties of compounds that undergo paracellular permeation?

3. How will passive diffusion permeability change as pH increases from 4.5 to 8 for: (a) basic compound, (b) acidic compound?

4. List important permeability barriers for drug discovery.

5. Which of the following structural modifications likely will improve permeability?: (a) change an amine to a methyl, (b) add a hydroxyl group, (c) remove a propyl group, (d) change a carboxylic acid to an ethyl ester, (e) change a carboxylic acid to a tetrazole.

6. For the following lead compared, what structural modifications could you make that might improve permeability?

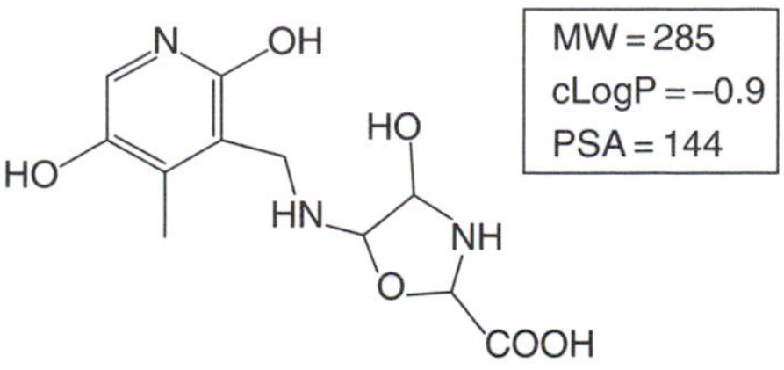

MW = 285
cLogP = −0.9
PSA = 144

7. Permeability is important for which of the following?: (a) absorption in intestine, (b) CYP metabolism, (c) BBB penetration, (d) dissolution in the intestinal lumen, (e) in vitro cell-based assay, (f) to reach intracellular targets in vivo.

8. Following are groups that could be added to a lead compound that is MW 300 and has ClogP 2.0. Rank them from lowest to highest predicted permeability of the product: (a) $-CH_3$, (b) $-OH$, (c) $-OCH_3$, (d) $-COOH$.

9. Following are groups that could be added to a lead compound that is MW 450 and has ClogP 4.5. Rank them from lowest to highest predicted permeability of the product: (a) $-C_6H_5$, (b) $-CH_3$, (c) $-C_3H_7$.

10. Following are groups that could be added to a lead compound that is MW 250 and has ClogP 0.0. Rank them from lowest to highest permeability of the product: (a) $-CH_3$, (b) $-C_6H_{11}$, (c) $-C_3H_7$.

References

1. Mandagere, A. K., Thompson, T. N., & Hwang, K.- K. (2002). Graphical model for estimating oral bioavailability of drugs in humans and other species from their Caco-2 permeability and in vitro liver enzyme metabolic stability rates. *Journal of Medicinal Chemistry*, *45*, 304–311.

2. Artursson, P. (2002). Prediction of drug absorption: Caco-2 and beyond. In *PAMPA 2002*: San Francisco, CA.

3. van de Waterbeemd, H., Smith, D. A., Beaumont, K., & Walker, D. K. (2001). Property-based design: optimization of drug absorption and pharmacokinetics. *Journal of Medicinal Chemistry*, *44*, 1313–1333.

4. Ellens, H., Eddy, E. P., Lee, C.- P., et al. (1997). In vitro permeability screening for identification of orally bioavailable endothelin receptor antagonists. *Advanced Drug Delivery Reviews*, *23*, 99–109.

5. Quan, M. L., Lam, P. Y. S., Han, Q., et al. (2005). Discovery of 1-(3'-aminobenzisoxazol-5'-yl)-3-trifluoromethyl-N-[2-fluoro-4-[(2'-dimethylaminomethyl)imidazol-1-yl]phenyl]-1H-pyrazole-5-carboxyamide hydrochloride (Razaxaban), a highly potent, selective, and orally bioavailable factor Xa inhibitor. *Journal of Medicinal Chemistry*, *48*, 1729–1744.

6. Liljebris, C., Larsen, S. D., Ogg, D., Palazuk, B. J., & Bleasdale, J. E. (2002). Investigation of potential bioisosteric replacements for the carboxyl groups of peptidomimetic inhibitors of protein tyrosine phosphatase 1B: Identification of a tetrazole-containing inhibitor with cellular activity. *Journal of Medicinal Chemistry*, *45*, 1785–1798.

7. Andersen, H. S., Olsen, O. H., Iversen, L. F., et al. (2002). Discovery and SAR of a novel selective and orally bioavailable nonpeptide classical competitive inhibitor class of protein-tyrosine phosphatase 1B. *Journal of Medicinal Chemistry*, *45*, 4443–4459.

8. Palanki, M. S. S., Erdman, P. E., Gayo-Fung, L. M., et al. (2000). Inhibitors of NF-kB and AP-1 gene expression: SAR studies on the pyrimidine portion of 2-chloro-4-trifluoromethylpyrimidine-5-[N-(3',5'-bis(trifluoromethyl)phenyl)carboxamide]. *Journal of Medicinal Chemistry*, *43*, 3995–4004.

9. Cheng, M., De, B., Almstead, N. G., et al. (1999). Design, synthesis, and biological evaluation of matrix metalloproteinase inhibitors derived from a modified proline scaffold. *Journal of Medicinal Chemistry*, *42*, 5426–5436.

10. Blanchfield, J. T., Dutton, J. L., Hogg, R. C., et al. (2003). Synthesis, structure elucidation, in vitro biological activity, toxicity, and Caco-2 cell permeability of lipophilic analogues of α-conotoxin MII. *Journal of Medicinal Chemistry*, *46*, 1266–1272.

11. Goodwin, J. T., Conradi, R. A., Ho, N. F. H., & Burton, P. S. (2001). Physicochemical determinants of passive membrane permeability: Role of solute hydrogen-bonding potential and volume. *Journal of Medicinal Chemistry*, *44*, 3721–3729.

12. Di, L., Kerns, E. H., Fan, K., McConnell, O. J., & Carter, G. T. (2003). High throughput artificial membrane permeability assay for blood-brain barrier. *European Journal of Medicinal Chemistry*, *38*, 223–232.

13. Kerns, E. H., Di, L., Petusky, S., Farris, M., Ley, R., & Jupp, P. (2004). Combined application of parallel artificial membrane permeability assay and Caco-2 permeability assays in drug discovery. *Journal of Pharmaceutical Sciences, 93*, 1440–1453.

14. Di, L., & Kerns, E. H. (2006). Application of physicochemical data to support lead optimization by discovery teams. In R. T. Borchardt, E. H. Kerns, M. J. Hageman, D. R. Thakker, & J. L. Stevens (Eds.), *Optimizing the drug-like properties of leads in drug discovery*. New York: Springer, AAPS Press.

Part 3
Disposition, Metabolism, and Safety

Transporters

Overview

▶ *Membrane transporters increase the influx and efflux of substrate compounds.*

▶ *Transporters are found in many tissues in vivo.*

▶ *P-glycoprotein efflux in the blood–brain barrier, cancer cells, and intestine is a iability for some compounds.*

Membrane transporters are responsible for two important permeability mechanisms, active uptake and efflux. Carrier mediated transport can contribute significantly to the pharmacokinetics characteristics of a compound.[1,2] Structures can be modified to reduce the deleterious effects of efflux. The possibility of improving absorption or BBB permeation by structure design to enhance uptake transport is a future opportunity.

9.1 Transporter Fundamentals

Passive diffusion is the predominant mechanism for the permeation of drugs throughout the body. A compound must have favorable physicochemical properties (i.e., lipophilicity, hydrogen bonds, molecular weight) to undergo passive diffusion. Many endogenous biochemical compounds that are necessary for life do not have physicochemical properties that allow sufficient passive diffusion, so there are trans-membrane transporters that greatly enhance their permeability. Examples of transporters in the intestine are shown in Table 9.1. For many biochemicals to function properly their concentrations must be significantly higher within a cell compared to the surrounding extracellular fluid. Some compounds, such as bile salts, must be exported to the bile from hepatocytes. Specific transporters move their

TABLE 9.1 ▶ Transporters Affecting Gastrointestinal Absorption of Some Drugs

---

Uptake
Oligopeptide transporters (PEPT1, PEPT2)
Organic anion transporters (OATP1, OAT1, OAT3)
Organic cation transporters (OCT1)
Bile acid transporters (NTCP)
Nucleoside transporters
Vitamin transporters
Glucose transporters (GLUT1)
Efflux
P-glycoprotein (Pgp, MDR1)
Breast cancer resistance protein (BCRP)

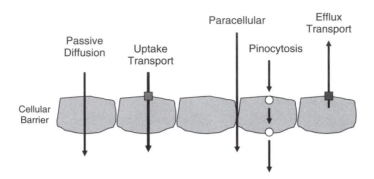

Figure 9.1 ▶ Uptake and efflux transporters contribute to net permeability.

substrates against a concentration gradient to enhance their accumulation. Uptake (import) transporters provide necessary nutrients and other compounds with physiological functions to tissues that would otherwise not have sufficient concentration for their physiological roles. The process is often referred to as active transport. Other transporters enhance the movement of compounds out of a cell. Efflux (export) transporters assist the bulk movement of compounds. An example is the efflux of potentially toxic xenobiotics (e.g., drugs) from the endothelial cells of the blood-brain barrier by P-glycoprotein (Pgp) before they reach sensitive brain cells. Transporters use energy input, such as from ATP, to perform their function. The roles of uptake and efflux transporters are illustrated in Figure 9.1. A particular transporter is expressed on only one surface of the cell (apical or basolateral). This results in the directional movement of substrates, for example from the blood stream into the bile. There is overlap in the substrate specificity of transporters, which may result in a cooperative effect.

Transporters affect drug pharmacokinetics. New transporters have typically been identified through cloning techniques. Their natural function, substrate specificity, kinetics, expression, and implications for drug development are an active area of research.

9.2 Transporter Effects

Transporters can affect the ADME/Tox characteristics of a compound. Transport occurs when the drug contains a moiety that has a similar moiety to the natural substrate of a transporter, or if it has structural elements that facilitate binding to a transporter with wide substrate specificity (e.g., Pgp). Here are a few examples of how transporters affect ADME/Tox:

▶ Uptake transporters enhance the absorption of some drug molecules in the intestine.

▶ Efflux transporters on the luminal (apical) surface of gastrointestinal epithelial cells oppose the absorption of some molecules.

▶ Transporters assist the uptake of some molecules into hepatocytes to enhance metabolic and biliary clearance.

▶ Efflux transporters oppose the distribution of some drugs from the bloodstream into organs, such as the brain.

▶ Uptake transporters enhance the distribution of certain drugs into some organs.

▶ Elimination of many drugs and metabolites is enhanced by active secretion in the nephrons of the kidney.

▶ Co-administered drugs can compete for a transporter for which they both have affinity, resulting in drug-drug interactions (DDI) and modification of the pharmacokinetics of one of the compounds.

Owing to the finite number of transporter protein molecules on the cell surface, they can be saturated if the concentration of a substrate is high enough. As the concentration of substrate increases, the flux of molecules increases and then levels off as the maximum capacity of the transporters is reached. Above this concentration the flux is the same. Saturation of drug uptake and efflux transporters in the intestine is observed when the luminal concentration following oral dosing exceeds the saturation concentration of the transporter. Passive diffusion permeation, by contrast, does not saturate.

Many commercial drugs are substrates for transporters. Some of these are shown in Table 9.2.[3-5] The effect of the transporters depends on the tissue and substrate concentration. For example, efflux transporters in the blood-brain barrier can exclude some drugs from distributing into the brain, whereas, in the intestine, efflux transporters appear to have less influence on absorption. This is because the drug concentration in the intestine after oral dosing is high (mM) and saturates the efflux transporters, whereas the drug concentration in the blood, thus at the BBB, is much lower (μM) and does not saturate the efflux transporters.

Transporters can be rate limiting in ADME/Tox processes. While passive diffusion is the predominant permeability mechanism for many compounds, transporters can greatly enhance or reduce total permeability for some compounds at certain membranes.

TABLE 9.2 ▶ Examples of Drugs with Active Uptake Transport

Oligopeptide transporters (PEPT1)

Captopril	
Enalapril (R=H) Enalaprilat (R=C₂H₅)	
Ampicillin (R=H) Pivampicillin (R=C₆H₁₁O₂)	
Cephalexin Also: Ceftibuten Cefoxitin Acyclovir Valacyclovir	

Continued

TABLE 9.2 ▶ *Continued*

Large neutral amino acid transporter (LAT1)

L-dopa

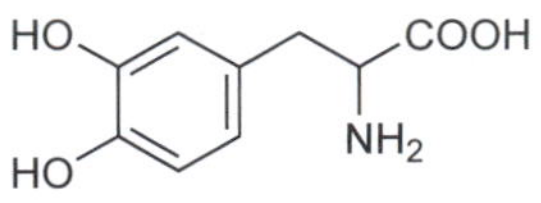

Methyldopa

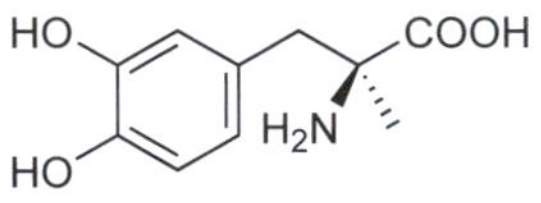

Melphalan

Also:

Phenylalanine

Baclofen

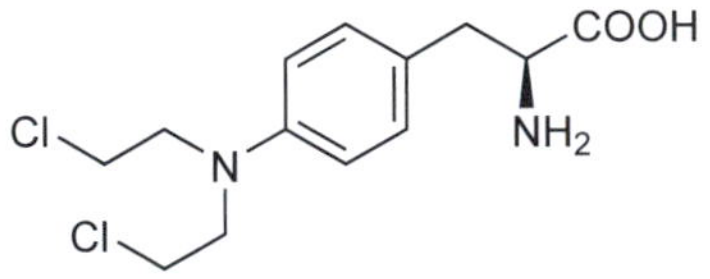

Monocarboxylic acid transporter (MCT1)

Salicylic acid

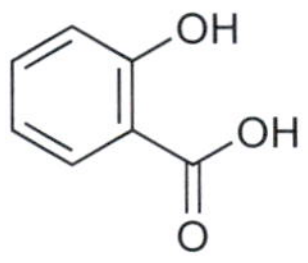

Pravastatin

Also:
Lovastatin acid
Simvastatin acid

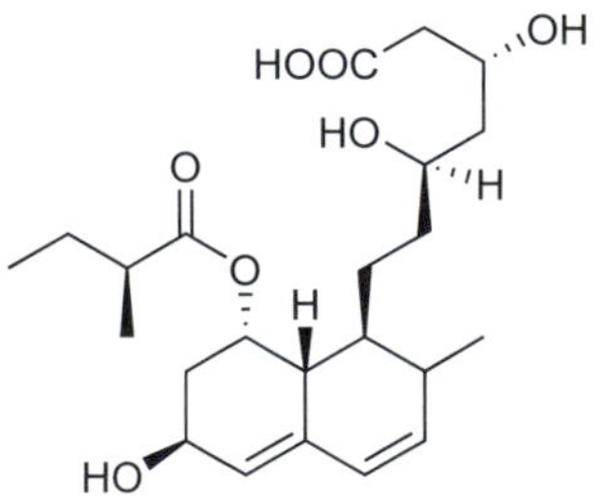

Organic anion transporter polypeptide (OATP1)

Fexofenadine

Also:
Enalapril
Temocaprilat
Deltrophin II

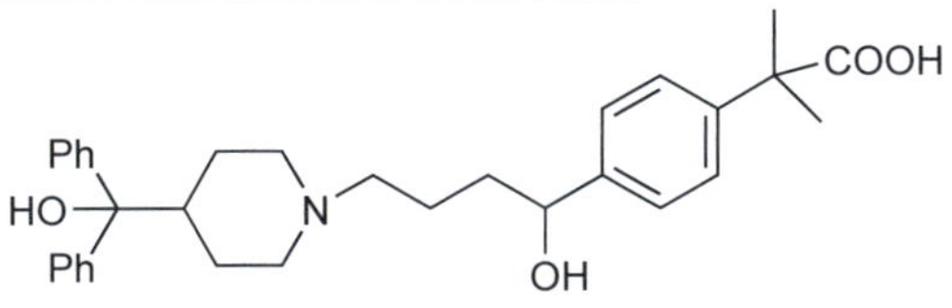

Transporters are found at barrier membranes throughout the body. Many of the transporters that are known to affect drug ADME/Tox are listed in Table 9.3 and shown in Figure 9.2. The letter abbreviations are complicated and some transporters have several letter abbreviations. The following sections discuss some of the effects of transporters in different barriers.

TABLE 9.3 ▶ Compilation of Many Transporters and Their ADME Functions

Organ/Barrier	Direction	Function	Transporter	Reference
Intestine/ Epithelial Cells	Absorption	Lumen to Blood	PEPT1	[6]
			ISBT	[6]
			OATP2B1	[6]
			MRP1	[7]
			OSTα	
			OSTβ	
			MRP3	[7]
			OATP1B1	[17]
	Secretion	Blood to Lumen; Epithelium to Lumen	Pgp	[6, 7]
			MRP2	[6, 7]
			BCRP	[6, 7]
Liver/ Hepatocytes	Excretion	Hepatocyte to Bile	Pgp	[6, 7]
			MDR2	
			MDR3	[6]
			BSEP	[6]
			MRP2	[6, 7, 17]
			BCRP	[6, 7]
	Excretion	Blood to Hepatocyte	OATPB	[6,17]
			OATP1B1 (OATPC)	[6, 17]
			OATP1B3 (OATP8)	[6, 17]
			OCT1	[6]
			OAT2	[6]
			NTCP	[6, 17]
			PGT	
			PEPT1	
	Retention/ Recirculation	Hepatocyte to Blood	MRP1	[6, 7]
			MRP3	[6, 7, 17]
			MRP4	[6]
Kidney/Renal Epithelial Cells	Excretion	Blood to Renal Epithelium	OAT1	[6, 17]
			OAT2 [6, 17]	
			OAT3	[6, 17]
			OAT4c1	[6]
			OCT1	[6]
			OCT2	[6]
			OCT3	[6]
	Excretion	Renal Epithelium to Urine	OAT4	[6, 17]
			OAT-K1	[6]
			OAT-K2	[6]
			Pgp	[6]
			MRP2	[6]
			MRP4	[6]

Continued

TABLE 9.3 ▶ *Continued*

Organ/Barrier	Direction	Function	Transporter	Reference
	Reabsorption	Urine to Renal Epithelium	PEPT1	[6]
			PEPT2	[6]
			OATP1	[17]
	Reabsorption	Renal Epithelium to Blood	MRP1	[6]
			MRP3	[6]
			MRP6	[6]
Blood-Brain Barrier	Into Brain	BBB Endothelium to Brain	MCT1	[18]
			OATP1	[17]
			GLUT1	[18]
			GLUT2	
			SGLT1	
			LAT1	[18]
			CAT1	[18]
			CNT2	[18]
			CHTX	[18]
			NBTX	[18]
			OATP3	[18]
	Elimination from Brain	BBB Endothelium to Blood	Pgp	[18, 28]
			MRP	[18, 28]
			BCRP	[18]
			OATs	[18]
			OATPs	[18]
			EAATs	[18]
			TAUT	[18]

9.2.1 Transporters in Intestinal Epithelial Cells

Transporters in the small intestine can modify absorption of some compounds. This can affect absorption if the substrate compound has low passive permeability and may not be obvious if a compound has high passive permeability. In intestinal epithelial cells, transporters are involved with:

▶ absorptive uptake (from the gastric lumen through epithelial cells and into blood)

▶ efflux (from the epithelial cell membrane back into the gastric lumen)

▶ secretory efflux (from the blood, into the epithelial cells, and into the gastric lumen)

Absorptive uptake increases the concentration of drug in the blood, while secretory efflux lowers the concentration in blood.

9.2.2 Transporters in Liver Hepatocytes

Transporters are important in hepatic clearance. They enhance metabolism by facilitating uptake into hepatocytes, where the molecules encounter metabolizing enzymes. They

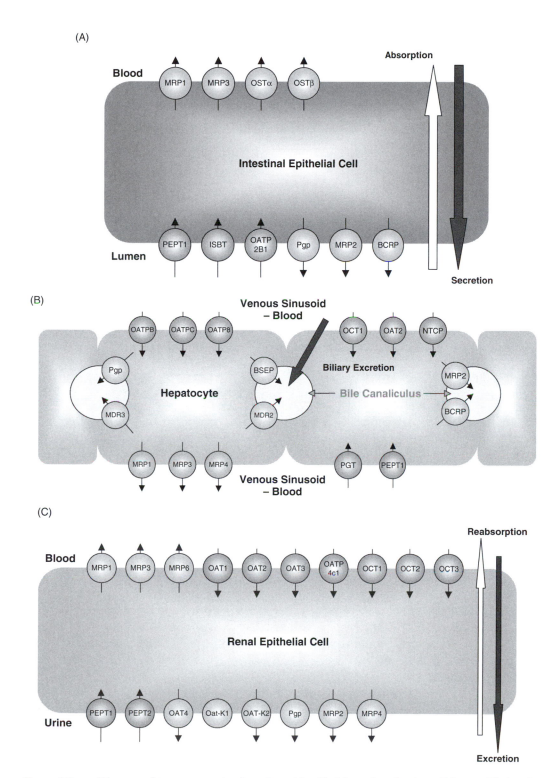

Figure 9.2 ▶ Diagram of transporters that have been identified in various barrier cell layers. Their roles in transporting drugs continues to be investigated. A) Intestinal epithelium, B) Liver hepatocytes, and C) Kidney epithelial cells[6].

enhance clearance by increasing the flux of the compound and metabolite molecules into the bile canaliculus or back into the blood for renal clearance. Transport rate is more important than metabolic rate in the clearance of some drugs.[6] In liver hepatocytes, transporters are involved with:

- ► hepatic uptake (from the blood into the hepatocyte)

- ► biliary clearance (from the hepatocyte into the bile canaliculus)

- ► hepatocyte efflux (from the hepatocyte into the blood)

9.2.3 Transporters in Kidney Epithelial Cells

In the renal epithelial cells of the nephron, transporters enhance renal clearance for some compounds, which results in a lower compound concentration in the blood. Some compounds are transported from the urine back into the blood. In the kidney, transporters are important for:

- ► tubular secretion (from blood through renal epithelial cells to urine)

- ► reabsorption (from urine through renal epithelial cells to blood)

9.2.4 Transporters in Blood–Brain Barrier Endothelial Cells

Chapter 10 discusses BBB permeation. Transporters are involved in keeping compounds out of the brain, but uptake transporters increase the distribution of some compounds into the brain. The important BBB functions of transporters are:

- ► efflux (from the BBB endothelial cells back into the blood)

- ► uptake (from the blood, through the BBB endothelial cells and into the brain)

Examples of commercial drugs for which transporter effects on ADME/Tox have been studied are compiled by Shitara, et al.[6]

The following sections discuss some of the transporters that affect drug permeability. The most important efflux transporter for discovery project teams, in general, is Pgp. It is useful for discovery scientists to understand the characteristics and function of Pgp. The effects of other transporters may be recognized (from unexplained pharmacokinetics parameters) during the course of a discovery project. When a project has a lead series that is active *in vitro*, but does not achieve sufficient *in vivo* exposure due to efflux, they might decide to modify the structure to reduce efflux or enhance uptake. For these reasons, the following sections provide introductory information on many of the most important transporters.

9.2.5 Consequences of Chirality on Transporters

Stereoselectivity is observed for transporters. Table 9.4 lists examples of drugs for which transporters have enantiomeric selectivity. This has an effect on any barrier in vivo for which the particular transporter has a significant role.

TABLE 9.4 ▶ Examples of Stereoselectivity of Transporters

Drug	Enantiomeric ratio	Transporter
Methotrexate	40 (L)	Dipeptide
Cephalexin	>100 (D)	Dipeptide
Dopa	>100 (L)	Amino acid

9.3 Efflux Transporters

Efflux transporters facilitate the export of compounds from the cell. These transporters belong to the ATP-binding cassette (ABC) family.[7]

9.3.1 P-glycoprotein (MDR1, ABCB1) [Efflux]

P-glycoprotein (Pgp) is the most widely known efflux transporter to discovery scientists because it can have a great effect on the success of some drug discovery projects. It is a 170 KD protein with 1280 amino acids and 12 trans-membrane segments (Figure 9.3).[8] Pgp is a member of the ATP Binding Cassette (ABC) family of transporters, of which over 50 are known and have natural transporter functions. Pgp has also been referred to as multi-drug resistant protein 1. Its gene is know as MDR1 or ABCB1. A drug molecule attaches to the binding domain of Pgp, which appears to be within the bilayer membrane. Then two ATPs, bound to the ATP binding regions, become hydrolyzed and induce a conformation change to open a pathway for the drug molecule to pass through into the extracellular fluid.[9]

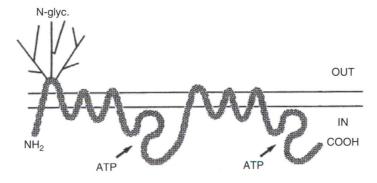

Figure 9.3 ▶ Schematic diagram of Pgp and its 12 transmembrane segments. Used with permission from [8].

Pgp was initially identified as a major cause of resistance by cancer cells to multiple drugs (e.g., paclitaxel, etoposide) having a variety of structures. With initial chemotherapeutic treatment, many of the cells in tumors died, but some cells lived and continued to grow. These cells were found to express Pgp, or another efflux transporter, which effluxes the cytotoxic cancer drug from the cells, thus allowing them to survive. Oncology research programs have dealt with Pgp for decades. A major oncology discovery strategy has been to test lead compounds for their ability to overcome multidrug resistance in cell lines that are highly drug resistant through the expression of high levels of Pgp and other efflux gransporters

It was also discovered that Pgp is present in many tissues of the body. Pgp is abundant in cell barriers that have a protective function, such as:

▶ blood-brain barrier

▶ small and large intestine

▶ liver

▶ kidney

▶ adrenal gland

▶ pregnant uterus

Pgp is expressed on the luminal surface of gastrointestinal epithelial cells, where it can reduce the total permeability of Pgp substrates. It has been shown to mediate the secretion of some drugs (e.g., digoxin) from the blood into the gastric lumen. In the liver and kidney, Pgp enhances drug and metabolite clearance to the bile and urine, respectively. Pgp attenuates penetration of some compounds into the brain, uterus, testes and other tissues. Pgp knock out animals have been developed. Pgp substrates typically have increased absorption, reduced excretion, increased toxicity, and increased distribution to protected tissues in knock out animals. Efflux by Pgp is a major challenge for some discovery projects, because it affects ADME processes, resulting in reduced exposure of the compound to the therapeutic target.

Efflux appears to have a greater relative effect when the drug concentration on the luminal surface is low (see Section 9.2). For example, the drug concentration circulating in the blood stream and exposed to the luminal surface of the BBB is much lower than the concentration on the luminal surface of intestinal epithelial cells after oral dosing. Therefore, Pgp may have little effect on the oral absorption of a particular Pgp substrate drug at high oral doses, but it may have a major effect on its brain penetration. It is also common to observe a greater effect of Pgp efflux on total permeability when the compound has low passive diffusion compared to high passive diffusion, because passive diffusion can dominate the process.

The substrate specificity for Pgp is very broad. Compounds ranging from a molecular weight of 250 to 1850 are known to be transported by Pgp. There is even evidence that Pgp can efflux peptides Aβ40 and Aβ42 with 40 and 42 amino acids, respectively, which are involved in Alzheimer's disease. Substrates may be aromatic, non-aromatic, linear or circular. The charges on the substrate molecules can be basic, uncharged, zwitterionic, or negatively charged. Some substrates are hydrophobic and some are amphipathic.

It is important to remember that the Pgp binding and efflux of a compound differs between species, owing to the differences in the protein's sequence. *In vivo* Pgp data from one species (e.g., mdr1a knock out mouse) may not properly predict effects in another species (e.g., human). In the same manner, *in vitro* Pgp data from MDR1-MDCKII (transfected with human Pgp-producing gene) cell monolayer efflux assay may not translate well to mouse or rat efficacy species pharmacokinetics.

Owing to the major potential effect on BBB permeability, Pgp efflux has been of particular interest in CNS discovery projects. Industry Pgp research has been greatly motivated by the need to deliver compounds to brain targets.

9.3.1.1 Rules for Pgp Efflux Substrates

As for other properties, rules are useful for the initial assessment of a compound, based on its structure. Rules for Pgp are referred to as "rule of 4". A compound is more likely to be a Pgp substrate if its structure has[10]:

▶ $N + O \geq 8$

▶ $MW > 400$

▶ Acid with $pKa > 4$

A compound is more likely to be a Pgp non-substrate if its structure has:

▶ N+O ≤ 4

▶ MW < 400

▶ Base with pKa < 8

Increasing numbers of hydrogen bond acceptors (N+O) appear to confer increasing likelihood of Pgp efflux.[11] This may be because binding to Pgp occurs in the lipophilic membrane region. Also, hydrogen bonds afford energetic binding interactions. Another contributor to Pgp binding may be a structural motif involving two H-bond acceptors 4.6 Å apart or three H-bond acceptors 2.5 Å apart.[12]

9.3.1.2 Case Study of Pgp Efflux

For a project lead series, it was shown[11] that compounds with N+O = 4 had a 33% chance of being effluxed by Pgp, compounds with N+O = 6 had a 65% chance, and compounds with N+O = 8 or 9 had a 87% chance. This is consistent with an increasing risk of efflux with increasing hydrogen bond acceptors.

An example of the development of structure-efflux relationships for a lead series is shown in Figure 9.4.[11] The efflux effects of various substituents at two positions in the scaffold were rated for their influence on measured efflux. There was a trend with increasing hydrogen

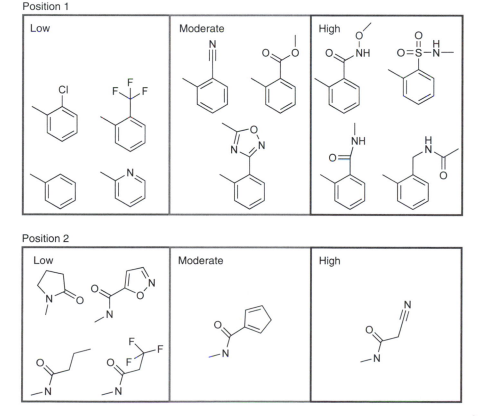

Figure 9.4 ▶ Influence of substituents at two positions on Pgp transport for a project lead series[11].

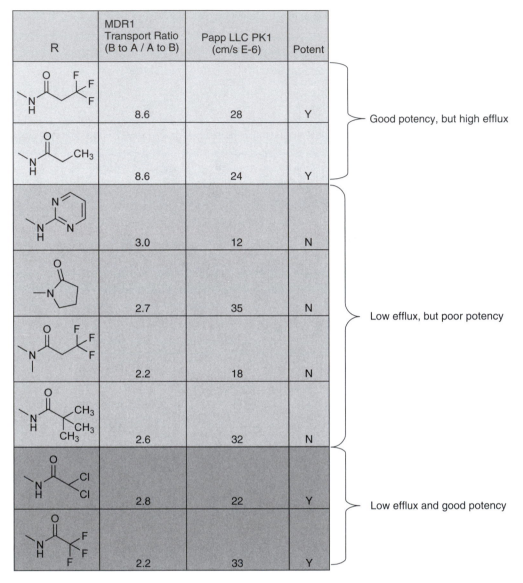

Figure 9.5 ▶ In this lead series, reduction of Pgp efflux was achieved while maintaining potency. Used with permission from [11].

bond acceptors. The aromatic amides were especially susceptible to efflux. Substitution of moieties was successful in reducing the Pgp efflux while maintaining potency (Figure 9.5).

9.3.1.3　Structure Modification Strategies to Reduce Pgp Efflux

Structure modification strategies have been successful in reducing Pgp efflux. First, try to identify hydrogen bond accepting atoms that are shown through structure-efflux relationship studies, or reasonable conjecture, to be involved in the Pgp binding. Then:

1. Introduce steric hindrance to the hydrogen bond donating atoms by:

　a. Attach a bulky group

　b. Methylate the nitrogen

2. Decrease H-bond acceptor potential

   a. Add an adjacent electron withdrawing group

   b. Replace or remove the hydrogen bonding group (e.g., amide)

3. Modify other structural features so that they may interfere with Pgp binding, such as adding a strong acid.

4. Modify the overall structure's Log P to reduce penetration into the lipid bilayer where binding to Pgp occurs.

Steric hindrance can be increased to reduce Pgp efflux. The example in Figure 9.6[13] is for a series of cancer drug candidates for the purpose of overcoming Pgp-induced resistance. A lower Pgp/no Pgp ratio indicates less difference between the resistant (Pgp) cells compared to the normal cells (no Pgp), because the compound is no longer a Pgp substrate. This compound series has increasingly hindered amines at R, resulting in reduced efflux.

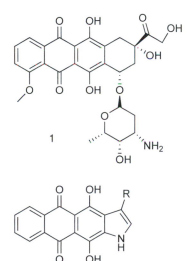

R	IC50 (uM) K562 (no Pgp)	IC50 (uM) K562i/S9 (with Pgp)	Pgp/no Pgp
1	0.2	1.5	8
CH$_2$NMe$_2$	1.2	12.5	10
—N͡NH	1.2	1.2	1
-N-quinuclidine	3.2	2.2	1

Figure 9.6 ► Increasing steric hindrance reduces Pgp efflux[13].

Increased acid strength reduces the Pgp substrate affinity for paclitaxel. The structure modification shown in Figure 9.7 introduced a carboxylic acid.[14] This imparted a 10 fold increase in brain penetration by reducing Pgp efflux.

A Pgp inhibitor can chemically knock out efflux, allowing compounds affected by Pgp efflux to reach higher levels at the therapeutic target. Pretreatment or co-dosing with a Pgp inhibitor has been performed for discovery projects as part of pharmacology proof of concept studies. This is also being investigated as a therapeutic strategy in the clinic,[15,16] but safety is a concern.

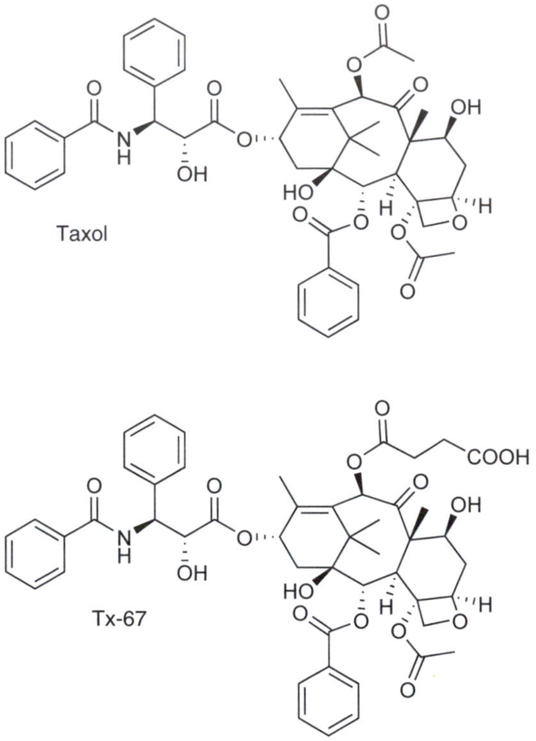

In Situ Rat Brain Perfusion $P_{app} \times 10^7$ cm/s	
Taxol	0.845
Tx-67	8.47

Figure 9.7 ▶ Pgp efflux at the BBB was decreased by adding a carboxylic acid moiety [14].

9.3.2   Breast Cancer Resistance Protein (BCRP, ABCG2) [Efflux]

BCPR efflux transporter was identified from chemotherapeutically resistant breast tumor cells. It is expressed normally in many tissues, such as placenta, hepatocytes, and small intestine. BCRP appears to be naturally involved in the efflux of porphyrins and their metabolites. Its role in the elimination of topotecan was demonstrated, and it appears to affect the disposition of several other drugs.[17]

9.3.3   Multidrug Resistance Protein 2 (MRP2, ABCC2) [Efflux]

MRP2 is an efflux transporter that came to light because it can also contribute to cancer multidrug resistance.[17] It has also been termed cMOAT for multispecific organic anion transporter. MRP2 transports glutathione, glucuronide, and sulfate conjugates of lipophilic compounds and some unconjugated compounds. It is expressed in intestinal epithelial cells, where it opposes absorption of substrates, and the canalicular membrane of hepatocytes and on renal tubule cells (kidney), where it enhances elimination of substrates. MRP1 through MRP9 have also been characterized.

9.3.4   Efflux Transporters in the BBB

Efflux transporters at the BBB include: Pgp, MRPs, BCRP, OATs, OATPs, EAATs (glutamic acid/acidic amino acids), and TAUT (taurine). These transporters export their substrates from the brain and BBB endothelial cells into the blood. They appear to work sequentially in concert: some of the transporters are on the abluminal (toward the brain tissue) membrane and some on the luminal (toward the blood) membrane.[18]

9.4 Uptake Transporters

Uptake transporters facilitate the permeation of compounds into cells.

9.4.1 Organic Anion Transporting Polypeptides (OATPs, SLCOs) [Uptake]

OATP1A2 (human) (a.k.a. OATP1, OATP-A) is found in the BBB (uptake), hepatocytes (uptake), and renal epithelium (reabsorption). It is known to transport organic anions (bile acids, steroid glucuronide conjugates, anionic dyes, thyroid hormones), as well as ouabain, cortisol, and large organic cations. It transports the drugs fexofenadine, enalapril, and temocaprilat, N-methyl quinidine, DPDPE and deltrophin II.[17,19]

OATP1B1 (human) (a.k.a. OATP2, OATP-C, LST1 [liver specific transporter 1]) is expressed in the liver (and may be in intestine). It has a similar substrate specificity to OATP1A2 and can also transport eicosanoids, benzylpenicillin, methotrexate, rifampin, pravastatin, rosuvastatin, and cerivastatin.[17] There are many other members of the OATP family that have been found in humans and rodents (e.g., OATP3 is found in the kidney, OATP9 in liver transports cardiac glycosides, OATP-K1 in kidney transports methotrexate).

9.4.2 Di/Tri Peptide Transporters (PEPT1, PEPT2) [Uptake]

These transporters enhance the uptake of dipeptides and tripeptides, but not individual amino acids or tetrapeptides.[20] A proton is co-transported with the substrate. Hydrophobicity increases PEPT1 binding and aromatic residues are preferred. Examples in Figures 9.8[21] and 9.9[22] show that prodrugs are transported by PEPT1 when natural amino acid, valine, is attached as a promoiety. Absorption of these prodrugs increased through both passive diffusion and active uptake. PEPT1 is known to transport β-lactam antibiotics and other drugs that contain peptides, as shown in Table 9.2. Peptide transporters have been reviewed.[23]

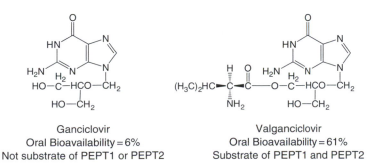

Ganciclovir
Oral Bioavailability = 6%
Not substrate of PEPT1 or PEPT2

Valganciclovir
Oral Bioavailability = 61%
Substrate of PEPT1 and PEPT2

Figure 9.8 ► Valganciclovir: Enhanced oral absorption by PEPT1 and PEPT2 peptide transporters.

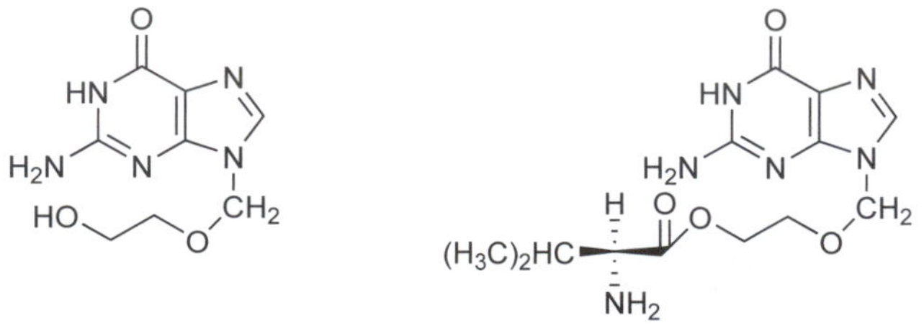

Acyclovir (Zovirax)
Not substrate of PEPT1 or PEPT2

Valacyclovir (Valtrex)
Substrate of PEPT1 and PEPT2
Oral bioavailability
3-5 fold higher than Acyclovir

Figure 9.9 ▶ Valacyclovir: Enhanced oral absorption by PEPT1 and PEPT2 peptide transporters.

9.4.3 Organic Anion Transporters (OATs) [Uptake]

OATs enhance renal clearance of some drugs and drug metabolites by uptake from capillary blood vessels into renal tubule cells.[17,24] OAT1 is known to transport β-lactam antibiotics, NSAIDs, antivirals, AZT, acyclovir, and many other drugs. Other members of the family are: OAT2 through 4, OAT-K1 through 2.

9.4.4 Organic Cation Transporter (OCT) [Uptake]

OCT1 through 3 and OCTN1 through 2 enhance transport into the urine in kidney.[24] Increasing hydrophobicity enhances binding to OCT.

9.4.5 Large Neutral Amino Acid Transporter (LAT1) [Uptake]

LAT1 is present in the apical membrane of the endothelial cells of the BBB. It transports amino acids, such as leucine and phenylalanine. It also transports the drugs L-DOPA, methyl DOPA, daclofen, and melphalan.

9.4.6 Monocarboxylic Acid Transporter (MCT1) [Uptake]

MCT1 is expressed on the apical membrane of the endothelial cells of the BBB and epithelial cells of the intestine. It is involved in uptake of acids. It enhances uptake of salicylic acid, pravastatin, lovastatin, simvastatin acid,[25] and probenecid .[18] MCTs have been reviewed.[26]

9.4.7 Other Uptake Transporters

▶ **Glucose Transporter (GLUT1) [Uptake]** is present in the apical membrane of the endothelial cells of the BBB and is involved in uptake.

▶ **Bile Salt Export Pump (BSEP, ABCB11) [Efflux]** is naturally involved in the export (efflux) of bile salts from hepatocytes into bile.

▶ **Sodium Dependent Taurocholate Co-transporting Polypeptide (NCTP) [Uptake]** assists the enterohepatic circulation of bile acids by transporting bile acids from the blood into hepatocytes, where they are secreted into the bile canaliculus. It may indirectly affect the role of bile acids on nuclear hormone receptors PXR and FXR for regulation of CYP expression and cholesterol metabolism.

▶ **Uptake transporters in the BBB include:** GLUT1, LAT1, MCT1, CAT1 (cationic amino acids), CNT2 (nucleosides), CHT (choline), and NBT (nucleobase).[18] These enhance the uptake of their substrates into the brain from the blood.

9.4.8 Structure Modification Strategies for Uptake Transporters

Uptake transporters are an attractive option for enhancing the permeability of compounds that are active *in vitro*, but whose passive diffusion is low.[4, 18, 26−29] Uptake transporters enhance the uptake of many drugs; however, this has often been discovered after the design phase in discovery. This strategy may represent an opportunity in future drug design.

Such an approach would utilize traditional SAR approaches, with informed structure modifications. These would be checked using *in vitro* assays for the specific transporter (see Chapter 27). The relationship of the structural modification to enhanced carrier mediated transport would guide further modifications or decisions to test the pharmacokinetics or tissue uptake *in vivo*.

Problems

(Answers can be found in Appendix I at the end of the book.)

1. Transporters are involved in which of the following: a) GI absorption of nutrients, b) BBB efflux, c) BBB uptake of some drugs, d) GI passive diffusion, e) GI efflux, f) renal secretion, g) GI hydrolysis, h) uptake into hepatocytes, i) biliary clearance.

2. At higher drug concentration, transporters: a) are most effective, b) may be saturated.

3. Which of the following can be affected by transporters: a) absorption, b) distribution, c) metabolism, d) excretion

4. What is the most consistently important transporter in drug discovery and why?

5. Which of the following compounds is more likely to be a substrate for Pgp:

Compound	MW	Ionization	H-Bond acceptors	H-Bond donors	PSA
A	350	pKa = 3	4	1	55
B	520	pKa = 9	10	5	140
C	400	pKa = 4	3	3	60
D	470	pKa = 8	8	2	75

6. What structural modifications may reduce Pgp efflux?

7. Pgp efflux of a compound can be proven in vivo using what?

8. Of the following transporters, which are efflux- and which are uptake-transporters: a) OATP1A2, b) BCRP, c) PEPT1, d) LAT1, e) MRP2, f) MCT1?

9. A particular transporter is found on which cell membrane: a) apical only, b) both apical and basolateral, c) either apical or basolateral, depending on the cell type, d) basolateral only?

10. Which of the following structure classes might have their oral absorption increased by active transport: a) amino acids, b) antibiotics, c) carboxylic acids, d) vitamins, e) di- and tri-peptides?

11. Pgp transporters are present in which of these cells: a) intestinal epithelium, b) blood-brain barrier, c) liver, d) kidney, e) skin?

12. Pgp causes which of the following: a) reduced blood-brain barrier penetration, b) cancer cell drug resistance, c) increased bioavailability?

13. Rank the following groups for increasing potential for Pgp transport:

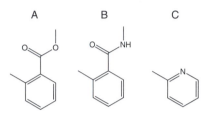

References

1. Ho, R. H., Tirona, R. G., Leake, B. F., et al. (2006). Drug and bile acid transporters in rosuvastatin hepatic uptake: Function, expression, and pharmacogenetics. *Gastroenterology, 130*, 1793–1806.

2. Kunta, J. R., & Sinko, P. J. (2004). Intestinal drug transporters: In vivo function and clinical importance. *Current Drug Metabolism, 5*, 109–124.

3. De Vrueh, R. L. A., Smith, P. L., & Lee, C.- P. (1998). Transport of L-valine-acyclovir via the oligopeptide transporter in the human intestinal cell line, Caco-2. *Journal of Pharmacology and Experimental Therapeutics, 286*, 1166–1170.

4. Walter, E., Kissel, T., & Amidon, G. L. (1996). The intestinal peptide carrier: A potential transport system for small peptide derived drugs. *Advanced Drug Delivery Reviews, 20*, 33–58.

5. Tamai, I., & Tsuji, A. (1996). Carrier-mediated approaches for oral drug delivery. *Advanced Drug Delivery Reviews, 20*, 5–32.

6. Shitara, Y., Horie, T., & Sugiyama, Y. (2006). Transporters as a determinant of drug clearance and tissue distribution. *European Journal of Pharmaceutical Sciences, 27*, 425–446.

7. Chan, L. M. S., Lowes, S., & Hirst, B. H. (2004). The ABCs of drug transport in intestine and liver: Efflux proteins limiting drug absorption and bioavailability. *European Journal of Pharmaceutical Sciences, 21*, 25–51.

8. Schinkel, A. H. (1999). P-Glycoprotein, a gatekeeper in the blood-brain barrier. *Advanced Drug Delivery Reviews, 36*, 179–194.

9. Hennessy, M., & Spiers, J. P. (2007). A primer on the mechanics of P-glycoprotein the multidrug transporter. *Pharmacological Research, 55*, 1–15.

10. Didziapetris, R., Japertas, P., Avdeef, A., & Petrauskas, A. (2003). Classification analysis of P-glycoprotein substrate specificity. *Journal of Drug Targeting, 11*, 391–406.

11. Hochman, J., Mei, Q., Yamazaki, M., et al. (2006). Role of mechanistic transport studies in lead optimization. In R. T. Borchardt, E. H. Kerns, M. J. Hageman, D. R. Thakker, & J. L. Stevens (Eds.), *Optimizing the "drug-like" properties of leads in drug discovery* (pp. 25–48). New York: Springer.

12. Seelig, A., & Landwojtowicz, E. (2000). Structure-activity relationship of P-glycoprotein substrates and modifiers. *European Journal of Pharmaceutical Sciences, 12*, 31–40.

13. Shchekotikhin, A. E., Shtil, A. A., Luzikov, Y. N., Bobrysheva, T. V., Buyanov, V. N., & Preobrazhenskaya, M. N. (2005). 3-Aminomethyl derivatives of 4,11-dihydroxynaphtho[2,3-f]indole-5,10-dione for circumvention of anticancer drug resistance. *Bioorganic & Medicinal Chemistry, 13*, 2285–2291.

14. Rice, A., Liu, Y., Michaelis, M. L., Himes, R. H., Georg, G. I., & Audus, K. L. (2005). Chemical modification of paclitaxel (Taxol) reduces P-glycoprotein interactions and increases permeation across the blood-brain barrier in vitro and in situ. *Journal of Medicinal Chemistry, 48*, 832–838.

15. Teodori, E., Dei, S., Scapecchi, S., & Gualtieri, F. (2002). The medicinal chemistry of multidrug resistance (MDR) reversing drugs. *Farmaco, 57*, 385–415.

16. Breedveld, P., Beijnen, J. H., & Schellens, J. H. M. (2006). Use of P-glycoprotein and BCRP inhibitors to improve oral bioavailability and CNS penetration of anticancer drugs. *Trends in Pharmacological Sciences, 27*, 17–24.

17. Glaeser, H., & Kim, R. B. (2006). The relevance of transporters in determining drug disposition. In R. T. Borchardt, E. H. Kerns, M. J. Hageman, D. R. Thakker, & J. L. Stevens (Eds.), *Optimizing the "drug-like" properties of leads in drug discovery* (pp. 423–460). New York: Springer.

18. Pardridge, W. M. (2007). Blood-brain barrier delivery. *Drug Discovery Today, 9*, 605–612.

19. Kim, R. B. (2002). Transporters and xenobiotic disposition. *Toxicology, 181–182*, 291–297.

20. Vig, B. S., Stouch, T. R., Timoszyk, J. K., et al. (2006). Human PEPT1 pharmacophore distinguishes between dipeptide transport and binding. *Journal of Medicinal Chemistry, 49*, 3636–3644.

21. Sugawara, M., Huang, W., Fei, Y.- J., Leibach, F. H., Ganapathy, V., & Ganapathy, M. E. (2000). Transport of valganciclovir, a ganciclovir prodrug, via peptide transporters PEPT1 and PEPT2. *Journal of Pharmaceutical Sciences, 89*, 781–789.

22. Ganapathy, M. E., Huang, W., Wang, H., Ganapathy, V., & Leibach, F. H. (1998). Valacyclovir: A substrate for the intestinal and renal peptide transporters PEPT1 and PEPT2. *Biochemical and Biophysical Research Communications, 246*, 470–475.

23. Herrera-Ruiz, D., & Knipp, G. T. (2003). Current perspectives on established and putative mammalian oligopeptide transporters. *Journal of Pharmaceutical Sciences, 92*, 691–714.

24. Dresser, M. J., Leabman, M. K., & Giacomini, K. M. (2001). Transporters involved in the elimination of drugs in the kidney: Organic anion transporters and organic cation transporters. *Journal of Pharmaceutical Sciences, 90*, 397–421.

25. Enerson, B. E., & Drewes, L. R. (2003). Molecular features, regulation, and function of monocarboxylate transporters: Implications for drug delivery. *Journal of Pharmaceutical Sciences, 92*, 1531–1544.

26. Sai, Y., & Tsuji, A. (2004). Transporter-mediated drug delivery: Recent progress and experimental approaches. *Drug Discovery Today, 9*, 712–720.

27. Majumdar, S., Duvvuri, S., & Mitra, A. K. (2004). Membrane transporter/receptor-targeted prodrug design: Strategies for human and veterinary drug development. *Advanced Drug Delivery Reviews, 56*, 1437–1452.

28. Sun, H., Dai, H., Shaik, N., & Elmquist, W. F. (2003). Drug efflux transporters in the CNS. *Advanced Drug Delivery Reviews, 55*, 83–105.

29. Ho, R. H., & Kim, R. B. (2005). Transporters and drug therapy: Implications for drug disposition and disease. *Clinical Pharmacology & Therapeutics, 78*, 260–277.

Chapter 10

Blood–Brain Barrier

Overview

▶ *Blood–brain barrier (BBB) is restrictive for some compounds owing to P-glycoprotein efflux, absence of paracellular permeation, and limited pinocytosis.*

▶ *Brain exposure is assessed in terms of BBB permeability or brain/plasma partition.*

▶ *Brain exposure is enhanced by reducing H-bonds, molecular weight, P-glycoprotein efflux, metabolism, and plasma protein binding, or by increasing Log P.*

The pharmaceutical treatment of central nervous system (CNS) disorders is the second largest area of therapy, following cardiovascular disease.[1] U.S. sales for CNS drugs exceeded $53 billion in 2002 to 2003 (Figure 10.1). CNS disorders are five of the top 10 causes of disability. Stroke is the third leading cause of death and costs the economy $40 billion annually. Fifteen million people suffer from Alzheimer's disease, which is the second most expensive disease to the economy at $100 billion annually. Many brain diseases do not have satisfactory treatments. Clearly, CNS disorders are an important current and future priority for the pharmaceutical industry.

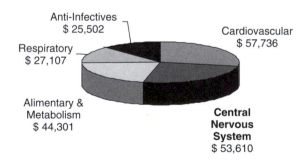

Figure 10.1 ▶ Central nervous system (CNS) disorders are the second largest pharmaceutical therapeutic area. U.S. pharmacy purchases September 2002 to August 2003 are shown.

In order for CNS drugs to penetrate to the brain tissue, they must pass through the blood–brain barrier (BBB). Many of the compounds that otherwise would be effective in treating CNS diseases are excluded from reaching a sufficient concentration in the brain tissue and producing the desired therapeutic effect (Figure 10.2). It has been estimated that only 2% of the possible CNS therapeutic compounds can pass the BBB.[2]

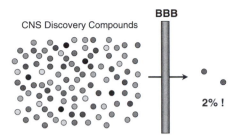

Figure 10.2 ▶ The BBB excludes as much as 98% of potentially beneficial drugs and is a major challenge for CNS therapy.

10.1 BBB Fundamentals

The term *BBB permeation* is widely used in CNS discovery projects. However, it is important to recognize that the goal is "brain penetration," the *exposure of compound to the therapeutic target in the brain. BBB permeation* is a major factor in brain penetration, and a closer look indicates that it is the sum of multiple mechanisms at the BBB (see Section 10.1.1). In addition, *brain distribution* mechanisms (e.g., metabolism, protein binding) also affect brain penetration of drugs (see Section 10.1.2). As with intestinal absorption, various compounds have different mechanisms, or combination of mechanisms, that limit their brain penetration. These depend on the compound properties and target location (e.g., membrane, cytoplasm, brain region). Therefore, it is important that discovery scientists be aware of the many brain penetration mechanisms and determine the mechanism(s) that best correlates with diagnosing and optimizing brain penetration of their leads.

The BBB is associated with the microcapillary blood vessels that run throughout the brain in close proximity to brain cells (Figure 10.3). These vessels naturally provide the nutrients and oxygen needed by the CNS cells and carry away waste. Over 400 miles of blood microcapillaries are present in the brain, with a surface area of approximately 12 m^2.[2]

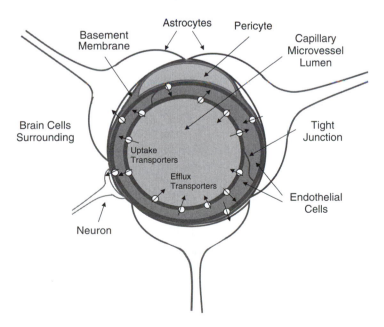

Figure 10.3 ▶ Schematic diagram of a cross-section of a brain capillary microvessel that constitutes the blood–brain barrier.

The BBB consists of the endothelial cells that form a monolayer lining the inner surface of the capillaries. CNS drugs must permeate through the endothelial cells to penetrate to the brain cells. The endothelial cells are associated with astrocyte and pericyte cells, which do not resist drug penetration but apparently can modify endothelial cell characteristics.

10.1.1 BBB Permeation Mechanisms

The BBB forms a permeation barrier that is more limiting to compound penetration than are most other membrane barriers in the body. Mechanisms affecting BBB permeation are shown in Figure 10.4 and include the following:

- ▶ Restrictive physicochemical characteristics that limit passive diffusion
- ▶ High efflux activity
- ▶ Lack of "leaky" paracellular permeation and capillary wall fenestrations
- ▶ Limited pinocytosis
- ▶ Metabolism within endothelial cells
- ▶ Uptake transport

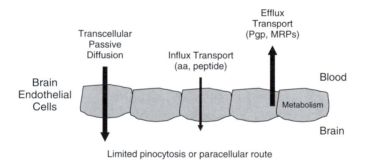

Figure 10.4 ▶ Permeation mechanisms at the BBB can limit compound exposure to brain cells. Used with permission from [20].

Transport of drugs into the brain is predominantly by transcellular passive diffusion. The same physicochemical properties that affect the permeation of compounds in the GI and other biological membranes also affect BBB permeation. In fact, they are more restrictive for the BBB. In addition, the profile of phospholipids in the BBB endothelial cells tends to have significant negatively charged polarity head groups, which opposes acids. These important limitations are discussed in Section 10.3.

BBB endothelial cells express P-glycoprotein (Pgp) on the apical surface. Efflux is a major limitation to BBB permeation for some compounds. Pgp efflux excludes molecules before they can reach brain cells. Therefore, an important strategy for increasing brain exposure of these compounds is to reduce efflux by Pgp. Endothelial cells also express breast cancer resistance protein (BCRP) and multidrug resistance protein 1 (MRP1) through MRP6, which efflux some compounds.[3] The roles of these efflux transporters are being investigated, but Pgp clearly is the transporter of greatest concern for neuroscience discovery projects.[4,5]

Efflux potential is assessed using an in vitro cell monolayer permeability assay (e.g., MDR1-MDCKII, Caco-2) that expresses Pgp. The term *efflux ratio* (ER) is defined as the permeability in the efflux direction divided by the permeability in the influx direction (Figure 10.4), as modeled using the in vitro assay (see Section 27.2.1). Generally, ER > 3 indicates significant efflux. Pgp knockout mice also are used for verification of the in vitro conclusions (see Section 27.3.1). A survey of successful CNS drugs indicated ER < 1 for 22%, 1–3 for 72%, and > 3 for 6%.[6] This suggests that many commercial CNS drugs are subject to low-to-moderate levels of Pgp efflux that can be overcome with clinical dosing. However, high Pgp efflux (ER > 3) is uncommon in commercial CNS drugs and should be avoided.

Paracellular permeation is drastically limited in the BBB because the endothelial cells form tight junctions. Endothelial cells of other capillaries in the body do not form such tight junctions and they have fenestrations, which are leaky sections in the vessel. Pinocytosis in BBB endothelial cells is limited.

Metabolism (phases I and II) has been observed in BBB endothelial cells. This structurally modifies compounds before they can reach brain tissue. The role of metabolism at the BBB likely is small.

BBB permeation is enhanced for a few compounds that are substrates for uptake transporters on the endothelial cells. These transporters naturally facilitate the uptake of nutrients (e.g., amino acids, peptides, glucose) and other endogenous compounds. A small number of commercial drugs partially or predominantly penetrate the BBB by active transport. Uptake enhancement is most commonly discovered by serendipity.

10.1.2 Brain Distribution Mechanisms

Several mechanisms limit the access of compounds to brain cells by affecting the distribution of compound to or within the brain (Figure 10.5):

▶ Metabolic clearance

▶ Plasma protein binding

▶ Nonspecific binding to proteins and lipids in brain tissue

▶ Clearance of compound from the extracellular fluid (ECF) into the blood and cerebrospinal fluid (CSF)

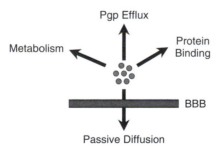

Figure 10.5 ▶ Mechanisms affecting compound exposure to brain cells are complex. (Reprinted with permission from [21]).

High hepatic clearance limits exposure of compounds to the brain.[7] It restricts compound from systemic circulation. Therefore, an important strategy for increasing brain exposure is to reduce hepatic clearance. For the example shown in Figure 10.6, in vitro assays indicated that passive diffusion (predicted by PAMPA-BBB, see Section 28.2.1.1) of the compound was good (CNS+), and there was no Pgp efflux. However, the compound did not penetrate into the brain in vivo to a significant amount. The reason was found to be a high rate of

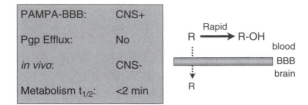

Figure 10.6 ▶ Example of a CNS project compound for which rapid metabolism resulted in low brain exposure.

liver metabolism ($t_{1/2}$ < 2 minutes) that greatly reduced the blood concentration and limited access of the compound to the brain.

Plasma protein binding limits penetration to the brain. As shown in Figure 10.7, only free unbound drug permeates the BBB. If the compound is highly bound to plasma protein and the on/off kinetics are moderate to slow, then little free drug is available to penetrate into the brain tissue. However, one must be careful in applying in vitro plasma protein binding data. Figure 10.8 shows that bound drug can release in brain microvessel circulation in vivo much more than with *in vitro* assays.[7] Plasma protein binding is useful in retrospectively diagnosing the causes of low in vivo brain penetration.

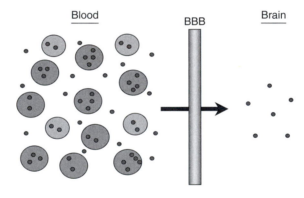

Figure 10.7 ▶ High plasma protein binding limits BBB permeation.

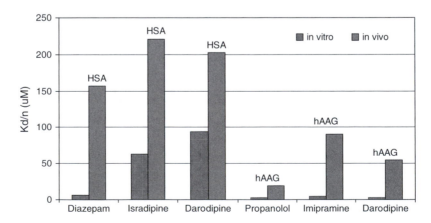

Figure 10.8 ▶ Plasma protein binding in vitro versus in vivo. Bound drugs release in brain microcirculation in vivo. (Modified from [9]).

Drug molecules that permeate the BBB are subject to *nonspecific binding* in the brain. The "free-drug hypothesis" suggests that a compound's efficacy is determined by the free (unbound) drug concentration in the brain and that binding restricts compound access to the therapeutic target. In this case, analysis of total brain tissue following dosing may indicate good total brain concentration; however, much of the compound may be restricted by nonspecific binding from interacting with the therapeutic target. Measurements of unbound percent in brain for commercial drugs[3] range widely (0.07%–52%). Therefore, some compounds are active in the brain despite high levels of nonspecific binding. This suggests that nonspecific binding correlates with pharmacological activity at some times but not at others. Compound in ECF is cleared into the blood and the CSF. If a compound has low BBB permeation, this clearance may limit the concentration in the ECF.[7] Figure 10.9 summarizes the major mechanisms that affect compound exposure in the brain.

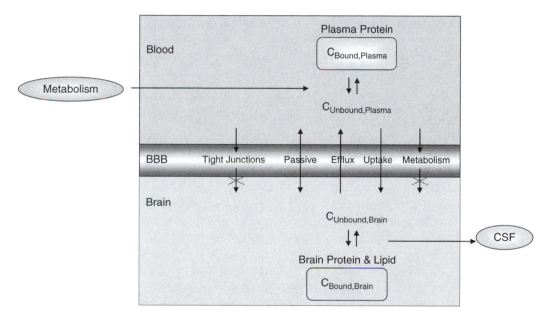

Figure 10.9 ▶ Compilation of major mechanisms affecting unbound (free) compound concentration in the CNS ECF ($C_{Unbound,Brain}$). $C_{Bound,Brain}$, concentration of compound bound nonspecifically to brain tissue components; $C_{Bound,Plasma}$, concentration of compound bound to plasma proteins; $C_{Unbound,Plasma}$, concentration unbound in blood; *CSF*, cerebrospinal fluid.

10.1.3 Brain–CSF Barrier

A second interface between the blood and the brain is the choroid plexus (Figure 10.10). The BBB interfaces the blood and the ECF of the brain. The choroid plexus interfaces the blood and the CSF and forms the blood–cerebrospinal fluid barrier (BCSFB).The BCSFB is not considered an effective route for drug delivery to the brain because (1) the surface area of the BBB is 5,000-fold larger than the BCSFB, (2) there is little mixing of the CSF components with the ECF, (3) the CSF flows very fast away from the brain tissue toward the arachnoid villi, and (4) CSF is turned over every 5 hours.

Some of the compound in the ECF is cleared into the CSF. Therefore, CSF contains material that penetrated the BBB. CSF has been sampled to study compound penetration into brain.[6] In some cases, CSF concentration correlates with ECF concentration, but it should not be assumed that, in general, compound concentration in CSF is the same as in ECF.[7]

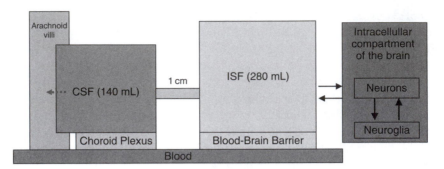

Figure 10.10 ► The BBB is a more effective route for drugs to reach the brain tissue. The BBB has 5,000-fold the surface area of the BCSFB, and CSF flows rapidly away from the brain tissue.[9,22]

10.1.4 Interpreting Data for Brain Penetration

Methods for brain penetration are discussed in Chapter 28. An understanding of how it was generated and its limitations benefits interpretation of these data as discussed in Sections 10.1.4.1 and 10.1.4.2.

10.1.4.1 B/P, Log BB, K_p, and $K_{p,free}$ Indicate the Extent of Brain Distribution

The most commonly used brain penetration data are from in vivo pharmacokinetic (PK) studies, which produce a ratio of drug in brain to drug in plasma or blood. This value is often termed *B/P*; its log is *Log BB*.

The B/P of most prescribed CNS drugs is > 0.3.[6] B/P $= 0.3$ means that the total compound concentration in the brain is 30% of the total compound concentration in the plasma. For many companies, B/P > 0.3 is used as a minimum guideline for CNS discovery projects. Compounds with B/P < 0.1 penetrate poorly into the brain, yet some commercial CNS drugs have B/P < 0.1.

B/P is actually a partition coefficient between total brain tissue and total plasma (just as Log P is a partition coefficient between octanol and water). It is a useful indicator of brain distribution; however, it has limitations as a sole indicator of brain exposure, as discussed below.

The term B/P is sometimes used loosely and is calculated from various data:

► AUC_{brain}/AUC_{plasma}

► $Concentration_{brain}/Concentration_{plasma}$ at a single time point or at C_{max}

► $Concentration_{brain}/Concentration_{plasma}$ at steady state (also termed K_p)

It is important that discovery scientists understand how the B/P values for their compounds were determined in order for the implications and limitations of the value to be properly understood and applied. If B/P is derived at a single time point, it may not accurately reflect brain penetration because C_{max}, t_{max}, or $t_{1/2}$ for a compound can vary between brain and plasma. Thus, B/P at a single time point can vary with time. Also, B/P frequently varies with dosing level, route of administration, vehicle, and species. For example, if there is significant efflux, the B/P value increases with dose as the efflux transporters are increasingly saturated. Other scenarios can affect B/P, so the project team should think through the effects of other properties of their compounds.

Another limitation of B/P is that it is calculated from total drug in the plasma and brain. A compound with a high B/P value and high nonspecific brain binding has a low free concentration in the brain. By comparison, a compound with a low B/P and low nonspecific brain binding may actually have a higher free concentration in the brain than the first example. William Pardridge, a noted expert on the BBB, warned against relying heavily on B/P values because they may mislead project teams.[7]

B/P is also limited because it is the result of multiple mechanisms. It does not give medicinal chemists specific guidance on what can be modified in the structure to improve brain penetration. Findings of low B/P values should be followed with investigation of specific mechanisms using in vitro assays to uncover the limiting mechanism(s). The in vitro assay(s) then can be used to monitor new analogs that are synthesized for the purpose of improving the specific mechanism.

It is important to distinguish brain/plasma partitioning (i.e., brain distribution) from BBB permeability. These are distinctly different, and both are significant. The structure modifications strategies for improving distribution differ from those for improving permeability.

A useful ratio is free compound in brain to free compound in plasma[8]:

$$K_{p,free} = C_{Unbound,Brain} / C_{Unbound,Plasma},$$

where $C_{Unbound,Brain}$ = free compound concentration in the brain ECF available to interact with brain cells, and $C_{Unbound,Plasma}$ = free compound concentration in the blood available to permeate the BBB. It is apparent that one strategy for increasing the brain free concentration is to increase the plasma concentration.

10.1.4.2 P_{app}, PS, and $t_{\frac{1}{2}eq.in}$ Indicate the Rate of BBB Permeability

In vitro, in situ, and in vivo experiments can provide BBB permeability (P). Pardridge[9] demonstrated the value of the permeability surface area coefficient (PS), especially in pharmacokinetic studies. Liu[8] proposed use of "the half-life to reach equilibrium between free drug in brain and plasma" ($t_{\frac{1}{2}eq.in}$). This value is the result of BBB permeability and brain free compound concentration and is not dependent on plasma free compound concentration.

10.2 Effects of Brain Penetration

BBB penetration is necessary for CNS drugs.[10] However, drugs for therapy in peripheral tissues may cause side effects in the brain; therefore, it is desirable that they have minimal penetration into the brain (Figure 10.11). It may be possible to reduce CNS side effects of peripheral drugs by making structural modifications that reduce the brain penetration.

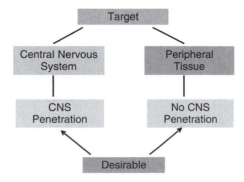

Figure 10.11 ▶ It is necessary for CNS drugs to penetrate into the brain, but it is undesirable for drugs treating diseases in the peripheral tissues to penetrate into the brain, where they can cause CNS side effects.

10.3 Structure–BBB Penetration Relationships

Physicochemical properties greatly affect the passive transcellular BBB permeation of compounds. Studies have shown the following key structural properties for discovery of CNS drugs[9,11–14]:

▶ Hydrogen bonds (acceptors and donors)

▶ Lipophilicity

▶ Polar surface area (PSA)

▶ Molecular weight (MW)

▶ Acidity

These properties are more restrictive at the BBB than at most other membrane barriers in the body. As a group, commercial CNS drugs compared to non-CNS drugs have fewer hydrogen bond donors, higher Log P, lower PSA, and fewer rotatable bonds.[12] A set of physicochemical BBB rules was first proposed by Pardridge.[13] The structure should have the following:

▶ H-bonds (total) < 8–10

▶ MW < 400–500

▶ No acids

Spraklin[3] suggests H-bond donors < 2 and H-bond acceptors < 6. This is in agreement with general consensus that H-bond donors are more limiting than H-bond acceptors.

Another set of BBB rules was compiled by Clark[14] and Lobell et al.[15] The structure should have the following:

▶ N + O < 6

▶ PSA < 60–70 Å^2

▶ MW < 450

▶ Log D = 1–3

▶ ClogP – (N + O) > 0

These rules are useful for evaluating BBB permeability prior to synthesis, assessing compounds being brought into a project (e.g., alliance partner), diagnosing poor in vivo brain penetration, and guiding which structural modifications might best improve BBB permeation of a discovery lead series.

Indomethacin is an example of poor BBB permeability by acids (Figure 10.12). CNS drugs tend to contain a basic amine, such as trifluoroperizine. Positively charged amines favorably interact with the negatively charged head groups of phospholipids at the BBB (Figure 10.13). Approximately 75% of the most prescribed CNS drugs are basic, 19% are neutral, and 6% are acids.[11] The amine functional group often is important for CNS activity; however, the basicity also assists BBB influx.

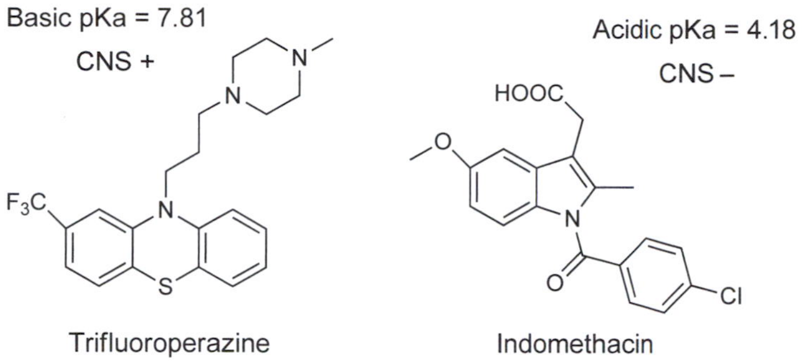

Basic pKa = 7.81
CNS +
Acidic pKa = 4.18
CNS –
HOOC
O
N
F3C
S
O
Cl
Trifluoroperazine
Indomethacin

Figure 10.12 ▶ Acids poorly permeate the BBB (CNS–), whereas bases generally penetrate much better (CNS+). (Reprinted with permission from [21]).

R2
R1–NH+
R3
R2
R1–N
R3

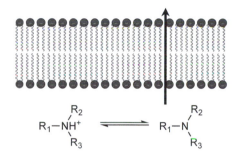

Figure 10.13 ▶ Amines have favorable interaction with predominantly negatively charged phospholipids head groups at the BBB.

It has been noted that in silico quantitative structure–activity relationship descriptors developed using B/P (Log BB) distribution datasets are similar to those developed using BBB permeability (Log PS) datasets.[16] Thus, the fundamental structural modifications (i.e., H-bonding, lipophilicity, MW, acidity) for brain penetration are consistent.

10.4 Structure Modification Strategies to Improve Brain Penetration

Strategies for modifying structures to improve brain penetration are listed in Table 10.1.

TABLE 10.1 ▶ Structure Modification Strategies for Brain Penetration Improvement

Structure modification	Section
Reduce P-glycoprotein efflux	10.4.1
Reduce hydrogen bonds	10.4.2
Increase lipophilicity	10.4.3
Reduce molecular weight	10.4.4
Replace carboxylic acid groups	10.4.5
Add an intramolecular hydrogen bond	10.4.6
Modify or select structures for affinity to uptake transporters	10.4.7

10.4.1 Reduce Pgp Efflux

Pgp efflux is the most significant limitation to BBB permeation of some discovery lead series. It is important to assess Pgp efflux early for a series. Structure–efflux relationships can be established by running series examples in an in vitro Pgp assay (see Chapter 28). These relationships will indicate which portions of the molecule might be modified to attempt efflux reduction. Other strategies for structural modifications to reduce Pgp efflux are discussed in Section 9.3.1.3.

10.4.2 Reduce Hydrogen Bonds

Reducing the total number of hydrogen bonds will increase BBB permeation, especially H-bond donors. The effect of reducing the number of total hydrogen bonds on BBB permeation is shown in Figure 10.14 for a series of steroids.[9] As the number of total hydrogen bond donors and acceptors decreases, BBB permeation increases. Functional groups that form hydrogen bonds can be removed, substituted, or blocked to enhance BBB permeation.

An example from a discovery CNS project is shown in Figure 10.15. Reduction of one hydrogen bond donor by blocking with a methyl group greatly increased brain penetration.

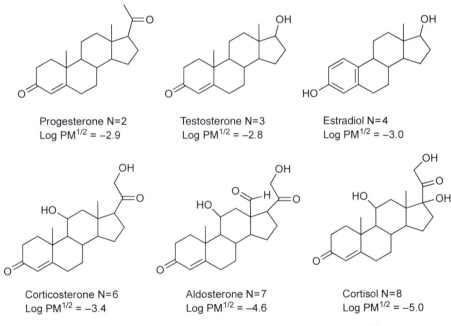

Progesterone N=2
Log PM$^{1/2}$ = –2.9

Testosterone N=3
Log PM$^{1/2}$ = –2.8

Estradiol N=4
Log PM$^{1/2}$ = –3.0

Corticosterone N=6
Log PM$^{1/2}$ = –3.4

Aldosterone N=7
Log PM$^{1/2}$ = –4.6

Cortisol N=8
Log PM$^{1/2}$ = –5.0

Figure 10.14 ▶ Effects of H-bonding on BBB permeation.[9]

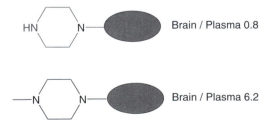

Brain / Plasma 0.8

Brain / Plasma 6.2

Figure 10.15 ▶ Example from a discovery CNS project in which the brain penetration was enhanced (B/P) with the reduction of one H-bond donor.

10.4.3 Increase Lipophilicity

Increasing lipophilicity will increase BBB permeation.[9] For example, addition of one methyl group to morphine produces codeine, which has a 10-fold higher BBB permeation (Figure 10.16). Addition of two acetyl groups to produce heroin further increases BBB permeation because of increased lipophilicity, despite increasing the number of H-bond acceptors. Nonpolar groups (e.g., methyl) also can be added to a molecule to enhance lipophilicity.

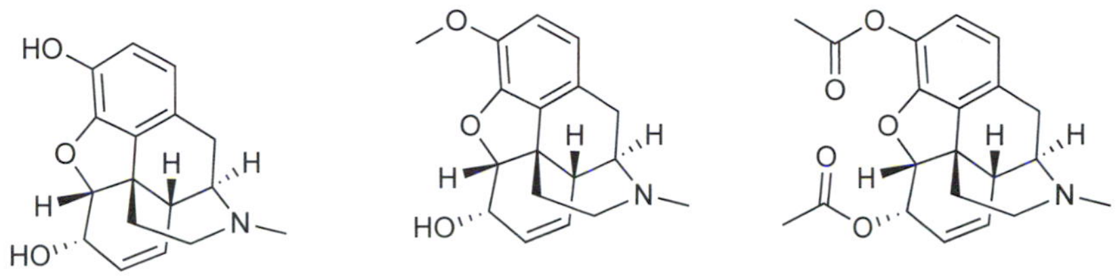

Morphine

Codeine
10x ↑ BBB

Heroin
100x ↑ BBB

Figure 10.16 ▶ Effects of lipophilicity on BBB permeation. (Reprinted with permission from [21]).

It should be noted that increasing lipophilicity to improve BBB permeation also can have a negative effect on BBB penetration. According to the "pharmacokinetic rule,"[13] increasing lipophilicity also increases nonspecific binding in the brain. This reduces free compound concentration in ECF at brain cells and decreases activity. Therefore, it is important to balance lipophilicity for optimizing permeability and minimizing brain binding.

10.4.4 Reduce MW

If groups on the structure can be removed without greatly impairing activity, then removing them can be beneficial. This reduces molecular size and improves permeation through lipid bilayer membranes.

10.4.5 Replace Carboxylic Acid Groups

Elimination of an acidic group will increase BBB permeation. An example of carboxylic acid replacement is shown in Figure 10.17.[17]

10.4.6 Add an Intramolecular Hydrogen Bond

An intramolecular hydrogen bond will increase BBB permeation. It reduces the total number of hydrogen bonds with water that must be broken for BBB permeation. In the example shown in Figure 10.18,[18] an amine was added that was reported to form two intramolecular hydrogen bonds and enhanced brain penetration.

10.4.7 Modify or Select Structures for Affinity to Uptake Transporters

Membrane transporters enhance the BBB permeation of some drugs. Carrier-mediated transporters have been advocated as a means of enhancing the penetration of compounds with poor passive BBB permeation.[19] For example, LAT1 (large neutral amino acid transporter) enhances the brain uptake of L-dopa and gabapentin.[8] Other transporters on the luminal side of the BBB endothelial cells are GLUT1 (glucose), MCT1 (monocarboxylic acids),

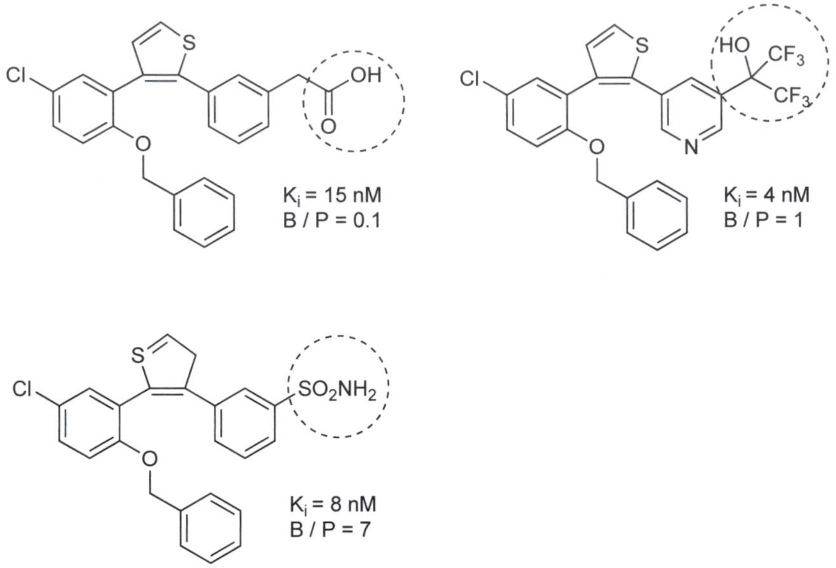

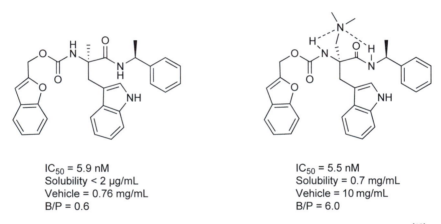

Figure 10.17 ▶ Replacement of carboxylic acid group in an EP1 receptor antagonist series improved brain penetration.[17]

Figure 10.18 ▶ Addition of an intramolecular hydrogen bond increases brain penetration.[18] (Reprinted with permission from [21]).

CAT1 (cationic amino acids), and CNT2 (nucleosides). The transporter uptake strategy has not been widely implemented but remains an intriguing prospect.

▢ Problems

(Answers can be found in Appendix I at the end of the book.)

1. The BBB consists of: (a) a membrane between the skull and the brain, (b) an impermeable membrane surrounding brain cells, (c) the endothelial cells of brain capillaries, (d) membranes that surround each section of the brain.

2. Why is Pgp efflux much more important at the BBB than at the GI endothelium?

3. Uptake of most drugs at the BBB occurs by which mechanism?: (a) Pgp efflux, (b) uptake transport, (c) metabolism, (d) paracellular, (e) passive diffusion, (f) endocytosis, (g) plaque formation.

4. Which of the following compounds are likely to have poor brain penetration and why?

Compound	MW	Ionization	H-bond acceptors	A-bond donors	PSA
A	350	$pK_a = 9$, base	4	1	55
B	520	$pK_a = 5$, acid	10	5	140
C	600	$pK_a = 8$, base	3	3	60
D	470	$pK_a = 8$, base	8	2	75

5. How could the following compound be structurally modified to improve BBB permeation by passive diffusion?

MW = 519.6
cLogP = −0.76
PSA = 136

6. Which of the following molecular properties are *not* favorable for BBB permeation?: (a) MW < 450, (b) PSA > 60–70 Å^2 (c) Solubility > 50 µM, (d) Log D < 1, (e) (N + O) > 5, (f) ClogP – (N + O) < 0?

7. If low brain concentrations of drug or low B/P are observed in vivo, which of the following approaches can be used to diagnose the cause?: (a) physicochemical/molecular "rules" for brain penetration, (b) thermodynamic solubility, (c) Pgp efflux ratio, (d) metabolic stability, (e) plasma protein binding, (f) CYP inhibition, (g) Pgp knockout animals (lacking Pgp).

8. Which of the following reduce brain penetration?: (a) CSF, (b) Pgp efflux, (c) tight BBB endothelial cell junctions, (d) uptake transporters, (e) high plasma protein binding, (f) carboxylic acid group on molecule, (g) intramolecular H-bonds in molecule.

9. Which of these structural modifications are likely to increase brain penetration?: (a) replace –COOH with –SO$_2$NH$_2$, (b) increase hydrogen bond donors, (c) increase lipophilicity, (d) increase intramolecular hydrogen bonding, (e) reduce hydrogen bond donors, (f) increase MW, (g) reduce PSA.

References

1. *IMS Health, Drug Monitor.*

2. Pardridge, W. M. (2001). Crossing the blood-brain barrier: are we getting it right? *Drug Discovery Today*, 6, 1–2.

3. Maurer, T. S., DeBartolo, D. B., Tess, D. A., & Scott, D. O. (2005). Relationship between exposure and nonspecific binding of thirty-three central nervous system drugs in mice. *Drug Metabolism and Disposition*, 33, 175–181.

4. Graff, C. L., & Pollack, G. M. (2004). Drug transport at the blood-brain barrier and the choroid plexus. *Current Drug Metabolism, 5*, 95–108.

5. Golden, P. L., & Pollack, G. M. (2003). Blood-brain barrier efflux transport. *Journal of Pharmaceutical Sciences, 92*, 1739–1753.

6. Doran, A., Obach, R. S., Smith, B. J., Hosea, N. A., Becker, S., Callegari, E., et al. (2005). The impact of P-glycoprotein on the disposition of drugs targeted for indications of the central nervous system: Evaluation using the MDR1A/1B knockout mouse model. *Drug Metabolism and Disposition, 33*, 165–174.

7. Pardridge, W. M. (2004). Log(BB), PS products and in silico models of drug brain penetration. *Drug Discovery Today, 9*, 392–393.

8. Liu, X., & Chen, C. (2005). Strategies to optimize brain penetration in drug discovery. *Current Opinion in Drug Discovery & Development, 8*, 505–512.

9. Pardridge, W. M. (1995). Transport of small molecules through the blood-brain barrier: biology and methodology. *Advanced Drug Delivery Reviews, 15*, 5–36.

10. Reichel, A. (2006). The role of blood-brain barrier studies in the pharmaceutical industry. *Current Drug Metabolism, 7*, 183–203.

11. Liu, X. (2006). Factors affecting total and free drug concentration in the brain. In *AAPS Conference: critical issues in discovering quality clinical candidates*. Philadelphia, PA.

12. Doan, K. M.M., Humphreys, J. E., Webster, L. O., Wring, S. A., Shampine, L. J., Serabjit-Singh, C. J., et al. (2002). Passive permeability and P-glycoprotein-mediated efflux differentiate central nervous system (CNS) and non-CNS marketed drugs. *Journal of Pharmacology and Experimental Therapeutics, 303*, 1029–1037.

13. Pardridge, W. M. (1998). CNS drug design based on principles of blood-brain barrier transport. *Journal of Neurochemistry, 70*, 1781–1792.

14. Clark, D. E. (2003). In silico prediction of blood-brain barrier permeation. *Drug Discovery Today, 8*, 927–933.

15. Lobell, M., Molnar, L., & Keseru, G. M. (2003). Recent advances in the prediction of blood-brain partitioning from molecular structure. *Journal of Pharmaceutical Sciences, 92*, 360–370.

16. Clark, D. E. (2005). Computational prediction of blood-brain barrier permeation. *Annual Reports in Medicinal Chemistry, 40*, 403–415.

17. Ducharme, Y., Blouin, M., Carriere, M.-C., Chateauneuf, A., Cote, B., Denis, D., et al. (2005). 2, 3-Diarylthiophenes as selective EP1 receptor antagonists. *Bioorganic & Medicinal Chemistry Letters, 15*, 1155–1160.

18. Ashwood, V. A., Field, M. J., Horwell, D. C., Julien-Larose, C., Lewthwaite, R. A., McCleary, S., et al. (2001). Utilization of an intramolecular hydrogen bond to increase the CNS penetration of an NK1 receptor antagonist. *Journal of Medicinal Chemistry, 44*, 2276–2285.

19. Pardridge, W. M. (2005). The blood-brain barrier: bottleneck in brain drug development. *NeuroRx: the Journal of the American Society for Experimental NeuroTherapeutics, 2*, 3–14.

20. Di, L., Kerns, E. H., Fan, K., McConnell, O. J., & Carter, G. T. (2003). High throughput artificial membrane permeability assay for blood-brain barrier. *European Journal of Medicinal Chemistry, 38*, 223–232.

21. Di, L., & Kerns, E. H. (2006). Application of physicochemical data to support lead optimization by discovery teams. In R. T. Borchardt, E. H. Kerns, M. J. Hageman, D. R. Thakker, & J. L. Stevens, eds. *Optimizing the drug-like properties of leads in drug discovery,* New York: Springer, AAPS Press.

22. Audus, K. L. (2002). Overview of the blood-brain and blood-fluid barriers of the central nervous system. In *Designing Drugs to Minimize or Maximize Exposure to the Brain*. Residential School on Medicinal Chemistry, Drew University: Princeton, NJ.

Chapter 11

Metabolic Stability

Overview

▶ *Metabolism is the enzymatic modification of compounds to increase clearance.*

▶ *It is a determinant of oral bioavailability, clearance, and half-life in vivo.*

▶ *Metabolism occurs predominantly in the liver, and some may occur in the intestine.*

▶ *Metabolic stability is increased by structure modifications that block or sterically interfere with metabolic sites or withdraw electrons.*

Drugs encounter formidable challenges to their stability in vivo. Unfortunately, this imposes significant limitations on the structures of drugs. Structures that are highly active in vitro may not make good drugs because they are susceptible to metabolism in the body. Most drug discovery project teams encounter stability limitations for their lead series. Many pharmacologically interesting molecules must be passed over because they are not sufficiently stable. This and other chapters on stability issues aim to inform discovery scientists about the causes of instability and guide medicinal chemists through successful strategies of structure modification to improve the lead series stability.

An overview of in vivo stability challenges suggests a diverse ensemble of chemical and enzymatic reactions poised to attack various moieties of the molecule (Figure 11.1).[1] In the gut, a molecule risks *intestinal decomposition*, owing to various pHs and enzymatic hydrolysis reactions. As the molecule moves through the gut wall, enzymes can initiate *intestinal metabolism*. Molecules that reach the portal vein are immediately carried to the liver where they encounter diverse *hepatic metabolism* reactions. Molecules that survive the liver encounter *plasma decomposition* by hydrolytic enzymes in the blood.

In addition to in vivo degradation, stability issues are encountered in drug discovery laboratories. Compounds can encounter *chemical decomposition* due to environmental challenges in the laboratory. It has become apparent that compounds also can undergo in vitro *decomposition* in biological experiments, which confuses the structure–activity relationship (SAR). Medicinal chemists should be aware of all potential in vivo and in vitro stability problems and ensure that their compounds are tested and structurally stabilized against these various challenges. A scheme of in vitro assays that can assess a wide range of stability issues in drug discovery is shown in Figure 11.2. Chapters 12 and 13 discuss the challenges faced as a result of chemical and enzymatic decomposition (e.g., solution and plasma stability). This chapter focuses on metabolic stability, which is widely considered one of the most significant challenges of drug discovery. Methods for predicting stability are discussed in Chapters 29 through 31.

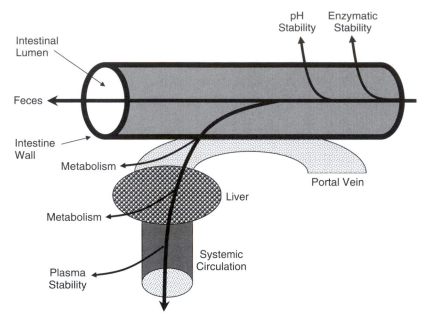

Figure 11.1 ▶ Stability challenges following oral administration.

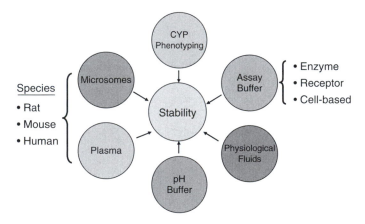

Figure 11.2 ▶ Scheme for in vitro assessment of diverse stability challenges during drug discovery. (Reprinted with permission from [29].)

▢ 11.1 Metabolic Stability Fundamentals

Drug metabolism often is referred to as *biotransformation*, and metabolism reactions have been divided into two phases. Phase I reactions are modifications of the molecular structure itself, such as oxidation or dealkylation. Phase II reactions are additions (conjugations) of polar groups to the molecular structure. Sometimes they are considered sequential: first addition of an attachment point (e.g., hydroxyl) and then addition of a large polar moiety (e.g., glucuronic acid). However, a compound need not undergo a phase I reaction before a phase II reaction if it already has a functional group that is susceptible to conjugation. Both of these types of reactions produce more polar products that have higher aqueous solubility, so they are more readily excreted from the body via bile and urine. Metabolism increases clearance, reduces exposure, and is a major cause of low bioavailability. The result is a lower concentration of a drug at the therapeutic target than without metabolism. Medicinal chemists often must make structural modifications to lead compounds during drug discovery in an attempt to reduce metabolism.

If the drug is metabolized prior to reaching systemic circulation, it is said to have undergone *presystemic* or *first-pass metabolism*. First-pass metabolism clears absorbed drug before it reaches systemic circulation. First-pass metabolism can occur in both the gut and the liver. By excluding exogenous compounds (e.g., drugs) from the bloodstream by first-pass metabolism and then further reducing drug concentration by metabolism of a portion of the circulating drug with each pass through the liver, metabolic stability is an underlying root property that affects exposure and bioavailability.

Differences in metabolism are observed among species. The metabolic profiles from different species vary in the metabolite structures and their production rates.

11.1.1 Phase I Metabolism

Phase I reactions consist of several mechanisms that modify the compound structure. These include oxidation and reduction.

Several different enzyme families catalyze these reactions. The most prominent are called *monooxygenases*, which include (a) the cytochrome P450 (CYP) family (Figure 11.3) and (b) the flavine monooxygenase (FMO) family. A generalized reaction for monooxygenases is:

$$R - H + O_2 + NADPH + H^+ \rightarrow R - OH + NADP^+ + H_2O.$$

Monooxygenases are bound to the endoplasmic reticulum (ER) inside cells and are found in high abundance in hepatocytes. The membrane association of the enzymes is consistent with the correlation of compound metabolism to lipophilicity.

The reaction is catalyzed in CYPs by a heme group containing an iron atom at the active site (Figure 11.3). In CYPs, the iron atom binds oxygen and transfers it to the drug molecule via a series of reactions (Figure 11.4). NADPH provides the electrons for reducing Fe^{III} to Fe^{II} via a second coupled enzyme (NADPH-cytochrome P450 reductase). The CYP family consists of over 400 isozymes. CYP enzymes are found in mammals, insects, plants, yeasts, and bacteria. In mammals, they are distributed in liver, kidney, lung, intestine, colon, brain, skin, and nasal mucosa. The different amino acid sequences of the isozymes result in different binding affinity for compound classes. The rate of the reaction is a result of (1) the affinity of the compound for binding to the CYP enzyme and (2) the reactivity of the position(s) on the molecule that are brought into close proximity with the heme group. In FMOs, a flavine group at the active site catalyzes the reaction. NADPH directly reduces the flavine.

A compound can be metabolized by more than one enzyme or isozyme. If metabolism of a drug at one enzyme is blocked by substrate saturation of the enzyme or by structural modification, then metabolism at another enzyme with weaker binding or lower reactivity may become a more favorable route. This process is called *metabolic switching*. Drug–drug interactions can occur if two or more drugs are coadministered, and one drug inhibits the metabolism of a second drug at a particular isozyme, leading to toxic effects. Drug–drug interactions are discussed in Chapter 15.

Common phase I metabolic reactions are shown in Figure 11.5. These examples indicate many of the potential metabolic reactions that discovery compounds can undergo. The most commonly observed metabolism is hydroxylation of aliphatic and aromatic carbon atoms.

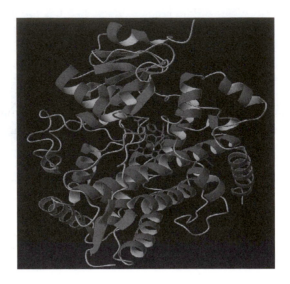

Figure 11.3 ▶ Structure of human cytochrome P450 3A4 with heme and inhibitor metyrapone.[30] (Drawing courtesy Kristi Fan.) (see Plate 2)

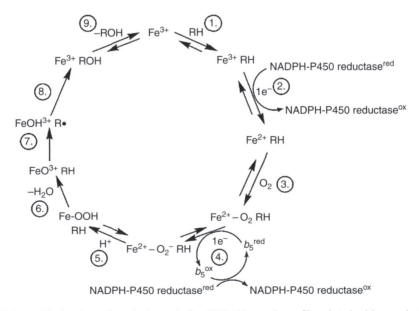

Figure 11.4 ▶ Mechanism of catalytic cycle for CYP450 reactions. (Reprinted with permission [31].)

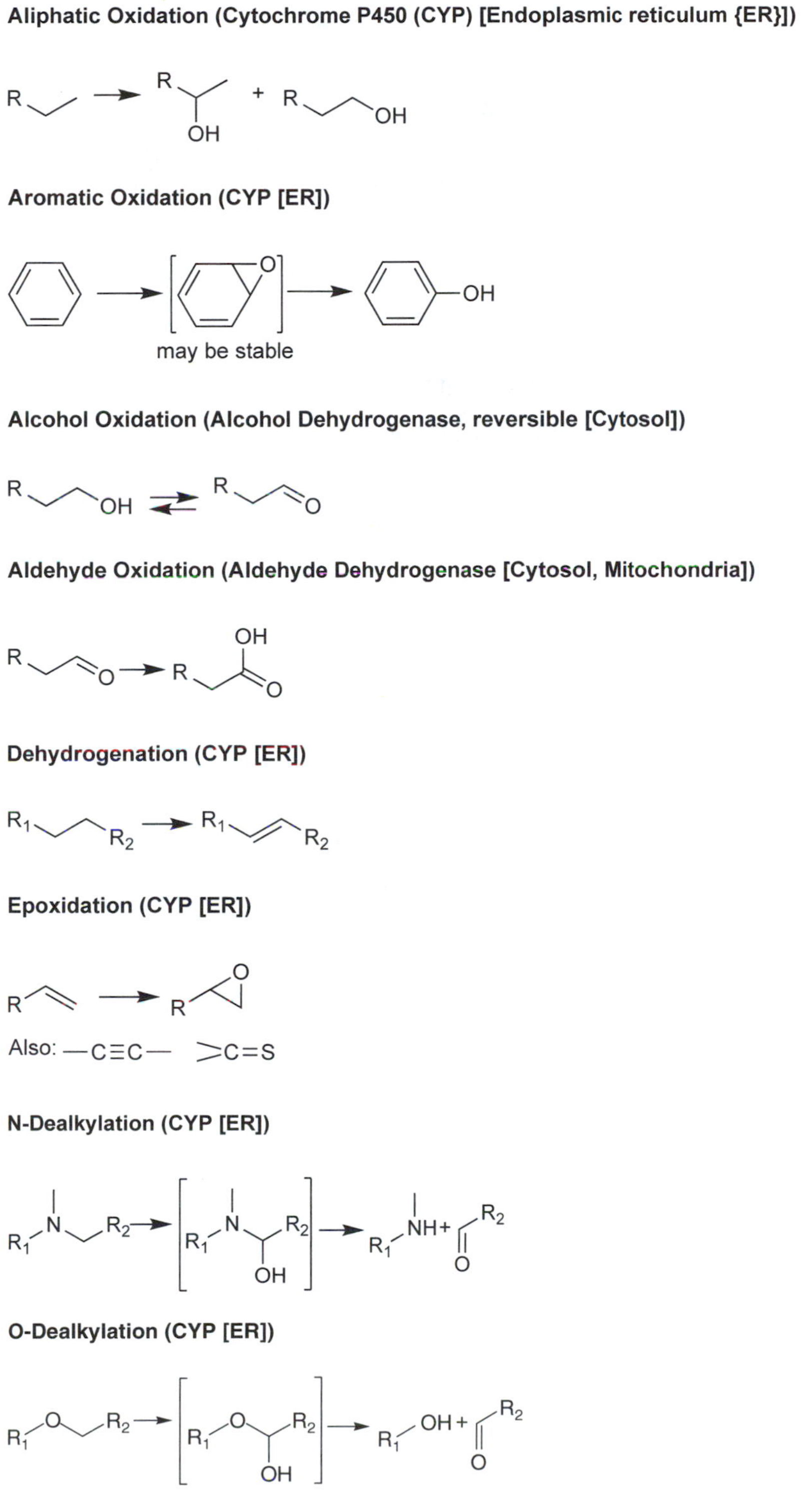

Aliphatic Oxidation (Cytochrome P450 (CYP) [Endoplasmic reticulum {ER}])

Aromatic Oxidation (CYP [ER])

may be stable

Alcohol Oxidation (Alcohol Dehydrogenase, reversible [Cytosol])

Aldehyde Oxidation (Aldehyde Dehydrogenase [Cytosol, Mitochondria])

Dehydrogenation (CYP [ER])

Epoxidation (CYP [ER])

Also: $-C\equiv C-$ $>C=S$

N-Dealkylation (CYP [ER])

O-Dealkylation (CYP [ER])

Figure 11.5 ▶ Examples of major phase I metabolic reactions.

S-Dealkylation (CYP [ER])

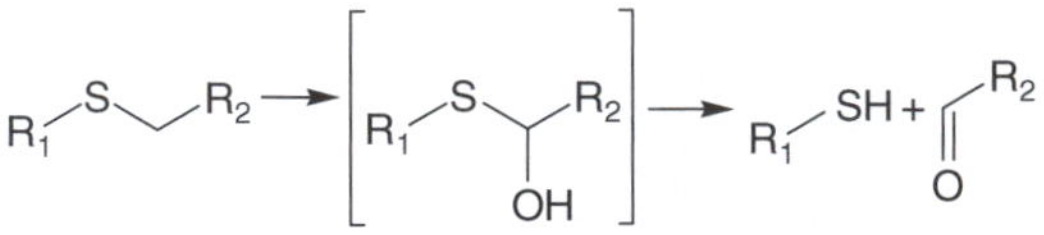

$$R_1\text{—S—}R_2 \rightarrow \left[R_1\text{—S—}\underset{OH}{R_2} \right] \rightarrow R_1\text{—SH} + \underset{O}{R_2}$$

Oxidative Deamination (Monoamine- & Diamine-Oxidases [Mitochondria])

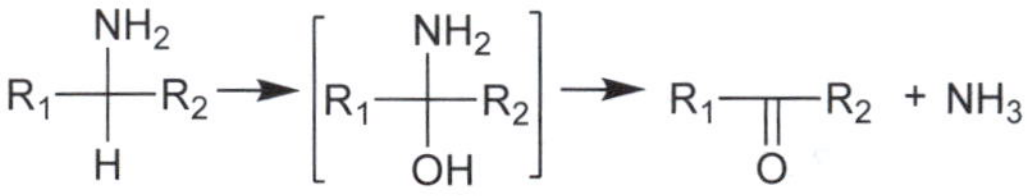

$$R_1\underset{H}{\overset{NH_2}{|}}R_2 \rightarrow \left[R_1\underset{OH}{\overset{NH_2}{|}}R_2 \right] \rightarrow R_1\underset{O}{—}R_2 + NH_3$$

N-Oxidation (Flavin Monooxygenase (FMO) [ER])

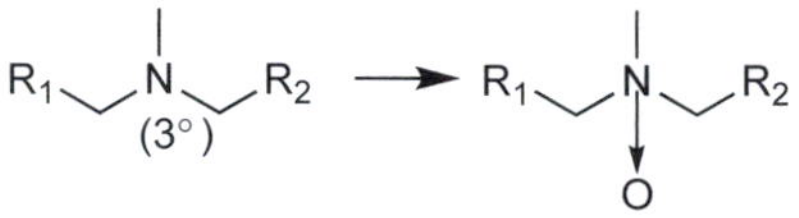

$$R_1\text{—N(3°)—}R_2 \rightarrow R_1\text{—N—}R_2 \ (\downarrow O)$$

N-Hydroxylation (CYP [ER])

$$R_1\overset{H}{—N—}R_2 \rightarrow R_1\overset{OH}{—N—}R_2$$

S-Oxidation (FMO [ER])

$$R_1\text{—S—}R_2 \rightarrow R_1\overset{O}{—S—}R_2 \rightarrow R_1\overset{O\ O}{—S—}R_2$$

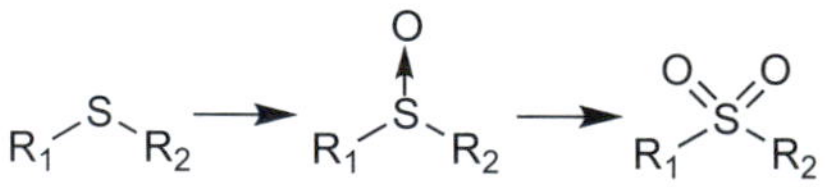

Cyclic Amines to Lactams (Aldehyde oxidase)

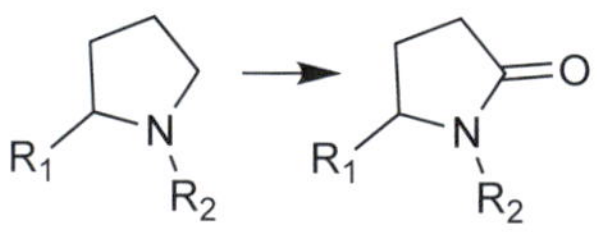

Reductions

$$Ph\text{—}NO_2 \rightarrow Ph\text{—}NH_2 \quad \text{([NADPH-CYP450 reductase [ER] and Nitroreductase [cystosol])}$$

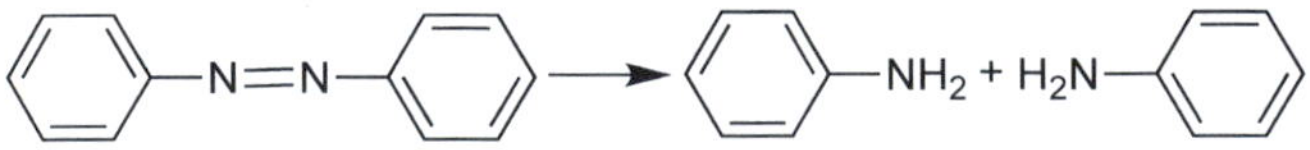

$$Ph\text{—N=N—}Ph \rightarrow Ph\text{—}NH_2 + H_2N\text{—}Ph \quad \text{(Azoreductase)}$$

$$R_1\underset{R_2}{\overset{O}{—}} \rightarrow R_1\underset{R_2}{\overset{OH,H}{—}} \quad \text{(R_2 may be H, Alcohol Dehydrogenase, [cytosol])}$$

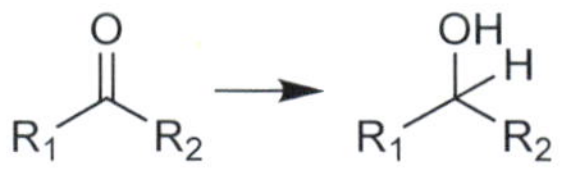

Figure 11.5 ► *Continued*

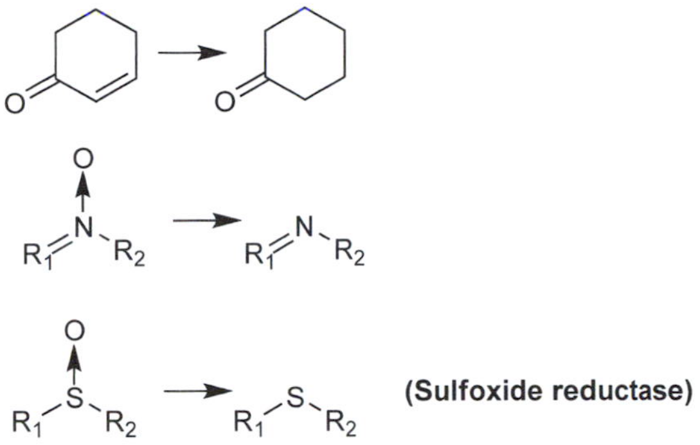

Figure 11.5 ► *Continued*

Aliphatic hydroxyls may be further converted to aldehydes and carboxylic acid. Carbon atoms adjacent to nitrogen atoms are susceptible to oxidation, which leads to dealkylation to form the amine and aldehyde. In the same manner, carbon atoms adjacent to ether oxygens or sulfides can be oxidized and lead to dealkylation. Nitrogen atoms of amines can be oxidized to *N*-oxides. Sulfur atoms of sulfides can be oxidized to sulfoxides and sulfones. Mechanisms of biotransformation reactions are discussed by Magdalou et al.[2] Some software products provide useful predictions of the site of metabolism (see Chapter 29).

11.1.2 Phase II Metabolism

Phase II metabolism is the addition of polar moieties to the molecule. Common additions are shown in Figure 11.6. Glucuronic acid can be added to aromatic or aliphatic hydroxyls, and occasionally carboxylic acids and amines by UDP-glucuronosyltransferases (UGTs) to form glucuronide metabolites. Sulfate can be added to aromatic, aliphatic, or hydroxylamine

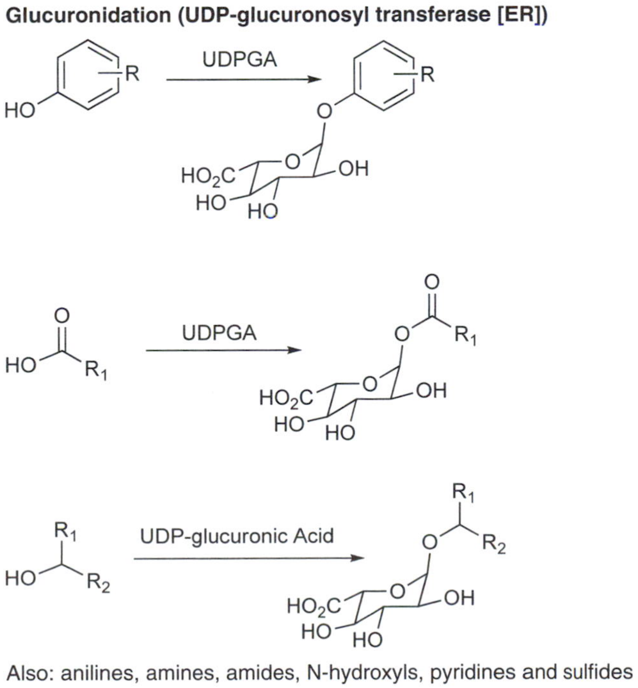

Glucuronidation (UDP-glucuronosyl transferase [ER])

Also: anilines, amines, amides, N-hydroxyls, pyridines and sulfides

Figure 11.6 ► Examples of major phase II metabolism reactions.

Carbamic Acid Glucuronidation

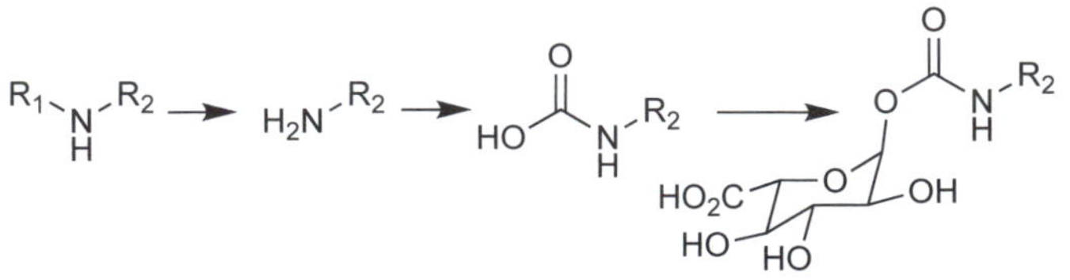

Sulfation (Sulfotranserase [cytosol])

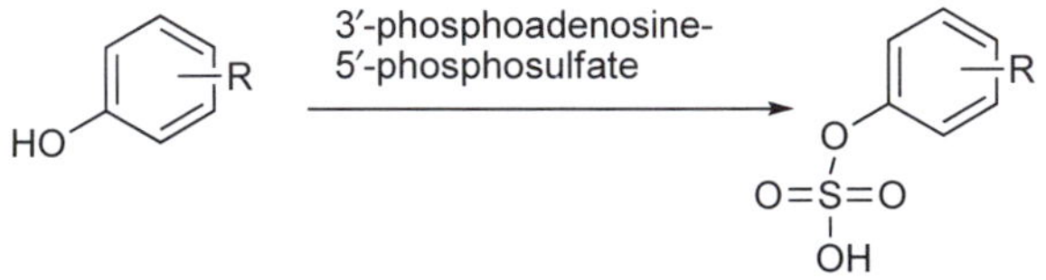

3′-phosphoadenosine-
5′-phosphosulfate

Acetylation (N-Acetyltransferases [cytosol])

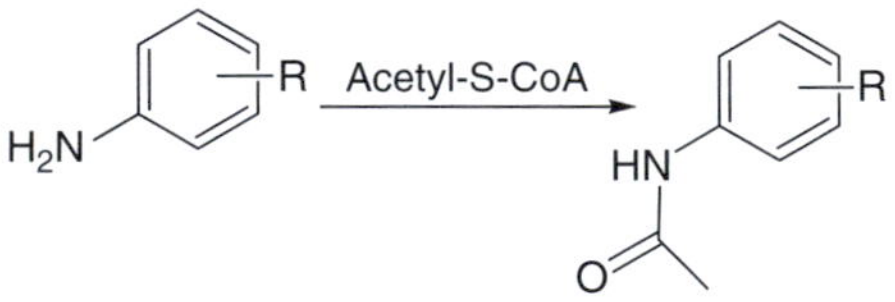

Acetyl-S-CoA

Also: 1°, 2° Amines, Hydrazines, Hydrides; R-NH-OH → -NH-O-COCH$_3$

Glycination [mitochondria]

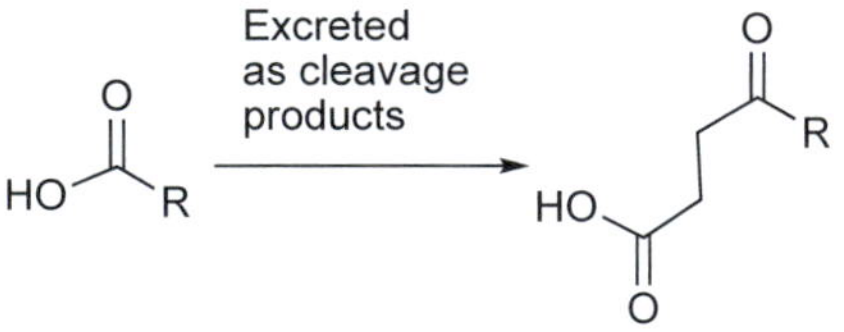

Excreted
as cleavage
products

Also: other amino acid additions (e.g., taurine, glutamine)

Glutathione Conjugation (Glutathione-S-transferases [cytosol])

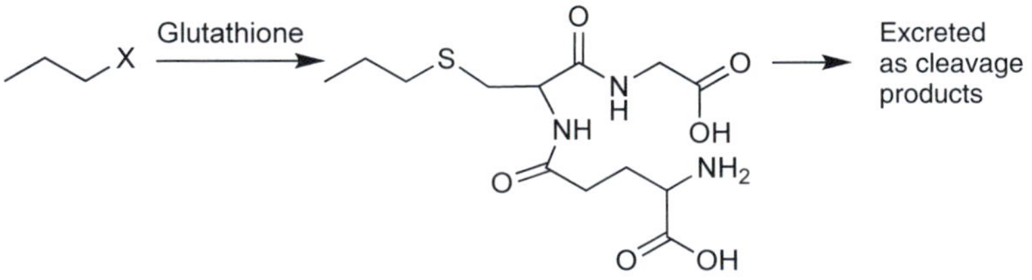

Glutathione

Excreted
as cleavage
products

X: Halogen, Electron-deficient double bond, or epoxide

Figure 11.6 ▶ *Continued*

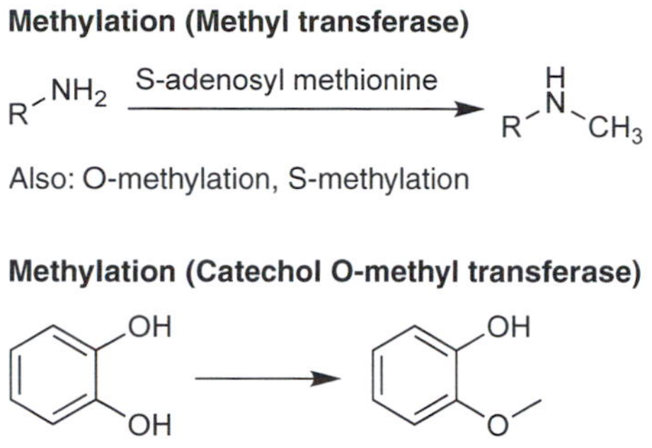

Methylation (Methyl transferase)

$$R-NH_2 \xrightarrow{\text{S-adenosyl methionine}} R-\overset{H}{\underset{}{N}}-CH_3$$

Also: O-methylation, S-methylation

Methylation (Catechol O-methyl transferase)

Figure 11.6 ▶ *Continued*

hydroxyls by sulfotransferases to form sulfate metabolites. This is a rapid, but saturable, reaction. Glutathione can be added to reactive electrophiles (nucleophilic displacement) or to an electron-deficient double bond (nucleophilic addition) by glutathione-*S*-transferases to form glutathione conjugates. This is a major detoxification mechanism for reactive xeno-biotics and reactive metabolites. Amines can be acetylated by *N*-acetyltransferases to form amides. Various amino acids, such as glycine, can be added to form conjugates.

The greatly increased hydrophilicity of phase II conjugates enhances their elimination in the bile and urine.

11.2 Metabolic Stability Effects

Metabolic stability affects pharmacokinetics (PK), as shown in Figure 11.7.[3] Metabolic stability has an inverse relationship with clearance (Cl). A decrease in metabolic stability leads to an increase in clearance. Cl and volume of distribution (V_d) directly affect the PK half-life ($t_{1/2} = 0.693 \times V_d/Cl$), which determines how often the dose must be administered. Cl and absorption, which is determined by intestinal permeability and solubility, directly affect oral bioavailability (F). F determines how much drug must be administered.

This is illustrated with compounds from a discovery project lead series in Table 11.1. The compounds having short in vitro metabolic stability $t_{1/2}$ tend to have high in vivo clearance (Cl) and low oral bioavailability (%F).

An example of different metabolic stabilities of analogs with the same core template is shown in Table 11.2[4]. Structural differences at R_1, R_2, and R_3 had significant effects on metabolic stability. In addition, the bioavailability increased as the percent metabolized decreased.

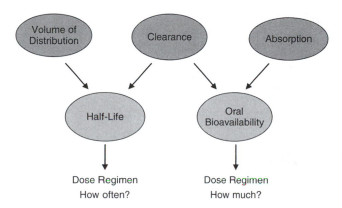

Figure 11.7 ▶ Effects of metabolic stability on pharmacokinetics. (Reprinted with permission from [3].)

TABLE 11.1 ▶ Example Compounds from a Discovery Project Lead Series Demonstrating Relationship Between In Vitro Metabolic Stability (t₁/₂) and In Vivo Cl and F

Compound	In vitro $t_{1/2}$ (min)	In vivo Cl (ml/min/kg)	% F Rat
1	5	53	3
2	6	55	8
3	7	49	15
4	14	18	20
5	>30	14	41

TABLE 11.2 ▶ Relationship Among Structure, In Vitro Metabolic Stability, and Oral Bioavailability

Compd	R_1	R_2	R_3	% Metabolized (S9, 1h)			Oral Bioavailability
				Rat	Dog	Human	% F in Rat
A	H	H		96	99	66	14
B	F	H		88	99	63	3
C	H	H		99	99	94	7
D	F	CF$_3$		7	4	5	40
E	F	F		7	90	21	NA
F	F	F		15	46	43	58

11.3 Structure Modification Strategies for Phase I Metabolic Stability

Before beginning structure modification to improve stability, it is useful to elucidate the specific site of metabolism. In the past, it was necessary to predict the most likely sites of metabolism and then make analogs to counteract possible metabolism at those sites. However, higher throughput methods for metabolite structure elucidation have been developed and are discussed in Chapter 29.

Several strategies for increasing the metabolic stability of compounds have been successful. These strategies are listed in Table 11.3 for phase I metabolism and Table 11.4 for phase II metabolism. The modifications for phase I metabolism are based on two key characteristics of metabolic reactions: (a) binding of the compound to the metabolic enzyme and (b) reactivity of the site on the molecule that is adjacent to the reactive heme of CYP or reactive site of another metabolic enzyme. Structural changes that reduce compound binding or reactivity at the labile site will increase metabolic stability. These strategies are not always 100% successful, owing to metabolic switching.

TABLE 11.3 ► Structure Modifications Strategies for Phase I Metabolic Stability Improvement

Structure modification	Section
Block metabolic site by adding fluorine	11.3.1
Block metabolic site by adding other blocking groups	11.3.2
Remove labile functional group	11.3.3
Cyclization	11.3.4
Change the ring size	11.3.5
Change chirality	11.3.6
Reduce lipophilicity	11.3.7
Replace unstable groups	11.3.8

11.3.1 Block Metabolic Site By Adding Fluorine

The strategy of blocking the site of hydroxylation is shown in Figure 11.8. The blocking group is less reactive than the hydrogen that was in the site in the metabolized analog. Fluorine is the most commonly used blocking group. Of the possible blockers, it seems to have the least effect on the size of the molecule.

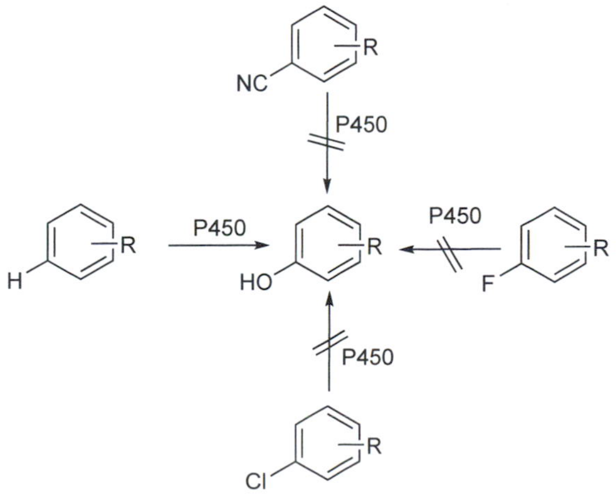

Figure 11.8 ► Blocking the site of metabolism by adding fluorine, chlorine, or nitrile.

Figure 11.9 illustrates the addition of fluorine at the site of metabolism.[5] Hydroxylation of the 5′ position of pyrimidine is a common metabolic reaction for buspirone. By adding

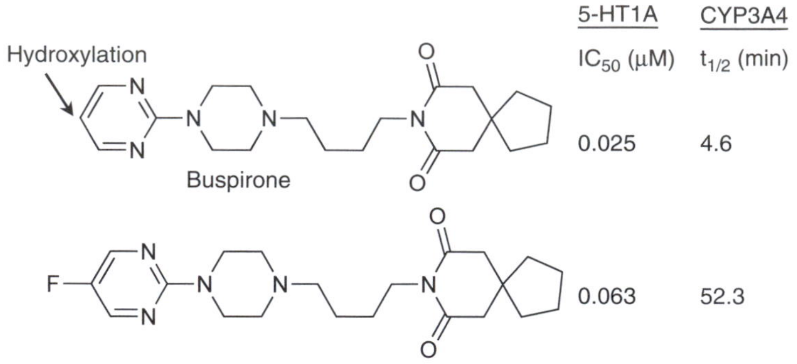

Figure 11.9 ► Fluorination of buspirone at the 5′ position of pyridine blocks metabolism and increases the in vitro $t_{1/2}$.

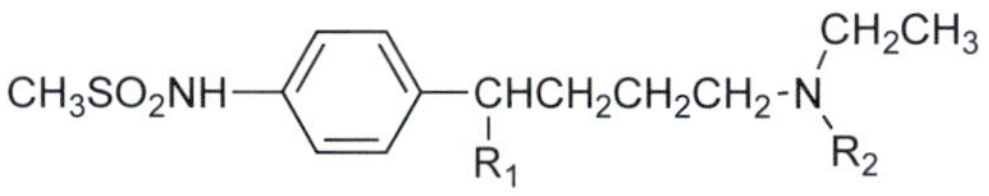

R1	R2	Relative Metabolic Stability
(R)-OH	$(CH_2)_6CH_3$	1
OH	$(CH_2)_6CH_2F$	4.2
OH	$(CH_2)_6CHF_2$	>20
OH	$(CH_2)_5CHF_2$	>20
(S)-OH	$(CH_2)_5C(CH_3)_2F_2$	>20
H	$(CH_2)_6CH_2F$	1
H	$(CH_2)_5CH(F)CH_3$	3
H	$(CH_2)_5C(CH_3)_2F$	3

Figure 11.10 ► Substitution of fluorine atoms in R2 increased the metabolic stability.

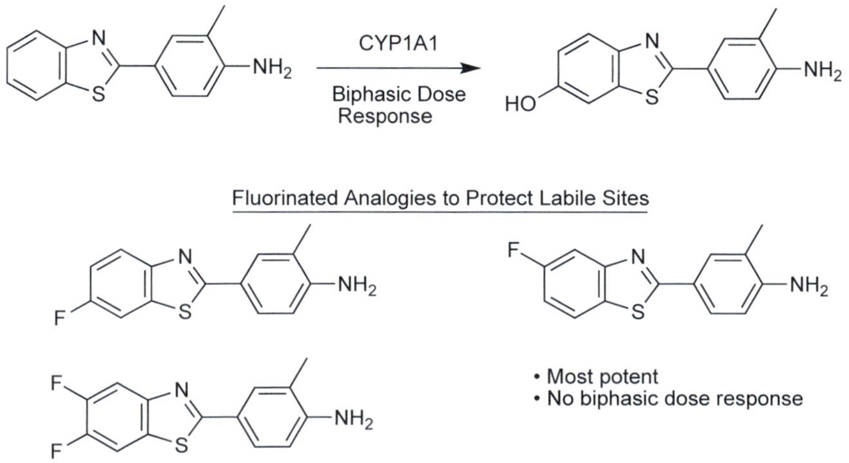

Fluorinated Analogies to Protect Labile Sites

• Most potent
• No biphasic dose response

Figure 11.11 ► Metabolism caused a biphasic dose response. Blocking the metabolic site eliminated the biphasic dose response.

fluorine at this site, the in vitro $t_{1/2}$ with CYP3A4 isozyme increases from 4.6 to 52 minutes, without a major loss of activity. CYP3A4 is the major metabolic enzyme for buspirone.

Various fluorine analogs of ibutilide at the R_2 alkyl moiety are shown in Figure 11.10.[6] Replacement of one or two hydrogens with fluorine atoms increased the relative metabolic stability by 4- and 20-fold, respectively.

The compound in Figure 11.11 demonstrated a biphasic dose response due to metabolism by CYP1A1.[7] Fluorinated analogs blocked metabolism and did not exhibit the biphasic dose response.

11.3.2 Block Metabolic Site By Adding Other Blocking Groups

Groups other than fluorine can be added at the labile site. Substitution with chlorine is illustrated in Figure 11.12. The pharmacokinetic half-life was lengthened from 6 to 33 hours by substitution of the methyl group on tolbutamide with chlorine to make chlorpropamide.

For the compound shown at the top of Figure 11.13, the benzylic methine was found to be a labile site.[8] Additions here resulted in greater metabolic stability and increased in vivo exposure (i.e., area under the curve [AUC]) while improving the activity.

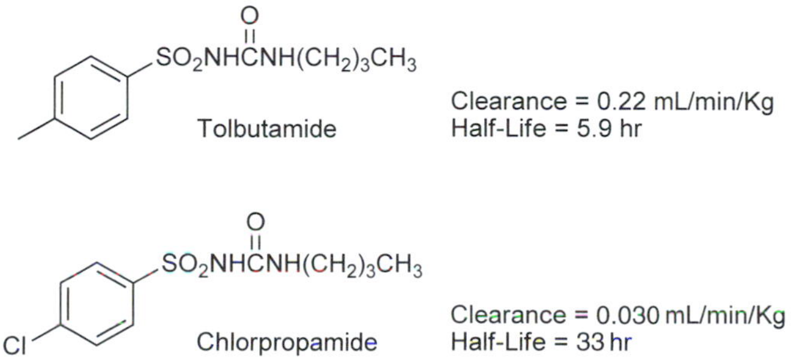

Clearance = 0.22 mL/min/Kg
Half-Life = 5.9 hr

Clearance = 0.030 mL/min/Kg
Half-Life = 33 hr

Figure 11.12 ► Substitution of methyl in tolbutamide with chlorine to make chlorpropamide greatly increased stability.

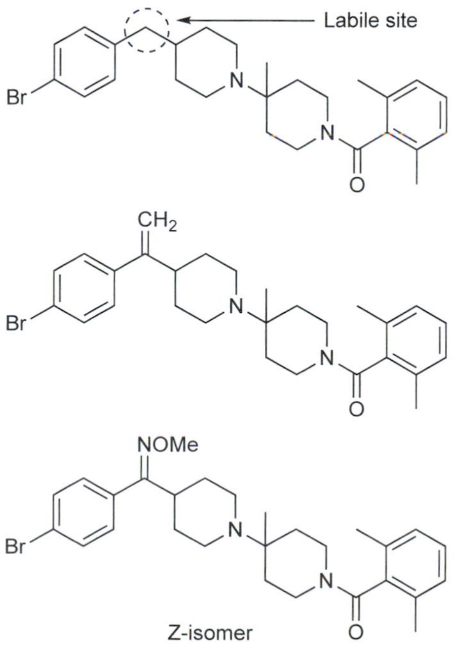

	Ki (nM)	IC$_{50}$ (nM)	AUC (p.o.) (h.µg/mL)
	66	10	0.04
	8	1.0	0.59
	2	1.3	1.2

Figure 11.13 ► Blockage of a labile site improved metabolic stability and oral exposure.

More bulky aliphatic groups can also be added adjacent to the labile site. This provides steric hindrance, thus reducing access of the metabolizing enzyme active site to the labile compound site. For example, modification of the methoxy in metoprolol to the cyclopropylmethoxy in betaxolol (Figure 11.14) reduced *O*-dealkylation by CYP2D6 and increased bioavailability.[9]

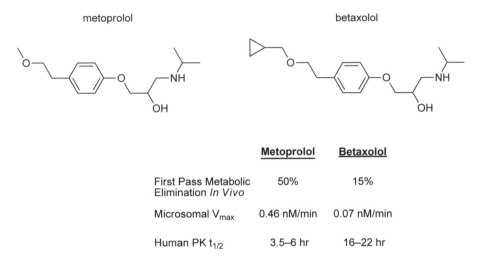

	Metoprolol	**Betaxolol**
First Pass Metabolic Elimination *In Vivo*	50%	15%
Microsomal V_{max}	0.46 nM/min	0.07 nM/min
Human PK $t_{1/2}$	3.5–6 hr	16–22 hr

Figure 11.14 ► *O*-Dealkylation of the methoxy of metoprolol was reduced by modification to the cyclopropylmethoxy to make betaxolol, which has improved metabolic stability.

11.3.3 Remove Labile Functional Group

Removing labile groups can improve metabolic stability.[10] For the compound in Figure 11.15, removal of the methoxy methyl increased stability. Further removal of the *N*-propylene group increased stability greatly while maintaining activity. These sites are reactive for dealkylation in the original molecule.

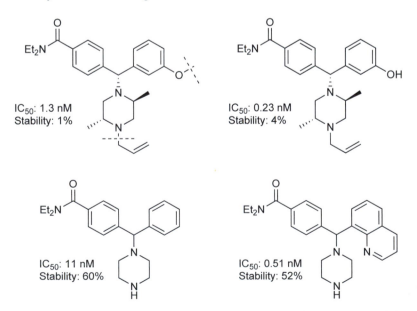

Figure 11.15 ► Removal of labile groups improved metabolic stability (rat, percent remaining after 1-hour incubation).

For the upper compound in Figure 11.16, the methoxy group was found to be metabolically unstable.[11] Removal of the methoxy and substitution with an amide improved the metabolic stability and resulted in a greater C_{max}.

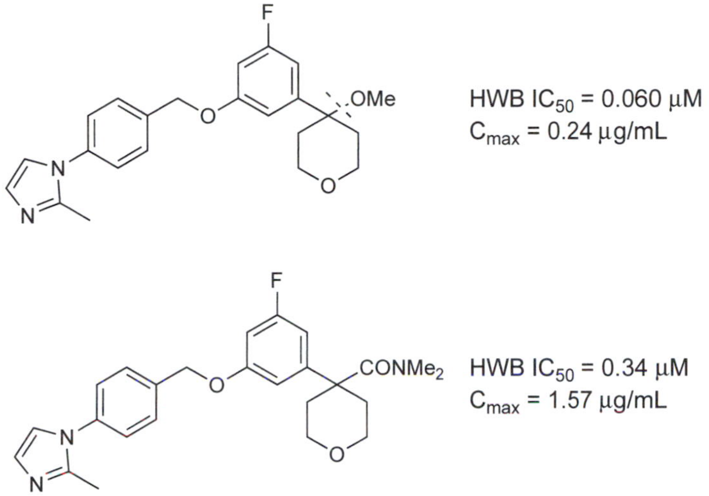
F
OMe
O
N
N
O
HWB IC50 = 0.060 μM
Cmax = 0.24 μg/mL
F
CONMe2
O
N
N
O
HWB IC50 = 0.34 μM
Cmax = 1.57 μg/mL

Figure 11.16 ► Remove labile group and replace to improve the metabolic stability (monkey, oral 5 mg/kg dose).

11.3.4 Cyclization

Metabolism of labile groups can be reduced by incorporation into a cyclic structure.[12] For example, the compound in Figure 11.17 was susceptible to *N*-demethylation. Incorporation of the methyl group into a cyclic structure preserved the functionality for the sake of activity while improving metabolic stability.

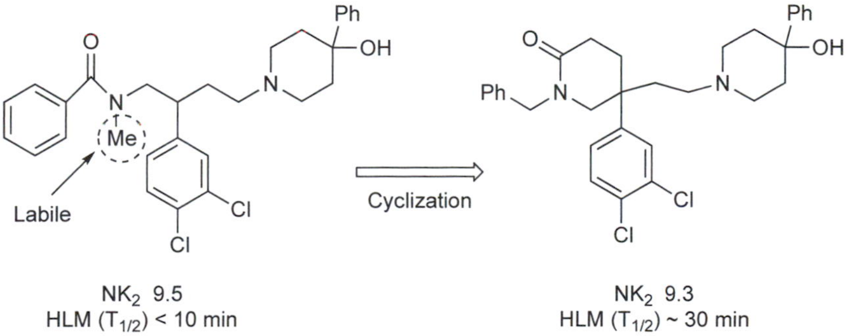
O
N
Me
Labile
N
Ph
OH
Cl
Cl
Cyclization
O
Ph
N
N
Ph
OH
Cl
Cl
NK2 9.5
HLM (T1/2) < 10 min
NK2 9.3
HLM (T1/2) ~ 30 min

Figure 11.17 ► Cyclization strategy to improve metabolic stability in human liver microsomes (HLM) while maintaining activity (NK2).

11.3.5 Change Ring Size

The size of attached rings can be changed to modify metabolic stability. For the example in Figure 11.18, reducing the ring size increased the metabolic stability ($t_{1/2}$ in human liver microsomes).[12]

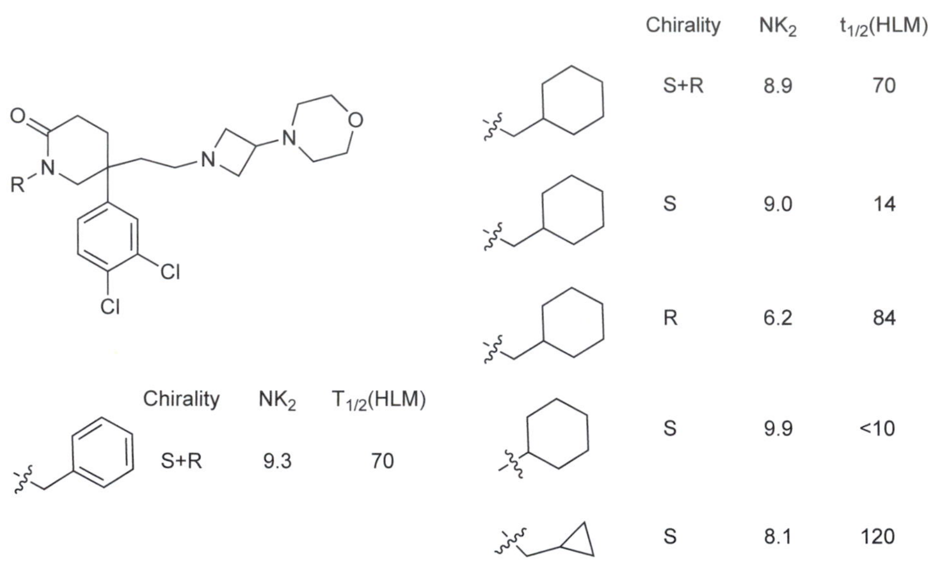

	Chirality	NK$_2$	t$_{1/2}$(HLM)
	S+R	8.9	70
	S	9.0	14
	R	6.2	84
	S	9.9	<10
	S	8.1	120

Chirality	NK$_2$	T$_{1/2}$(HLM)
S+R	9.3	70

Figure 11.18 ► Reduction of ring size and modification of chirality can increase metabolic stability.

11.3.6 Change Chirality

The chirality of attached groups can affect metabolic stability. Changes in the chirality of the series analogs in Figure 11.18 lead to a significant improvement in metabolic stability.[12] This suggests that the enantiomers bound differently to the metabolic enzyme.

11.3.7 Reduce Lipophilicity

A decrease in lipophilicity often will improve metabolic stability. This reduces binding to metabolic enzymes, which typically have lipophilic binding pockets. The example in Figure 11.19 demonstrates how reduced Log D can lead to improved metabolic stability.[13]

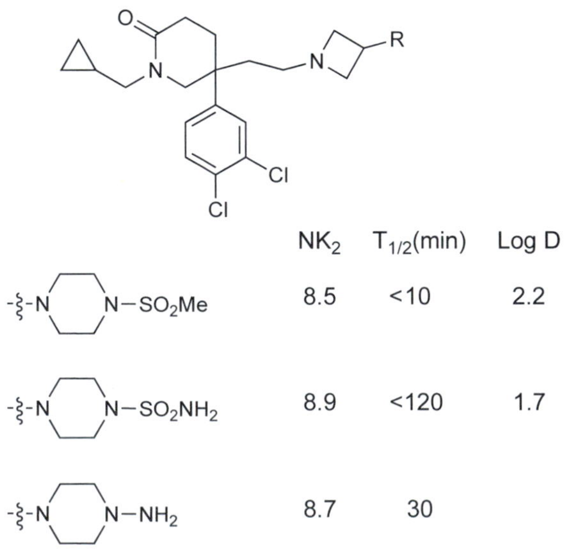

	NK$_2$	T$_{1/2}$(min)	Log D
-N N—SO$_2$Me	8.5	<10	2.2
-N N—SO$_2$NH$_2$	8.9	<120	1.7
-N N—NH$_2$	8.7	30	

Figure 11.19 ► Reducing the lipophilicity of the R group improved metabolic stability.

11.3.8 Replace Unstable Groups

Groups that cause considerable metabolism of the lead can be substituted to achieve improved stability. The compounds in Figure 11.20 that contained a piperidine group were metabolically unstable. Substitution with a piperazine greatly improved metabolic stability[13].

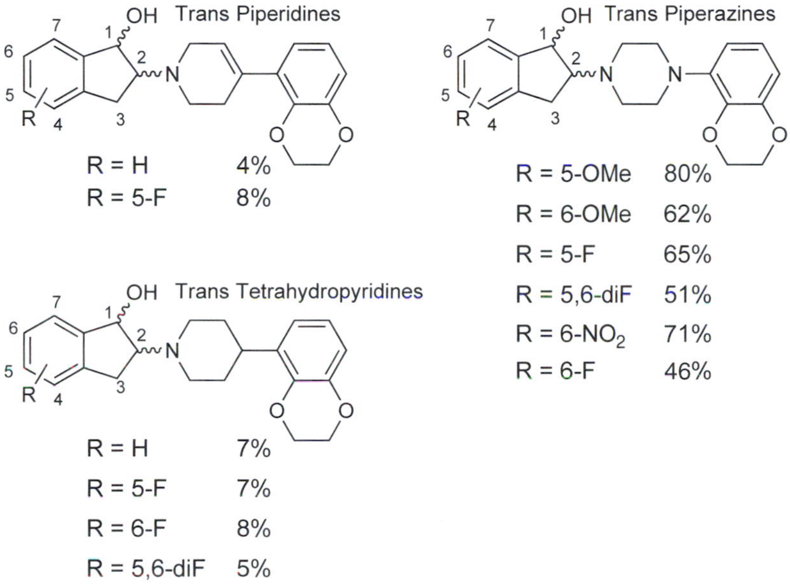

Trans Piperidines

R = H   4%
R = 5-F   8%

Trans Piperazines

R = 5-OMe   80%
R = 6-OMe   62%
R = 5-F   65%
R = 5,6-diF   51%
R = 6-NO$_2$ 71%
R = 6-F   46%

Trans Tetrahydropyridines

R = H   7%
R = 5-F   7%
R = 6-F   8%
R = 5,6-diF   5%

Figure 11.20 ► Substitution of piperazine for piperidine improved the metabolic stability for this series, resulting in increased %F.

The metabolic stability of tiamulin (Figure 11.21) was improved by substitution of the side chain with the carbamate-containing side chain.[14] The new compound has excellent broad-spectrum antibacterial activity and 10-fold slower metabolism.

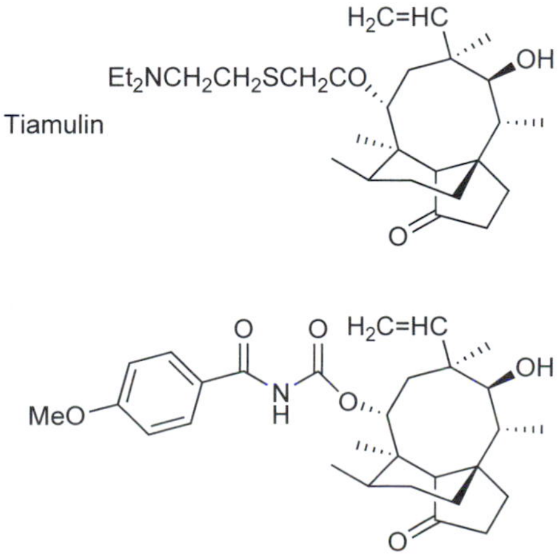

Tiamulin

Et$_2$NCH$_2$CH$_2$SCH$_2$CO,

Figure 11.21 ► Substitution of the carbamate side chain improved the metabolic stability while maintaining good antibacterial activity.

11.4   Structure Modification Strategies for Phase II Metabolic Stability

Phase II reactions can be reduced through structure modification (Table 11.4).

TABLE 11.4 ▶ Structure Modifications Strategies for Phase II Metabolic Stability Improvement

Structure modification	Section
Introduce electron-withdrawing groups or steric hindrance	11.4.1
Change phenolic hydroxyl to cyclic urea or thiourea	11.4.2
Change phenolic hydroxyl to prodrug	11.4.3

11.4.1   Introduce Electron-Withdrawing Groups and Steric Hindrance

Addition of electron withdrawing groups on the aromatic ring will reduce the rate of glucuronidation of a phenol. Figure 11.22 shows an example of the addition of a chlorine atom to the phenol adjacent to the hydroxyl. This modification reduced glucuronidation by reducing the reactivity of the phenol and by steric hindrance.

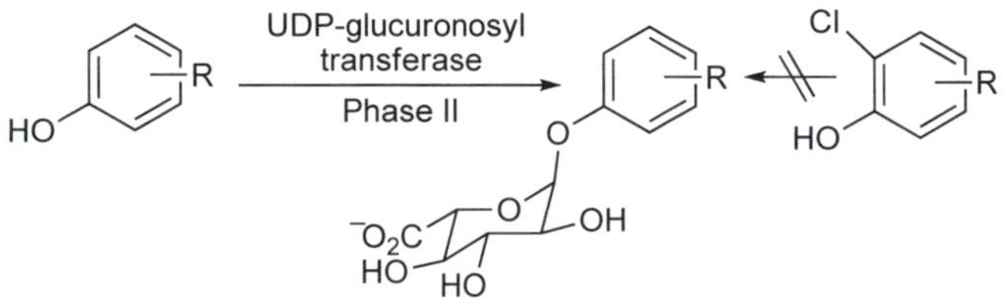

Figure 11.22 ▶ Addition of a halogen to a phenolic compound reduces glucuronidation by electron withdrawal and steric hindrance.

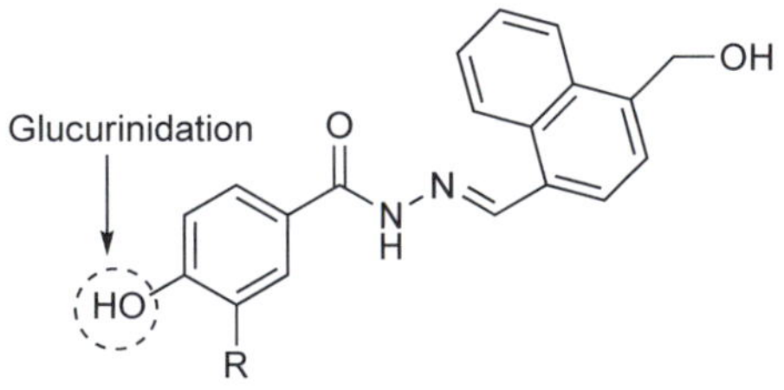

R	hGluR binding affinity (nM)	Metabolism Clearance (pmol/min/mg)
Cl	41	75
F	29	89
2,3-di-Cl	54	103
CN	30	37

Figure 11.23 ▶ Addition of a cyano group to the phenolic compound reduced glucuronidation while increasing activity.

Glucurinidation
O
N
H
N
HO
R
O
N

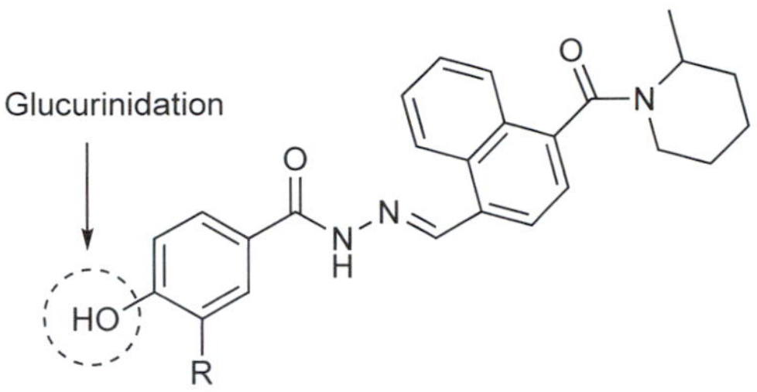

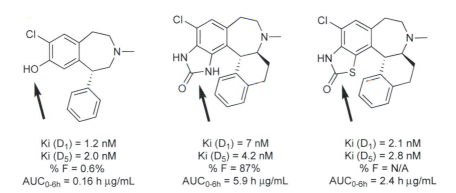

R	hGluR binding affinity (nM)	Metabolism Clearance (pmol/min/mg)
Cl	8	267
CN	12	65

Figure 11.23 ▸ *Continued*

Addition of a cyano group to the phenolic ring adjacent to the hydroxyl will also reduce glucuronidation. Figure 11.23 shows examples of two phenolic groups whose glucuronidation was reduced while improving target activity.[15]

11.4.2 Change Phenolic Hydroxyl to Cyclic Urea or Thiourea

A phenolic group can be replaced with an isostere to reduce glucuronidation. The lead on the left in Figure 11.24 was modified by the addition of a cyclic urea or thiourea. The metabolic stability increased (as indicated by the improved F and AUC) while the activity was maintained.[16]

Cl
HO
N—
Ki (D1) = 1.2 nM
Ki (D5) = 2.0 nM
% F = 0.6%
AUC0-6h = 0.16 h µg/mL
Cl
HN
NH
O
N—
Ki (D1) = 7 nM
Ki (D5) = 4.2 nM
% F = 87%
AUC0-6h = 5.9 h µg/mL
Cl
HN
S
O
N—
Ki (D1) = 2.1 nM
Ki (D5) = 2.8 nM
% F = N/A
AUC0-6h = 2.4 h µg/mL

Figure 11.24 ▸ Isosteric replacement of the phenolic hydroxyl improved metabolic stability against glucuronidation.

11.4.3 Change Phenolic Hydroxyl to Prodrug

A phenolic hydroxyl can be modified to a prodrug (Figure 11.25).[17] The pro-moiety is slowly hydrolyzed to release the free phenol, after it has bypassed first-pass metabolism.

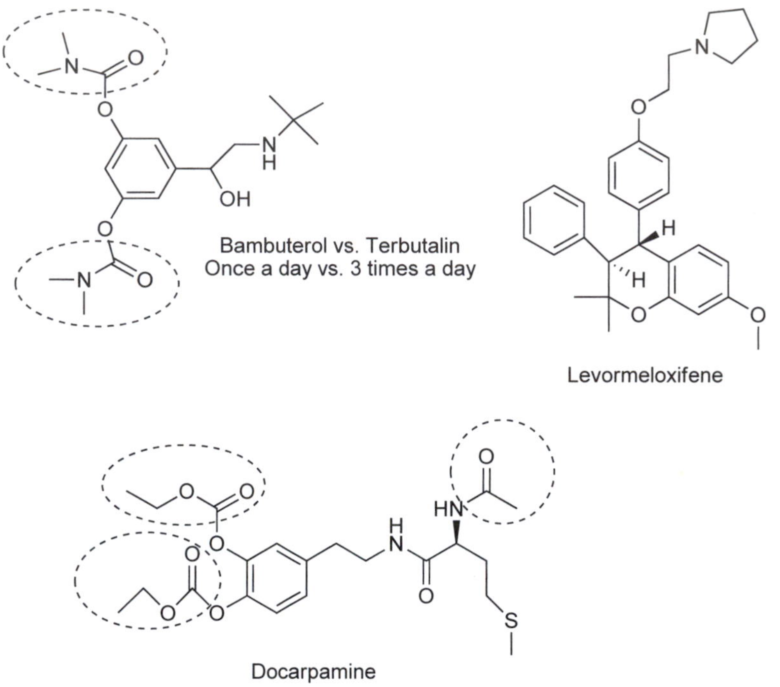

Bambuterol vs. Terbutalin
Once a day vs. 3 times a day

Levormeloxifene

Docarpamine

Figure 11.25 ▶ Phenols can be modified to form prodrugs. The free phenol is slowly released.

11.5 Applications of Metabolic Stability Data

In vitro metabolic stability data are often used to:

▶ Guide structure modifications to improve stability

▶ Select the optimal compound(s) for in vivo PK or activity testing

▶ Prospectively predict in vivo PK performance

▶ Retrospectively diagnose the root causes of poor in vivo PK

Metabolic stability plays a major role in drug clearance. Schemes for the application of metabolic clearance to in vivo PK clearance and bioavailability are shown in Figure 11.26.[3]

Metabolic stability is commonly measured in vitro during drug discovery as soon as a new compound is synthesized. This provides feedback that alerts the project team to metabolic limitations and provides data to guide metabolic stability improvement through structural modifications.

The rate of metabolism of a compound by different metabolizing enzymes is useful to know. During the lead optimization phase, this information can be combined with knowledge of the particular enzyme's substrate specificity. This can guide structural modifications to reduce metabolism. During late discovery, knowledge of the primary metabolizing enzymes can suggest issues with drug–drug interactions (see Chapter 15). When the in vitro experiment studies which CYP isozyme(s) metabolizes the compound, the assay

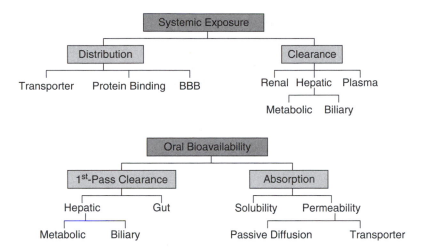

Figure 11.26 ► Schemes for the diagnosis of in vivo PK performance using metabolic clearance. (Reprinted with permission from [3].)

is termed *CYP phenotyping*. This and other metabolic stability assays are discussed in Chapter 29.

Metabolic stability data also can be beneficial in optimizing in vivo PK performance. Improvement of microsomal stability typically leads to lower in vivo Cl and higher F (Table 11.1).

Another example of the application of metabolic stability to a discovery project is shown in Figure 11.27. The compounds in the lead series for this project initially had low metabolic stability. After 3 months of structural optimization, metabolic stability was greatly improved. The keys to this success were as follows:

► High-throughput microsomal stability method that assayed hundreds of compounds per month for each of the company's many projects

► Fast turnaround of data (1–2 weeks)

► Company commitment to parallel optimization of activity and properties

Start, $t_{1/2}$ (min)		3 months later, $t_{1/2}$ (min)		
Rat	Mouse	Rat	Mouse	Human
5	10	>30	12	>30
7	7	>30	29	>30
5	5	20	10	18
7	8	>30	14	>30
3	2	12	30	>30
8	5	6	10	>30
5	3	>30	13	>30

Figure 11.27 ► Data from a discovery project show that the compounds initially had low metabolic stability. Structural modifications resulted in improved metabolic stability in a short time.

A comparison of the current approach versus the traditional approach is shown in Figure 11.28. The new screening paradigm allows teams to use metabolic stability data to make informed decisions.

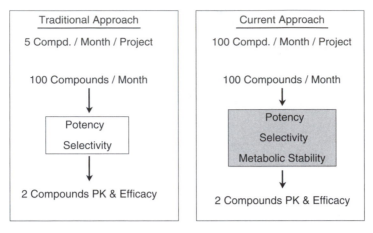

Figure 11.28 ▶ New approach includes metabolic stability for 100 compounds per month for each project, which enables informed decisions. (Reprinted with permission from [32].)

The crucial nature of metabolic stability and the relatively low cost of in vitro stability assays can be leveraged to assist with more expensive processes in discovery. Metabolic stability of two compounds from a series is shown Figure 11.29. The *R, S,* and racemic compounds in the top table all had the same microsomal stability. For the second compound in the table, the desirable *S*-enantiomer was less stable than the undesirable *R*-enantiomer. These have been consistently observed for this series of compounds. Therefore, using metabolic stability data from the racemate will not underestimate microsomal stability of the desirable enantiomer. Hence, the team uses microsomal stability data along with

Half-Life in minutes

Compounds	Rat	Mouse	Human	
Racemate	3	5	18	
S	5	4	16	Same
R	3	6	18	

Compounds	Rat	Mouse	Human	
Racemate	9	6	11	
S	2	3	3	R > S
R	21	7	19	

Figure 11.29 ▶ Strategy for triaging chiral compounds using an initial evaluation by the less expensive microsomal stability assay.

potency and selectivity to triage compounds for scaleup, chiral separation, and animal testing (Figure 11.30). In vitro metabolic stability data can be used to help teams make an informed decision on which compounds to take to the more expensive studies (e.g., chirals resolution).

In vitro microsomal stability data can be used to plan in vivo efficacy study designs. The compound in Figure 11.31 had a short microsomal $t_{1/2}$. Therefore, the in vivo study was planned for a short time window, when there would be a maximum of compound

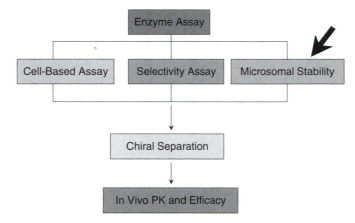

Figure 11.30 ▶ An efficient discovery project strategy for using microsomal stability for selection of compounds prior to expensive chiral separation and in vivo testing. (Reprinted with permission from [32].)

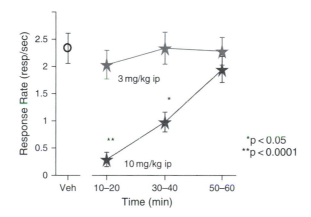

Figure 11.31 ▶ Microsomal stability indicated low stability, so the in vivo efficacy study time window was set to be short in this behavioral model. Efficacy was observed at 10–20 minutes and 30–40 minutes, whereas insignificant efficacy would have been observed if the window was 50–60 minutes.

concentration. The compound showed significant in vivo efficacy at early time points, which would have been missed if the time window had been 50 to 60 minutes.

An "intrinsic clearance" can be calculated using the in vitro microsomal stability. This is the predicted Cl due to hepatic metabolic reactions. The method for calculating intrinsic clearance was developed by Obach.[18]

$$Cl'_{int} = (0.693/t_{1/2 \text{ microsomal}}) * (\text{mL incubation/mg microsomal protein})$$

$$* (\text{mg microsomal protein/g liver}) * (\text{g liver/kg body weight}).$$

For example, if:

▶ $t_{1/2 \text{ microsomal}} = 15$ minutes

▶ microsomal protein concentration $= 0.5$ mg/mL

▶ mg microsomal protein/g liver $= 45$ mg/g

▶ g liver/kg body weight $= 20$ g/kg (human).

Then:

$$Cl'_{int} = (0.693/15\,min) * (1\,mL/0.5\,mg\ protein) * (45\,mg\ protein/g)$$
$$* (20\,g/kg\ weight) = 83\,mL/min/kg.$$

Cl'_{int} can be useful in predicting the PK of a compound. This must be used carefully because CYP metabolism may not be the only mechanism of clearance for a compound. Clearance also may be affected by extramicrosomal metabolism, renal clearance, biliary extraction, and hydrolysis in plasma or intestine. Examples of the classification of clearance values are given in Table 11.5.

TABLE 11.5 ▶ **Example of PK Clearance Classifications**

Species	HBF	Low Cl (20% HBF)	High Cl (80% HBF)
Mouse	90	18	72
Rat	55	11	44
Monkey	44	9	35
Human	21	4	17

All values are given as mL/min/kg. Cl, Clearance; HBF, hepatic blood flow.

11.6 Consequences of Chirality on Metabolic Stability

Chirality can greatly affect the metabolic stability of compounds. This is because enantiomers bind with different affinity and orientation to metabolizing enzymes. This affects both phase I and II metabolic enzymes. Up to 75% of all chiral drugs show stereoselective metabolism. Differential hepatic clearances are shown for several drugs in Table 11.6. Verapamil clearance shows difference between enantiomers. This is further exemplified

TABLE 11.6 ▶ **Stereospecific Hepatic Metabolism**

Drugs	Clearance (L/min)		
	(R)	(S)	Ratio
Propanolol (IV)	1.21	1.03	1.2
Propanolol (PO)	2.78	1.96	1.4
Verapamil (IV)	0.80	1.40	1.8
Verapamil (PO)	1.72	7.46	4.3
Warfarin (PO)	0.23	0.33	1.4
Propafenone (PO)	13.5	32.6	2.4

by the large stereospecific effect between routes of administration, owing to first-pass metabolism following oral administration. Stereoselective metabolism of various compounds is shown in Figures 11.32, 11.33, and 11.34.[19−22] Another example is terbutaline, which has a 1.9 enantiomeric ratio (+/−) for metabolism in the small intestine.

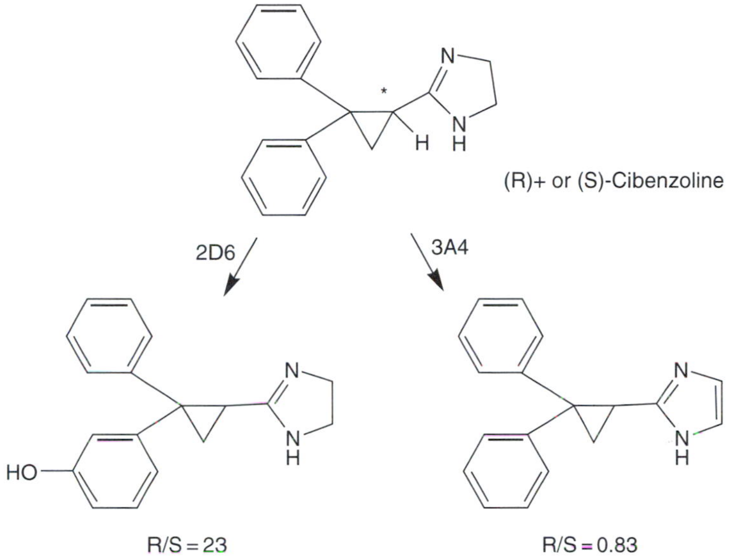

Figure 11.32 ► Stereoselective metabolism of cibenzoline.[19]

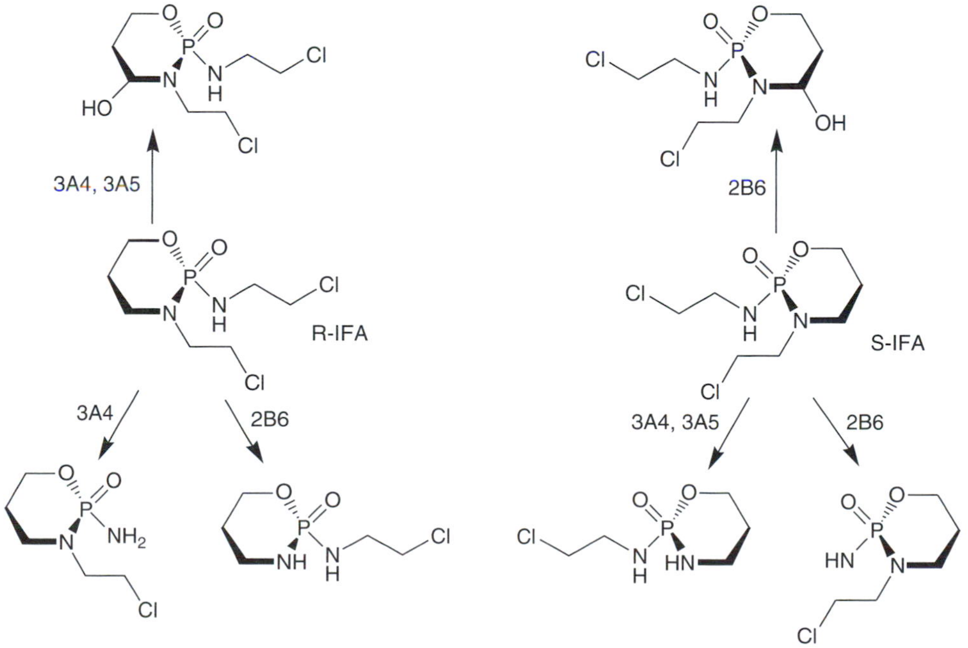

Figure 11.33 ► Stereoselective pathway of ifosfamide metabolism.[20,21]

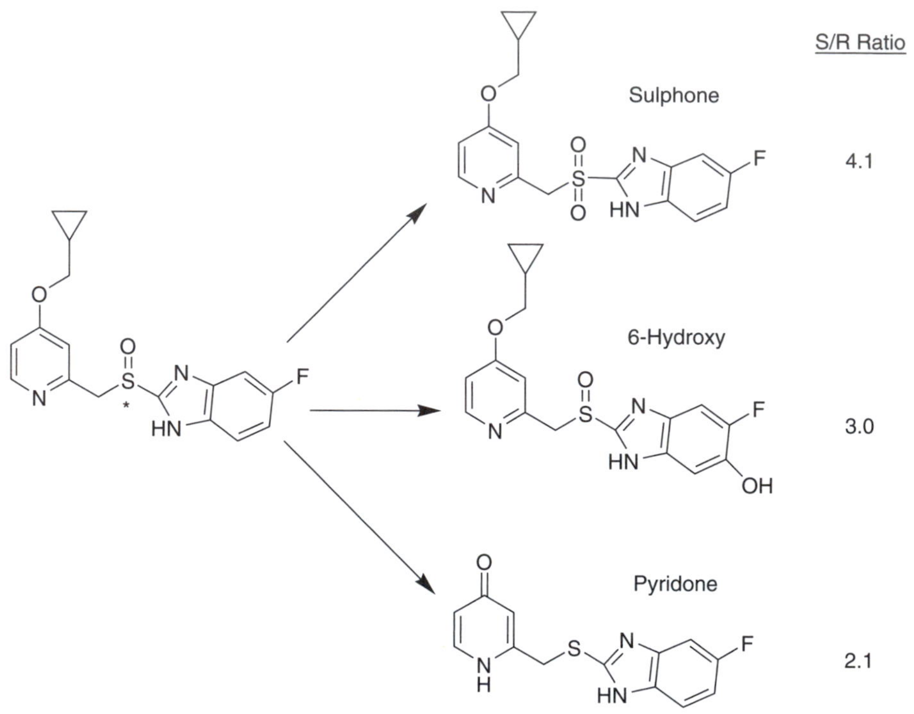

S/R Ratio

Sulphone   4.1

6-Hydroxy   3.0

Pyridone   2.1

Figure 11.34 ► Stereoselective metabolism by human liver microsomes of a substituted benzimidazole.[22]

11.7 Substrate Specificity of CYP Isozymes

Often a drug is metabolized by more than one CYP isozyme. However, different CYP isozymes have differences in the characteristics of the drugs they tend to metabolize. These differences are listed in Table 11.7.[23] Insights on the structural features that enable binding and the likely sites of oxidation are useful in designing molecules with enhanced metabolic stability.

TABLE 11.7 ► **Characteristics of CYP Isozyme Substrates**

CYP	Range of Log P	Other characteristics	Typical substrate
3A4	0.97 to 7.54	Large molecules	Nifedipine
2D6	0.75 to 5.04	Basic (Ionized)	Propranolol
2C9	0.89 to 5.18	Acidic (Nonionized)	Naproxen
1A2	0.08 to 3.61	Planar amines and amides	Caffeine

11.7.1 CYP1A2 Substrates

Example CYP1A2 substrates are shown in Figure 11.35. CYP1A2 tends to catalyze the metabolism of planar amines and amides.

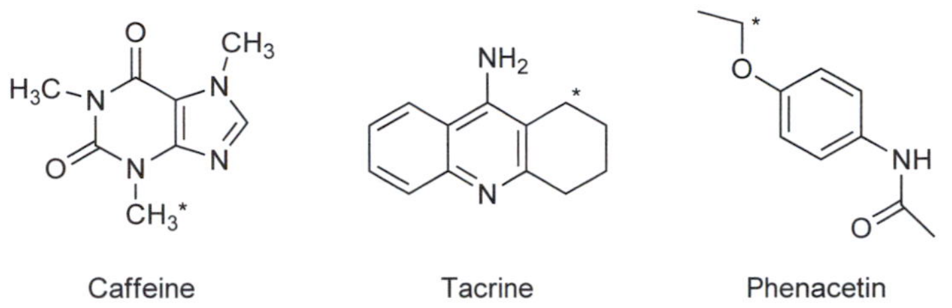

Caffeine   Tacrine   Phenacetin

Figure 11.35 ▶ Example CYP1A2 substrates. The major site of metabolism is labeled.

11.7.2 CYP2D6 Substrates

CYP2D6 tends to metabolize medium size basic amines.[24] The SAR for the active site is shown in Figure 11.36. CYP2D6 substrates have:

▶ ≥1 basic nitrogen atom

▶ Flat hydrophobic area (e.g., planar aromatic) at or near the site of oxidation

▶ 5–7 Å from the basic nitrogen (charged at pH 7.4) to the site of oxidation

▶ Negative molecular electronic potential above the planar part

The cation binds strongly to the anionic aspartic acid in the active site.

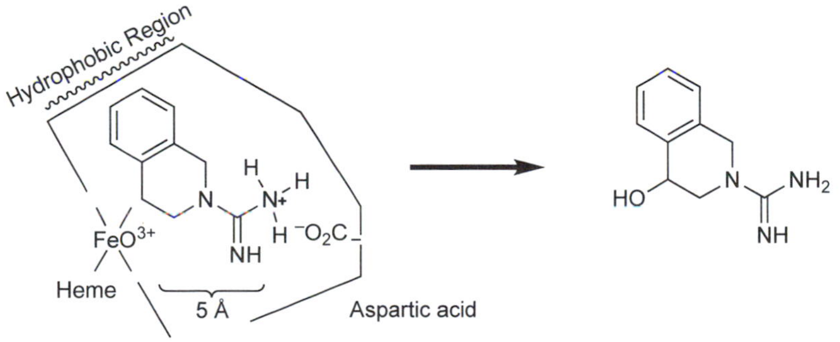

Figure 11.36 ▶ SAR of the active site for CYP2D6. The site of oxidation is 5–7 Å from the basic nitrogen.[24]

Examples of CYP2D6 substrates are shown in Figure 11.37. A large number of drugs have basic nitrogen atoms. For this reason, CYP2D6 metabolizes about 30% of commercial drugs despite its relatively low abundance in the liver (~2%). This low abundance may be saturable and result in a nonlinear increase in drug concentration with dose. Another problem is that about 7% to 10% of the white population lacks CYP2D6. For these individuals, drugs that are metabolized primarily by CYP2D6 are not cleared as fast and can build up to toxic concentrations.

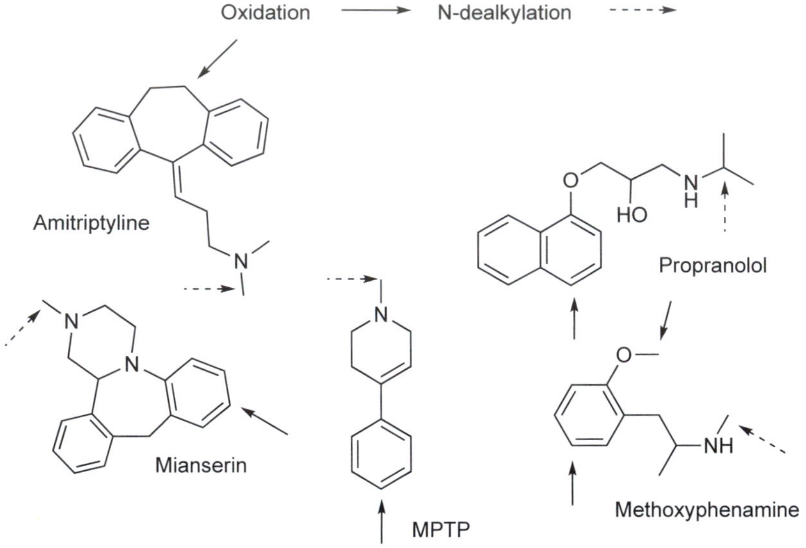

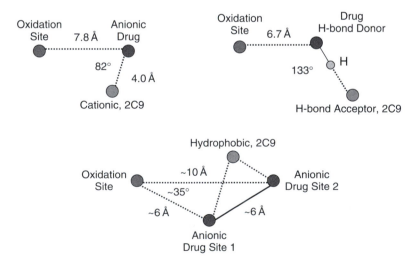

Figure 11.37 ▶   Examples of CYP2D6 substrates. The typical distance between the basic nitrogen and the site of oxidation is 5–7 Å. The major sites and types of metabolism are labeled.[33]

11.7.3 CYP2C9 Substrates

Substrates for CYP2C9 are less strictly defined. They are characterized as having the following[25–27]:

▶ Large dipole or negative charges

▶ Oxygen-rich (with a carboxylic acid (e.g., nonsteroidal antiinflammatory drugs, sulfonamide, alcohol)), hydrogen-bond acceptors

▶ Aromatic ring or lipophilic interaction

The SAR of the active site is shown in Figure 11.38 and indicates the relationships of atoms in the active site. Examples of CYP2C9 substrates are shown in Figure 11.39.

Figure 11.38 ▶   SAR/active site for CYP2C9, indicating substrate specificity.[27]

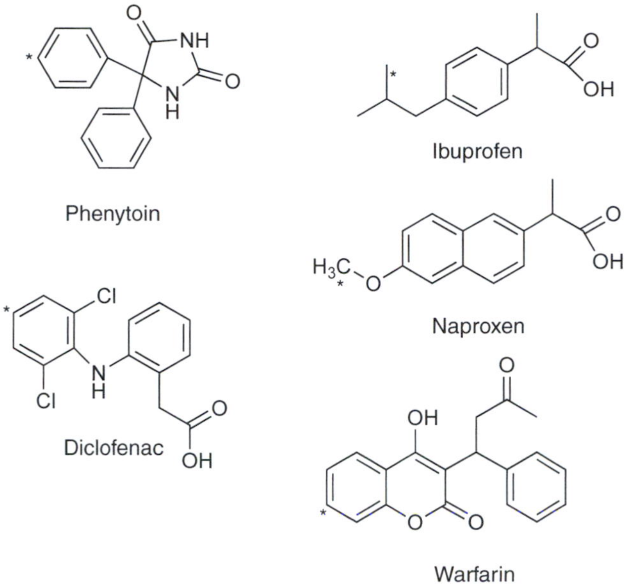
Phenytoin
Ibuprofen
Naproxen
Diclofenac
Warfarin

Figure 11.39 ▸ Examples of CYP2C9 substrates.[25–27] Compounds typically are carboxylic acids or are oxygen-rich (hydrogen-bond acceptors). The major site of metabolism is labeled.

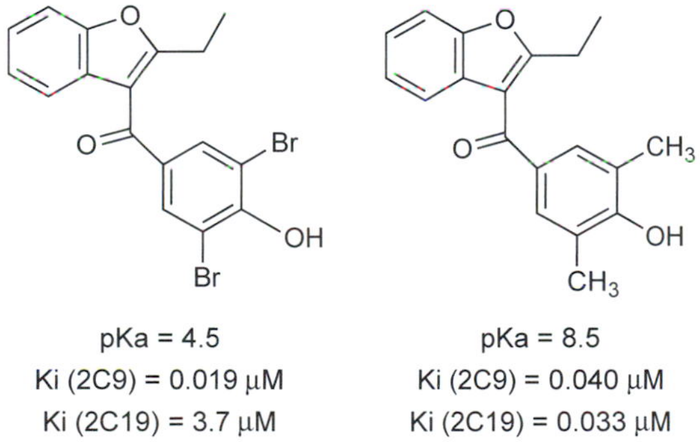
pKa = 4.5
Ki (2C9) = 0.019 μM
Ki (2C19) = 3.7 μM
pKa = 8.5
Ki (2C9) = 0.040 μM
Ki (2C19) = 0.033 μM

Figure 11.40 ▸ CYP2C9 and CYP2C19 have similar binding specificity. CYP2C19 prefers neutral substrates.[28]

CYP2C9 and CYP2C19 have 91% sequence homology. CYP2C9 has both neutral and anion binding sites. CYP2C19 prefers neutral substrates. The examples in Figure 11.40 [28] show how acidic compounds bind more weakly to CYP2C19.

Problems

(Answers can be found in Appendix I at the end of the book.)

1. List four types or locations of decomposition/metabolism reactions in living systems.
2. Metabolic reactions make the molecule: (a) more chemically stable, (b) more permeable, (c) more polar, (d) less toxic?

3. Two important aspects of metabolic reactions are: (a) compound solubility, (b) binding to the metabolic enzyme, (c) rotatable bonds, (d) reactivity at molecule position adjacent to the active site?

4. Define metabolic switching.

5. List some major phase I metabolic reactions.

6. List some major phase II metabolic reactions.

7. In vitro metabolic stability is an important property in determining which of the following PK parameters?: (a) C_O, (b) volume of distribution, (c) clearance, (d) t_{max}, (e) bioavailability.

8. The following compound undergoes the major metabolic reactions shown. How could the compound be structurally modified to improve metabolic stability?

9. Which sites on the following molecule might undergo phase II metabolism? What reactions might occur?

10. What can metabolic stability data be used for?: (a) select compounds for PK studies with highest chance of good bioavailability, (b) guide synthetic modification for stability

improvement, (c) diagnose cause of low bioavailability, (d) guide synthetic modification to increase permeability, (e) select compounds for more expensive in vivo efficacy studies.

11. Which of the following structural modifications may increase metabolic stability?: (a) add F or Cl at a site of metabolism, (b) remove labile groups, (c) add hydroxyl groups, (d) cyclization at labile site, (e) reduce ring size, (f) add steric hindrance at labile site, (g) add lipophilic groups.

References

1. Rowland, M., & Tozer, T. N. (1995). *Clinical pharmacokinetics: concepts and application*s. Philadelphia: Lippincott Williams & Wilkins.

2. Magdalou, J., Fournel-Gigleux, S., Testa, B., Ouzzine, M., & Nencki, M. (2003). Biotransformation reactions. In Camille Georges Wermuth (ed.), *Practice of Medicinal Chemistry* (2nd ed., pp. 517–543). Amsterdam: Elsevier Academic Press.

3. van de Waterbeemd, H., & Gifford, E. (2003). ADMET in silico modelling: towards prediction paradise? *Nature Reviews Drug Discovery 2*, 192–204.

4. Wring, S. A., Silver, I. S., & Serabjit-Singh, C. J. (2002). Automated quantitative and qualitative analysis of metabolic stability: a process for compound selection during drug discovery. *Methods in Enzymology 357*, 285–295.

5. Tandon, M., O'Donnell, M.-M., Porte, A., Vensel, D., Yang, D., Palma, R., et al. (2004). The design and preparation of metabolically protected new arylpiperazine 5-HT1A ligands. *Bioorganic & Medicinal Chemistry Letters 14*, 1709–1712.

6. Hester, J. B., Gibson, J. K., Buchanan, L. V., Cimini, M. G., Clark, M. A., Emmert, D. E., et al. (2001). Progress toward the development of a safe and effective agent for treating reentrant cardiac arrhythmias: synthesis and evaluation of ibutilide analogues with enhanced metabolic stability and diminished proarrhythmic potential. *Journal of Medicinal Chemistry 44*, 1099–1115.

7. Hutchinson, I., Jennings, S. A., Vishnuvajjala, B. R., Westwell, A. D., & Stevens, M. F. G. (2002). Antitumor benzothiazoles. 16. Synthesis and pharmaceutical properties of antitumor 2-(4-aminophenyl)benzothiazole amino acid prodrugs. *Journal of Medicinal Chemistry 45*, 744–747.

8. Palani, A., Shapiro, S., Josien, H., Bara, T., Clader, J. W., Greenlee, W. J., et al. (2002). Synthesis, SAR, and biological evaluation of oximino-piperidino-piperidine Amides. 1. Orally bioavailable CCR5 receptor antagonists with potent anti-HIV activity. *Journal of Medicinal Chemistry 45*, 3143–3160.

9. Manoury, P. M., Binet, J. L., Rousseau, J., Lefevre-Borg, F. M., & Cavero, I. G. (1987). Synthesis of a series of compounds related to betaxolol, a new b1-adrenoceptor antagonist with a pharmacological and pharmacokinetic profile optimized for the treatment of chronic cardiovascular diseases. *Journal of Medicinal Chemistry 30*, 1003–1011.

10. Plobeck, N., Delorme, D., Wei, Z.-Y., Yang, H., Zhou, F., Schwarz, P., et al. (2000). New diarylmethylpiper-azines as potent and selective nonpeptidic d-opioid receptor agonists with increased in vitro metabolic stability. *Journal of Medicinal Chemistry 43*, 3878–3894.

11. Mano, T., Okumura, Y., Sakakibara, M., Okumura, T., Tamura, T., Miyamoto, K., et al. (2004). 4-[5-Fluoro-3-[4-(2-methyl-1H-imidazol-1-yl)benzyloxy]phenyl]-3,4,5,6-tetrahydro-2H-pyran-4-carboxamide, an orally active inhibitor of 5-lipoxygenase with improved pharmacokinetic and toxicology characteristics. *Journal of Medicinal Chemistry 47*, 720–725.

12. MacKenzie, A. R., Marchington, A. P., Middleton, D. S., Newman, S. D., & Jones, B. C. (2002). Structure-activity relationships of 1-alkyl-5-(3,4-dichlorophenyl)-5-{2-[(3-substituted)-1-azetidinyl]ethyl}-2-piperidones. 1. Selective antagonists of the neurokinin-2 receptor. *Journal of Medicinal Chemistry 45*, 5365–5377.

13. Peglion, J.-L., Goument, B., Despaux, N., Charlot, V., Giraud, H., Nisole, C., et al. (2002). Improvement in the selectivity and metabolic stability of the serotonin 5-HT1A ligand, S 15535: A series of cis- and trans-2-(arylcycloalkylamine) 1-indanols. *Journal of Medicinal Chemistry 45*, 165–176.

14. Brooks, G., Burgess, W., Colthurst, D., Hinks, J. D., Hunt, E., Pearson, M. J., et al. (2001). Pleuromutilins. Part 1. The identification of novel mutilin 14-carbamates. *Bioorganic & Medicinal Chemistry 9*, 1221–1231.

15. Madsen, P., Ling, A., Plewe, M., Sams, C. K., Knudsen, L. B., Sidelmann, U. G., et al.). Optimization of alkylidene hydrazide based human glucagon receptor antagonists. Discovery of the highly potent and orally available 3-cyano-4-hydroxybenzoic acid [1-(2,3,5,6-tetramethylbenzyl)-1H-indol-4-ylmethylene]hydrazide. *Journal of Medicinal Chemistry 45*, 5755–5775.

16. Wu, W.-L., Burnett, D. A., Spring, R., Greenlee, W. J., Smith, M., Favreau, L., et al. (2005). Dopamine D1/D5 receptor antagonists with improved pharmacokinetics: Design, synthesis, and biological Evaluation of phenol bioisosteric analogues of benzazepine D1/D5 antagonists. *Journal of Medicinal Chemistry 48*, 680–693.

17. Ettmayer, P., Amidon, G. L., Clement, B., & Testa, B. (2004). Lessons learned from marketed and investigational prodrugs. *Journal of Medicinal Chemistry 47*, 2393–2404.

18. Obach, R. S. (1999). Prediction of human clearance of twenty-nine drugs from hepatic microsomal intrinsic clearance data: An examination of in vitro half-life approach and nonspecific binding to microsomes. *Drug Metabolism and Disposition 27*, 1350–1359.

19. Niwa, T., Shiraga, T., Mitani, Y., Terakawa, M., Tokuma, Y., & Kagayama, A. (2000). Stereoselective metabolism of cibenzoline, an antiarrhythmic drug, by human and rat liver microsomes: possible involvement of CYP2D and CYP3A. *Drug Metabolism and Disposition 28*, 1128–1134.

20. Lu, H., Wang, J. J., Chan, K. K., & Philip, P. A. (2006). Stereoselectivity in metabolism of ifosfamide by CYP3A4 and CYP2B6. *Xenobiotica 36*, 367–385.

21. Roy, P., Tretyakov, O., Wright, J., & Waxman, D. J. (1999). Stereoselective metabolism of ifosfamide by human P-450S 3A4 and 2B6. Favorable metabolic properties of R-enantiomer. *Drug Metabolism and Disposition 27*, 1309–1318.

22. Abelo, A., Andersson, T. B., Bredberg, U.S., Skanberg, I., & Weidolf, L. (2000). Stereoselective metabolism by human liver CYP enzymes of a substituted benzimidazole. *Drug Metabolism and Disposition 28*, 58–64.

23. Lewis, D. F.V, & Dickins, M. (2002). Substrate SARs in human P450s. *Drug Discovery Today 7*, 918–925.

24. ter Laak, A. M., Vermeulen, N. P.E, & de Groot, M. J. (2002). Molecular modeling approaches to predicting drug metabolism and toxicity. In A. D. Rodrigues (ed.). *Drug-drug interactions* (pp. 505–548). New York: Marcel Dekker.

25. Rao, S., Aoyama, R., Schrag, M., Trager, W. F., Rettie, A., & Jones, J. P. (2000). A refined 3-dimensional QSAR of cytochrome P450 2C9: computational predictions of drug interactions. *Journal of Medicinal Chemistry 43*, 2789–2796.

26. de Groot, M. J., Alex, A. A., & Jones, B. C. (2002). Development of a combined protein and pharmacophore model for cytochrome P450 2C9. *Journal of Medicinal Chemistry 45*, 1983–1993.

27. de Groot, M. J., & Ekins, S. (2002). Pharmacophore modeling of cytochromes P450. *Advanced Drug Delivery Reviews 54*, 367–383.

28. Locuson, C. W., Suzuki, H., Rettie, A. E., & Jones, J. P. (2004). Charge and substituent effects on affinity and metabolism of benzbromarone-based CYP2C19 inhibitors. *Journal of Medicinal Chemistr, 47*, 6768–6776.

29. Di, L., Kerns, E. H., Hong, Y., & Chen, H. (2005). Development and application of high throughput plasma stability assay for drug discovery. *International Journal of Pharmaceutics 297*, 110–119.

30. Williams, P. A., Cosme, J., Vinkovic, D. M., Ward, A., Angove, H. C., Day, P. J., et al. (2004). Crystal structures of human cytochrome P450 3A4 bound to metyrapone and progesterone. *Science 305*, 683–686.

31. Guengerich, F. P., & Johnson, W. W. (1997). Kinetics of ferric cytochrome P450 reduction by NADPH-cytochrome P450 reductase: rapid reduction in the absence of substrate and variations among cytochrome P450 systems. *Biochemistry 36*, 14741–14750.

32. Di, L., & Kerns, E. H. (2005). Application of pharmaceutical profiling assays for optimization of drug-like properties. *Current Opinion in Drug Discovery & Development 8*, 495–504.

33. De Groot, M. J., Ackland, M. J., Horne, V. A., Alex, A. A., & Jones, B. C. (1999). A novel approach to predicting P450 mediated drug metabolism. CYP2D6 catalyzed N-dealkylation reactions and qualitative metabolite predictions using a combined protein and pharmacophore model for CYP2D6. *Journal of Medicinal Chemistry 42*, 4062–4070.

Chapter 12

Plasma Stability

Overview

► *Compound decomposition can be catalyzed in plasma by hydrolytic enzymes.*

► *Increased clearance can occur for hydrolyzable substrate compounds.*

► *Plasma stability increases with steric hindrance, electron-withdrawing groups, or replacement with a less reactive group.*

Compounds with certain functional groups can decompose in the bloodstream. Unstable compounds often have high clearance and short $t_{1/2}$, resulting in poor in vivo pharmacokinetics (PK) and disappointing pharmacological performance. Plasma degradation clearance can be overlooked if discovery project teams focus on microsomal stability. Microsomal enzymes are different than plasma enzymes. Stability in liver microsomes does not imply stability in plasma. Instability in plasma also can cause erroneous PK assay results if the compound degrades in the plasma sample after it is taken from the animal or if quantitative analysis standards are prepared in plasma and degrade. Pharmaceutical companies typically do not develop clinical candidates that are unstable in plasma, unless they are prodrugs or antedrugs. Therefore, it is important for discovery scientists to anticipate and assess this issue early. Plasma degradation can be used to advantage in the development of prodrugs and antedrugs.

12.1 Plasma Stability Fundamentals

Blood contains a large number of hydrolytic enzymes, such as cholinesterase, aldolase, lipase, dehydropeptidase, alkaline and acid phosphatase.[1] The amount of each enzyme is dependent on species, disease state, gender, age, and race.[2] If the compound has affinity for one of these enzymes and it has a hydrolyzable group in the right position, it can be decomposed in the plasma. Many such groups are used to enhance the compound's pharmacological activity at the target protein; thus, medicinal chemists may be reluctant to remove or replace them. However, hydrolysis in plasma can be a major cause of compound clearance, and pharmacologically efficacious concentrations may not be achievable in vivo. For this reason, it is important to assess the liability of potentially unstable moieties in the project's lead series during an early discovery stage and either modify or deprioritize the series before a large amount of effort is expended on activity optimization.

Several functional groups are susceptible to plasma degradation and include the following:

► Ester

► Amide

▶ Carbamate

▶ Lactam

▶ Lactone

▶ Sulfonamide

Leads containing these groups, especially peptides and peptide mimetics, should be tested for plasma stability.

12.1.1 Consequences of Chirality on Plasma Stability

Plasma stability is affected by chirality, owing to the differential binding of enantiomers to plasma enzymes. For example, in Figure 12.1 the hydrolysis rate constant for O-acetyl propranolol is affected by the stereochemistry.[3]

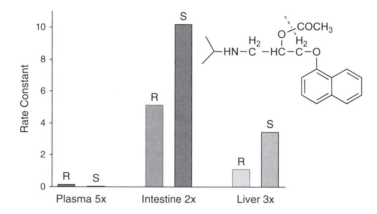

Figure 12.1 ▶ Hydrolysis rate constant of O-acetyl propranolol is affected by chirality.[3]

12.2 Effects of Plasma Stability

Medicinal chemists can take advantage of plasma reactions as part of a prodrug approach. In order to enhance permeation or metabolic stability, a prodrug strategy can be applied. Typically, esters are prepared. The prodrug enhances permeability or metabolic stability so that high concentrations of the prodrug reach the bloodstream. In the blood, a hydrolytic enzyme cleaves the prodrug to release the active drug. Prodrugs are discussed further in Chapter 39.

Antedrugs ("soft drugs") are the opposite of prodrugs.[4,5] These drugs are active locally but rapidly degrade to an inactive compound once they reach the bloodstream. The purpose of this action is to reduce side effects by minimizing the systemic toxicity of the drug. Examples of antedrugs are shown in Figure 12.2.[5] Ciclesonide works locally at the lung for treatment of asthma and chronic obstructive pulmonary disease. Fluocortin-butyl is applied topically as an antiinflammatory agent. Loteprednol etabonate is used as local treatment of eye inflammation. These are esters for local delivery and are readily inactivated systemically through hydrolysis.

In selecting project compounds for in vivo study, plasma stability data can be used to indicate which compounds are most likely to be stable and succeed. For lead series that have a liability for stability, these data can assist selection of the most stable compounds.

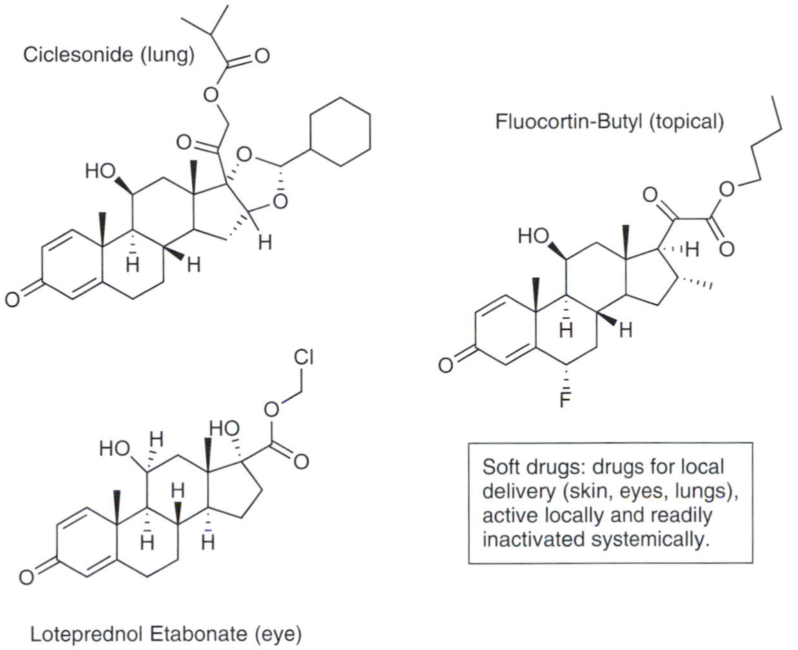

Ciclesonide (lung)

Fluocortin-Butyl (topical)

Loteprednol Etabonate (eye)

Soft drugs: drugs for local delivery (skin, eyes, lungs), active locally and readily inactivated systemically.

Figure 12.2 ► Antiinflammatory antedrugs (soft drugs).

Plasma stability can vary greatly among species, which makes difficult the prediction of human clinical outcomes of prodrugs from animal data. The plasma stability of the compound in Figure 12.3 increased in stability according to the following species order: rat < dog < human.[6] Typically, compounds are less stable in rodents than in humans.

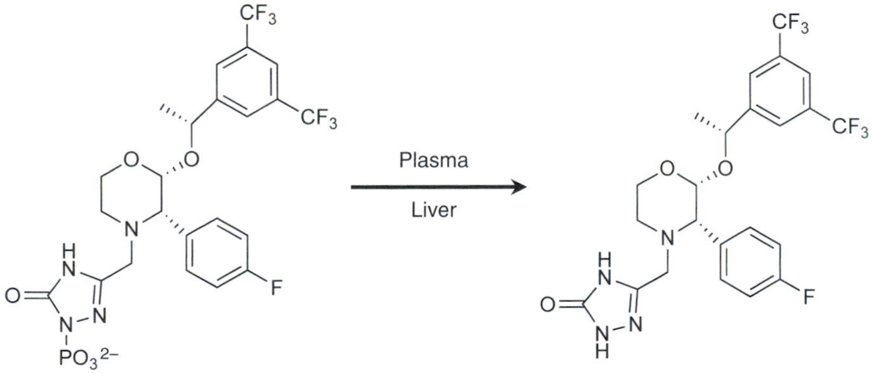

Time	Rat (µM)		Dog (µM)		Human (µM)	
(min)	Prodrug	Parent	Prodrug	Parent	Prodrug	Parent
0	23	0.6	26	1.2	26	0.8
15	18	5.8	26	1.8	27	1.0
30	13	8.0	25	2.4	26	1.1
60	6.4	11	24	3.9	26	1.4
120	1.9	15	20	6.0	26	2.3

Figure 12.3 ► Species differences in plasma stability.

12.3 Structure Modification Strategies to Improve Plasma Stability

Strategies for improving plasma stability are listed in Table 12.1.

TABLE 12.1 ▶ Structure Modifications Strategies for Plasma Stability Improvement

Structure modification	Section
Substitute an amide for an ester	12.3.1
Increase steric hindrance	12.3.2
Add electron-withdrawing groups to decrease for antedrug	12.3.3
Eliminate the hydrolyzable group	

12.3.1 Substitute an Amide for an Ester

Amides are more stable against plasma hydrolysis than are esters in the same position. For example, in Figure 12.4, the ester had a half-life <1 minute.[7] An amide at that position increased the half-life to 69 hours. The amide retained the activity of the ester. An ether or amine at the same position, although very stable, greatly reduced activity.

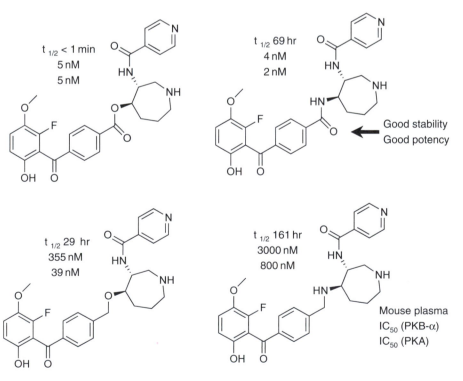

Figure 12.4 ▶ Substitution of an amide for an ester greatly increased plasma stability and maintained activity. (Reprinted with permission from [8].)

12.3.2 Increase Steric Hindrance

Addition of steric hindrance near a hydrolyzable group can increase plasma stability.[8] The examples in Figures 12.5 and 12.6 demonstrate the stability enhancements achieved with addition of consecutively more bulky *R*-groups attached to the lactam carbonyl. Increasing steric hindrance resulted in longer half-lives in plasma.

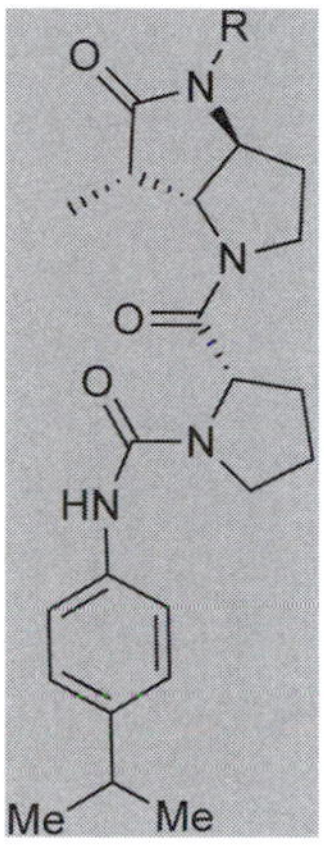

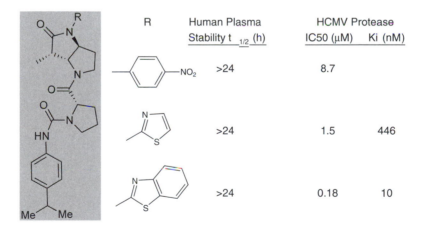

R	Human Plasma Stability t $_{1/2}$ (h)	HCMV Protease IC50 (µM)	Ki (nM)
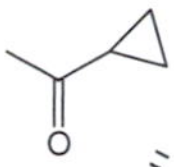	0.5	0.2	2.4
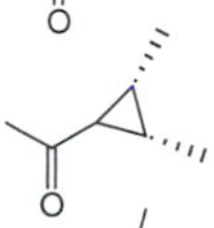	1.5	1.8	
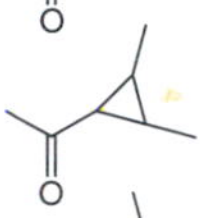	6	0.3	16
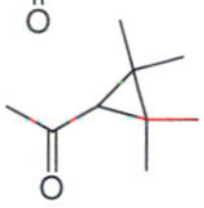	16	>20	

Figure 12.5 ► Increased steric hindrance of the lactam carbonyl increased plasma stability. (Reprinted with permission from [8].)

R	Human Plasma Stability t $_{1/2}$ (h)	HCMV Protease IC50 (µM)	Ki (nM)
—NO₂	>24	8.7	
	>24	1.5	446
	>24	0.18	10

Figure 12.6 ► Increased steric hindrance of the lactam carbonyl increased plasma stability. (Reprinted with permission from [8].)

12.3.3 Electron-Withdrawing Groups Decrease Plasma Stability for Antedrug

Antedrugs are purposely modified to decrease their plasma stability. This causes the drugs to be cleared rapidly systemically in order to reduce side effects. Figure 12.7 shows an example of modifying the phosphonamide ester of matrix metalloproteinase inhibitors to reduce plasma stability.[4] Including electron-withdrawing groups increased the positive charge on the phosphorous atom and increased the rate of hydrolysis.

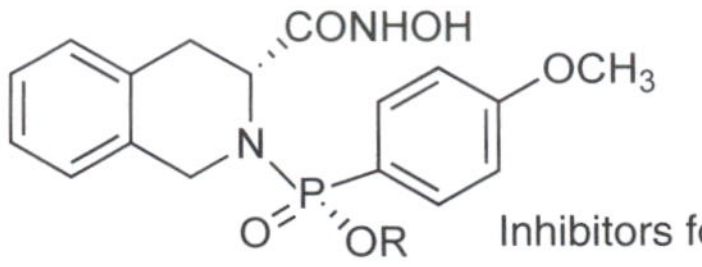

Inhibitors for MMP

R	Human Plasma Stab. %Remaining @ 60 min	IC50 (μM) HB-EGF	AR
Et	100	0.23	0.35
CH_2CH_2F	96	0.18	0.47
CH_2CHF_2	0 ($t_{1/2}$ ~ 10 min)	0.51	1.47
CH_2CF_3	0 ($t_{1/2}$ < 1 min)	0.73	0.95
$CH_2CH_2CF_3$	99	0.31	0.97

Figure 12.7 ► Introduce electron-withdrawing group to decrease plasma stability and increase clearance to avoid adverse effects using antedrug approach.

12.4 Applications of Plasma Stability Data

Plasma stability data are used for many purposes in drug discovery.[9] Some of those applications are discussed here.

12.4.1 Diagnose Poor In Vivo Performance

Sometimes the cause of poor in vivo performance of a compound (e.g., low area under the curve, short $t_{1/2}$, high clearance) cannot be attributed to low metabolic stability. The clearance may be due in part to low plasma stability if the compound has a group(s) that is susceptible to plasma enzyme hydrolysis. Plasma stability data can be used as a custom assay to diagnose this possible cause of poor in vivo performance. When a compound is not stable in plasma, clearance higher than hepatic blood flow can sometimes be observed.

12.4.2 Alert Teams to a Liability

Use of plasma stability as a general screen can alert teams to labile structural motifs. This allows them to recognize problems early and to fix the problem or deprioritize the compound series.

12.4.3 Prioritize Compounds for In Vivo Animal Studies

Information on the plasma stability of compounds adds to the ensemble of data for making informed decisions by teams. This information is used to prioritize compounds for in vivo studies of pharmacology and PK.

For some series of compounds, plasma stability differentiates between compounds and is an effective tool in selecting compounds for further study. For example, Figure 12.8 shows the plasma stability results for 24 of more than 200 compounds in a series.[9] In general, all of the series compounds had similar properties. However, the plasma stability of certain compounds was clearly greater than others. It would not be wise to continue work on the unstable compounds when the stable compounds offer clear advantages. These plasma stability data were combined with other potency and property data to select a small number of compounds for in vivo studies.

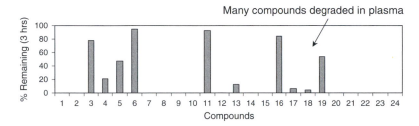

Figure 12.8 ▶ Plasma stability clearly differentiated series compounds for further study. (Reprinted with permission from [9].)

12.4.4 Prioritize Synthetic Efforts

The example in Figure 12.9 helped to prioritize synthetic efforts. The sulfide series was stable in plasma, but the sulfone series was unstable.[9] Based on this information, the sulfone series was terminated, and the team focused their effort on the sulfide series.

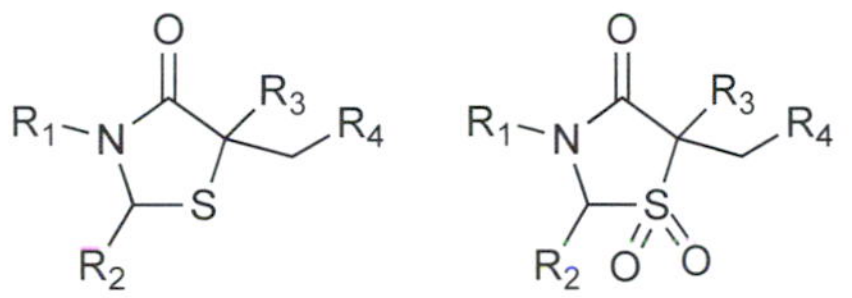

Stable in plasma   Rapid degradation
Termination of series

Figure 12.9 ▶ Plasma stability data provided key insight for continuing the series on the left and terminating the series on the right. (Reprinted with permission from [9].)

12.4.5 Screening of Prodrugs

If the discovery project team is taking a prodrug strategy, in vitro plasma stability can be measured to select a prodrug with the optimal properties. Plasma stability can be used to profile prodrugs. In the example in Table 12.2, a discovery team synthesized different diester prodrugs for the purpose of improving permeability to optimize oral bioavailability.[9] Compound 1 was too stable in plasma to be useful as a prodrug. Compound 6 was stable in gastric intestinal fluid and rapidly converted to diacid in plasma. This is a favorable profile

TABLE 12.2 ▶ Screening of Diester Prodrugs: Percent of Ester Observed After 3-Hour Incubation in Rat Plasma

Prodrugs	% Diester at 3 Hours	% Monoester at 3 Hours	% Diacid at 3 Hours
1	100	0	0
2	0.0	57.3	42.7
3	0.0	76.7	23.4
4	0.0	77.7	22.3
5	0.0	34.9	65.2
6	0.0	10.4	89.6
7	0.0	0.3	99.8

for a prodrug. In the same manner, antedrugs can be screened in vitro to select compounds for in vivo testing.

12.4.6   Guide Structural Modification

The structures of plasma degradation products often can be obtained using liquid chromatography/mass spectrometry. Figure 12.10 shows an example of a series containing a terminal carbamate and a cyclic carbamate.[9] Based on molecular weight, hydrolysis of the two carbamates could be easily distinguished. The terminal carbamate was found to be unstable, while the cyclic carbamate was stable. Synthetic modifications were undertaken to stabilize or replace the terminal carbamate.

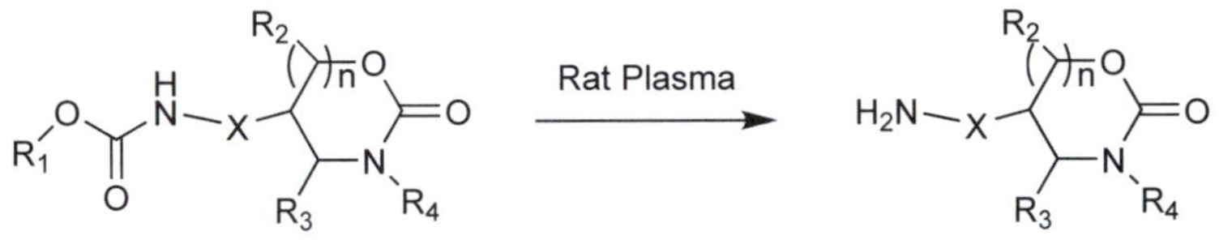

Figure 12.10 ▶ Identification of degradation products, in addition to quantitative determination of half-life, identified that the terminal carbamate was unstable but the cyclic carbamate was not. Information is used to guide structural modification. (Reprinted with permission from [9].)

Problems

(Answers can be found in Appendix I at the end of the book.)

1. Which of the following structures might be partially or completely unstable in plasma?

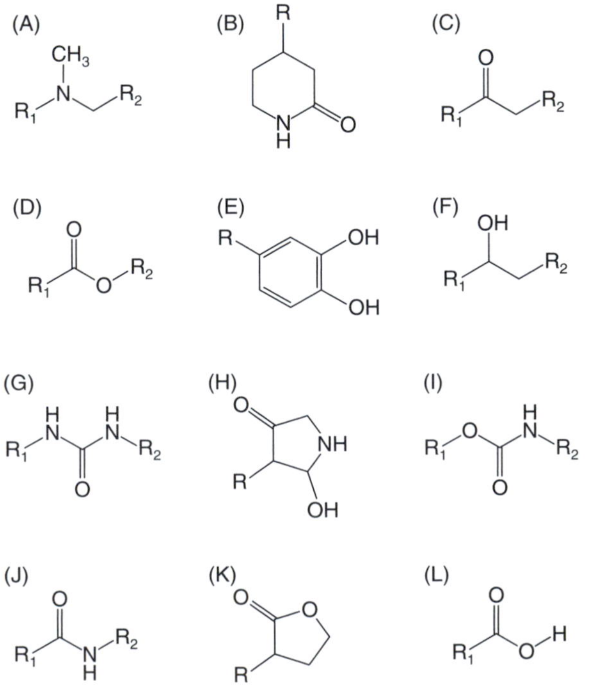

2. List two applications where plasma instability is advantageous.

3. What are some structural modifications that you could try in order to improve the plasma stability of the following compound?

4. Microsomes have hydrolysis activity. Are they useful for assessing potential plasma hydrolysis?

5. Which of the following groups should alert you to potential degradation due to plasma hydrolysis?: (a) phenyl, (b) carboxylic acid, (c) ester, (d) lactone, (e) trifluoromethyl, (f) carbamate, (g) amide.

References

1. Dittmer, D. S., & Altman, P. L. (1961). *Blood and other body fluids*. Washington, DC: Federation of American Societies for Experimental Biology.

2. Cook, C. S., Karabatsos, P. J., Schoenhard, G. L., & Karim, A. (1995). Species dependent esterase activities for hydrolysis of an anti-HIV prodrug glycovir and bioavailability of active SC-48334. *Pharmaceutical Research*, *12*, 1158–1164.

3. Yoshigae, Y., Imai, T., Horita, A., & Otagiri, M. (1997). Species differences for stereoselective hydrolysis of propranolol prodrugs in plasma and liver. *Chirality*, *9*, 661–666.

4. Sawa, M., Tsukamoto, T., Kiyoi, T., Kurokawa, K., Nakajima, F., Nakada, Y., et al. (2002). New strategy for antedrug application: Development of metalloproteinase inhibitors as antipsoriatic drugs. *Journal of Medicinal Chemistry*, *45*, 930–936.

5. Ettmayer, P., Amidon, G. L., Clement, B., & Testa, B. (2004). Lessons learned from marketed and investigational prodrugs. *Journal of Medicinal Chemistry*, *47*, 2393–2404.

6. Hale, J. J., Mills, S. G., MacCoss, M., Dorn, C. P., Finke, P. E., Budhu, R. J., et al. (2000). Phosphorylated morpholine acetal human neurokinin-1 receptor antagonists as water-soluble prodrugs. *Journal of Medicinal Chemistry*, *43*, 1234–1241.

7. Breitenlechner, C. B., Wegge, T., Berillon, L., Graul, K., Marzenell, K., Friebe, W.-G., et al. (2004). Structure-based optimization of novel azepane derivatives as PKB inhibitors. *Journal of Medicinal Chemistry*, *47*, 1375–1390.

8. Borthwick, A. D., Davies, D. E., Ertl, P. F., Exall, A. M., Haley, T. M., Hart, G. J., et al. (2003). Design and synthesis of pyrrolidine-5,5′-trans-lactams (5-oxo-hexahydropyrrolo[3,2-b]pyrroles) as novel mechanism-based inhibitors of human cytomegalovirus protease. 4. Antiviral activity and plasma stability. *Journal of Medicinal Chemistry*, *46*, 4428–4449.

9. Di, L., Kerns, E. H., Hong, Y., & Chen, H. (2005). Development and application of high throughput plasma stability assay for drug discovery. *International Journal of Pharmaceutics*, *297*, 110–119.

Chapter 13

Solution Stability

Overview

▶ *Compounds can decompose in solution owing to pH, dosing solution excipients, solution components, enzymes, light, oxygen, or temperature.*

▶ *Instability in solution can cause erroneous in vitro assay or in vivo pharmacokinetic results.*

▶ *Stability increases by replacing the unstable group, adding steric hindrance, or electron withdrawal.*

Stability of a compound in solution is necessary for success at each stage of discovery and for development of a successful drug product.[1–6] During discovery, compounds must be stable in biological assay buffers in order to produce accurate activity data in enzyme, receptor, and cell-based assays.[7,8] For in vivo oral dosing, compounds must be stable in the acidic, basic, and enzymatic conditions of the gastrointestinal (GI) tract.[9–13] Solution stability also is crucial for prodrugs, which must be stable under certain physiological solution conditions and then hydrolyze to release the active drug under other conditions.[14–17] Parenteral drugs must be stable in the presence of formulations containing various excipients.[18,19] Despite these potential solution stability obstacles, discovery project teams may overlook the assessment of solution stability. Solution instability frequently occurs and can confuse the structure–activity relationship (SAR) and reduce in vivo performance. It is prudent for discovery teams to fully assess the potential for chemical instability of their project compounds.

13.1 Solution Stability Fundamentals

Compounds are exposed to a wide range of solutions during drug discovery. These solutions include the following:

▶ Organic solvent stocks

▶ Aqueous buffers

▶ Bioassay buffers

▶ Dosing solutions

▶ GI tract

The specific conditions and components of each of these solutions can cause compound instability. Some of these challenges are as follows:

► pH

► Water

► Counter ions of salts

► Solution components (e.g., dithiotheital (DTT))

► Excipients

► Enzymes

► High-performance liquid chromatography modifiers (especially if concentrated after purification)

► Temperature

► Light

► Oxygen

It often is assumed that compounds are stable in stock solutions of organic solvents or in aqueous buffers. Reactions can occur unexpectedly. For example, paclitaxel epimerizes in solutions that are slightly basic. Organic or aqueous buffer solutions can expose compounds to decomposition reactions induced by ambient laboratory light, elevated temperature, and oxygen absorbed from the air.

Bioassay buffers can contain components that promote compound decomposition. pH can promote hydrolysis or hydrate formation. DTT can cause reduction or react as a nucleophile. Solution components can react with discovery compounds. Furthermore, throughout the lifetime of a discovery project, several different assays (binding, cell-based functional assay, selectivity assay, etc.) are used, and each assay buffer solution may introduce new conditions or components that cause compound decomposition. This results in different stability in different assay solutions.

Excipients of dosing solutions for oral, intraperitoneal, or IV administration may promote compound decomposition. For example, lactic acid-containing dosing solutions have a low pH and could potentially cause acid-catalyzed decomposition.

In the GI tract, compounds are exposed to a wide variety of pHs. The pH ranges from acidic in the stomach and upper intestine to basic in the colon. Also, a wide array of hydrolytic enzymes are present in the GI tract, such as pepsin and pancreatin. These enzymes have the natural function of digesting macromolecules to monomers for use as nutrients, but they can also bind and hydrolyze drug compounds.

Hydrolysis probably is the most common reaction causing instability during discovery. It can occur for compounds that contain an ester, amide, thiol ester, imide, imine, carbamic ester, acetal, alkyl chloride, lactam or lactone. For example, aspirin contains an ester that can hydrolyze in water.

Hydrolysis can be catalyzed in solution by acidic or basic conditions, depending on the functional group. The rate of the reaction depends on the functional group, attached chemical constituents, physicochemical nature of the solution, and chemical and enzymatic components in the solution. Other reactions in solution include oxidation (initiated by trace metal, light, or autoxidation), isomerization, dimerization, and racemization.

13.2   Effects of Solution Instability

Chemical instability in bioassay buffer solution reduces the compound concentration and produces decomposition products that themselves may be active. The lower concentration results in a lower apparent activity of the compound and produces erroneous SAR. Without stability data, the team may interpret stability differences between compounds as real activity differences. If team chemists know that a compound or series is not stable in solution, they can make chemical modifications to improve stability and produce reliable activity data. Alternately, if a buffer component or conditions of the assay are causing degradation and these conditions are not present in vivo, then it may be possible to modify the assay to be more accurate for the compound series.

Compounds can decompose as they are stored in organic solvent solutions. Oxygen, water, trace metals, and materials leached from the glass or plastic containers can react with the compound. Water is rapidly absorbed into dimethylsulfoxide (DMSO) when solutions are exposed to the air, especially with condensations when solutions are cool from refrigerator storage. Absorption of water and residual acidic additives (e.g., trifluoroacetic acid (TFA), formic acid) from chromatographic purification promotes degradation of compound libraries dissolved in DMSO during storage [20–23]. It was found that water is more important in causing compound degradation than oxygen in DMSO stock solution [23]. Light can induce reactions to certain sensitive compounds in a solution that is exposed to laboratory light.

Instability in vivo reduces the compound's PK performance and the in vivo pharmacological activity. This may be overlooked as a cause of low bioavailability and high clearance. The compound may not achieve a sufficiently high concentration to produce in vivo efficacy or achieve pharmacological proof of concept. In drug development, formulation strategies can be developed to enhance stability of dosage forms, which are rarely applied in drug discovery.

13.3   Structure Modification Strategies to Improve Solution Stability

The structure modifications that can improve solution stability depend on the conditions and functional group (Table 13.1). Strategies for overcoming enzymatic hydrolysis are similar to improvement of plasma stability (see Chapter 12).

TABLE 13.1 ▶ Structure Modifications Strategies for Solution Stability Improvement

Structure modification	Section
Eliminate or modify unstable group	13.3.1
Add an electron-withdrawing group	13.3.2
Isosteric replacement of labile functional group	13.3.3
Increase steric hindrance	13.3.4

13.3.1   Eliminate or Modify the Unstable Group

If the labile functional group does not contribute to binding at the therapeutic target's active site, it can be eliminated without significant loss of activity. Figure 13.1 shows how elimination of the acetal type group in artemisinin increased stability by 10-fold.[18] The lipoxin analogs in Figure 13.2 were modified to greatly improve solution stability, resulting in significant in vivo exposure improvement.[13]

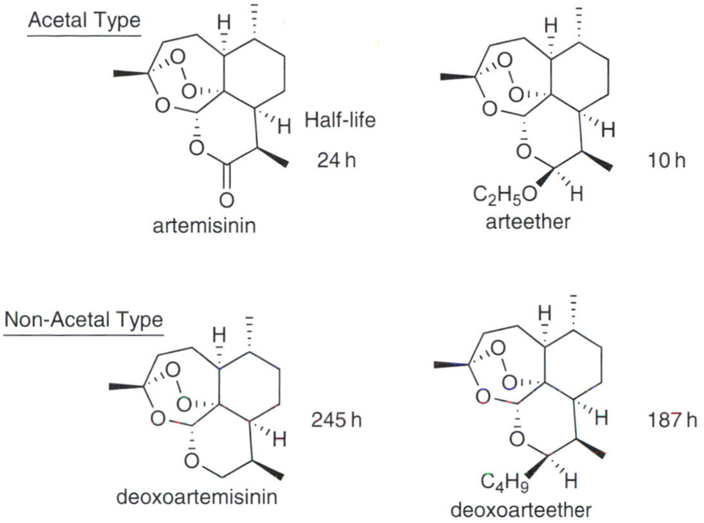

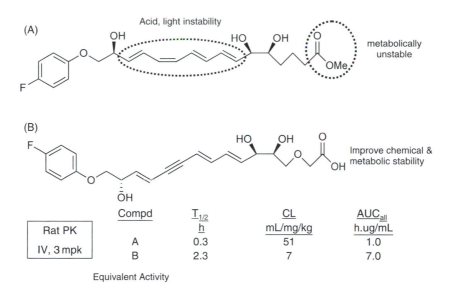

Figure 13.1 ► Improvement of acidic stability of artemisinin analogs. Conditions were pH 2 at 37°C.

Rat PK	Compd	$T_{1/2}$ h	CL mL/mg/kg	AUC_{all} h.ug/mL
IV, 3 mpk	A	0.3	51	1.0
	B	2.3	7	7.0

Equivalent Activity

Figure 13.2 ► Modification of the conjugated section and elimination of the ester greatly improved the stability and exposure of these lipoxin analogs.

13.3.2 Add an Electron-Withdrawing Group

Addition of an electron-withdrawing group adjacent to an epoxide can reduce the reaction rate and stabilize the compound. In Figure 13.3, a cyano group was added adjacent to the epoxide, resulting in a 50-fold improvement in stability.[10]

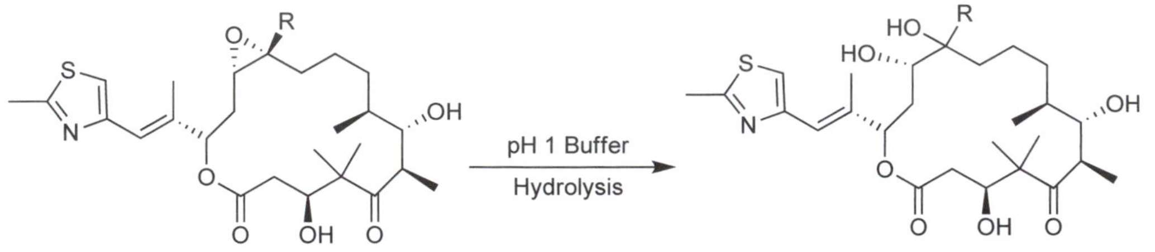

R	$EC_{0.01}$	IC_{50}	T_{95}
	(tubuline)	(HCT-116)	(5% Degrade)
Me (Epothilone B)	2.2 μM	4.4 nM	<0.2 hrs
CN	2.5 μM	4.1 nM	11 hrs

Electron-withdrawing R group slows down S_N1 hydrolysis

Figure 13.3 ▶ Addition of an electron-withdrawing cyano group reduced hydrolysis of the adjacent epoxide.

13.3.3 Isosteric Replacement of Labile Functional Group

If the functional group makes a significant contribution to the therapeutic target binding, it is important to know what aspects of the group contribute to active site interactions and attempt to find an isosteric replacement that aids binding without contributing to instability.

13.3.4 Increase Steric Hindrance

Introducing steric hindrance near the site of the reaction can reduce access of the labile functional group and increase stability. Figure 13.4 shows how stability of imidazolines improved at pH 7.4 by introducing bulky substituents adjacent to the labile group.[8] The stability of an amide was enhanced by adding steric hindrance near the amide nitrogen (Figure 13.5).[7]

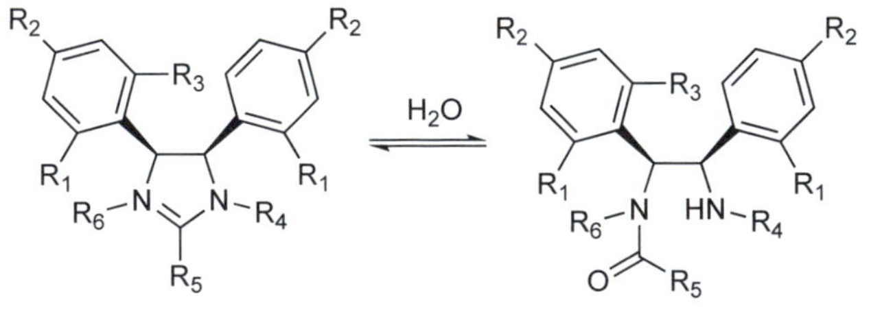

Compds	R_1	R_2	R_3	R_4	R_5	R_6	half-life ($t_{1/2}$, h)	k_{obs} (h^{-1})	Rel. Act. (% 1uM)
1	Cl	OCH_3					4.58	0.198	
2		OH					6.36	0.122	0
3	Cl	OH					5.94	0.15	67
4	Cl	OH	Cl				13.41	0.052	112
5	Cl	OH			C_2H_4OH		stable	stable	20
6	Cl	OH		C_2H_5			154.03	0.004	103
7	Cl	OH		C_2H_5		C_2H_5	stable	stable	60

Figure 13.4 ▶ Stability of imidazolines at pH 7.4. Stability improved with increase in steric hindrance.

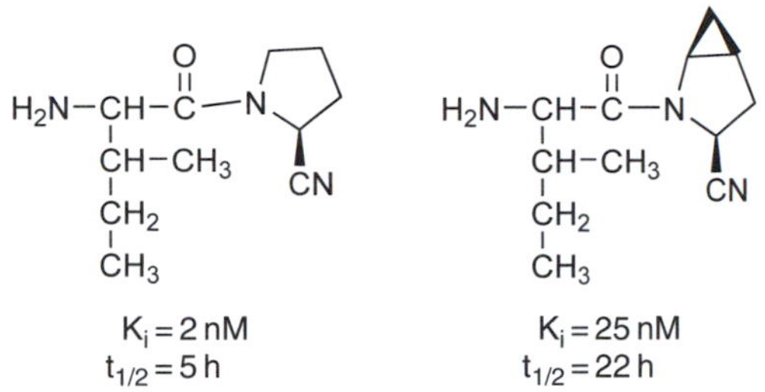

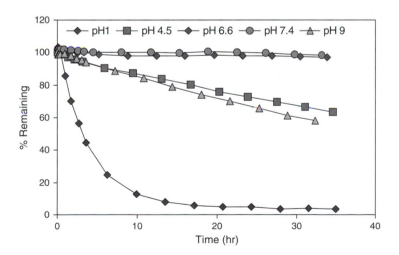

Figure 13.5 ► Dipeptidyl peptidase IV (DPP-IV) inhibitors. Chemical stability in pH 7.2 at 39.5°C. Introducing steric hindrance enhanced stability.

13.4 Applications of Solution Stability Data

Solution stability data can be used in drug discovery for many purposes.

Provide an Early Alert to Liabilities

Early knowledge that a lead series has stability limitations is valuable to a discovery team. Otherwise, the team may synthesize many series analogs that turn out to be wasted effort. Also, biological and property testing may be carried out and result in confusing conclusions and SAR. In Figure 13.6, the pH stability of a β-lactam compound is profiled using the methodology described in Chapter 31. The compound is not stable at low or high pH values. Profiling of solution stability typically is performed during the early exploratory phase to help prioritize different chemical series and alert teams to potential issues in later stages. Information on the stability of compounds at different pH values contributes to selecting the best bioassay conditions, designing synthetic strategies, developing optimal formulations, and predicting oral absorption.[14–17]

Figure 13.6 ► Stability pH profile of a β-lactam compound. (Reprinted with permission from [4].)

Selection of Conditions for Compound Purification

Early access to information on the stability of a compound at various pHs can suggest conditions for purification at which it is stable. The compound in Figure 13.7 degraded at pH 1. Therefore, acidic conditions (e.g., TFA, formic acid) should be avoided during sample purification.

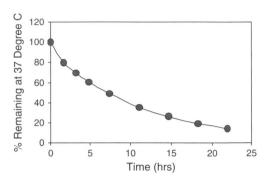

Buffers	% Remaining at 24 hrs at 37° C
pH 1	14
pH 4.5	96
pH 6.6	101
pH 7.4	100
pH 9	99

Figure 13.7 ▶   Screening of compound stability at different pH values.

Develop Structure–Stability Relationships

Solution stability testing of several series analogs can provide structure–stability relationships. This can indicate possible modifications to the series that improve stability.

Diagnose Poor In Vitro Bioassay Performance

A compound will not produce reliable activity in vitro if it is unstable in the bioassay buffer. The assay conditions can be altered to allow for accurate activity measurement. Figure 13.8 shows an example of bioassay buffer screening of 96 compounds. They all were from the same lead series and contained a labile functional group that was susceptible to nucleophilic attack by water. Many compounds were hydrated and rearranged. Stability screening in a 96-well format can quickly evaluate a large number of compounds to diagnose their bioassay stability.

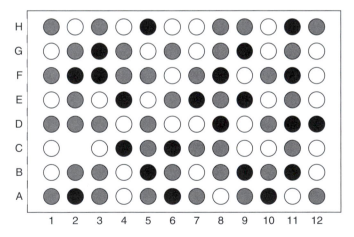

Figure 13.8 ▶   The stability of 96 compounds in biological assay buffer can be rapidly profiled to indicate which series analogs are stable (white), moderately stable (gray), or unstable (black).

Diagnose Poor In Vivo Performance

Poor PK or pharmacological performance can occur in vivo if the compound is not stable in the GI system. Low pH and enzymatic hydrolysis in the stomach or intestine can reduce in vivo exposure (area under the curve). Little of the dose is absorbed if the compound is degraded in the stomach or intestine. In vitro testing with simulated gastric fluid and simulated intestinal fluid can help determine if a compound is not stable in the stomach or intestine, respectively (see Chapter 31.

Prioritize Compounds for In Vivo Animal Studies

Knowledge of the stability of compounds in simulated physiological fluids can help evaluate whether they should be tested in vivo. Comparative stability provides a foundation for decisions on which compounds to dose in vivo. An example of screening four compounds for stability in simulated physiological fluids is shown in Figure 13.9. Compounds 1 and 3 are predicted to be the most stable in the GI tract.

Compounds	Buffers	Incubation Time at 37° C	% Remaining
1	SGF	24 hr	106
	SIF	24 hr	102
2	SGF	24 hr	100
	SIF	1 hr	42
3	SGF	24 hr	102
	SIF	24 hr	100
4	SGF	15 min	0
	SIF	15 min	0

SGF: Simulated Gastro Fluid, pH 1.2 + Pepsin
SIF: Simulated Intestinal Fluid, pH 6.8 + Pancreatin

Figure 13.9 ▶ Compounds can be assayed for solution stability in simulated gastric fluid (SGF) and simulated intestinal fluid (SIF) prior to oral dosing.

Structure Elucidation of Solution Stability Products Guides Synthetic Optimization

Evidence for the structures of solution degradants can be readily obtained using liquid chromatography/mass spectometry. This can guide synthetic modifications at the site of the reaction.

Problems

(Answers can be found in Appendix I at the end of the book.)

1. In which of the following solutions might a compound be unstable?: (a) stomach fluid, (b) enzyme assay media, (c) high-throughput screening buffer, (d) pH 7.4 buffer, (e) animal gavage dosing solution, (f) ethanol stock solution, (g) cell assay buffer.

2. If a compound is unstable in solution, wouldn't it be best to eliminate it from further study?

References

1. Di, L., Kerns, E. H., Chen, H., & Petusky, S. L. (2006). Development and application of an automated solution stability assay for drug discovery. *Journal of Biomolecular Screening*, *11*, 40–47.

2. Kerns, E. H., & Di, L. (2003). Pharmaceutical profiling in drug discovery. *Drug Discovery Today*, *8*, 316–323.

3. Kerns, E. H., & Di, L. (2006). Accelerated stability profiling in drug discovery. In B. Testa, S. D. Kramer, H. Wunderli-Allenspach, & G. Folkers (Eds.), *Pharmacokinetic profiling in drug research: biological, physicochemical and computational strategies* (pp. 281–306). Zurich: Wiley.

4. Kerns, E. H., & Di, L. (2006). Chemical stability. In: *Comprehensive medicinal chemistry, vol. 5* (pp. 489–507). Philadelphia: Elsevier.

5. Kibbey, C. E., Poole, S. K., Robinson, B., Jackson, J. D., & Durham, D. (2001). An integrated process for measuring the physicochemical properties of drug candidates in a preclinical discovery environment. *Journal of Pharmaceutical Sciences, 90*, 1164–1175.

6. Shah, K. P., Zhou, J., Lee, R., Schowen, R. L., Elsbernd, R., Ault, J. et al. (1994). Automated analytical systems for drug development studies. I. A system for the determination of drug stability. *Journal of Pharmaceutical and Biomedical Analysis, 12*, 993–1001.

7. Magnin, D. R., Robl, J. A., Sulsky, R. B., Augeri, D. J., Huang, Y., Simpkins, L. M., et al. (2004). Synthesis of novel potent dipeptidyl peptidase IV inhibitors with enhanced chemical stability: Interplay between the N-terminal amino acid alkyl side chain and the cyclopropyl group of a-aminoacyl-L-cis-4,5-methanoprolinenitrile-based inhibitors. *Journal of Medicinal Chemistry, 47*, 2587–2598.

8. von Rauch, M., Schlenk, M., & Gust, R. (2004). Effects of C2-alkylation, N-alkylation, and N,N′-dialkylation on the stability and estrogen receptor interaction of (4R,5S)/(4S,5R)-4,5-bis(4-hydroxyphenyl)-2-imidazolines. *Journal of Medicinal Chemistry, 47*, 915–927.

9. Chang, C.-W.T., Hui, Y., Elchert, B., Wang, J., Li, J., & Rai, R. (2002). Pyranmycins, a novel class of amino glycosides with improved acid stability: The SAR of D-pyranoses on ring III of pyranmycin. *Organic Letters, 4*, 4603–4606.

10. Regueiro-Ren, A., Leavitt, K., Kim, S.-H., Hoefle, G., Kiffe, M., Gougoutas, J. Z., et al. (2002). SAR and pH stability of cyano-substituted epothilones. *Organic Letters, 4*, 3815–3818.

11. Posner, G. H., Paik, I.-H., Sur, S., McRiner, A. J., Borstnik, K., Xie, S., et al. (2003). Orally active, antimalarial, anticancer, artemisinin-derived trioxane dimers with high stability and efficacy. *Journal of Medicinal Chemistry, 46*, 1060–1065.

12. Chong, Y., Gumina, G., Mathew, J. S., Schinazi, R. F., & Chu, C. K. (2003). L-2′,3′-Didehydro-2′,3′-dideoxy-3′-fluoronucleosides: Synthesis, anti-HIV activity, chemical and enzymatic stability, and mechanism of resistance. *Journal of Medicinal Chemistry, 46*, 3245–3256.

13. Guilford, W. J., Bauman, J. G., Skuballa, W., Bauer, S., Wei, G. P., Davey, D., et al. (2004). Novel 3-oxa lipoxin A4 analogues with enhanced chemical and metabolic stability have anti-inflammatory activity in vivo. *Journal of Medicinal Chemistry, 47*, 2157–2165.

14. Song, Y., Schowen, R. L., Borchardt, R. T., & Topp, E. M. (2001). Effect of "pH" on the rate of asparagine deamidation in polymeric formulations: "pH"-rate profile. *Journal of Pharmaceutical Sciences, 90*, 141–156.

15. Fubara, J. O., & Notari, R. E. (1998). Influence of pH, temperature and buffers on cefepime degradation kinetics and stability predictions in aqueous solutions. *Journal of Pharmaceutical Sciences, 87*, 1572–1576.

16. Zhou, M., & Notari, R. E. (1995). Influence of pH, temperature, and buffers on the kinetics of ceftazidime degradation in aqueous solutions. *Journal of Pharmaceutical Sciences, 84*, 534–538.

17. Muangsiri, W., & Kirsch, L. E. (2001). The kinetics of the alkaline degradation of daptomycin. *Journal of Pharmaceutical Sciences, 90*, 1066–1075.

18. Jung, M., Lee, K., Kendrick, H., Robinson, B. L., & Croft, S. L. (2002). Synthesis, stability, and anti-malarial activity of new hydrolytically stable and water-soluble (+)-deoxoartelinic acid. *Journal of Medicinal Chemistry, 45*, 4940–4944.

19. Akers, M. J. (2002). Excipient-drug interactions in parenteral formulations. *Journal of Pharmaceutical Sciences, 91*, 2283–2300.

20. Bowes, S., Sun, D., Kaffashan, A., Zeng, C., Chuaqui, C., Hronowski, X., et al. (2006). Quality assessment and analysis of Biogen Idec compound library. *Journal of Biomolecular Screening, 11*, 828–835.

21. Kozikowski, B. A., Burt, T. M., Tirey, D. A., Williams, L. E., Kuzmak, B. R., Stanton, D. T., et al. (2003). The effect of freeze/thaw cycles on the stability of compounds in DMSO. *Journal of Biomolecular Screening, 8*, 210–215.

22. Kozikowski, B. A., Burt, T. M., Tirey, D. A., Williams, L. E., Kuzmak, B. R., Stanton, D. T., et al. (2003). The effect of room-temperature storage on the stability of compounds in DMSO. *Journal of Biomolecular Screening, 8*, 205–209.

23. Cheng, X., Hochlowski, J., Tang, H., Hepp, D., Beckner, C., Kantor, S., et al. (2003). Studies on repository compound stability in DMSO under various conditions. *Journal of Biomolecular Screening, 8*, 292–304.

Chapter 14

Plasma Protein Binding

Overview

▶ *Compounds can bind to albumin, α_1-acid glycoprotein, or lipoproteins in blood.*

▶ *Binding reduces free drug in solution for penetration into tissue to reach the therapeutic target or to the liver and kidney for elimination.*

A majority of pharmaceutical treatment strategies use the bloodstream to deliver the drug to disease targets. Most delivery routes, such as oral dosing, intravenous injection (IV), intraperitoneal (IP), intramuscular (IM), subcutaneous (SC), and transdermal, transfer drug into the bloodstream and distribute it to the different tissues. Once the drug molecules are in the bloodstream, they can bind to a variety of blood constituents, including red blood cells, leukocytes and platelets, as well as proteins such as albumin, α_1-acid glycoprotein (AGP), lipoproteins, erythrocytes and alpha-, beta- and gamma-globulins.[1] Plasma proteins can adsorb a significant percentage of drug molecules. Binding to plasma protein can affect the pharmacokinetics (PK) of the drug substance in tissues and blood as well as the dosing regimen for the drug product.

Plasma is the liquid portion of blood (not including cells). Plasma for in vitro studies is obtained when fresh blood is collected in the presence of anticoagulant (e.g., heparin) and centrifuging to remove blood cells. Plasma proteins remain with the liquid. Serum is plasma from which clotting factors (e.g., fibrinogen) have been removed. Serum is collected without anticoagulant. PK studies typically collect plasma, thus retaining both the protein-bound and unbound drug, but discarding cell-bound drug.

14.1 Plasma Protein Binding Fundamentals

Drug compound molecules that are dissolved in the blood are in equilibrium with plasma proteins.[1] The interaction of drug molecules with plasma proteins is electrostatic and hydrophobic. The binding usually is rapid, with an average equilibrium time of 20 ms. The available binding sites on plasma proteins can be saturated. Plasma protein binding (PPB) is reversible. It can vary among species. Plasma protein concentrations can vary in different disease states or with age.[2–4] Figure 14.1 shows species differences in PPB. Ceftriaxone has very different protein binding for human and dog plasma, with 4.6-fold higher percent bound in human.[5] The percent of zamifenacin unbound in human is 10- and 20- fold lower than in dog and rat, respectively. This results in lower clearance in human and C_{max} (total of drug bound to plasma proteins and unbound) in human that is 40 and 74 times higher than in dog and rat.[6] It is important to determine PPB across different species to establish safety margins for human exposure and doses for clinical trials.[6] Subtle concentration variations occur between human individuals. These variations can sometimes affect PK. In most cases, however, changes in free drug concentration do not have a significant impact on

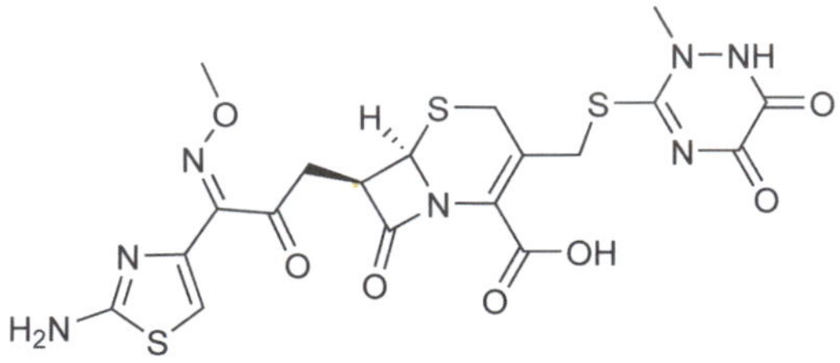

Ceftriaxone
Human Plasma Protein Binding = 90.8% Bound
Dog Plasma Protein Binding = 19.6% Bound

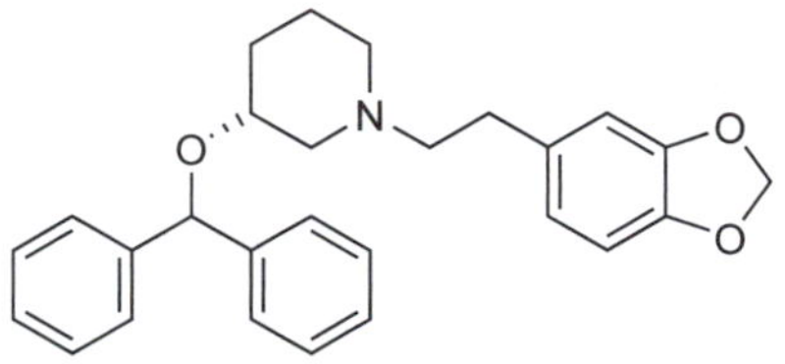

Zamifenacin
Human Plasma Protein Binding = 0.01% Unbound
Dog Plasma Protein Binding = 0.10% Unbound
Rat Plasma Protein Binding = 0.20% Unbound
C_{max} in Humans: 40 and 74 Times Higher than in Dog and Rat, Respectively

Figure 14.1 ▶ Species differences in plasma protein binding.[5]

clinical exposure and, therefore, do not lead to changes in pharmacological or toxicological responses.[7–9]

Two plasma proteins are most responsible for binding of drug molecules: albumin, which in humans is called human serum albumin (HSA), and AGP. Some lipophilic drugs also bind to plasma lipoproteins (very-high-density lipoprotein [VHDL], high-density lipoprotein [HDL], low-density lipoprotein [LDL], very-low-density lipoprotein [VLDL]).

HSA primarily binds strongly to organic anions (e.g., carboxylic acids, phenols), but it also can bind to basic and neutral drugs. It is the most abundant protein in plasma (60% of the total plasma protein), with a concentration of 500 to 750 μM (35–50 mg/mL). Each molecule circulates throughout the body about once every minute, but of this minute it spends only 1 to 3 seconds in any particular capillary, where it can exchange transported substances with the neighboring cells.[10,11] HSA has 585 amino acids in a single polypeptide chain and a molecular weight (MW) of 66.5 kDa. An x-ray structure was obtained in 1989. HSA has at least six primary binding sites of high specificity. The most common binding sites are I and II.[12] A large number of secondary binding sites with low affinity are nonsaturable. For example, one albumin molecule can weakly bind up to 30 imipramine molecules. Binding is primarily by hydrophobic interactions. The primary natural functions of HSA are to maintain blood pH and osmotic pressure and to transport molecules throughout the body.

AGP primarily binds basic drugs (e.g., amines). It also binds hydrophobic compounds (e.g., steroids). Its concentration in the blood is 15 μM (0.5–1.0 mg/mL). In some disease states it can reach a concentration of 3 mg/mL. AGP consists of 181 amino acids in a single polypeptide chain and has a MW of 44 kDa. It has a very high carbohydrate content (45%) and a very acidic isoelectric point around 3. AGP has one binding site per molecule, which binds compounds primarily by nonspecific hydrophobic interactions. Its primary function is to carry steroids throughout the body.

Acidic drugs that bind to HSA can be classified into different types based on their binding characteristics (Table 14.1). These types demonstrate the wide range of binding modes of drugs to HSA. Class I drugs, typified by warfarin and diazepam, bind tightly to HSA. There are one to three binding sites per molecule, depending on the compound, and they are saturable. Class II drugs are exemplified by indomethacin, which binds moderately to HSA and has six binding sites per HSA molecule. Phenytoin is an example of a class III drug. It has weak HSA binding but many binding sites per molecule.

TABLE 14.1 ▶ Classification of Acidic Drugs for HSA Binding

Types of drugs	I	II	III
Reference drugs	Warfarin diazepam	Indomethacin	Phenytoin
Binding proteins	HSA	HSA	HSA
Binding processes	Saturable	Saturable and nonsaturable	Nonsaturable
Association constant (M^{-1})	$10^4 - 10^6$	$10^3 - 10^5$	$10^2 - 10^3$
Binding sites per molecule	1 to 3	6	Many

Basic and neutral drugs that bind to HSA are shown in Table 14.2. Class IV is typified by digitoxin, which binds to HSA and is not saturable. Class V basic drugs, such as erythromycin, bind to HSA and can be saturated. Basic drug imipramine typifies class VI and can bind to HSA, AGP, and lipoproteins (HDL, LDL, VLDL). Lipoproteins tend to bind lipophilic basic and neutral compounds (e.g., probucol, etretinate). Highly water-soluble compounds typically are highly unbound (e.g., caffeine, ketamine).

TABLE 14.2 ▶ Classification of Non-ionized and Basic Drugs

Types of drug	IV	V	VI
Reference drugs	Digitoxin	Erythromycin	Imipramine
pK_a	–	8.8	9.5
Binding protein	HSA (NS)	HSA (NS)	HSA (NS), α_1-AGP (S), HDL (NS), LDL (NS), VLDL (NS)
Drug plasma saturation	No	Possible	Possible

14.1.1 Consequences of Chirality on PPB

Chirality affects binding of compounds to plasma proteins. Table 14.3 shows examples of stereoselectivity in PPB. This differential PPB affects compound distribution, metabolism, and renal clearance (see Section 14.2). The clinical significance is greater with higher binding and disease state. The stereoselectivity of binding also varies with animal species.

TABLE 14.3 ▶ Examples of Stereoselectivity in Plasma Protein Binding

	% Unbound		Enantiomeric
Drug	+ Enantiomer	− Enantiomer	ratio
Propranolol	12	11	1.1
Warfarin	1.2	0.9	1.3
Disopyramide	27	39	1.4
Verapamil	6.4	11	1.7
Indacrinone	0.3	0.9	3.0

14.2 PPB Effects

PPB impacts the PK of a drug and exposure to the therapeutic target. This section discusses various PPB effects so that drug discovery project teams can consider all of the possibilities. However, the effects of PPB can vary among compound series and project. Fraction unbound in plasma does not always correlate to in vivo PK parameters. It is advisable to consider the effects for each project and lead series in the context of the other properties of the series. Some organizations take the strategy of considering PPB as a primary property of interest, while others use PPB retrospectively to diagnose poor PK or in vivo pharmacological data.

According to the "free drug hypothesis," the drug–plasma protein complex cannot permeate through cell membranes by passive transcellular or paracellular permeation. Only free drug passes through membranes to reach tissues, and only free drug molecules are available for liver metabolism and renal excretion. Therefore, PPB can confine the compounds in the bloodstream and limit penetration of drug molecules to the target tissue to produce pharmacological effects, into other body tissues, and into clearance organs (e.g., liver, kidney).

There are two complementary factors of PPB:

▶ Extent of binding at equilibrium (expressed as percent bound or percent unbound in plasma [$f_{u,plasma}$], or equilibrium dissociation constant K_d)

▶ Rate of association and dissociation (expressed as association and dissociation rate constant k_a and k_d)

In vivo, these factors (Figure 14.2) affect the absorption, distribution, metabolism, and excretion (ADME) of drug molecules (Table 14.4).[10,13,14] If the drug molecules are highly bound (low percent unbound) and tightly bound (slow dissociation) to plasma proteins, the effects of PPB can be as follows:

▶ Retain drug in plasma compartment

▶ Restrict distribution of drug into target tissue (reduce volume of distribution [V_d])

▶ Decrease metabolism, clearance, and prolong $t_{1/2}$

▶ Limit brain penetration

▶ Require higher loading doses but lower maintenance doses

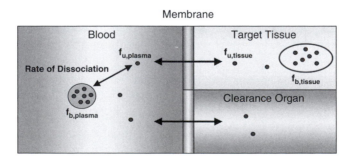

Figure 14.2 ▶ Penetration of drug into the target tissue or clearance organ depends on the fraction unbound in plasma ($f_{u,plasma}$).

TABLE 14.4 ► Effect of Equilibrium and Rate Constant on In Vivo ADME Properties

% Bound (equilibrium)	k_d (Dissociation rate constant)	In vivo ADME effects
High	Slow	Restrictive
High	Fast	Permissive
Low	Slow	Permissive
Low	Fast	Permissive

Thus, compounds that have high percent bound and slow dissociation may not be able to exert their therapeutic effect before being removed from the site of action by the flow of blood.[10]

On the other hand, if the compound has fast kinetics (high dissociation rate), PPB might not limit compound distribution into tissues, metabolism, excretion, or brain penetration, even if it has apparently high percent bound. Thus, high binding to plasma protein (high percent bound) alone does not itself determine the consequences of plasma binding; the on/off rate of binding can act as a major determining factor.[10]

14.2.1 Impact of PPB on Distribution

PPB can have either a "restrictive" or a "permissive" (nonrestrictive) effect on drug disposition. Examples of drugs for which PPB is restrictive or permissive are given in Table 14.5. For example, furosemide has 4% free drug, which is restrictive of the volume of distribution ($V_d = 0.2$ L/kg). Imipramine has 5% free drug but is permissive of the volume of distribution ($V_d = 30$ L/kg).

TABLE 14.5 ► Restrictive and Permissive Effects of PPB on Drug Disposition

Drug	Free drug in plasma (%)	Volume of distribution (L/kg)
Restrictive		
Furosemide	4	0.2
Ibuprofen	<1	0.14
Nafcillin	10	0.63
Warfarin	<1	0.1
Permissive		
Desipramine	8	40
Imipramine	5	30
Vinblastine	30	35
Vincristine	30	11

PPB also can be restrictive of BBB permeation. Binding keeps the compound in the bloodstream, resulting in reduced permeation (Figure 14.3). Teams often use in vitro PPB in vitro measurements to estimate this effect. However, as shown in Figure 10.8, in vitro measurements of PPB may underestimate the release of drug molecules in the microcirculation environment of brain microvessels.[15] On/off rate also is important.[14]

PPB also affects V_d (see Chapter 19). V_d is determined as follows:

$$V_d = V_{plasma} + V_{tissue}(f_{u,plasma}/f_{u,tissue}),$$

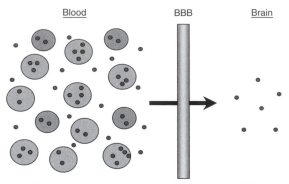

Free drug molecules permeate through the BBB

Figure 14.3 ▶   High PPB can restrict BBB permeation.

where V_{plasma} = volume of plasma in the body, V_{tissue} = volume of tissue in the body, $f_{u,plasma}$ = fraction unbound in the plasma, and $f_{u,tissue}$ = fraction unbound in the tissue. A couple of examples illustrate the possible scenarios:

▶ If PPB is high, $f_{u,plasma}$ is low, resulting in a low V_d.

▶ If PPB is low, $f_{u,plasma}$ is high, resulting in a high V_d.

▶ If nonspecific binding in tissue is high, $f_{u,tissue}$ is low, resulting in a high V_d.

▶ If nonspecific binding in tissue is low, $f_{u,tissue}$ is high, resulting in a low V_d.

Therefore, in addition to PPB, tissue binding also determines V_d. The ratio $f_{u,plasma}/f_{u,tissue}$ is a balance between the nonspecific binding in the tissue and PPB.[16] As this ratio increases, V_d is greater (more drug in tissue). For V_d much greater than 1 L/kg, high nonspecific tissue binding occurs. High binding to tissue proteins and lipids increases V_d.

14.2.2   Effect of PPB on Clearance

High PPB can be *restrictive* or *permissive* of liver extraction. Table 14.6 shows the case of propranolol, with >90% PPB but is *permissive* of >90% liver extraction. However, warfarin, with 99% PPB, is *restrictive* of liver extraction (<0.3%). Rate constants play an important role in liver clearance.[13]

TABLE 14.6 ▶   **Restrictive and Permissive Effects of PPB on Liver Extraction**

Drug	Bound drug in plasma	Liver extraction	Consequence
Propranolol	>90%	>90%	Permissive
Warfarin	>99%	<0.3%	Restrictive

14.2.3   Effect of PPB on Pharmacology

Pharmacology can be affected by PPB. Enzyme inhibition can be reduced if the compound is bound to plasma proteins. For example, inhibition of HIV protease by VX-478 is reduced two-fold if 45% HSA is added to the in vitro incubation media. PPB reduces vascular receptor occupancy because there are no diffusion barriers for vascular receptors. For example, the activity of angiotensin II is decreased by adding albumin. PPB directly reduces antimicrobial activity, which has not been seen with other groups of agents.

A study of 1,500 frequently prescribed drugs showed 43% of compounds have >90% PPB.[17] PPB shows no significant difference among therapeutic areas: CNS, inflammation, and renal/cardiovascular. An exception is perhaps antiinflammatory drugs, which have a higher percentage (26%) of high protein binding (>99%), and many of them are acids. It is surprising that many CNS drugs possess high PPB characteristics. The most striking finding of this study is that chemotherapeutics, including antibiotic, antiviral, antifungal, and anticancer drugs, have a high percentage (77%) of low binding drugs. In designing drugs for chemotherapeutics, low PPB appears to be advantageous.

14.3 PPB Case Studies

Nonsteroidal antiinflammatory drugs (NSAIDs) are extensively bound to HSA, with a high binding constant. Plasma binding is higher than tissue binding. For this reason, NSAIDs have low tissue distribution.[18]

Christ and Trainor[19] discuss the addition of HSA and AGP to in vitro cell culture assays for HIV reverse transcriptase inhibitors in discovery. The shift in IC_{90} is indicative of the effect of PPB on free compound concentration and has been called a *shift assay*. This experiment was used as part of the strategy for selecting clinical candidates with higher free drug concentration for treatment of mutant HIV strains.

Webster et al.[20] correlated the in vitro IC_{50} that causes hERG blocking (see Chapter 16) to free drug concentration in the plasma that causes QT prolongation (increases the QT ECG interval) and torsades de pointes (TdP) arrhythmia (Figure 14.4). There is a strong correlation, but IC_{50} values overestimate the free drug concentration that causes QT.

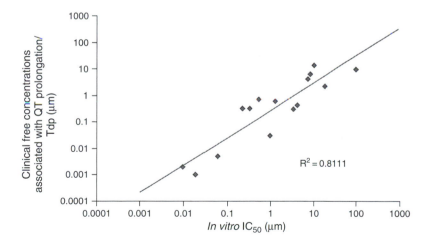

Figure 14.4 ▶ Relationship of in vitro IC_{50} for hERG blocking and clinical free plasma concentration causing QT prolongation and torsades de pointes (TdP). (Reprinted with permission from [20].)

14.4 Structure Modification Strategies for PPB

Quantitative structure–activity relationship studies have evaluated the physicochemical properties that correlate with PPB.[21] Increasing lipophilicity has the greatest effect on increasing

PPB. This analysis suggests various structure modifications that can be performed to *reduce PPB* (Table 14.7). Synthetic strategies for these structural modifications are found in several other chapters.

> **TABLE 14.7 ▶ Structure Modification Strategies to Reduce PPB in Order of Highest to Lowest Potential Effect**
>
> **Structure Modification Strategy**
>
> Reduce lipophilicity (Log P for acids, Log $D_{7.4}$ for nonacids)
> Reduce acidity (increase pK_a of the acid)
> Increase basicity (increase pK_a of the base)
> Reduce nonpolar area
> Increase PSA (increasing PSA increases hydrogen bonding)

Urien et al.[16] have suggested that PBB can be used to advantage in restricting the distribution of drugs into general tissues throughout the body where they can cause side effects. It is suggested, as in the case of cetirizine (zwitterion with moderate lipophilicity), that high plasma binding and low tissue binding would restrict the compound to the blood. This may be a useful strategy when the target is in close proximity to the bloodstream.

14.5 Strategy for PPB in Discovery

In general, the prospective use of PPB data for predicting in vivo PK and pharmacodynamics in drug discovery can be misleading. Many commercial drugs have high (>99%) PPB. PPB may be restrictive or permissive for penetration into tissues. This is because PPB is modified by the extent of plasma binding, the plasma protein–drug dissociation rate, and the extent of nonspecific tissue binding. PPB can increase the PK $t_{1/2}$ (by keeping the compound in the blood and restricting clearance), but it also can restrict exposure to the therapeutic target (by reducing penetration into tissues). PPB alone can be either a positive or a negative aspect of a compound.

However, PPB can be useful, retrospectively, as part of an ensemble of in vitro diagnostic tests to understand the impact of PPB on PK or pharmacological effects. Only when PPB is placed into context with PK parameters can valuable insight be gained into the disposition of the molecule.[6]

14.6 Red Blood Cell Binding

Drug molecules can bind to red blood cells (RBCs). In PK studies, RBCs are immediately removed by centrifugation of blood samples to produce plasma from which drug concentration is quantitated. Therefore, drug bound to RBCs will not be measured in the sample. If in vivo PK studies show only low plasma concentrations or unexplained clearance, especially from IV dosing, red blood cell binding could be one of the causes.

Problems

(Answers can be found in Appendix I at the end of the book.)

1. Would high plasma protein binding of a compound (e.g., 99.9%) and low dissociation rate tend to increase or decrease each of the following, compared to a compound with moderate plasma protein binding (e.g., 50%) and moderate dissociation rate?: (a) metabolic

clearance, (b) tissue concentration, (c) tissue distribution, (d) blood concentration, (e) renal clearance, (f) PK half-life, (g) pharmacological effect for non-bloodstream target, (h) brain penetration.

2. List three plasma proteins to which drugs bind.

3. Why is a shift in activity sometimes observed in vitro when albumin is added to the assay media?

4. For the following compounds, list structure modifications and structural examples that can be synthesized in an attempt to reduce plasma protein binding.

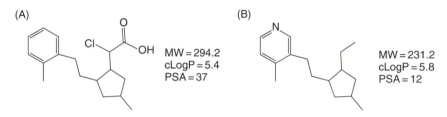

(A)

MW = 294.2
cLogP = 5.4
PSA = 37

(B)

MW = 231.2
cLogP = 5.8
PSA = 12

References

1. Smith, D. A., Van de Waterbeemd, H., & Walker, D. K. (2001). *Pharmacokinetics and metabolism in drug design.* Weinheim, Germany: Wiley-VCH.

2. Grandison, M. K., & Boudinot, F. D. (2000). Age-related changes in protein binding of drugs: implications for therapy. *Clinical Pharmacokinetics, 38,* 271–290.

3. Kosa, T., Maruyama, T., & Otagiri, M. (1998). Species differences of serum albumins: II. Chemical and thermal stability. *Pharmaceutical Research, 15,* 449–454.

4. Kosa, T., Maruyama, T., & Otagiri, M. (1997). Species differences of serum albumins: I. Drug binding sites. *Pharmaceutical Research, 14,* 1607–1612.

5. Kratochwil, N. A., Huber, W., Mueller, F., Kansy, M., & Gerber, P. R. (2004). Predicting plasma protein binding of drugs: revisited. *Current Opinion in Drug Discovery & Development, 7,* 507–512.

6. van de Waterbeemd, H., Smith, D. A., Beaumont, K., & Walker, D. K. (2001). Property-based design: optimization of drug absorption and pharmacokinetics. *Journal of Medicinal Chemistry, 44,* 1313–1333.

7. Benet, L. Z., & Hoener, B.-A. (2002). Changes in plasma protein binding have little clinical relevance. *Clinical Pharmacology & Therapeutics, 71,* 115–121.

8. Rolan, P. E. (1994). Plasma protein binding displacement interactions: why are they still regarded as clinically important? *British Journal of Clinical Pharmacology, 37,* 125–128.

9. Sansom, L. N., & Evans, A. M. (1995). What is the true clinical significance of plasma protein binding displacement interactions? *Drug Safety, 12,* 227–233.

10. Talbert, A. M., Tranter, G. E., Holmes, E., & Francis, P. L. (2002). Determination of drug-plasma protein binding kinetics and equilibria by chromatographic profiling: Exemplification of the method using L-tryptophan and albumin. *Analytical Chemistry, 74,* 446–452.

11. Guyton, A. C. (1996). *Textbook of medical physiology* (9th ed.). Philadelphia: WB Saunders.

12. Ascenzi, P., Bocedi, A., Notari, S., Fanali, G., Fesce, R., & Fasano, M. (2006). Allosteric modulation of drug binding to human serum albumin. *Mini-Reviews in Medicinal Chemistry, 6,* 483–489.

13. Weisiger, R. A. (1985). Dissociation from albumin: a potentially rate-limiting step in the clearance of substances by the liver. *Proceedings of the National Academy of Sciences of the United States of America, 82,* 1563–1567.

14. Robinson, P. J., & Rapoport, S. I. (1986). Kinetics of protein binding determine rates of uptake of drugs by brain. *American Journal of Physiology, 251,* R1212–R1220.

15. Pardridge, W. M. (1995). Transport of small molecules through the blood-brain barrier: biology and methodology. *Advanced Drug Delivery Reviews, 15,* 5–36.

16. Urien, S., Tillement, J.-P., & Barre, J. (2001). The significance of plasma-protein binding in drug research. In *Pharmacokinetic Optimization in Drug Research: Biological, Physicochemical, and Computational Strategies, [LogP2000, Lipophilicity Symposium], 2nd, Lausanne, Switzerland, March 5–9, 2000,* pp. 189–197.

17. Kratochwil, N. A., Huber, W., Muller, F., Kansy, M., & Gerber, P. R. (2002). Predicting plasma protein binding of drugs: a new approach. *Biochemical Pharmacology, 64,* 1355–1374.

18. Tillement, J. P., Houin, G., Zini, R., Urien, S., Albengres, E., Barre, J., et al. (1984). The binding of drugs to blood plasma macromolecules: recent advances and therapeutic significance. *Advances in Drug Research, 13,* 59–94.

19. Christ, D. D., & Trainor, G. L. (2004). Free drug! The critical importance of plasma protein binding in new drug discovery. *Biotechnology: Pharmaceutical Aspects, 1,* 327–336.

20. Webster, R., Leishman, D., & Walker, D. (2002). Towards a drug concentration effect relationship for QT prolongation and torsades de pointes. *Current Opinion in Drug Discovery & Development 5,* 116–126.

21. Fessey, R. E., Austin, R. P., Barton, P., Davis, A. M., & Wenlock, M. C. (2006). The role of plasma protein binding in drug discovery. In *Pharmacokinetic Profiling in Drug Research: Biological, Physicochemical, and Computational Strategies, [LogP2004, Lipophilicity Symposium], 3rd, Zurich, Switzerland, Feb. 29–Mar. 4, 2004,* pp. 119–141.

Cytochrome P450 Inhibition

Overview

▶ *Drug–drug interactions can occur when two drugs are coadministered and compete for the same enzyme.*

▶ *In cytochrome P450 (CYP) inhibition, one drug ("perpetrator") binds to the isozyme and the other drug ("victim") is excluded from metabolism, thus increasing to a toxic concentration.*

▶ *Irreversible binding inactivates CYP and is termed mechanism-based inhibition.*

▶ *CYP inhibition can cause withdrawal from clinical use or restrictive labeling for a drug.*

Many patients receive more than one drug at a time, and physicians must be careful to avoid drug–drug interactions (DDI). DDI is the interference of one drug with the normal metabolic or pharmacokinetic behavior of a coadministered drug. DDI typically occurs by competition at a specific protein, such as at a metabolizing enzyme, involved in ADME processes. A monograph discusses DDI in detail.[1] A major DDI concern is cytochrome P450 (CYP) inhibition. (Other DDI issues are discussed in Section 15.6.) CYP inhibition has caused withdrawal from clinical use or restricted use of some major drugs. Because of its effects on clearance and half-life, CYP inhibition has become an important concern with the Food and Drug Administration (FDA) and at pharmaceutical companies. CYP inhibition now is assessed for a lead series from the earliest stages of the discovery project and can cause a lead series to be diminished in priority if the issue is uncorrectable. Medicinal chemists often can modify the structure to reduce CYP inhibition.

15.1 CYP Inhibition Fundamentals

CYP inhibition is illustrated in Figure 15.1. When a single drug is administered to a patient, it undergoes normal metabolism at one or more of the CYP isozymes and is eliminated at a predictable clearance (Cl) rate. This clearance is used for calculations of patient pharmacokinetics, dosage levels, and frequency. When the same drug is administered with a second drug that competes for binding at the CYP isozyme that metabolizes the first drug, then the clearance of the first drug is reduced. This can be described as follows:

$$Cl_{int}(i) \approx Cl_{int}/[1 + (I/K_i)],$$

where Cl_{int} = normal intrinsic clearance (without inhibitor), $Cl_{int}(i)$ = intrinsic clearance in the presence of the inhibitor, I = inhibitor concentration at the CYP isozyme, and K_i = inhibitor constant for an isozyme.[2] The result is a reduced clearance rate. Cl decreases

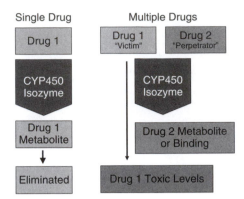

Figure 15.1 ▶ Administration of drug 1 alone allows a normal rate of metabolism at a particular CYP isozymes. Coadministration of drug 1 and drug 2 can result in drug 2 being an inhibitor of drug 1. Lack of drug 1 clearance can cause drug 1 to increase to a toxic concentration.

with increasing inhibitor concentration. With the reduced clearance, the concentration (e.g., C_{max}) or exposure (i.e., area under the curve [AUC]) of the first drug increases to a higher level than expected. This high concentration may be toxic or cause side effects. The CYP inhibitor is termed the *perpetrator,* and the compound whose metabolism is inhibited is termed the *victim.*

In order to recognize the impact of CYP inhibition, it is useful to understand the CYP isozymes. Many CYP isozymes have been discovered, and their individual contributions to drug metabolism are becoming better understood. The major isozymes present in human liver microsomes (HLM) are shown in Figure 15.2. Some of the major isozymes are the 3A family (28% of total CYP protein) and 2C family (18%).

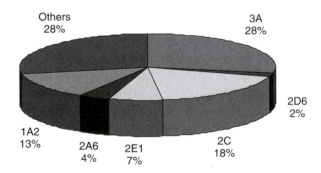

Figure 15.2 ▶ CYP isoforms in human liver microsomes and their relative abundances.[22]

The percentages of drugs metabolized by different CYP isozymes are shown in Figure 15.3 and summarized in Table 15.1.[3] Although 3A constitutes 28% of the human liver CYP, it is responsible for metabolism of 50% of all drugs. Even more remarkable, 2D6, which constitutes only 2% of the human liver CYP, is responsible for metabolism of 30% of all drugs. 2D6 primarily metabolizes basic amine compounds, of which there are many drugs. Thus, an inhibitor of 3A4 or 2D6 poses significant risk when coadministered with many drugs.

A CYP inhibitor perpetrator itself may be metabolized by the isozyme and, therefore, be a competitive substrate with the victim compound for the isozyme. Alternatively, the CYP inhibitor may bind to the isozyme and itself not be metabolized but inhibit other compounds from binding and being metabolized.

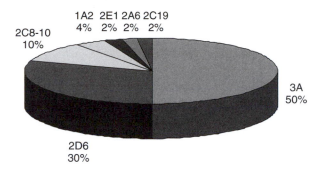

Figure 15.3 ▶ Percentage of drugs metabolized by different CYP isoforms.[22]

TABLE 15.1 ▶ **Summary of Important CYP Isozymes**

Isozyme	Distribution in HLM	Drugs metabolized	Comments
3A family	28%	50%	Most abundant
2D6	2%	30%	Polymorphic, 5% of white males lack isozyme
2C family	18%	10%	Polymorphic
1A2	13%	4%	Enzyme induction

15.2 Effects of CYP Inhibition

An example of CYP inhibition is coadministration of erythromycin and terfenadine (Figure 15.4). Erythromycin inhibits the metabolism of terfenadine by CYP3A4 isozyme. Terfenadine increases to abnormal concentrations and can cause prolongation of the cardiac QT interval and trigger torsades de pointes arrhythmia (see Chapter 16). Coadministered erythromycin is also known to inhibit the metabolism of cyclosporin, carbamazepine, and midazolam.[4]

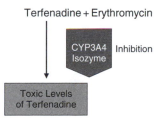

Figure 15.4 ▶ When erythromycin and terfenadine are coadministered, erythromycin inhibits terfenadine metabolism at CYP3A4 and can result in cardiac arrhythmia.

Pharmaceutical companies must test each clinical candidate for its potential to inhibit CYP isozymes. They also must determine what will happen if the drug's metabolism is inhibited by other drugs. Methods used in CYP inhibition testing are discussed in Chapter 32. In general the assays are conducted by coincubating the CYP isozyme, test compound, and a substrate compound, whose rate of metabolism is well established. Reduction in the rate of metabolism of the substrate is evidence for CYP inhibition by the test compound at the particular isozyme. During initial in vitro CYP inhibition assessment, often at a test compound concentration of $3\,\mu M$, the following guidelines are generally useful in recognizing the potential risk of DDI:

▶ <15% inhibition @ $3\,\mu M$ or $IC_{50} > 10\,\mu M$ CYP inhibition *low*

▶ 15%–50% inhibition @ $3\,\mu M$ or $3\,\mu M < IC_{50} < 10\,\mu M$ CYP inhibition *moderate*

▶ >50% inhibition @ $3\,\mu M$, $IC_{50} < 3\mu M$ CYP inhibition *high*

Some groups are more conservative and use more stringent guidelines:

▶ $IC_{50} > 100\,\mu M$ CYP inhibition *low*

▶ $10\,\mu M < IC_{50} < 100\,\mu M$ CYP inhibition *moderate*

▶ $IC_{50} < 10\,\mu M$ CYP inhibition *high*

Each organization decides on its own guidelines. This can be determined by using the data obtained from the in-house CYP inhibition assay(s) for known CYP inhibitors from the scientific literature or by using industry consensus.

A useful decision-making tool for discovery project teams is to plot the therapeutic target IC_{50} of project compounds versus the IC_{50} for CYP inhibition.[5] Use of a tool such as Spotfire allows differentiation of compound classes to visualize the trends of different lead series (Figure 15.5).

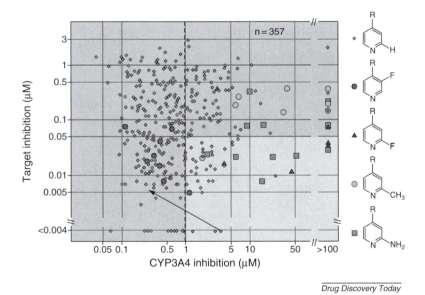

Drug Discovery Today

Figure 15.5 ▶ A plot of activity versus CYP inhibition is useful for multivariable decision making by discovery project teams. Here the CYP3A4 inhibition versus target potency plot shows which pyridine substitutions were effective in reducing CYP3A4 inhibition while maintaining target inhibition. The optimal area for activity with minimal CYP3A4 inhibition is the light area in the bottom right corner. Such plots are useful for all properties. (Reprinted with permission from [5].)

It is important to pay careful attention to compounds having moderate-to-high CYP inhibition. Three options in this case are to attempt structure modification to reduce the inhibition, reduce the priority of that particular lead series, or discuss the ramifications with clinical metabolism experts.

The magnitude of the CYP inhibition problem in vivo is determined by many parameters, in addition to $IC_{50,}$ which should be discussed among drug metabolism experts and the project team scientists. Compound concentration (C_{max}) in vivo relative to K_i is very important. The isozyme that is inhibited is important. For example, a drug that inhibits 3A4, which metabolizes 50% of drugs, has many more drugs it could interact with than if it inhibited 2C19, which metabolizes a low percentage of drugs. Another parameter is plasma protein binding of inhibitor, which reduces its exposure to the liver. Conversely, the inhibitor concentration in liver may be higher than in plasma. For example, hydroxyzine concentration in liver is 10 times the plasma concentration. For drugs that might be inhibited by coadministered drugs, it is useful to consider whether they might switch to another CYP isoform(s) for metabolism if their normal primary isozyme is inhibited, or whether clearance by other routes (renal, other enzyme systems) might increase.

In later stages of discovery[6], an assessment is made of K_i versus drug concentration levels in vivo following dosing. A PhARMA working group published a consensus summary of DDI assessment. Among its findings are that reversible CYP inhibition assessments can be guided by the CYP inhibition results (K_i), and the C_{max} expected at the highest human clinical dose. C_{max} is the highest inhibitor concentration reached in the blood in vivo and is an approximation of the inhibitor concentration (I) at the isozymes in the liver. The guidelines are as follows:

▶ $C_{max}/K_i < 0.1$ CYP inhibition *not likely*

▶ $0.1 < C_{max}/K_i < 1$ CYP inhibition *possible*

▶ $C_{max}/K_i > 1$ CYP inhibition *likely*

These ratios are used in planning early clinical DDI studies.

Data on CYP inhibition is included in regulatory submissions to the FDA. In vitro CYP450 inhibition data showing negligible inhibition can be used to conclude that the compound lacks DDI potential and requires no clinical DDI studies. If in vitro data indicate potential DDI, they are used to plan specific clinical DDI experiments with human volunteers during clinical development.[7] PhARMA has agreed on guidelines for conducting in vitro and in vivo DDI studies. The package insert for a drug contains information on DDI. It states whether the compound is a CYP inhibitor or if its pharmacokinetics is significantly affected by another drug that is an inhibitor. Therefore, strong DDI potential can greatly affect the marketability of the drug. It is good for a new clinical candidate to be metabolized by multiple CYP isozymes, to reduce its potential metabolic inhibition by another drug.

15.3 CYP Inhibition Case Studies

CYP inhibition can lead to toxic effects in patients. Several commercial drugs have been voluntarily withdrawn from the market because of the effects of CYP inhibition (Table 15.2). Terfenadine (Seldane)[8] and cisapride (Propulsid) can produce fatal torsades de pointes

TABLE 15.2 ▶ Drug Withdrawals due to CYP Inhibition

Drug	Generic name	Date voluntarily withdrawn
Posicor	Mibefradil dihydrochloride	June 1998
Seldane	Terfenadine	February 1998
Hismanal	Astemizole	June 1999

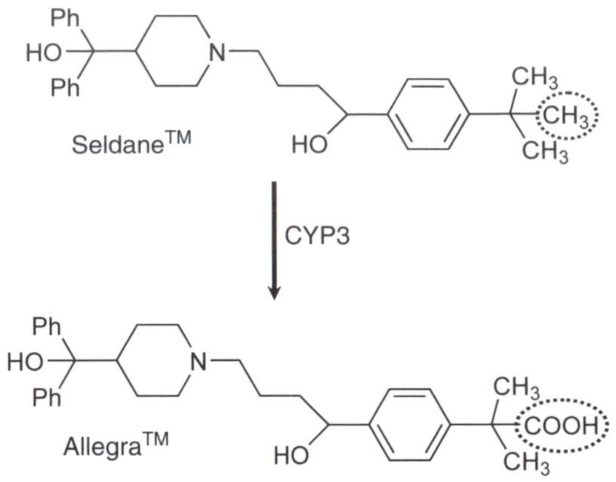

Figure 15.6 ► Seldane was withdrawn from the market because of high C_{max} values when coadministered with a CYP3A4 inhibitor. It was replaced by its active metabolite Allegra.

arrhythmia (see Section 16.2) when their metabolism is inhibited by a coadministered drug. Seldane was voluntarily removed from the market, and its active metabolite Allegra was developed and introduced for clinical use (Figure 15.6). Clinical use of cimetidine (Tagamet) diminished when it was found to cause DDI in the clinic.[9] When cimetidine and disopyramide are coadministered, the plasma C_{max} of disopyramide is greatly increased compared to individual administration.[10]

Reversible competitive inhibition of a CYP isozyme is dependent on the same factors that enhance binding of any ligand to an enzyme active site. These include specific interactions with active site moieties (e.g., lipophilic) and molecular shape.[4] In addition, the presence of a lone electron pair on the inhibitor appears to enhance binding to the heme group of the CYP enzyme. The increased binding energy is on the order of 6 kcal/mol.[11] For example, cimetidine contains an imidazole group and is a CYP3A4 and 2D6 inhibitor, whereas ranitidine has no imidazole group and does not inhibit disopyramide metabolism. Quinoline groups, as contained in quinidine (CYP2D6 inhibitor) and ellipticine (CYP1A inhibitor),[12] and pyridines, as contained in indinavir (CYP3A4 inhibitor),[13] also appear to interact with the heme during competitive binding.

15.3.1 Consequences of Chirality on CYP Inhibition

Chirality affects CYP inhibition. Two examples are shown in Figure 15.7. The (+)-isomer (3R, 5S) of fluvastatin (Figure 15.7A), an antilipidemic drug, more strongly inhibits the CYP2C9 isozyme than its enantiomer.[14] For the structure in Figure 15.7B, the

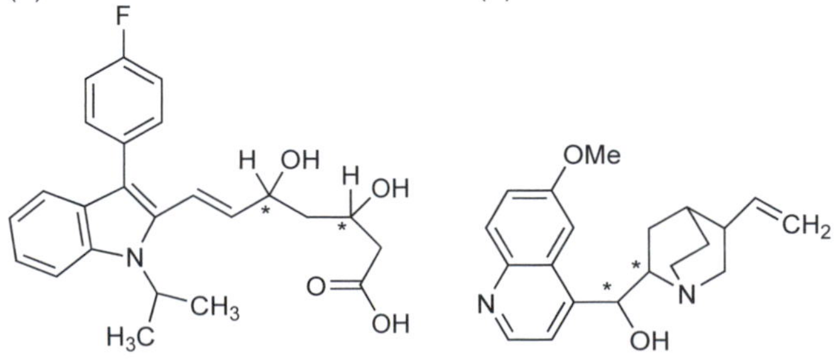

Figure 15.7 ► Stereoselective cytochrome P450 inhibition is observed for **(A)** fluvastatin[14] and **(B)** (+)quinidine and (−)quinine.[15]

$(+)$ enantiomer is quinidine and the $(-)$ enantiomer is quinine. Quinidine is a strong inhibitor of CYP2D6, while quinine has no effect on CYP2D6 metabolism.[15]

15.4 Structure Modification Strategies to Reduce CYP Inhibition

Modification of the structure of the lead series can reduce the IC_{50} of CYP inhibition. In the pyridinyloxazole series shown in Figure 15.8, the IC_{50} values for inhibition of three CYP isozymes were greatly increased with structure modification, while not affecting the activity or selectivity.[16]

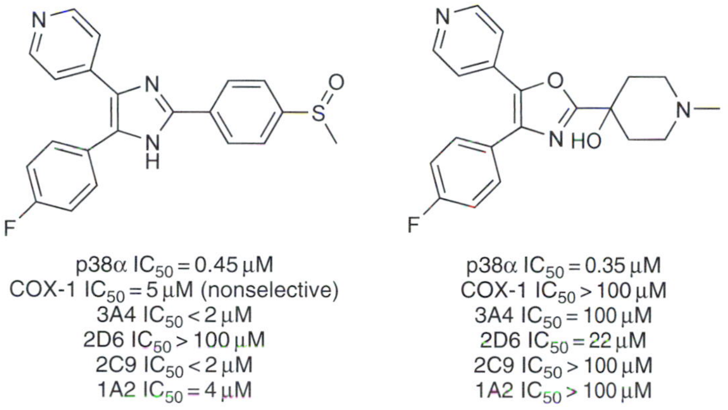

p38α $IC_{50} = 0.45\,\mu M$
COX-1 $IC_{50} = 5\,\mu M$ (nonselective)
3A4 $IC_{50} < 2\,\mu M$
2D6 $IC_{50} > 100\,\mu M$
2C9 $IC_{50} < 2\,\mu M$
1A2 $IC_{50} = 4\,\mu M$

p38α $IC_{50} = 0.35\,\mu M$
COX-1 $IC_{50} > 100\,\mu M$
3A4 $IC_{50} = 100\,\mu M$
2D6 $IC_{50} = 22\,\mu M$
2C9 $IC_{50} > 100\,\mu M$
1A2 $IC_{50} > 100\,\mu M$

Figure 15.8 ▶ Pyridinyloxazole series modification successfully reduced IC_{50} for inhibition of three CYP isozymes without reducing activity and selectivity.

In a sodium channel-blocking project, CYP2D6 inhibition was reduced by structural modification (Figure 15.9). At the same time, the activity of the series was maintained or improved.[17]

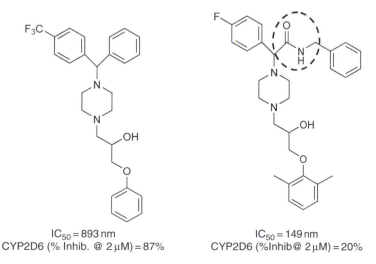

$IC_{50} = 893\,nm$
CYP2D6 (% Inhib. @ $2\,\mu M$) = 87%

$IC_{50} = 149\,nm$
CYP2D6 (%Inhib @ $2\,\mu M$) = 20%

Figure 15.9 ▶ CYP2D6 inhibition was reduced while sodium channel-blocking activity was improved by structure modification of this series. The circled modification was primarily responsible for reducing the CYP2D6 inhibition.

A structural series for a G-protein–coupled receptor (GPCR) target showed significant CYP2D6 inhibition (Figure 15.10). This was overcome by structure modification while maintaining the GPCR agonism.[18]

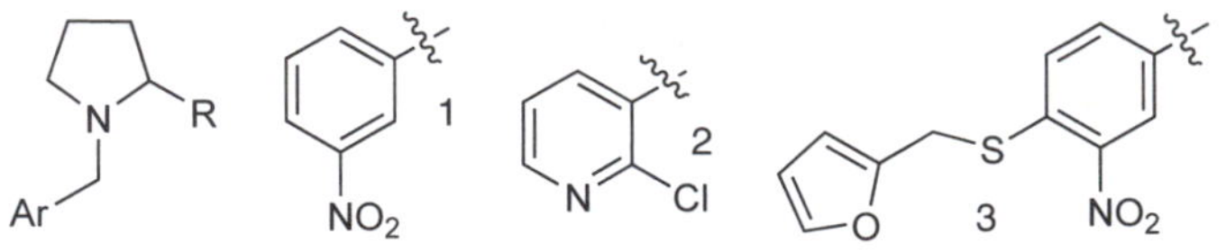

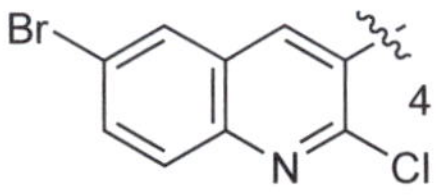

Compd	GPCR IC$_{50}$ (uM)	CYP2D6 IC$_{50}$ (uM)	Selectivity Ratio
1	0.33	<0.05	<0.15
2	0.22	0.02	0.09
3	0.22	2.2	10
4	0.19	22	116

Figure 15.10 ► CYP2D6 inhibition was reduced while maintaining GPCR agonism by structural modifications. Compound 1 was the original lead, which had high CYP2D6 inhibition.

Riley et al.[11] studied CYP3A4 inhibition of pyridine-containing drugs and suggested general rules for predicting CYP3A4 inhibition by nitrogen heterocycle-containing drugs (i.e., triazoles, pyridines, imidazoles, quinolines, thiazoles). Reducing the interaction of the nitrogen loan pair with the 3A4 heme group was beneficial. Riley's rules and related work[5] provide guidance for structure modification (Table 15.3) to reduce CYP3A4 inhibition.

TABLE 15.3 ► **Structure Modification Strategies to Reduce CYP Inhibition**

Structure Modification Strategy
Decrease the lipophilicity (Log D$_{7.4}$) of the molecule
Add steric hindrance to the heterocycle para to the nitrogen
Add an electronic substitution (e.g., halogen) that reduces the pK_a of the nitrogen

☐ 15.5 Reversible and Irreversible CYP Inhibition

CYP inhibition has two general modes. The mode most commonly considered is *reversible inhibition*, in which the inhibitor binds to the CYP enzyme and then releases in a reversible binding scheme. The second mode is *irreversible inhibition*. One irreversible mechanism is the formation of a covalent bond between the inhibitor metabolite and the enzyme (often with the heme). For example, reactive intermediates generated from spironolactone can react with the heme group or with the protein chain. Another irreversible mechanism is tight quasi-irreversible binding. For example, oxidation of the tertiary amine in the amino sugar ring of erythromycin generates a nitroso metabolite that complexes with the CYP3A4 heme.[4] Irreversible binding has been called *mechanism-dependent* (inhibition or *mechanism-based inhibition*). It results in permanent inactivation of the enzyme molecule *in vivo*[19].

Irreversible inhibition is diagnosed by time dependency of inhibition. With reversible inhibition, IC$_{50}$ should be the same for any incubation time. However, with irreversible inhibition, IC$_{50}$ decreases with incubation time. This is because the fraction of deactivated enzyme increases with reaction time, resulting in fewer functional enzyme molecules, making IC_{50} appear to be lower if the starting enzyme concentration is used to calculate IC_{50}. An example of irreversible inhibition diagnosis is shown in Figure 15.11.[20] For R $=$ H in the compound series and for troleandomycin, IC_{50} decreases with increasing incubation time. For the modification of R to F and for ketoconazole, IC_{50} is the same at each incubation time.

Compound	Time-Dependent Inhibition (IC_{50}, µM)			
	5 min	15 min	30 min	45 min
R = H	96	62	33	22
R = F	18	21	19	19
Troleandomycin	61	33	20	16
Ketoconazole	0.016	0.013	0.017	0.022

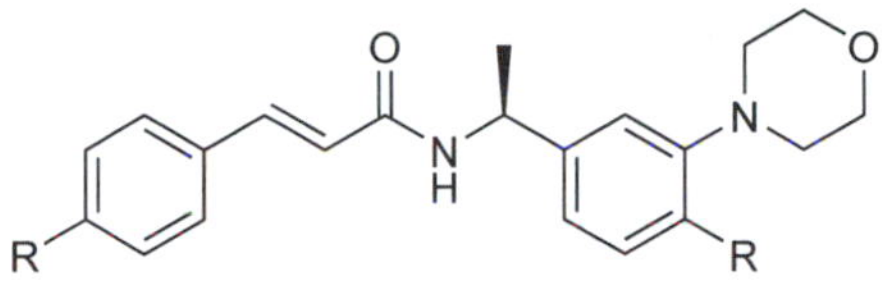

Figure 15.11 ▶ Examples of reversible and irreversible (red arrows) CYP inhibition. For the structure shown, when R is H irreversible CYP inhibition is observed, as indicated by the decrease in IC_{50} with incubation time. Modification of R to F eliminated the mechanism-dependent inhibition. Examples are also shown for troleandomycin, which exhibits time-dependent inhibition, and ketoconazole, which does not.

Irreversible inhibition is also diagnosed by dialysis or gel filtration. If inhibition is eliminated by dialysis or filtration of the small molecule inhibitor away from the enzyme, then the inhibition is reversible. However, continuation of inhibition after dialysis is evidence of irreversible inhibition.

Irreversible inhibition also may be diagnosed by preincubating the isozymes with inhibitor and cofactor (i.e., NADPH) and then incubating with the test compound. If the inhibitor or inhibitor metabolite is irreversibly bound to the isozymes, then IC_{50} will be lower (apparently more inhibitory) with preincubation versus no preincubation.

The consequences of irreversible inhibition can vary with the conditions. They depend on what fraction of the isozyme is irreversibly inactivated and the replenishment rate of the isozymes.

15.6   Other DDI Issues

The discussion thus far has focused on the important issue of discovery candidates acting as perpetrators by inhibition of the metabolism of other drugs (victims). There are other mechanisms by which a new clinical candidate emerging from drug discovery might have DDI with an existing commercial drug. These mechanisms include (1) the candidate as a victim to a perpetrator commercial drug, (2) the candidate as a victim or perpetrator of DDI at a transporter, and (3) the candidate as a victim or perpetrator of metabolic enzyme induction.

15.6.1   Candidate as Victim to a Metabolism Inhibition Perpetrator

Some drugs in clinical application might act as a perpetrator and inhibit a key metabolic enzyme for the candidate (victim), causing the candidate's concentration to increase to toxic levels. One way to avoid this is to select a candidate that is metabolized by multiple metabolic isozymes. For such candidates, if a perpetrator inhibits one metabolic pathway, then other pathways will take on a greater portion of the candidate's metabolic clearance. For candidates that are cleared primarily by one isozyme, inhibition of that isozyme likely will lead to an increase in the candidate's concentration. For this reason, the candidate's metabolic stability at specific CYP isozymes and other metabolic enzymes are tested during drug discovery in a process called *metabolic phenotyping* (see Chapter 29). It is preferable for candidates to be metabolized by multiple isozyme pathways. One isozyme should not metabolize more than 50% of the candidate.[1,21] Interestingly, coadministration of an inhibitor also has been purposely performed in the clinic to enhance the blood concentration of drug that is expensive or in short supply.

15.6.2 Candidate as a Victim or Perpetrator at a Transporter

In addition to metabolic enzymes, DDI can occur at transporters. For example, if a candidate for a peripheral disease is a P-glycoprotein (Pgp) substrate, coadministration of a drug that inhibits Pgp can increase the candidate's exposure in protected tissues (e.g., brain, placenta) by inhibiting Pgp at the blood–tissue barrier. This also may cause increased absorption by inhibiting Pgp in the intestine, or it may reduce clearance by inhibiting Pgp active secretion in the nephron or bile canaliculus.

15.6.3 Candidate as a Victim or Perpetrator of Metabolic Enzyme Induction

Some candidates or drugs induce the production of metabolic enzymes and cause DDI. The candidate drug might be repeatedly coadministered or closely after a drug that induces a metabolic enzyme that clears a majority of the candidate. In this case, the candidate will be the victim of elevated enzyme levels, resulting in an enhanced rate of candidate metabolism and lower candidate exposure. This can reduce the pharmacological effect of the candidate, and the disease will not be treated with a concentration of candidate that produces the desired pharmacological effect. A candidate that is primarily metabolized by one enzyme should not be coadministered with a drug that induces that enzyme. Conversely, the candidate may be a metabolic enzyme inducer and perpetrate the enhanced metabolism of another (victim) drug. Such candidates often will fail to advance further in the development process.

▌ Problems

(Answers can be found in Appendix I at the end of the book.)

1. For initial CYP inhibition screening, a useful goal is an IC_{50} greater than what concentration?

2. For human studies, K_i should be what? At what concentration is there likely to be CYP inhibition?

3. Why was Seldane removed from the market?

4. What is the difference between reversible and mechanism-based CYP inhibition? How can you distinguish these mechanisms?

5. How might you modify the following structure to reduce CYP inhibition?

MW = 333.5
PSA = 29
cLogP = 6.0

6. What is the risk associated with CYP inhibition?: (a) a coadministered drug is metabolized too quickly, (b) a compound is not stable, (c) a coadministered drug is not metabolized quickly enough, (d) an isozyme may be induced.

7. Should CYP inhibition be used to estimate metabolic stability?

References

1. Rodrigues, A. D. (2002). *Drug-Drug Interactions*. New York: Marcel Dekker.

2. Kunze, K. L., & Trager, W. F. (1996). Warfarin-Fluconazole. III. A rational approach to management of a metabolically based drug interaction. *Drug Metabolism and Disposition, 24*, 429–435.

3. Shimada, T., Yamazaki, H., Mimura, M., Inui, Y., & Guengerich, F. P. (1994). Interindividual variations in human liver cytochrome P-450 enzymes involved in the oxidation of drugs, carcinogens and toxic chemicals: studies with liver microsomes of 30 Japanese and 30 Caucasians. *Journal of Pharmacology and Experimental Therapeutics, 270*, 414–423.

4. Yan, Z., & Caldwell, G. W. (2001). Metabolism profiling, and cytochrome P450 inhibition & induction in drug discovery. *Current Topics in Medicinal Chemistry, 1*, 403–425.

5. Zlokarnik, G., Grootenhuis, P. D. J., & Watson, J. B. (2005). High throughput P450 inhibition screens in early drug discovery. *Drug Discovery Today, 10*, 1443–1450.

6. Bjornsson, T. D., Callaghan, J. T., Einolf, H. J., Fischer, V., Gan, L., Grimm, S., et al. (2003). The conduct of in vitro and in vivo drug-drug interaction studies: a pharmaceutical and manufacturers of America (PhRMA) perspective. *Drug Metabolism and Disposition, 31*, 815-832.

7. Obach, R. S., Walsky, R. L., Venkatakrishnan, K., Gaman, E. A., Houston, J. B., & Tremaine, L. M. (2006). The utility of in vitro cytochrome P450 inhibition data in the prediction of drug-drug interactions. *Journal of Pharmacology and Experimental Therapeutics, 316*, 336–348.

8. Honig, P. K., Wortham, D. C., Zamani, K., Conner, D. P., Mullin, J. C., & Cantilena, L. R. (1993). Terfenadine-ketoconazole interaction. Pharmacokinetic and electrocardiographic consequences. *Journal of the American Medical Association, 269*, 1513–1518.

9. Sedman, A. J. (1984). Cimetidine-drug interactions. *American Journal of Medicine, 76*, 109–114.

10. Jou, M. J., Huang, S. C., Kiang, F. M., Lai, M. Y., & Chao, P. D. (1997). Comparison of the effects of cimetidine and ranitidine on the pharmacokinetics of disopyramide in man. *Journal of Pharmacy and Pharmacology, 49*, 1072–1075.

11. Riley, R. J., Parker, A. J., Trigg, S., & Manners, C. N. (2001). Development of a generalized, quantitative physicochemical model of CYP3A4 inhibition for use in early drug discovery. *Pharmaceutical Research, 18*, 652–655.

12. Tassaneeyakul, W., Birkett, D. J., Veronese, M. E., McManus, M. E., Tukey, R. H., Quattrochi, L. C., et al. (1993). Specificity of substrate and inhibitor probes for human cytochromes P450 1A1 and 1A2. *Journal of Pharmacology and Experimental Therapeutics, 265*, 401–407.

13. Boruchoff, S. E., Sturgill, M. G., Grasing, K. W., Seibold, J. R., McCrea, J., Winchell, G. A., et al. (2000). The steady-state disposition of indinavir is not altered by the concomitant administration of clarithromycin. *Clinical Pharmacology and Therapeutics, 67*, 351–359.

14. Transon, C., Leemann, T., & Dayer, P. (1996). In vitro comparative inhibition profiles of major human drug metabolizing cytochrome P450 isoenzymes (CYP2C9, CYP2D6 and CYP3A4) by HMG-Co, A., reductase inhibitors. *European Journal of Clinical Pharmacology, 50*, 209–215.

15. Otton, S. V., Crewe, H. K., Lennard, M. S., Tucker, G. T., & Woods, H. F. (1988). Use of quinidine inhibition to define the role of the sparteine/debrisoquine cytochrome P450 in metoprolol oxidation by human liver microsomes. *Journal of Pharmacology and Experimental Therapeutics, 247*, 242–247.

16. Revesz, L., Di Padova, F. E., Buhl, T., Feifel, R., Gram, H., Hiestand, P., et al. (2000). SAR of 4-hydroxypiperidine and hydroxyalkyl substituted heterocycles as novel p. 38 Map kinase inhibitors. *Bioorganic & Medicinal Chemistry Letters, 10*, 1261–1264.

17. Ashwell, M. A., Lapierre, J.-M., Kaplan, A., Li, J., Marr, C., & Yuan, J. (2004). The design, preparation and SAR of novel small molecule sodium (Na+) channel blockers. *Bioorganic & Medicinal Chemistry Letters, 14*, 2025–2030.

18. Biller, S., Custer, L., Dickinson, K. E., Durham, S. K., Gavai, A. V., Hamann, L. G., et al. (2004). In R. T. Borchardt, E. H. Kerns, C. A. Lipinski, D. R. Thakker, & B. Wang (Eds.), The Challenge of Quality in Candidate Optimization. *Pharmaceutical profiling in drug discovery for lead selection* (pp. 413–429). Arlington, VA: AAPS Press.

19. Fontana, E., Dansette, P. M., & Poli, S. M. (2005). Cytochrome P 450 enzymes mechanism based inhibitors: common sub-structures and reactivity. *Current Drug Metabolism, 6*, 413–454.

20. Wu, Y.-J., Davis, C. D., Dworetzky, S., Fitzpatrick, W. C., Harden, D., He, H., et al. (2003). Fluorine substitution can block CYP3A4 metabolism-dependent inhibition: Identification of (S)-N-[1-(4-fluoro-3-morpholin-4-ylphenyl)ethyl]-3-(4-fluorophenyl)acrylamide as an orally bioavailable KCNQ2 opener devoid of CYP3A4 metabolism-dependent inhibition. *Journal of Medicinal Chemistry, 46*, 3778–3781.

21. Obach, S. (2007). The kinetics and pharmacokinetics of drug interactions: induction, inhibition (victims and perpetrators). In *MARM2007*. Collegeville, PA.

22. Clarke, S. E., & Jones, B. C. (Eds.). (2002). *Human cytochromes P450 and their role in metabolism-based drug-drug interactions*. New York: Marcel Dekker.

hERG Blocking

Overview

▶ *Certain compounds block the cardiac K^+ (hERG) ion channel and induce arrhythmia.*

▶ *The safety margin for hERG is $IC_{50}/C_{max, unbound} > 30$.*

▶ *hERG blocking might be decreased by reducing the basicity, reducing lipophilicity, and removing oxygen H-bond acceptors.*

hERG has rapidly emerged as an important safety issue in drug discovery. Awareness about the details of hERG and its impact are growing among discovery scientists. Recent efforts have quickly integrated assessment of hERG risk and solutions into discovery.

hERG is a gene that codes for a cardiac potassium ion channel. If this channel is blocked, a mechanism is initiated that can lead to cardiac arrhythmia. This arrhythmia proceeds to fatality in a small portion of the patient population. In the past, such arrhythmias were observed only after the drug was approved by the Food and Drug Administration (FDA) and was used by a large population of patients. Since the mechanism of this arrhythmia was elucidated, the FDA has carefully reviewed new drug applications for this potential problem. FDA approval or continued clinical use of a drug for a low-risk medical condition (e.g., allergy) is hard to justify if the drug may cause arrhythmia. Drug candidates that are hERG blockers require large clinical trials with many patients in order to demonstrate safety, because arrhythmia caused by hERG channel blockage is a rare event. Thus, the cost of clinical development is elevated if a compound has a potential hERG liability. Therefore, drug companies study this potential problem during discovery. In recent years, hERG blocking has been one of the leading causes for withdrawal from the market of drugs approved by the FDA. Other drugs with this problem have remained on the market, but their use has been severely restricted. Examples of these drugs are shown in Figure 16.1.

16.1 hERG Fundamentals

The full name of the hERG gene is "human ether-a-go-go related gene." The protein product of hERG is the inner pore-forming portion of a critical membrane bound potassium (K^+) channel in heart muscle tissue. It forms a tetramer, with each monomer having six transmembrane regions. It is controlled by voltage (membrane potential) and gates the flow of K^+ ions out of the cell. Movement of K^+ ions across the cell membrane creates the rapidly activating delayed rectifier K^+ current called I_{Kr}.

The potassium channel is part of the ensemble of ion channels that creates the cardiac action potential at the cellular level (Figure 16.2, *A*). The action potential is initiated with the opening of sodium (Na^+) channels. Na^+ ions flow quickly into the cell, causing rapid depolarization of the membrane potential from a resting state of about -90 mV to about

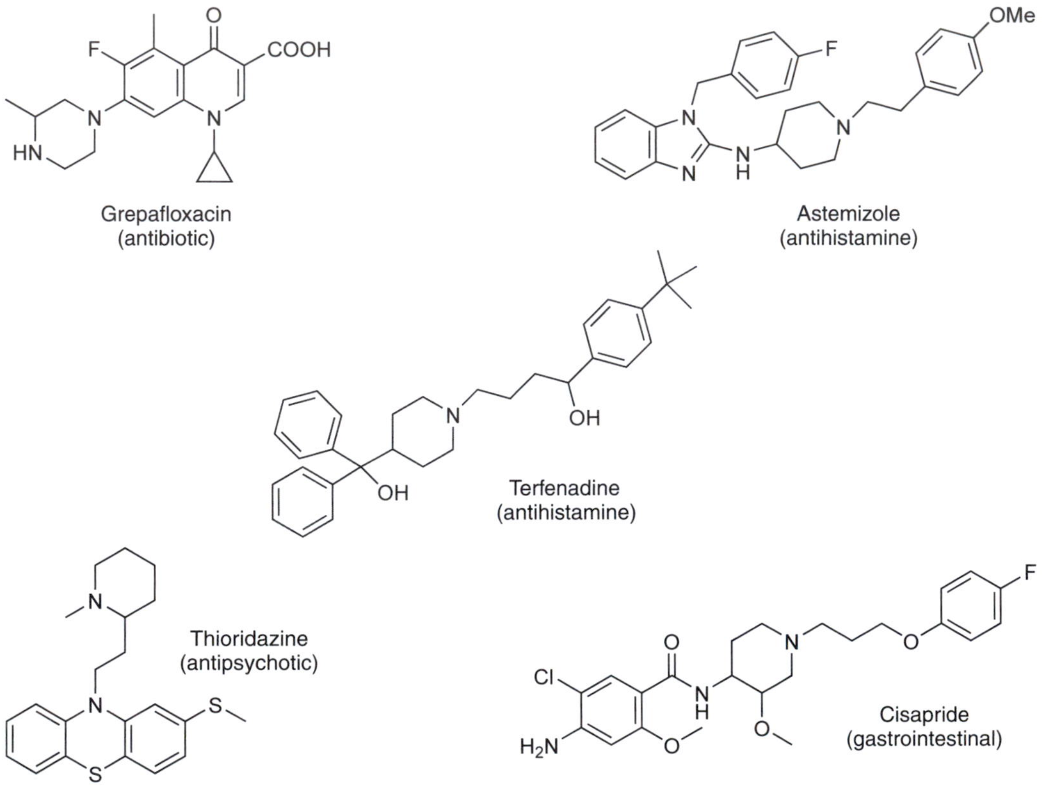

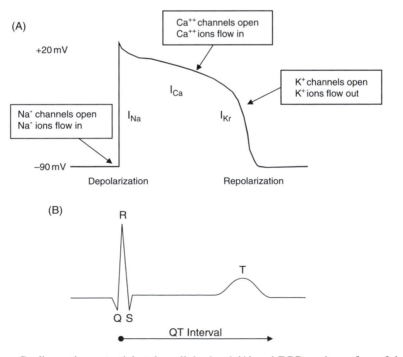

Grepafloxacin
(antibiotic)

Astemizole
(antihistamine)

Terfenadine
(antihistamine)

Thioridazine
(antipsychotic)

Cisapride
(gastrointestinal)

Figure 16.1 ► Commercial drugs that were withdrawn or had major labeling restrictions due to hERG blocking.

(A)

+20 mV

Ca++ channels open
Ca++ ions flow in

K+ channels open
K+ ions flow out

I_{Ca}

I_{Na}

I_{Kr}

Na+ channels open
Na+ ions flow in

−90 mV

Depolarization   Repolarization

(B)

R

T

Q S

QT Interval

Figure 16.2 ► Cardiac action potential at the cellular level (**A**) and ECG on the surface of the heart (**B**).

+20 mV (voltage inside the cell compared to outside). This depolarization is maintained by subsequent opening of calcium (Ca^{2+}) ion channels, allowing Ca^{2+} ions to flow into the cell. Repolarization to −90 mV occurs by opening of the potassium ion channels, allowing K^+ ions to move out of the cells. The hERG channel is the most important potassium channel for repolarization.

This action potential contributes to the overall electrical activity of the heart, which is measured using an electrocardiogram (ECG) on the surface of the heart tissue (Figure 16.2, *B*). On the ECG, the time from point Q to point T is called the *QT interval* (from depolarization to repolarization). A change in the action potential will change the ECG.

16.2 hERG Blocking Effects

If a compound binds within the hERG K^+ channel, it can obstruct the flow of K^+ ions out of the cell. This causes a slower outflow of K^+ ions, thus lengthening the time required to repolarize the cell (Figure 16.3). From the ECG, it can be seen that the T event is delayed, thus lengthening the QT interval (long QT [LQT]). LQT may trigger life-threatening torsades de pointes (TdP) arrhythmia (Figure 16.4). Although hERG blocking is a triggering factor for TdP, other physiological and genetic factors also increase the chances of LQT.[1] These factors include low serum K^+, slow heart rate, genetic factors (e.g., mutations affecting other ion channels), other cardiac conditions, coadministered drugs that also block hERG, coadministered drugs that inhibit metabolism (e.g., terfenadine), and gender. Some patients can have LQT without progressing to TdP, whereas others progress to TdP with only slight lengthening of the QT interval. TdP arrhythmia leads to ventricular fibrillation, which can cause sudden death. The involvement of the hERG channel in LQT is further supported by a naturally occurring inherited mutation in hERG that leads to LQT, TdP, and ventricular fibrillation.

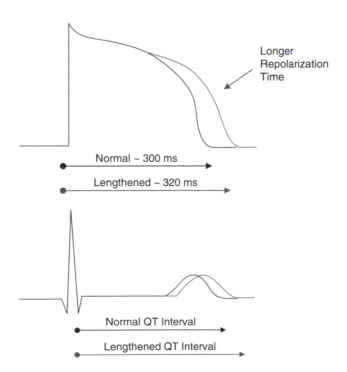

Figure 16.3 ▶ hERG potassium channel blocking lengthens the time until repolarization, resulting in LQT.

Figure 16.4 ▶ Torsades de pointes arrhythmia ECG.

The number of drugs that induce TdP is estimated to be higher than the number that induce more rare arrhythmias. TdP was reported to occur in patients taking quinidine (1%–3%), sotalol (1%–5%), dofetilide (1%–5%), and ibutilide (12.5%).[2] Arrhythmia was produced in $1:10^5$ to 10^6 persons taking antihistamines.[1] The incidence of arrhythmia was about 1:50,000 for patients taking terfenadine.[3]

Most drugs that cause LQT and trigger TdP are hERG blockers. However, not all hERG blockers produce TdP. This makes prediction and risk assessment difficult. A three-step approach to assessment of hERG blocking and TdP-induced arrhythmia often is used during drug discovery: (1) in silico structural alert, (2) in vitro hERG channel-blocking screening using a patch-clamp method, and (3) in vivo ECG. Methods for measuring hERG blocking and studying TdP are discussed in Chapter 34.

The safety margin for hERG blocking typically is evaluated as the ratio between the hERG IC_{50} (half maximal inhibition concentration) and $C_{max,unbound}$ (maximal human blood concentration of drug not bound to plasma proteins). A large value (often cited as >30) for the ratio hERG $IC_{50}/C_{max,unbound}$ is recommended. This is based on the experimental observation that for compounds with a ratio <30, 95% produce TdP and only 5% do not, whereas for a ratio >30, 15% produce TdP and 85% do not produce. Another safety margin is the time of QT interval lengthening. Regulatory agencies express major concern about the drug candidate when LQT exceeds an additional 5 ms compared to normal.[4]

Compounds that prolong the QT interval by >20 ms may be approvable if they are for important therapeutic indications, there is a strong benefit compared to the risk, or there is a feasible way to manage the risk. However, these drugs require special restrictive labeling, and their commercial impact is reduced. They may consume considerable resources during development and still not be approved. A human QT study is likely to be required during clinical development and is expensive.[4]

As with many other types of property data, a significant hERG response is not the sole criterion for terminating a compound. hERG-blocking data should be used in combination with all other data and project considerations in a holistic decision-making process.

16.3 hERG Blocking Structure–Activity Relationship

The amino acid residues in the hERG K^+ channel to which blocking drugs bind have been studied by means of single-site mutations.[1] Binding has been isolated to the central cavity of the channel, with Tyr652 and Phe656 of the pore helix being most important in binding. π-stacking appears to occur if the drug has an aromatic group. Alternately, interaction can occur with nonaromatic hydrophobic substructures in the drug. Cation–π interaction occurs between a basic nitrogen in the drug and Tyr652. Interaction with drugs appears to occur mostly when the K^+ channel is in the open position. Trapping of a drug molecule in the central cavity appears to be enhanced in the hERG channel, compared to other ion channels, by certain sequence differences that increase the hERG channel cavity volume, even in the closed state.

Studies agree on several structural features that are common to binding in the hERG channel:

▶ A basic amine (positively ionizable, pK_a >7.3)

▶ Hydrophobic/lipophilic substructure(s) (ClogP >3.7)

▶ Absence of negatively ionizable groups

▶ Absence of oxygen H-bond acceptors

In one hERG structure–activity relationship (SAR) model, the basic nitrogen is the top of a pyramid, with three or four hydrophobic substructures at the other loci, thus forming a plug of the channel.[1] hERG SAR is an active area of research, and increased data are expected to improve the SAR understanding for hERG.

16.4   Structure Modification Strategies for hERG

Initial suggestions for improving the structures of discovery leads to reduce hERG blocking follow the indications of the SAR and are listed in Table 16.1.

TABLE 16.1 ▶   Structure Modifications to Reduce hERG Blocking

Structure Modification Strategy
Reduce the pK_a (basicity) of the amine
Reduce the lipophilicity and number of substructures in the binding region
Add acid moiety
Add oxygen H-bond acceptors
Rigidify linkers

hERG molecular models continue to be improved. These guide structural modifications at points of interaction of the compound and the inner surface of the hERG channel.

Problems

(Answers can be found in Appendix I at the end of the book.)

1. hERG is the gene for what protein?

2. What is the function of the hERG protein?

3. What is LQT?

4. What is TdP?

5. How common in the population is TdP that is triggered by LQT?

6. What safety margin can be used in drug discovery for hERG blocking?

7. Where do most hERG blocking drugs bind?: (a) ATP binding site, (b) hinge region, (c) within the channel cavity, (d) at the allosteric site.

8. Which of the following structural features are favorable toward hERG blocking?: (a) low lipophilicity, (b) carboxylic acid, (c) secondary amine, (d) lipophilic moiety, (e) oxygen H-bond acceptors.

9. What structural modifications might be tried to reduce hERG blocking of the following structure?

MW = 310
PSA = 24
cLogP = 4.9

10. Compounds that cause hERG blocking are at risk for causing which of the following?: (a) K^+ channel opening, (b) myocardial infarction, (c) arrhythmia, (d) metabolic inhibition, (e) QT interval shortening.

References

1. Sanguinetti, M. C., & Mitcheson, J. S. (2005). Predicting drug-hERG channel interactions that cause acquired long QT syndrome. *Trends in Pharmacological Science*, 26:119–124.

2. Dorn, A., Hermann, F., Ebneth, A., Bothmann, H., Trube, G., Christensen, K., et al. (2005). Evaluation of a high-throughput fluorescence assay method for HERG potassium channel inhibition. *Journal of Biomolecular Screening*, 10:339–347.

3. Honig, P. K., Wortham, D. C., Zamani, K., Conner, D. P., Mullin, J. C., & Cantilena, L. R. (1993) Terfenadine-ketoconazole interaction. Pharmacokinetic and electrocardiographic consequences. *Journal of the American Medical Association*, 269:1513–1518.

4. Levesque, P. (2004) Predicting drug-induced qt interval prolongation. In *American Chemical Society, Middle Atlantic Regional Meeting*, 2004, Piscataway, NJ.

Additional Reading

1. Vaz, R. J., Li, Y., & Rampe, D. (2005). Human ether-a-go-go related gene (HERG): a chemist's perspective. *Progress in Medicinal Chemistry*, *43*, 1–18.

2. Aronov, A. M. (2005). Predictive *in silico* modeling for hERG channel blockers. *Drug Discovery Today*, 10:149–155.

3. Finlayson, K. (2004). Acquired QT interval prolongation and HERG: implications for drug discovery and development. *European Journal of Pharmacology*, 500:129–142.

4. Redfern, W. S (2003). Relationship between preclinical cardiac electrophysiology, clinical QT interval prolongation and torsade de pointes for a broad range of drugs: evidence for a provisional safety margin in drug development. *Cardiovascular Research*, 58:32–45.

Chapter 17

Toxicity

Overview

▶ *Toxicity remains a significant cause of attrition during development.*

▶ *Many toxic outcomes are possible, including carcinogenicity, teratogenicity, reproductive toxicity, cytotoxicity, and phospholipidosis.*

▶ *Toxic mechanisms include reactive metabolites, gene induction, mutagenicity, oxidative stress, and autoimmune response.*

▶ *The safety window is the concentration range between efficacious response and toxic response.*

Safety is one of the highest goals of pharmaceutical companies. Drugs provide a means of improving the quality and length of patients' lives, but they must minimize possible deleterious side effects and harm. Toxicology studies assess potential toxicities of a drug candidate and determine how drug therapy can be managed to minimize patient risk. As with the pharmacologically beneficial effects of drugs, toxic effects follow a dose–response relationship. Therefore, it is necessary to maximize the safety window (therapeutic index) between efficacy and toxicity (Figure 17.1).

In past years, toxicity studies were performed only during the clinical development phase. However, toxicity remains a major cause of drug candidate attrition during preclinical and clinical development (see Figure 2.4). The Kola-Landis study documented 20% to 30% attrition due to toxicity and clinical safety. A KMR study reported 44% toxicity attrition.[1]

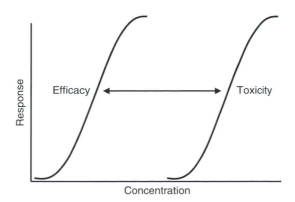

Figure 17.1 ▶ It is important to maximize the safety window (therapeutic index) of a drug candidate.

Moreover, many drugs have been withdrawn from the market due to human toxicity observed during clinical use in the wider population.

All companies perform Food and Drug Administration (FDA)-prescribed in vivo toxicity studies prior to clinical studies. Most companies also have addressed key safety issues, such as cytochrome P450 (CYP) inhibition and hERG blocking, with in vitro screens during drug discovery. However, these are only two of the potential toxicity challenges that drugs encounter. As with other absorption, distribution, metabolism, excretion, and toxicity (ADME/Tox properties), early screening and correction of toxicity issues can reduce preclinical and clinical attrition. This also can improve the efficiency of drug discovery by reducing the time and effort spent on lead series that are terminated later due to toxicity. In silico and in vitro assays can indicate potential toxicities during discovery. Another opportunity for improving efficiency is the early validation of each discovery target for target-based toxicity associated with modulating the target.[2]

Toxicity studies and safety optimization in discovery follow the same process as other ADME/Tox properties. In silico tools and in vitro higher-throughput screens indicate potential toxicity mechanisms. If performed early, these data are used as part of lead selection. Indications of toxicity from screens are followed with advanced studies that provide definitive data. The data are used to prioritize compounds and lead series and to guide structure optimization to reduced toxicity during discovery. Finally, toxicity data usually are included in the data package that is used to select clinical candidates.

Toxicity is a very broad topic. The purpose of this chapter is to introduce discovery scientists and students to key toxicity concepts and terms. Examples of chemical mechanisms of toxicity are described. The reader is directed to reviews on toxicity and mechanisms for in-depth study.[3–5] Methods for toxicity assessment during drug discovery are discussed in Chapter 35.

17.1 Toxicity Fundamentals

Data from standard animal toxicity studies must be obtained for clinical candidates prior to investigational new drug (IND) filing. Animal studies are designed with consideration to the expected dosing plan for humans so as to distinguish chronic effects (regular dosing over extended time periods) from acute effects (single dose). The *maximum tolerated dose* (MTD), the maximum dose at which no toxic effects are observed, is determined, and dosing proceeds up to a level that produces a toxic effect that can be used as a marker during first in human phase I studies. Animal studies provide clinical researchers with the basis for estimating the safety window (*therapeutic index*), which is the range between the concentration that is predicted to cause human toxicity and the concentration that is predicted to produce human efficacy. For example, it may be determined as the dose at which 50% of individual animals demonstrate a particular toxic response (e.g., death, tumor formation, side effects; i.e., LD_{50}) divided by the dose at which 50% of individual animals demonstrate an efficacious response (i.e., ED_{50}). A large safety window improves the chances of the drug being safe in the clinic. Investigators must evaluate the benefit versus risk to patients of new chemical entities. Pharmacokinetic differences between animals and humans are taken into consideration. In many cases, some toxicity is manageable if the drug provides important benefits to patients. Toxic effects may be reversible by the body (e.g., liver damage) or they may be permanent (e.g., cancer, brain damage, teratogenicity). Differences in absorption, distribution, metabolism, and elimination between people are better tolerated with a wider window. Toxicity may be observed during phase I human studies that was not observed in animals during preclinical safety studies.

17.1.1 Toxicity Terms and Mechanisms

Toxicity can be caused by many different mechanisms. The biochemical mechanism that initiates toxicity might be related to the following:

▶ Therapeutic target

▶ Off-target

▶ Reactive metabolic intermediates

Unintended effects of modulating the therapeutic target can be toxic. These are referred to as *target-* or *mechanism-based toxicity*. The compound also can affect other enzymes, receptors, or ion channels in the body; these are referred to as *off-target*. The effects of target or off-target modulation can be observed as pharmacological (functional effect), pathological (lethal effect), or carcinogenic (producing cancer). Reactive metabolites produce undesirable effects by covalent reaction with endogenous macromolecules, which can lead to cell death, carcinogenicity, or immunotoxicity.

Some of the safety issues and terms of interest in drug discovery are as follows (with examples):

Acute toxicity is the toxic response from a single dose. *Chronic* toxicity is the toxic response from long term dosing.

Carcinogenicity is the production of cancer. There is often a long latency period from compound exposure to tumor appearance.

Cytotoxicity is the production of cell death, such as in liver cells (hepatotoxicity).

Genotoxicity is a mutagenic change of the DNA sequence or chromosome damage. It often leads to cancer.

Idiosyncratic reactions are immune-mediated effects having delayed onset and are caused by reactive metabolites. Proposed mechanisms involve cell damage and individual patient conditions, such as enzyme irregularities or polymorphism.

Immunotoxicity is an immune reaction (e.g., lymphoproliferation, autoimmunity) triggered by the compound (e.g., penicillin) or reaction of a metabolite with endogenous macromolecules to form an antigen.

Organ toxicity deteriorates the function of the organ (e.g., heart, liver, kidney, blood cells, brain).

Phospholipidosis is the accumulation of polar phospholipids in lysosomes as lamellated bodies in response to cationic amphiphilic drugs.

Reproductive toxicology is a change in reproductive systems, normal function, or sexual behavior.

Safety pharmacology studies the effects of a compound on normal physiological functions. Abnormal function is often caused by inhibition of enzymes or antagonism or agonism of receptors or ion channels.

Teratogenicity is embryo toxicity or abnormal development. (The effect of thalidomide on human embryo development is well known.)

17.1.2 Toxicity Mechanisms

17.1.2.1 Therapeutic Target Effects

Modulation of the therapeutic target can have undesirable effects in the target tissue or elsewhere in the body. These effects are observed in animal dosing studies or reports from human clinical use.

17.1.2.2 Off-Target Effects

The lead compound or series might inhibit an unintended enzyme or receptor in the body. This disrupts the normal function and can cause toxicity or side effects. If these are not severe and can be tolerated by the patient, they may be manageable in clinical use of the drug.

One example of off-target effects is drug–drug interaction (DDI), where a drug interferes with the body's absorption, distribution, metabolism, or elimination of a second coadministered drug. DDI can cause a buildup of the second compound to a concentration that causes toxic effects. This is discussed in Chapter 15. DDI also can occur at other metabolic enzymes and at transporters involved in elimination of a drug by the liver or kidney.

Inhibition can occur at a protein that is unrelated to the compound's intended mechanism. For example, inhibition of hERG is discussed in Chapter 16. Compounds having certain physicochemical and structural features can bind in the hERG K^+ ion channel and prevent influx of K^+ ions, resulting in lengthening of the QT interval, which may trigger a fatal arrhythmia.

Just as the lead compound has agonist or antagonist activity at the project's target, it also can have activity at another target in the body. The effects may sometimes be helpful, as other potential positive uses are discovered for the compound. However, the effects can disrupt a biochemical biosynthesis or signaling pathway.

Off-target effects usually are screened in drug discovery by sending the compound to an outside laboratory (e.g., Novascreen, MDS PanLabs). Such laboratories have an array of biochemical assays for diverse targets against which the compound can be rapidly screened for activity. Off-target effects also are revealed in animal dosing studies in which an array of normal physiological functions is monitored.

17.1.2.3 Reactive Metabolites

A major cause of drug toxicity is metabolic bioactivation.[4–11] In most cases, metabolism serves the important function of chemically modifying the compound to make it more polar and, thus, more readily cleared into the bile and urine. An unfortunate consequence of metabolism is that a reactive metabolite or intermediate can cause toxicity. The reactive intermediate or metabolite might be an electrophile, which covalently binds to endogenous nucleophiles (e.g., proteins, DNA). The resulting adduct does not function normally. It can cause DNA mutation, which can lead to cancer. Modifications of proteins can elicit an immune response. Often these reactions occur in the liver, leading to hepatotoxicity, but the reactive metabolites can also cause damage at distal sites. Glutathione is a ubiquitous molecule that acts as a scavenger of electrophiles. However, it does not capture all of the reactive metabolites, especially when the cell is in oxidative stress (see Section 17.1.2.5). Some drugs are activated to toxic or allergenic degradants by exposure to light in the skin (e.g., sulfonamides). Idiosyncratic drug reactions appear to be initiated by reactive metabolites.[6,12]

Excellent reviews are available on the numerous mechanisms of metabolic activation of diverse functional groups and structural classes.[4,5,12] Some compounds are activated to multiple metabolites, and proving which compound or intermediate is responsible for the toxic response can be quite difficult. Table 17.1 lists some substructures that have produced reactive metabolites resulting in toxicity. Medicinal chemists can use this type of information to plan lead series analog synthesis and anticipate issues that might arise if one of these substructures is used in lead series expansion.

TABLE 17.1 ► Partial List of Substructures that may Initiate Toxicity and Their Proposed Reactive Metabolites[4,5,7–10,16]a

Substructure	Proposed reactive metabolite
Aromatic amine	Hydroxyl amine, nitroso, quinone-imine (oxidative stress)
Hydroxyl amine	Nitroso (oxidative stress)
Aromatic nitro	Nitroso (oxidative stress)
Nitroso	Nitroso, diazonium ions (oxidative stress)
Alkyl halide	Acylhalide
Polycyclic aromatic	Epoxide
α,β-Unsaturated aldehyde	Michael acceptor
Carboxylic acid	Acyl glucuronide
Nitrogen-containing aromatic	Nitrenium
Bromo aromatic	Epoxide
Thiophene	S-oxide, epoxide
Hydrazine	Diazene, diazonium, carbenium ion
Hydroquinones	p-Benzoquinone
o- or p-Alkylphenols	o- or p-Quinone methide
Quinone	Quinone (oxidative stress)
Azo	Nitrenium
Furans	α,β-Unsaturated dicarbonyl
Pyrrole	Pyrrole oxide
Acetamide	Radical (oxidative stress)
Nitrogen mustard	Aziridium ions
Ethinyl	Ketene
Nitrosamine	Carbenium ion
Polyhalogenated	Radicals, carbene
Thioamide	Thiourea
Vinyl	Epoxide
Aliphatic amines	Iminium ion
Phenol	Quinine
Arylacetic, arylpropionic acids	
Imidazole	
Medium-chain fatty acids	

aThese have been termed *toxic alerts*.

Figures 17.2 through 17.6 are examples of substructures that undergo metabolic activation mechanisms that lead to toxicity. Aniline is oxidized by cytochrome P450 (CYP) to form hydroxyl amine and nitroso (Table 17.1). These electrophiles react with proteins to form adducts, which induce an immunologic response. One mechanism for this process is sulfation

Figure 17.2 ► Primary and secondary amines (especially aromatic amines) form an *N*-hydroxyl amine, which produces a reactive nitrenium intermediate and reacts with a nucleophile "X" (e.g., DNA) to cause genotoxicity.[4,10] Example: acetaminofluorene.

by sulfotransferase, followed by elimination to form the electrophilic nitrenium ion, which reacts with a nucleophile.[10] Hydroxyl amines also can be converted to quinones, etc., which cause oxidative stress.

Carboxylic acids can be glucuronidated in the liver by UDP-glucuronosyltransferase (Figure 17.3). The compound migrates around the glucuronic acid molecule, which then opens to form a reactive center and produces adducts to protein molecules.[13] Alternatively, the drug adducts to the protein following ring opening.

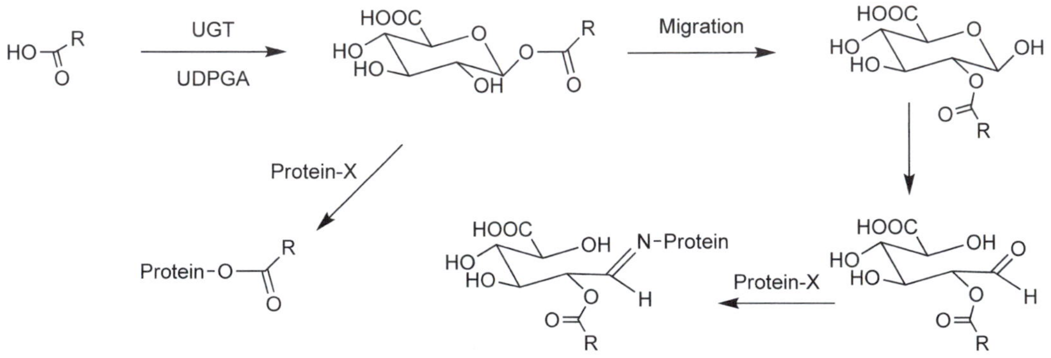

Figure 17.3 ▶ Carboxylic acids can be glucuronidated to form an acyl glucuronide, followed by rearrangement and reaction with a nucleophile (e.g., DNA, protein) to cause genotoxicity.[10] A proposed mechanism is shown (J. Wang, personal communication, 2006). X is $-NH_2$, $-OH$, or $-SH$. Examples: bromfenac, diclofenac.

Unsaturated bonds can be epoxidized by CYP (Figure 17.4). The epoxide readily reacts.[10] This mechanism can lead to adducting to DNA, which can cause mutation and trigger cancer.

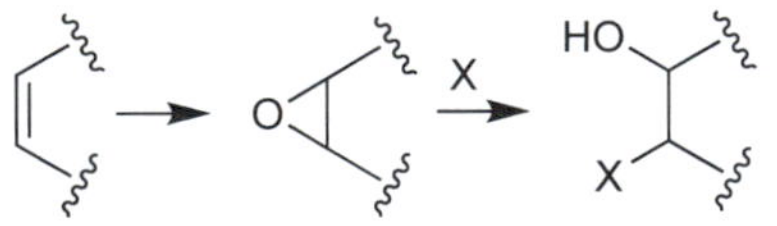

Figure 17.4 ▶ Epoxidation of unsaturated bonds leads to attachment to DNA and protein. Examples: aflatoxin B_1 with DNA (guanine), bromobenzene with protein (thiol groups).

Nitroaromatics are reduced to form reactive nitro radical, nitroso, nitroxyl radical, and aromatic N-oxide (Figure 17.5). These can induce oxidative stress (see 17.1.2.5). Thiophene is oxidized to sulfoxide, which can react with protein sulfhydryl groups (Figure 17.6).

The effects of reactive metabolites can be observed with in vivo dosing studies that include monitoring of normal physiological function and are followed by histopathological examination for abnormal tissue appearance. A large number of in vitro assays are available

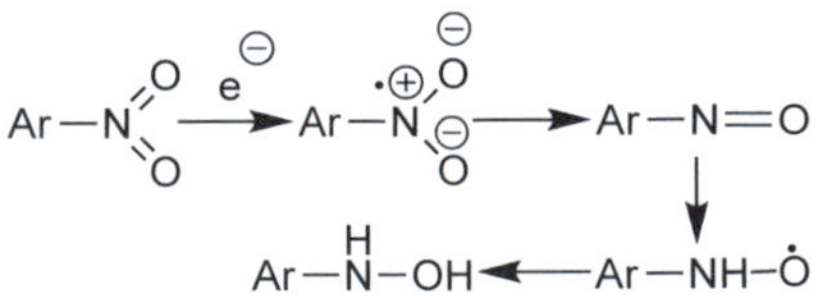

Figure 17.5 ▶ Nitroaromatics can be reduced to form several reactive intermediates: nitro radical, nitroso, nitroxyl radical, and aromatic N-oxide.

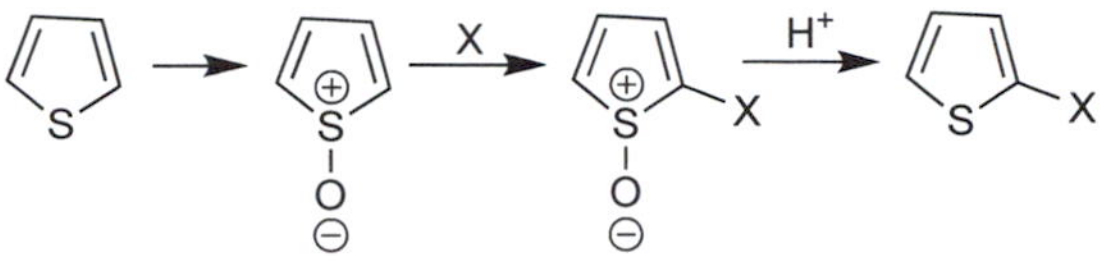

Figure 17.6 ▶ Thiophene oxidation is followed by attack on protein (sulfhydryl group) to produce covalent binding. Example: tienilic acid. Epoxide formation has also been proposed as a bioactivation mechanism.[4]

(see Chapter 35) to study the effects of mutation (e.g., Ames), chromosome damage (e.g., micronucleus), and cell toxicity (e.g., hepatotoxicity).

17.1.2.4 Gene Induction

A compound can induce the expression of a gene and its protein product. For example, a compound might activate pregnane X-receptor (PXR), which ultimately results in induction of higher levels of CYP3A4. The higher 3A4 levels result in faster metabolism of 3A4 substrates. Rifampicin can induce 3A4, which causes higher metabolism of coadministered oral contraceptives, resulting in their loss of efficacy.[14] Assays have been developed to indicate induction via PXR or aryl hydrocarbon receptor (AHR) activation (see Chapter 35).

17.1.2.5 Oxidative Stress

Some compounds undergo redox cycling (one electron reduction to the radical followed by reoxidation) within the cell and induce an oxidative environment through enhancing the process of oxygen reduction. Normally cells are maintained in a reducing environment by enzyme systems and glutathione. Oxidative stress results in the increase of free radicals and peroxides. These can abstract a hydrogen atom from lipids, glutathione, and DNA, resulting in cell damage and death. Examples of structures that induce oxidative stress are aromatic amines, aromatic nitros, and quinones.

17.2 Toxicity Case Studies

Matrix metalloproteinase (MMP) inhibitors have been studied clinically for treatment of degradation of collagen in arthritis, angiogenesis, and tumor growth. Unfortunately, MMP inhibitor candidates also cause musculoskeletal syndrome (MSS), a tendonitis-like fibromyalgia in humans. Studies have shown that the side effects are not caused by target-specific MMP-1 inhibition but more likely are due to nonselective inhibition of one or more other metalloproteinases. MMP inhibitor compounds that chelate zinc appear to be responsible. Research on new MMP inhibitors is focusing on nonchelators of zinc to reduce this off-target toxicity.[15]

Rofecoxib (Vioxx) was withdrawn from the market after a clinical study showed that patients receiving a 25-mg dose for 19 months had a 3.9-fold increase in thromboembolic adverse events (heart attack). The mechanism remains unclear[16] but may be related to the therapeutic target cyclooxygenase-2 (COX-2).

Troglitazone was withdrawn from the market in 2000 because of human hepatic failure. In vitro studies show that CYP3A4 oxidation of the chromane ring or the thiazolidinedione ring produces an electrophilic intermediate that covalently binds to protein.[4]

In high doses, acetaminophen causes liver damage. As shown in Figure 17.7, oxidation of acetaminophen by CYP produces *N*-acetyl-*p*-benzoquinoneimine (NAPQI), which reacts with nucleophiles, including sulfhydryl groups of proteins.

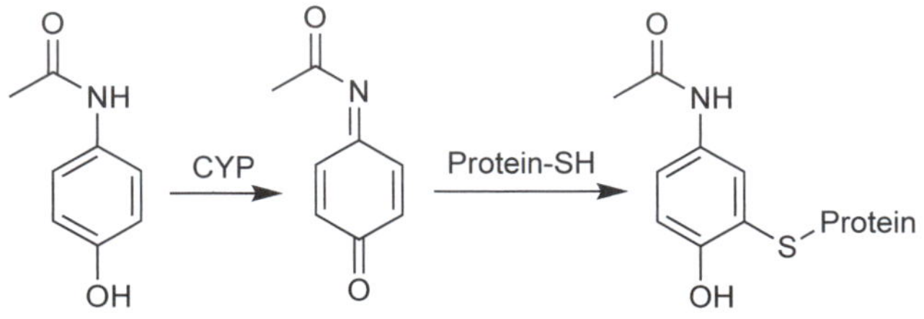

Figure 17.7 ▶ Proposed mechanism of acetaminophen metabolic activation and covalent binding to protein sulfhydryl groups.[3,4]

17.3 Structure Modification Strategies to Improve Safety

Medicinal chemists take several approaches to avoid or reduce toxicity during the discovery stage. The optimum time to obtain toxicity data for use in redesign of compound structures to reduce toxicity is during the active phase of analog synthesis in lead optimization. Structural strategies for toxicity include the following:

1. Avoid substructures that are known to induce toxic responses. During lead selection, compounds containing potentially toxic substructures (Table 17.1) can receive a lower priority.

2. Early synthetic modifications should be undertaken to remove any potentially toxic substructures from lead series.

3. Potentially toxic substructures should not be added to lead series structures during lead optimization. (If they are made for structure–activity relationship exploration purposes only, they may be acceptable.)

4. Perform reactive metabolite assays (as indicated in Chapter 35) to screen for potentially toxic compounds. In vitro assays range from trapping of glutathione conjugates to screening of DNA mutagenicity following S9 metabolic activation. Data indicating potential toxicity from these tests do not guarantee that toxicity will be observed in vivo, but they provide an early warning that can improve discovery efficiency.

5. Data indicating potential toxicity from in vitro assays can be further investigated by structure elucidation of the metabolites or trapped intermediate using spectroscopy. It can be difficult to definitely assign the toxicity to a specific metabolite; however, the data may point in a direction that can be followed as a possible hypothesis. Knowledge of reactive metabolite structures can suggest structural modifications to reduce metabolism.

6. Utilize the metabolite structural modification strategies discussed in Chapter 11 to attempt to reduce metabolic bioactivation.

Problems

(Answers can be found in Appendix I at the end of the book.)

1. Define therapeutic index (safety window). Is it preferable for this to be small or large?

2. How can reactive and unreactive metabolites be distinguished?

3. Which of the following moieties might themselves be or might form a reactive metabolite: a) aniline, b) thiophene, c) phenol, d) ester, e) Michael acceptor, f) propyl alcohol

4. Define oxidative stress.

5. How is gene induction toxic?

6. Why are off-target effects toxic?

References

1. KMR Group. (2003). *KMR benchmark survey* (pp. 98–101). Chicago, IL: KMR Group.

2. Car, B. D. (2006). Discovery approaches to screening toxicities of drug candidates. In *AAPS Conference: Critical issues in discovering quality clinical candidates.* American Society of Pharmaceutical Scientists, Philadelphia, PA, April 24–26, 2006.

3. Klaassen, C. D. Principles of toxicology and treatment of poisoning. In L. L. Brunton (Ed.). *Goodman and Gillman's the pharmacological basis of therapeutics* (11th ed.). New York: McGraw-Hill.

4. Macherey, A.-C., & Dansette, P. (2003). Chemical mechanisms of toxicity: basic knowledge for designing safer drugs. In C. G. Wermuth (Ed.). *The practice of medicinal chemistry* (2nd ed., pp. 545–560). Amsterdam: Elsevier Academic Press.

5. Kalgutkar, A. S., Gardner, I., Obach, R. S., Shaffer, C. L., Callegari, E., Henne, K. R., et al. (2005). A comprehensive listing of bioactivation pathways of organic functional groups. *Current Drug Metabolism, 6,* 161–225.

6. Uetrecht, J. (2003). Screening for the potential of a drug candidate to cause idiosyncratic drug reactions. *Drug Discovery Today 8,* 832–837.

7. Nassar, A.-E. F., Kamel, A. M., & Clarimont, C. (2004). Improving the decision-making process in structural modification of drug candidates: reducing toxicity. *Drug Discovery Today, 9,* 1055–1064.

8. Kazius, J., McGuire, R., & Bursi, R. (2005). Derivation and validation of toxicophores for mutagenicity prediction. *Journal of Medicinal Chemistry, 48,* 312–320.

9. Uetrecht, J. (2003). Bioactivation. In J. Lee, R. Obach and M. B. Fisher (Eds). Drug Metabolism Enzymes: Cytochrome P_{450} in Drug Discovery and Development. New York: Marcel Dekker.

10. Nelson, S. D. (2001). Molecular mechanisms of adverse drug reactions. *Current Therapeutic Research, 62,* 885–899.

11. Amacher, D. E. (2006). Reactive intermediates and the pathogenesis of adverse drug reactions: the toxicology perspective. *Current Drug Metabolism, 7,* 219–229.

12. Erve, J. C.L. (2006). Chemical toxicology: reactive intermediates and their role in pharmacology and toxicology. *Expert Opinion on Drug Metabolism & Toxicology, 2,* 923–946.

13. Georges, H., Jarecki, I., Netter, P., Magdalou, J., & Lapicque, F. (1999). Glycation of human serum albumin by acylglucuronides of nonsteroidal anti-inflammatory drugs of the series of phenylpropionates. *Life Sciences, 65,* PL151–PL156.

14. Li, A. P. (2001). Screening for human ADME/Tox drug properties in drug discovery. *Drug Discovery Today, 6,* 357–366.

15. Peterson, J. T. (2006). The importance of estimating the therapeutic index in the development of matrix metalloproteinase inhibitors. *Cardiovascular Research, 69,* 677–687.

16. Dogne, J.-M., Supuran, C. T., & Pratico, D. (2005). Adverse cardiovascular effects of the coxibs. *Journal of Medicinal Chemistry, 48,* 2251–2257.

Integrity and Purity

Overview

▶ *Data from testing a compound with the incorrect identity or low purity can affect the structure–activity relationship.*

▶ *The sample may have decomposed, been misidentified, or have significant impurities.*

The quality of materials used in drug discovery plays a crucial role in the success of discovery research. If there is an unknown problem with the quality of the material, the viability of the biological and absorption, distribution, metabolism, excretion, and toxicity (ADME/Tox) property studies will be in question or will confuse the discovery project team. For this reason, integrity and purity profiling of discovery compounds can be as important as other physicochemical and metabolic properties of the compounds.[1]

18.1 Fundamentals of Integrity and Purity

Integrity refers to whether the material is the same as the identity or structure that is recorded in the company's database. *Purity* refers to the percentage of the material that is the correct structure, as opposed to impurities in the material.

Medicinal chemists deal directly with integrity and purity for each new compound and batch that is synthesized. This is commonly accomplished using nuclear magnetic resonance (NMR) and mass spectrometry (MS) for structure verification and high-performance liquid chromatography (HPLC) for purity checking. Despite this early verification of the compound, there are many other stages of discovery where it is wise to requalify the materials.

18.2 Integrity and Purity Effects

Discovery experiments assume that test compounds have the correct identity and are relatively pure. Compounds come into discovery biology and ADME/Tox laboratories from many sources. It is prudent to verify the quality of compounds in order to produce an accurate structure–activity relationship (SAR).

The sources of test compounds include the following.

▶ Company repository provides the compound:

  ▶ Solid from a storage bottle

  ▶ Solution from a repository solution stock

  ▶ Solution from a well plate used for high-throughput screening (HTS)

▶ Company collaborator laboratory provides the compound as a solution from a well plate or vial that has been stored in the laboratory

▶ Partner research company provides the compound

▶ Purchased from an outside materials sourcing firm

▶ Solution or solid from a diversity array that was synthesized using parallel (combinatorial) synthesis

Each of these cases has the opportunity for integrity or purity errors. Of particular note are the age and storage conditions of the sample prior to reaching the assay laboratory, human errors in manual handling, and quality of the original structural and purity analysis. Discovery researchers should consider the quality of the material as a variable in the experiment.

The case of HTS is illustrative. For HTS, all the compounds from the corporate screening library are individually placed in solution and transferred to plates. These plates are stored and then retrieved and thawed for HTS runs. In this way, hundreds of thousands of structurally diverse compounds are biologically tested with the therapeutic target protein for activity. When activity is observed for a compound, this "hit" obtains greater interest by medicinal chemists as a possible lead. Verification of the hit identity and purity before further studies is valuable. Experience indicates that some compounds in screening libraries have inaccurate identity or low purity. Misidentification of the HTS hit compound can lead to the wrong SAR conclusions from the beginning of the project. An impurity in the solution can itself be active, whether or not the putative component is active. This can confuse and mislead the research team. In the same way, incorrect assignments of identity or poor purity of compounds from corporate repositories or external sources can lead to unproductive and time-consuming investigations.

In addition to problems with SAR, integrity and purity can affect assessment of the properties of compounds. Impurities can produce false-positive signals in property assay detection methods, such as UV plate readers, light scattering detectors, and fluorescence plate readers. An impurity could be the cause of a response in some assays, such as cytochrome P450 (CYP) inhibition. This will compromise the property results and lead to inaccurate structure–property relationships (SPR), which are used to select leads or improve the properties of leads.

There are several causes for problems with integrity and purity of some materials. Companies often collected compounds over many years. These compounds came from many origins. Some of the compounds were synthesized as long as 30 to 50 years ago and were stored under various conditions that were later found to be unsuitable. Compounds were collected from various companies that merged to form the present company. Compounds came from university laboratories and chemical companies that had various criteria for quality. They may have been synthesized as libraries by contract companies and never individually tested. Usually, the chemist who synthesized the compound is not accessible to the current discovery scientists, so no first-hand knowledge of the material is available.

Another source of error is misidentification. Mistakes in handling or labeling during transfer and weighing occur. Material can be cross-contaminated into another sample. In recent years, automated sample handling has reduced such errors, but manual handling is necessary for some steps.

Compounds may have been incorrectly characterized when they initially were placed in the collection. A spectrum may have been misinterpreted. The compound may have been synthesized before modern NMR and MS techniques were common. Significant levels of impurities may be present from starting materials or reaction by-products that went undetected.

Compounds may have decomposed. Storage in the solid form exposes the compound to oxygen and water from the air, which can react to form degradants. Counter-ions can react with the compound. Light and elevated temperature accelerate decomposition reactions.

Compounds can degrade in solutions that are stored in the laboratory prior to experiments. Water and oxygen from the air can dissolve in solutions. Cooling of solutions in the refrigerator can cause water from the air to condense and dissolve in the solution. Light from the laboratory can degrade certain compounds. Storage vials can catalyze reactions.

18.3 Applications of Integrity and Purity

It is useful for discovery scientists to check the purity and integrity of compounds at certain times. Here are a few examples of when verification is beneficial:

▶ Newly synthesized compounds, to assure that they are what was intended and that they have an acceptably low level of impurities

▶ Compounds that are being collected and formatted into plates for HTS screening, so that early discovery efforts in lead selection are not misled

▶ Compounds that were identified as potential actives from similarity searching of the corporate collection, so that crucial SAR assignments from a series are accurate

▶ HTS hits, to assure that they have not degraded, been misidentified, or been mislabeled

▶ Materials that are being approved for late discovery toxicology, pharmacology, or selectivity testing, to assure that the right compound is being tested and that any negative results (e.g., toxicity, activity at another receptor or enzyme) are not due to impurities

The different tasks require different methodologies, depending on the level of detail that is necessary to answer the research question and the cost that is reasonable at that point. In some cases, the integrity and purity examination may seem like overkill, but without it mistakes may be made if the compounds are assumed to be correct and pure.

18.3.1 Case Study

Popa-Burke et al.[2] provided an example of potential integrity and purity problems of compound libraries. Prior to placing compounds from vendors into a screening library, they were checked for identity using a rapid MS-based method. Approximately 10% of the compounds did not pass the identity check. Compounds from "historical" libraries had lower quality than recently synthesized compounds, suggesting potential degradation or initial misidentification. Furthermore, testing of compound solution concentration showed that the quantity of a compound could vary greatly from the intended concentration, thus compromising the IC_{50} measurements and rank ordering of compounds.

Without integrity and purity profiling, discovery experiments must proceed at risk. If compounds of poor integrity and purity slip into the project, the trends will be made less clear for concluding how the project should proceed.

Problems

(Answers can be found in Appendix I at the end of the book.)

1. What are the negative effects of low purity or inaccurate structural identity?

2. How might a sample have low purity or the wrong identity?

3. What techniques can be used for purity and integrity determination?: (a) HPLC, (b) NMR, (c) LC/MS.

4. Ensuring the integrity and purity of your HTS hits and project compounds does which of the following?: (a) ensures good drug-like properties, (b) avoids erroneous SAR.

References

1. Kerns, E. H., Di, L., Bourassa, J., Gross, J., Huang, N., Liu, H., et al. (2005). Integrity profiling of high throughput screening hits using LC-MS and related techniques. *Combinatorial Chemistry and High Throughput Screening*, 8, 459–466.

2. Popa-Burke, I .G., Issakova, O., Arroway, J. D., Bernasconi, P., Chen, M., Coudurier, L., et al. (2004). Streamlined system for purifying and quantifying a diverse library of compounds and the effect of compound concentration measurements on the accurate interpretation of biological assay results. *Analytical Chemistry*, 76, 7278–7287.

Pharmacokinetics

Overview

▶ *Pharmacokinetics (PK) studies the concentration time course of compounds and metabolites in vivo.*

▶ *Key parameters are volume of distribution, area under the curve, clearance, $t_{1/2}$, C_{max}, and bioavailability.*

▶ *PK data are heavily used by discovery teams and correlated to pharmacodynamics.*

19.1 Introduction to Pharmacokinetics

After a compound is administered, its concentration in the bloodstream and tissues changes with time, first increasing as it enters systemic circulation and then decreasing as it is distributed to tissues, metabolized, and eliminated. Pharmacokinetics (PK) is the study of the time course of compound and metabolite concentrations in the body.[1]

The PK parameters of a compound (e.g., clearance, half-life, volume of distribution) result from its physicochemical and biochemical properties (see Figure 2.1). These properties are determined by the structure of the compound and the physical and biochemical environment into which the compound is dosed. The dependence of PK parameters on these fundamental properties and on molecular structure is the reason we use property data to predict and improve the PK parameters of discovery compounds. The specific effects of individual physicochemical and biochemical properties on PK parameters are discussed in the chapters on each property and in Chapter 38.

When a compound is administered directly into the bloodstream by intravenous injection (IV), rapid circulation of the blood mixes the compound throughout the entire blood in minutes. Permeation of the compound from the blood capillaries into the tissues begins immediately and is called *distribution*. This rapidly reduces the concentration of compound in the blood during the distribution phase (Figure 19.1). The concentration of free compound in the tissues approaches equilibrium with the concentration of free compound in the blood.

At the same time, compound is being removed from the bloodstream, primarily by the liver and kidneys, in the process termed *elimination*. The elimination phase typically follows first-order kinetics. If the compound is rapidly eliminated, the rate of decrease during the elimination phase is high, and the compound concentration quickly approaches the baseline. As the free drug concentration in the blood drops, the compound permeates back out of the tissues and into the bloodstream.

The time course of compound concentrations is different with other routes of administration. When a compound is administered orally, it must first dissolve, permeate through the gastrointestinal membrane, and pass through the liver before it reaches systemic circulation. As a result, there is a time delay until the compound reaches its peak concentration in the blood (C_{max}). Once it reaches the bloodstream, it undergoes distribution to the tissues and

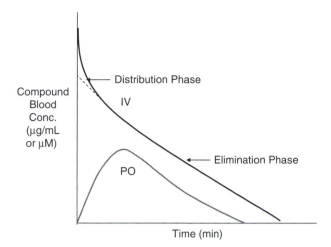

Figure 19.1 ▶ Hypothetical example of compound concentrations in blood over time following administration by intravenous (IV) and oral (PO) routes.

elimination. Compound exposure is indicated by the area under the time–concentration curve (area under the curve). Oral exposure usually is lower than IV exposure after dose normalization. This is because of the additional barriers (e.g., permeation, solubility, intestinal decomposition) that limit intestinal absorption as well as first-pass metabolism and biliary extraction that eliminate compound material before it reaches systemic circulation. These barriers are not encountered in IV dosing.

 ## 19.2 PK Parameters

Several key PK parameters are used in drug discovery, development, and clinical practice. These parameters are reported for discovery PK studies; thus, it is useful to understand their meaning and how they relate to the physicochemical, metabolic, and structural properties of the compound.

19.2.1 Volume of Distribution (V_d)

Volume of distribution (V_d) indicates how widely the compound is distributed in the body. Its units typically are L/kg or mL/kg of body weight. V_d is not a real measurable volume. Instead, it represents the apparent volume into which the compound is dissolved. V_d is a proportionality of compound concentration in plasma to total compound in the body throughout the time course (Figure 19.2).

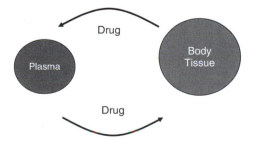

Figure 19.2 ▶ Volume of distribution (V_d) indicates the equilibrium established by the proportion of compound in the plasma to total compound in the body.

V_d is dependent on the properties of the compound:

▶ Compounds that are *highly and tightly bound to plasma protein* tend to be restricted to the bloodstream and do not enter the tissues in significant amounts. V_d is close to the volume of blood (approximately 0.07 L/kg).

▶ Compounds that are *hydrophilic* tend to be restricted to the bloodstream and do not enter the tissues in significant amounts. V_d is close to the volume of blood (approximately 0.07 L/kg).

▶ Compounds that are *moderately lipophilic and moderately bound to plasma protein and tissue components* tend to distribute evenly throughout the blood and tissues. V_d is in the range of the volume of body water (approximately 0.7 L/kg).

▶ Compounds that are *highly lipophilic* tend to bind to tissue components (e.g., proteins, lipids), and there is very low blood concentration. V_d exceeds body water volume (0.7 L/kg) and may reach levels as high as 200 L/kg.

Figure 19.3 shows these ranges of V_d. Examples of values of V_d for commercial drugs are listed in Table 19.1. Values range from the V_d of warfarin, which is primarily found in the bloodstream and binds tightly to plasma proteins, to those of imipramine and chloroquine, which are primarily found in the tissues and bind tightly to tissue components.

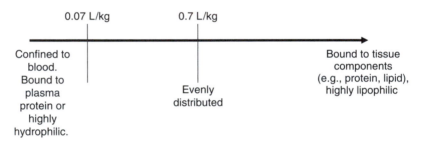

Figure 19.3 ▶ Diagnosing compound binding locations using V_d.

TABLE 19.1 ▶ **Values of V_d for Some Commercial Drugs**

Drug	V_d (L/kg)
Warfarin	0.11
Salicylic acid	0.14
Theophylline	0.5
Atenolol	0.7
Quinidine	2
Digoxin	7
Imipramine	30
Chloroquine	235

The binding of a compound to plasma protein versus tissue components is captured in the expression:

$$V_d = V_{blood} + V_{tissue} \bullet (f_{u,blood}/f_{u,tissue}),$$

where $f_{u,blood}$ = fraction of compound unbound in blood, and $f_{u,tissue}$ = fraction of compound unbound in tissue. This relationship explains the categories of compounds described above. If a compound is highly and tightly bound in blood, $f_{u,blood}$ is low and V_{blood} predominates. If a compound is highly bound in tissue, $f_{u,tissue}$ is low and V_{tissue} predominates. A high V_d does not guarantee high therapeutic target exposure ($f_{u,tissue}$), only that $f_{b,tissue}$ (bound drug in tissue) is high. Increasing tissue binding and increasing V_d are related to increasing lipophilicity.

Compound molecules are carried to the tissues by systemic circulation. Thus, compound distributes fastest to tissues that are highly perfused with blood (high blood flow). These tissues include brain, heart, lung, kidney, and liver. Compound distributes slowest to tissues that have lower perfusion, such as skeletal muscle, bone, and adipose tissue.

V_d is calculated in PK experiments using the following expression:

$$V_d = \text{Dose}/C_0,$$

where dose = compound dosed mass/animal mass (e.g., mg compound/kg body weight), and C_0 = initial blood concentration of compound after an IV dose. C_0 is determined by plotting the log of the blood concentrations versus (linear) time and extrapolating back to time zero.

19.2.2 Area Under the Curve (AUC)

Drug exposure is evaluated using the PK parameter area under the curve (AUC). This is the area under the compound blood concentration–time plot (Figure 19.4). Comparison of AUCs from lead series analogs provides a means to select the compounds that produce the highest exposure levels. It has been reported that, with a 10 mg/kg oral does, if the AUC from 0 to 6 hours is >500 ng•h/mL, then there is an 80% chance of an acceptable bioavailability (>20%F).[2] Structure–AUC (exposure) relationships also can be developed. Another use of AUC is to calculate other PK parameters, such as clearance and bioavailability (see Sections 19.2.3 and 19.2.5).

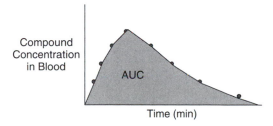

Figure 19.4 ► Area under the curve (AUC) is the area under the compound plasma concentration–time plot.

19.2.3 Clearance (Cl)

Another important PK parameter for discovery scientists is clearance (Cl). It indicates how rapidly the compound is extracted from systemic circulation (bloodstream) and eliminated. Clearance occurs primarily in two organs: kidney and liver.

A portion of cardiac output (i.e., blood flow) goes to the kidney (Figure 19.5). In the kidney, compounds and metabolites in blood are extracted into the urine by glomerular filtration and active secretion by transporters. This results in compound and metabolite elimination. Clearance specifically from the kidney can be determined in detailed PK studies and is termed *renal clearance* (Cl_R).

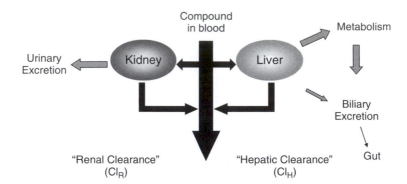

Figure 19.5 ▶ A significant portion of cardiac blood output flows to the kidney and liver, where a portion of the free compound in solution is extracted and cleared from the body.

The liver also receives a portion of the cardiac output. Compound permeates from blood into hepatocytes by passive diffusion and active transport. Within the hepatocytes, a variety of enzymatic reactions metabolize the compound. Metabolites and a portion of unchanged compound may be extracted into the bile ("biliary extraction") by passive diffusion or active transport into the bile canaliculus. Bile is stored in the gallbladder and excreted into the intestine, from which compounds and metabolites are eliminated in feces. Hepatic clearance (Cl_H) can be determined in detailed PK studies. Metabolites and unchanged compound can also exit the hepatocytes into blood by passive diffusion and active transport and be extracted by the kidney into urine. The amount of compound and metabolites that are extracted into bile or urine depend on their properties (e.g., passive diffusion, transporter affinity, metabolic stability).

The term *systemic clearance* (Cl_S) is used to indicate the total clearance of compound from all sources. Cl_S is the sum of Cl_R and Cl_H. Minor routes of clearance include saliva, sweat, and breath. The units used for clearance typically are mL/min/kg.

Clearance by an organ is determined by two factors: blood flow into the organ (Q) and extraction ratio by the organ (E; Figure 19.6), as follows:

$$Cl = Q \bullet E.$$

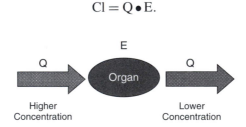

Figure 19.6 ▶ Extraction of compound by the liver or kidney is dependent on the blood flow to the organ (Q) and the extraction ratio of the organ (E).

Blood flow into the liver and kidney are well characterized in each species used for drug discovery PK studies. The total blood flow from the heart is also known and is termed *cardiac output*. The extraction ratio is the portion of compound removed by the organ with each pass of blood through the organ. For example, if cardiac output to the liver (Q_H) is 55 mL/min/kg and hepatic clearance (Cl_H) is measured as 25 mL/min/kg, then the hepatic extraction ratio (E_H) is 0.45 (i.e., 45% of the compound is removed with each pass through the liver).

E is constant for a specific compound, unless the capacity of the organ is saturated by high compound concentration in blood. The highest value for E is 1 (100%); thus, the highest value for Cl is the blood flow to the organ. Blood flow to organs of some discovery

TABLE 19.2 ▶ Blood Flow to Organs in Some Species Used in Drug Discovery

Species	Blood flow (mL/min/kg)		
	Hepatic	Renal (GFR)	Total
Mouse	90	15	400
Rat	55	5	300
Monkey	30	2	220
Dog	30	5	120
Human	20	2	80

GFR, Glomerular filtration rate.

species are listed in Table 19.2. The term *glomerular filtration rate* (GFR) is often used with regard to renal clearance. GFR is the flow rate of fluid passing by filtration from the glomerulus into the Bowman's capsule of all the nephrons in the kidney (typically in units of mL/min/kg body weight).

Clearance is calculated from intravenous (IV) PK studies using the dose and total compound exposure from the expression:

$$Cl = Dose/AUC_{IV}$$

Thus, if clearance is determined, it can be used to estimate the dose that must be administered to provide a level of compound exposure necessary for the therapeutic effect. For example, compounds with higher clearance require higher doses to achieve a certain exposure (AUC) level for in vivo activity. Metabolic stability is a major determinant of hepatic clearance. Thus, medicinal chemists have the opportunity to reduce clearance by structurally modifying the compound to reduce metabolism. Chemists can use in vitro metabolic stability half-life to correlate to Cl.

19.2.4 Half-life ($t_{1/2}$)

The time for the concentration of a compound in systemic circulation to reduce by half is termed *half-life* ($t_{1/2}$). Compound clearance typically follows first-order kinetics, and a plot of log compound concentration versus time indicates the elimination rate constant (k) from the slope (Figure 19.7). $t_{1/2}$ is calculated from k using the expression:

$$t_{1/2} = 0.693/k.$$

$t_{1/2}$ also can be calculated from V_d and Cl using the expression:

$$t_{1/2} = 0.693 \bullet V_d/Cl.$$

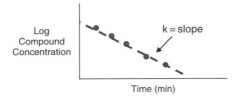

Figure 19.7 ▶ First-order rate constant (k) for elimination is obtained from a plot of log compound concentration versus time.

Thus, PK half-life is determined by Cl_H (metabolic stability and biliary clearance) Cl_R, and V_d. Chemists should not assume that in vitro metabolic stability half-life correlates to PK half-life. Half-life allows an estimation of how often the compound needs to be redosed to maintain the in vivo therapeutic concentration. Redosing typically is performed every 1 to 3 half-lives.

Another related PK parameter is mean residence time (MRT). This is the time for elimination of 63.2% of the IV dose.

19.2.5 Bioavailability (F)

One of the most commonly used PK parameters is bioavailability (F). It is the fraction of the dose that reaches systemic circulation unchanged. Less than 100% bioavailability typically results from incomplete intestinal absorption or first-pass metabolism. Secondary causes include enzymatic or pH-induced decomposition in the intestine or blood.

Bioavailability is determined by the following experiment. The compound is dosed IV, plasma samples are collected and analyzed, and AUC_{IV} is calculated. After a period of compound washout, or using different living subjects, the compound is dosed PO, plasma samples are collected and analyzed, and AUC_{PO} is calculated. Bioavailability is calculated from the expression:

$$\%F = (AUC_{PO}/AUC_{IV}) \bullet (Dose_{IV}/Dose_{PO}) \bullet 100\%.$$

Often a lower dose is given IV because of the high concentrations that result from IV dosing. Many companies have a goal of at least 20% oral bioavailability for advancement of a candidate to clinical trials. A compound with poor oral bioavailability can have significant patient variability if its metabolizing cytochrome P450 (CYP) isozyme is polymorphic in the population. Bioavailability is sometimes determined for other dosing routes versus IV, in order to diagnose issues such as absorption or first-pass metabolism. A compound with low bioavailability risks high patient-to-patient variability of blood concentrations, especially if the compound is metabolized by an enzyme having expression levels that vary greatly in the population.

19.3 Effects of Plasma Protein Binding on PK Parameters

The binding of compounds to plasma proteins is discussed in detail in Chapter 14. Compound molecules that are bound to plasma protein cannot permeate through membranes. Thus, the percentage of protein-bound compound molecules and their binding affinity determine how much of the compound permeates out of the blood capillaries and into the disease tissues for therapeutic action, other tissues for general distribution, and kidney and liver for clearance.

19.4 Tissue Uptake

Tissue uptake is important for the compound to reach the therapeutic target. Blood–organ barriers limit penetration into some tissues. For example, CNS drugs must penetrate into brain tissue through the blood–brain barrier (BBB; see Chapter 10). The penetration of cancer drugs into tumors may be reduced, compared to other tissues, by reduced blood flow and tumor morphology. Uptake into the target tissue is often measured to determine the exposure of the target to the compound.

19.5 Using PK Data in Drug Discovery

The major PK parameters used in drug discovery and their methods of determination are summarized in Table 19.3 and Figure 19.8. In addition to the PK parameters discussed in previous sections, Figure 19.8 shows the determination of the following parameters:

▶ C_{max}, the maximum compound concentration from oral dosing

▶ t_{max}, the time at which C_{max} is reached

Discovery scientists also examine these data in relation to IC_{50} or EC_{50} to estimate how long the plasma concentration is above the effective in vitro concentration. General classifications for PK parameters in discovery are suggested in Table 19.4. These often differ among companies and projects. Table 19.5 lists the PK parameters of some commercial drugs.[3] PK data for many more compounds are given in reference 3.

It is instructive to note in Table 19.5 the relationships of structural properties to PK performance. Compounds having properties that exceed the Lipinski rules tend to have low oral bioavailability and are administered by nonoral routes (e.g., paclitaxel, doxorubicin). Improvement of structural properties can improve bioavailability (e.g., cefuroxime vs cephalexin). High clearance may result in low bioavailability (e.g., buspirone). The prodrug valacyclovir provides improved PK performance for its active drug acyclovir.

This chapter has summarized the fundamental PK parameters used for drug discovery project team support, in order to assist the interpretation of PK data and planning of in vivo experiments for PK and efficacy studies. Chapter 37 discusses the measurement of PK parameters. Chapter 38 discusses how the compound's properties affect PK parameters so

TABLE 19.3 ▶ Pharmacokinetic Parameters: Definitions, Calculations, and Applications In Discovery

Pharmacokinetic parameter	Symbol	Description	Calculation	Application
Area under the curve	AUC	Area under the concentration vs time curve	Integrate area under curve	Estimate the level of exposure
Initial concentration	C_0	Initial blood concentration after IV dose	Extrapolate plot of log plasma concentration vs time back to zero time	Calculate V_d
Volume of distribution	V_d	Apparent volume in which compound is dissolved	$V_d = Dose/C_0$	Estimate how widely the compound is distributed in body; calculate half-life
Clearance	Cl	How rapidly compound is extracted from systemic circulation	$Cl = Dose/AUC_{IV}$	Calculate the dose needed to achieve a certain exposure (AUC); diagnose mechanisms of compound elimination
Elimination rate constant	k	First-order kinetics elimination rate	Slope of log plasma concentration vs time profile	Calculate $t_{1/2}$

Continued

TABLE 19.3 ► *Continued*

Pharmacokinetic parameter	Symbol	Description	Calculation	Application
Half-life	$t_{\frac{1}{2}}$	Time for blood concentration to reduce by half	$t_{\frac{1}{2}} = 0.693/kt_{\frac{1}{2}} = 0.693 \bullet V_d/Cl$	Calculate how frequently dose must be administered to maintain a therapeutic concentration
Bioavailability	%F	Fraction (percent) of dose reaching systemic circulation unchanged	$\%F = (AUC_{PO}/AUC_{IV}) \bullet (Dose_{IV}/Dose_{PO}) \bullet 100\%$	Diagnose mechanisms limiting compound exposure (e.g., solubility, permeability, first-pass metabolism)
Maximum concentration	C_{max}	Highest concentration of compound in systemic circulation following dose	Interpolate from plot of blood concentration–time profile	Evaluate if pharmacology is driven by C_{max}, time above IC_{50}, or AUC
Time of maximum drug concentration	t_{max}	Time of C_{max}	Interpolate from plot of blood concentration–time profile	Estimate any delay in reaching target tissue from bloodstream

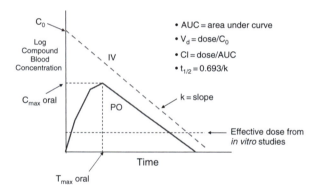

Figure 19.8 ► Pharmacokinetic parameters used in drug discovery and their determination.

TABLE 19.4 ► General PK Parameter Goals for Discovery Compounds

Pharmacokinetic parameter	Symbol	High	Low
Volume of distribution	V_d	>10 L/kg	<1 L/kg
Plasma clearance	Cl	Rat: >45 mL/min/kg Mouse: >70 mL/min/kg Human: >15 mL/min/kg	Rat: <10 mL/min/kg Mouse: <20 mL/min/kg Human: <5 mL/min/kg
Half-life	$T_{\frac{1}{2}}$	Rat: >3 h Mouse: >3 h Human: >8 h	Rat: <1 h Mouse: <1 h Human: <3 h
Oral bioavailability[2]	%F	>50%	<20%
Oral exposure (10 mpk)[2]	AUC	Rat >2,000 h•ng/mL	Rat <500 h•ng/mL
Time of maximum drug concentration	t_{max}	>3 h	<1 h

TABLE 19.5 ▶ **Pharmacokinetic Parameters of Selected Drug Compounds[a]**

Compound	Oral bioavailability (%)	Urinary excretion unchanged (%)	Plasma binding (%)	Cl (ml/min/kg)	Vd (L/kg)	$t_{1/2}$ (h)	t_{max} (h)	C_{max} (μg/mL)	Dose[b] (mg/kg)	Notes
Paclitaxel	Low	5	88–98	5.5	2.0	3	—	0.85	250 mg/m²	IV infusion, oncology
Doxorubicin	5	<7	76	666	682 L/m²	26	—	0.95	45 mg/m2	IV infusion, oncology
Buspirone	3.9	0.1	95	28.3	5.3	2.4	0.71	0.017	0.3	Oral, anxiolytic

Structure										
Cefuroxime	32	96	33	0.94	0.2	1.7	2–3	7–10	7.1 oral	Oral, antibiotic
Cephalexin	90	91	14	4.3	0.26	0.9	1.4	28	7.1 oral	Oral, antibiotic
Zolpidem	72	<1	92	4.5	0.68	1.9	1–2.6	0.076–0.14	0.14	Oral, sedative
Acetaminophen	88	3	<20	5	0.95	2	0.33–1.4	20	20 oral	Oral, analgesic
Propranolol	26	<0.5	87	16	4.3	3.9	1.5	0.049	11	Oral, 4 times per day β-adrenergic-hypertension

Continued

TABLE 19.4 ▶ *Continued*

Drug										
Atenolol	58	94	<5	2.4	1.3	6.1	3.3	0.28	0.7	Oral, β-adrenergic-hypertension
Amoxicillin	93	86	18	2.6	0.21	1.7	1–2	5	7 oral	Oral, antibiotic
Acyclovir	15–30	75	15	3.4	0.69	2.4	1.5–2	3.5–5.4 μM	5.7	Oral, 6 times per day, availability increases with dose, antiviral
Valacyclovir	Val: low Acy: 54%	Val: <1 Acy: 44	Val: 13.5–17.9 Acy: 22–33	—	—	Val: — Acy: 2.5	Val: 1.5 Acy: 1.9	Val: 0.56 Acy: 4.8	14	Oral prodrug for acyclovir, antiviral

^{a} Data from reference [3].
^{b} mg/kg dose is based on 70-kg average human weight.
Acy, Acyclovir; Val, valacyclovir.

that discovery scientists can diagnose the underlying properties that limit PK and modify the structure to achieve improved PK.

PK data are used throughout drug discovery, development, and clinical treatment. Animal PK data are very useful for discovery projects because they indicate compound behavior in an organism that is used as a model of the human body. Some of the useful applications of PK are as follows:

▶ Indicate if useful exposure levels can be achieved

▶ Determine the compound's bioavailability, a key indicator of PK performance

▶ Reveal PK limitations to prompt structural modifications for improved performance

▶ Show the dose necessary to produce a concentration at the therapeutic target at which a pharmacological response might be observed in vivo

▶ Compare lead series analogs to select compounds for more advanced studies

▶ Establish the relationship of PK parameters to pharmacodynamics (PD; in vivo activity), also known as PK/PD relationship

▶ Plan dose levels for efficacy and toxicity studies

▶ Estimate the safety window (therapeutic index; see Chapter 17)

▶ Measure compound concentration in the target disease tissue to determine penetration through organ barriers (e.g., BBB) and target exposure

▶ Evaluate potential for drug–drug interaction, based on compound concentration at therapeutic dosing levels

▶ Extrapolate animal PK parameters to humans to plan initial clinical dosing studies

▶ Meet company's criteria for advancement

In short, PK data provide key insights that are useful to discovery projects. Chemists optimize structures, pharmacologists plan biological experiments, and team leaders make informed decisions using the data.

Problems

(Answers can be found in Appendix I at the end of the book.)

1. Match the following PK parameters with their definitions:

PK parameter	Definition choices
Cl	a. percentage of the oral dose that reaches systemic circulation unchanged
C_{max}	b. rate a compound is removed from systemic circulation
V_d	c. time for the compound's concentration in systemic circulation to decrease by half
$t_{1/2}$	d. compound's exposure as determined by blood concentration over time
C_0	e. apparent volume into which the compound is dissolved
AUC	f. highest concentration reached in the blood
%F	g. initial concentration after IV dosing

2. Why does IV bolus administration of a compound have a higher initial concentration than PO administration?

3. What would: (a) enhance volume of distribution, (b) reduce volume of distribution?

4. A low AUC indicates exposure is: (a) high, (b) low?

5. Clearance occurs primarily in what organs?

6. Associate the following V_d values:
0.1, 1, 100,
with the following descriptions of distribution:
evenly distributed throughout the body, highly tissue bound, highly bloodstream restricted

7. Which of the following is a more preferred exposure (ng•h/mL)?: (a) AUC = 45, (b) AUC = 620/

8. Which of the following is a more preferred clearance (mL/min/kg)?: (a) Cl = 20, b) Cl = 60/

9. Which of the following is a more preferred half-life (h) for once-daily dosing?: (a) $t_{1/2} = 0.5$, (b) $t_{1/2} = 8$.

10. For the following experiments and data, calculate the bioavailabilities:

IV dose (mg/kg)	PO dose (mg/kg)	AUC PO (ng•h/mL)	AUC IV (ng•h/mL)	Bioavailability
1	10	500	500	
2	10	1000	500	
5	10	300	200	

11. What value of V_d (L/kg) indicates approximately equal distribution between blood and body tissue?: (a) 0.07, (b) 0.7, (c) 7, (d) 70, (e) 700.

References

1. Birkett, D. J. (2002). *Pharmacokinetics made easy.* Sydney, Australia: McGraw-Hill.

2. Mei, H., Korfmacher, W., & Morrison, R. (2006). Rapid in vivo oral screening in rats: reliability, acceptance criteria, and filtering efficiency. *AAPS Journal, 8,* E493–E500.

3. Thummel, K. E., Shen, D. D., Isoherranen, N., & Smith, H.E. Design and optimization of dosage regimens: pharmacokinetic data, Appendix II. In L.L. Brunton (Ed.). *Goodman and Gilman's the pharmacological basis of therapeutics* (11th ed.). New York: McGraw-Hill.

Chapter 20

Lead-like Compounds

Overview

▶ *Structure modification during activity optimization often increases H-bonds, molecular weight, and Log P, which deteriorates drug-like properties.*

▶ *Lead-like compounds have lower initial values for structural properties, allowing increases without becoming non–drug-like.*

▶ *Fragment screening typically provides more lead-like compounds for optimization.*

The ancient proverb says: "The house built on a foundation of sand will fall, but the house built on rock will prosper." For drug discovery, the foundation is the lead structure. If the foundation is strong, the project team can build a strong drug-like clinical candidate. If the foundation is weak, the team's effort may never advance a drug-like compound to development.

The "hits" that serve as starting places for leads come from high-throughput screening, virtual screening, natural ligands, natural products, and the scientific literature. In the "hit-to-lead" phase, it is important to include properties in the workflow and goals for lead selection. In this evaluation process, some effective concepts have been emerging:

▶ Lead-likeness

▶ Template conservation

▶ Triage

▶ Fragment-based screening.

The use of one or more of these concepts can increase the chances of success in discovering a strong lead-like structural foundation.

20.1 Lead-likeness

Shortly after introduction of the rule of 5, also known as the Lipinski rules,[1] the value of including property guidelines in the selection of leads was recognized. This was an important step forward in starting with a foundation of leads that are free from major liabilities that would later impede the accomplishment of a viable clinical candidate.

As experience with property guidelines accumulated, it was recognized that the lead optimization phase adds substructures onto the lead template to enhance target affinity and selectivity. Nonpolar groups are added to enhance binding to lipophilic pockets. Other groups are added to increase hydrogen bonding with the binding site. These modifications add lipophilicity, molecular weight (MW), and hydrogen bonding to the lead. This process can result in compounds that exceed the rule of 5 guidelines and have deleterious properties.

A proposed alternative was "lead-like properties,"[2] in which screening libraries are limited to compounds with:

▶ MW between 100 and 350

▶ Clog P between 1 and 3

By selecting leads that have lower MW, lower lipophilicity, and fewer hydrogen bonds, the eventual products of optimization are more likely to have acceptable drug-like properties. It also was proposed that lead-like structures are more likely to bind to the target protein "because they can more easily find a binding mode than larger drug-like molecules" that are commonly included in screening libraries.[2] A separate computational comparison indicated that, on average, lead compounds had lower values for structural properties than drug compounds, including 69-Da lower MW, one fewer ring, one fewer H-bond acceptor, two fewer rotatable bonds, 0.43 lower ClogP, and 0.97 lower $LogD_{7.4}$.[3]

A computational evaluation later suggested the following criteria for inclusion of compounds in lead-like screening libraries[4]:

▶ $MW \leq 460$

▶ $-4 \leq Log\ P \leq 4.2$

▶ Log of water solubility $(Log\ S_w) \leq -5$

▶ Rotatable bonds ≤ 10

▶ Rings ≤ 4

▶ Hydrogen-bond donors ≤ 5

▶ Hydrogen-bond acceptors ≤ 9

Lead-likeness also was suggested as extending to in vivo pharmacokinetic parameters and in vitro property criteria.[4] In many hit-to-lead programs, toxicity and pharmacokinetics are often not studied extensively. These could be used in the final selection of leads that will advance to optimization:

▶ Bioavailability $(\%F) \geq 30\%$

▶ Clearance (Cl) $< 30\,mL/min/kg$ in rat

▶ $0 \leq Log\ D_{7.4} \leq 3$

▶ Binding to cytochrome P450 isozymes $=$ low

▶ Plasma protein binding $\leq 99.5\%$

▶ Acute toxicity and chronic toxicity $=$ none (in therapeutic window)

▶ Genotoxicity, teratogenicity, carcinogenicity $=$ none (at dose 5–10 times therapeutic window)

20.2 Template Conservation

In many cases, a large portion of the lead structure is conserved throughout the lead optimization stage. Structure modifications during optimization are often added onto the lead template, thus retaining much of the original core structure of the lead. The properties associated with the core structure continue to be a primary component of the properties of the analogs and the eventual clinical candidate.

Examples of this are shown in Figure 20.1, and others are found in Proudfoot.[5] Many drugs retain a large portion of the lead core structure. This principle suggests that the greatest opportunity to "lock in" favorable properties is at the lead selection stage. In addition to selecting from among the screening "hits," often a small effort in synthetic modification at the lead selection stage can improve properties and provide a stronger lead for the optimization stage. If the lead coming into the optimization phase does not have good properties, the project team will be required to expend time and resources to improve properties and may never be able to accomplish this at a later time when there are many discovery objectives to complete in a short time. This would be like trying to go back and reconstruct the foundation after the house has already been built. The difficulty and time required are magnified compared to completing this process at the beginning. Also, there is a natural tendency in discovery to maximize target affinity through structure–activity relationship–guided structure design. If the lead already has good properties, then deterioration of properties for the sake of activity optimization still may result in a clinical candidate with acceptable drug-like properties.

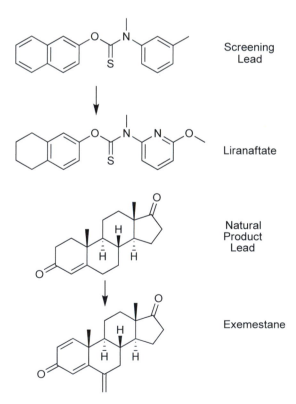

Screening Lead

Liranaftate

Natural Product Lead

Exemestane

Figure 20.1 ► The lead template often is conserved during lead optimization.[1]

20.3 Triage

Hopefully, the early screening activities of a project provide many hits for consideration. Inclusion of properties in the evaluation criteria, along with activity, selectivity, and novelty, is an important strategy for ensuring that leads will have strong properties. This is consistent with the emergence of "risk" as a major factor in drug discovery decision-making. Selecting compounds that have the greatest chances of success and downgrading compounds that have higher risk of failure is an efficient practice for a drug discovery enterprise.

This process of triage is aided by setting goals for each of the key criteria of the lead.[6] Figure 20.2 is an example of how exploratory medicinal chemists compare the activity, selectivity, and properties of potential leads. This disciplined evaluation process assists triage of hits, guidance of initial synthetic modifications for improvement, and selection of leads for optimization.

	Lead	Analog	Desired Profile
MW	330	445	<450
clogP	1.9	5.19	<4.0
IC50 (µM)	4.2	>20	<1.0 µM
Binding to target (STD, FP, Trp-Fl.)	X-ray		Yes (NMR, FP)
MIC			
B. subtilus	>200 µM	50 µM	<200 µM
S. aureus MRSA	>200 µM	25 µM	<200 µM
S. aureus ATCC	>200 µM	200 µM	<200 µM
S. pneumo +	>200 µM	25 µM	<200 µM
Selectivity: C. albicans (MIC µg/mL)	>200	>200	>10 fold
Aqueous Solubility (µg/ml @ pH 7.4)	>100	26.5	>60
Permeability (10⁻⁶ m/s @ pH 7.4)	0	0.15	>1
CYP 3A4 (% inhibition @ 3 µM)	11	7	<15
CYP 2D6 (% inhibition @ 3 µM)	0	1	<15
CYP 2C9 (% inhibition @ 3 µM)	NT	23	<15
Microsome stability (% remaining @ 30 min)	NT	NT	>80
Definable Series	Yes	Yes	Yes
Definable SAR	Yes	Yes	Yes

Figure 20.2 ▶ Example of goals used by Wyeth Research exploratory medicinal chemists for hit selection, initial structural modification, and lead selection in an acyl carrier protein synthase (AcpS) inhibitor project.[6] (see Plate 3)

20.4 Fragment-Based Screening

An emerging strategy in exploring for novel leads has been termed *fragment-based screening*. This approach is based on the theory that screening with larger structures that fit the very specific shape, electrostatic interactions, and hydrophobic contacts of the binding site of the target protein with appreciable affinity is a very low-percentage possibility. Instead, it is suggested that smaller, less complex compounds, or *fragments*, are more likely to bind to a portion of the binding site. From a fragment core, functionality can be added to enhance binding. Also, by selecting fragments that bind to different portions of the site and then joining them together with a "tether," it may be more likely to find a final lead that binds appreciably to the site. Although fragments bind with low affinity, tethered fragments forming a larger molecule will bind with greater affinity. An example of this method[7] is shown in Figure 20.3.

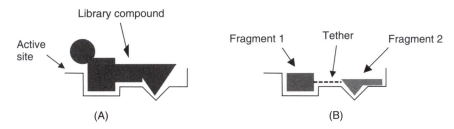

(A)   (B)

Figure 20.3 ▶ Fragment-based screening replaces the screening of large libraries containing complex structures to fit a specific active site (**A**) with the screening of a smaller library of smaller molecules that fit a portion of the active site and then are tethered together to form a lead with higher affinity (**B**).

When fragments bind, they have lower affinity ($IC_{50} \sim 50$ μM–1 mM)[7] but have a high efficiency of binding for their size.[8] Binding of fragments is difficult to detect with conventional biological assays; however, x-ray crystallography or NMR can be used to detect such weak binding fragments. The binding site and orientation in the pocket can be determined. These techniques are more expensive than conventional screening methods. Fortunately, it appears that a small library of fragments can produce a great deal of structural diversity because they are not encumbered by additional attached structure that would hinder binding. Also, cocktails of small molecules can be tested to accelerate the screening and to observe multiple binding locations. Excellent reviews have been published on fragment screening techniques.[9] library design,[10,11] and examples in drug discovery.[12–14]

The fragment approach nicely complements the goal of selecting leads with good properties. When large screening molecules, which are common in conventional screening libraries, bind to the target protein, they may have a considerable portion of the structure that is not involved in forming binding interactions. This extraneous structure adds to the MW, hydrogen bonds, and lipophilicity that detract from lead-like properties. On the other hand, fragments can minimize the extraneous structure and reduce the useless portions that would be detrimental to absorption. Also, a fragment screening library can be constructed from small molecules having good lead-like properties, allowing expanded lead or the tethered product of two fragments to be minimized to the structural features that are useful for effective binding and, ultimately, produce an active drug-like molecule.

Combination of the ideas of lead-likeness and fragment-based screening has produced guidelines for the properties of molecules used in fragment screening libraries. A set of rules for lead-like compounds has been proposed by Oprea as the "rule of 3"[15]:

▶ MW ≤300

▶ Clog P ≤3

▶ Rotatable bonds ≤3

▶ Hydrogen-bond donors ≤3

▶ Hydrogen-bond acceptors ≤3

▶ Polar surface area (PSA) ≤60 Å^2

20.5 Lead-like Compounds Conclusions

The well-studied lead-like criteria reinforce the principle that success in discovering drug-like clinical candidates is higher when starting with compounds that have structural properties with lower values, which allow the opportunity for structural modifications to enhance activity and selectivity without condemning the series to unacceptable pharmacokinetic performance. All discovery organizations should consider whether it is useful to include compounds in their screening libraries that have properties falling outside the range guidelines of lead-like, or even drug-like, compounds. The need for efficiency and success in drug discovery adds imperative to a disciplined workflow that bases drug discovery work on a firm foundation from which a strong clinical candidate ("house") can be built.

Problems

(Answers can be found in Appendix I at the end of the book.)

1. If leads are going to be structurally modified during lead optimization, why should compounds that exceed the "rule of 5" guidelines not be included as leads?

2. Evaluate the following compounds for their "lead-likeness" according to the "rule of 3" and list which properties are unfavorable:

A
MW = 474.6
PSA = 113
cLogP = 2.0

B
MW = 285.3
PSA = 144
cLogP = −0.9

C
MW = 316.4
PSA = 59
cLogP = 1.8

D
MW = 418
PSA = 142
cLogP = −2.4

3. To increase the chances of maintaining good drug-like properties throughout the optimization phase, start with a lead having which of the following "lead-like" properties?: (a) ClogP ≤ 3, (b) MW > 400, (c) PSA ≤ 60, (d) H-bond donors > 5.

References

1. Lipinski, C. A., Lombardo, F., Dominy, B. W., & Feeney, P. J. (1997). Experimental and computational approaches to estimate solubility and permeability in drug discovery and development settings. *Advanced Drug Delivery Reviews*, *23*, 3–25.

2. Teague, S. J., Davis, A. M., Leeson, P. D., & Oprea, T. (1999). The design of leadlike combinatorial libraries. *Angewandte Chemie, International Edition*, *38*, 3743–3748.

3. Oprea, T. I., Davis, A. M., Teague, S. J., & Leeson, P. D. (2001). Is there a difference between leads and drugs? A historical perspective. *Journal of Chemical Information and Computer Sciences*, *41*, 1308–1315.

4. Hann, M. M., & Oprea, T. I. (2004). Pursuing the leadlikeness concept in pharmaceutical research. *Current Opinion in Chemical Biology*, *8*, 255–263.

5. Proudfoot, J. R. (2002). Drugs, leads, and drug-likeness: an analysis of some recently launched drugs. *Bioorganic & Medicinal Chemistry Letters*, *12*, 1647–1650.

6. Ellingboe, J. (2005). *The application of pharmaceutical profiling data to lead identification and optimization.* Abstracts, 37th Middle Atlantic Regional Meeting of the American Chemical Society, New Brunswick, NJ, United States, May 22–25, 2005, GENE-231.

7. Carr, R., & Jhoti, H. (2002). Structure-based screening of low-affinity compounds. *Drug Discovery Today*, *7*, 522–527.

8. Carr, R. A. E., Congreve, M., Murray, C. W., & Rees, D. C. (2005). Fragment-based lead discovery: leads by design. *Drug Discovery Today*, *10*, 987–992.

9. Lesuisse, D., Lange, G., Deprez, P., Benard, D., Schoot, B., Delettre, G., et al. (2002). SAR and X-ray. A new approach combining fragment-based screening and rational drug design: application to the discovery of nanomolar inhibitors of Src SH2. *Journal of Medicinal Chemistry*, *45*, 2379–2387.

10. Jacoby, E., Davies, J., & Blommers, M. J.J. (2003). Design of small molecule libraries for NMR screening and other applications in drug discovery. *Current Topics in Medicinal Chemistry*, *3*, 11–23.

11. Schuffenhauer, A., Ruedisser, S., Marzinzik, A. L., Jahnke, W., Blommers, M., Selzer, P., & Jacoby, E. (2005). Library design for fragment based screening. *Current Topics in Medicinal Chemistry*, *5*, 751–762.

12. Verdonk, M. L., & Hartshorn, M. J. (2004). Structure-guided fragment screening for lead discovery. *Current Opinion in Drug Discovery & Development*, *7*, 404–410.

13. Gill, A., Cleasby, A., & Jhoti, H. (2005). The discovery of novel protein kinase inhibitors by using fragment-based high-throughput X-ray crystallography. *ChemBioChem*, *6*, 506–512.

14. Rishton, G. M. (2003). Nonleadlikeness and leadlikeness in biochemical screening. *Drug Discovery Today*, *8*, 86–96.

15. Congreve, M., Carr, R., Murray, C., & Jhoti, H. (2003). A "rule of three" for fragment-based lead discovery? *Drug Discovery Today*, *8*, 876–877.

Chapter 21

Strategies for Integrating Drug-like Properties into Drug Discovery

Overview

▶ *How and when property data are used can have great impact on a discovery project.*

▶ *Successful property strategies include assessing properties early, profiling properties rapidly, relating structures to properties, optimizing activity and properties together, using single-property assays, improving bioassays and interpretation, customizing assays for specific issues, and diagnosis.*

The integration of drug-like properties into the workflow of drug discovery deserves thought, creativity, and planning. Different companies may choose different approaches for integration of drug-like properties based on their resources, priorities, and organizational experience. The first decade of drug-like properties in discovery has indicated strategies that are beneficial. Many of these strategies are described in the following sections.

21.1 Assess Drug-like Properties Early

Property assessment can be performed early in the hit-to-lead process (Figure 21.1). This allows properties to be considered along with high-throughput screening (HTS) data, confirmatory in vitro bioassay data, and novelty. The property tools most appropriate for this stage are structural rules, in silico tools, and in vitro assays. If major liabilities are found that are unlikely to be improved by structure modification, then the structural series can be ranked lower in priority, and resources can be used for more promising series. If correctable liabilities are found, then studies can be planned for early structure modifications that attempt to minimize the problem(s) by modifying the lead before committing significant resources to the series. If acceptable or advantageous properties are found, then this series may be ranked higher. As discussed in Chapter 20, the structural template often is conserved during lead optimization, so it is important to advance a reliable core to the optimization phase. This approach emphasizes that absorption, distribution, metabolism, excretion, and toxicity (ADME/Tox) properties are important throughout the entire discovery process. This approach also improves efficiency, because deferring work on drug-like properties until later in discovery can result in series delay or failure after significant resources have already been invested in the series. The assessment of ADME/Tox properties at this stage should use methods having the appropriate level of specificity and resources, as discussed in Chapter 22, and need not be highly resource intensive.

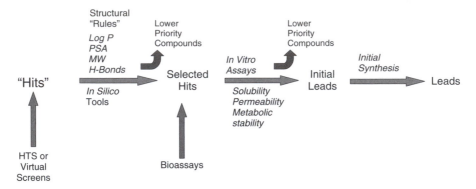

Figure 21.1 ▶ Early assessment of compound properties during the hit-to-lead phase can remove problematic compounds, augment lead selection using all available data, and help to plan initial synthetic modifications to rescue a series.

21.2 Rapidly Assess Drug-like Properties for All New Compounds

As discovery proceeds and new compounds are synthesized, properties can be rapidly assessed. This provides immediate feedback. If the new compound was synthesized to improve properties, rapid property measurement provides a check on the success of this approach. If the new compound was synthesized to improve activity or selectivity, then rapid property measurement provides a check on whether the properties were affected by the structural modification. Rapid property measurement allows teams to make decisions faster to improve properties. Further modifications can be planned in the same time period, thus providing increased chances of success.

21.3 Develop Structure–Property Relationships

Structure–property relationships (SPRs) can be developed for a series in the same manner as structure–activity relationships (SARs). SAR defines how structure modification at one moiety in the molecule affects activity. SPR defines how structure modification affects properties. An example of a hypothetical series is shown in Figure 21.2. The patterns

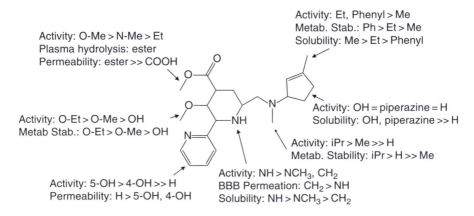

Figure 21.2 ▶ Develop structure–property relationships (SPR) to complement structure–activity relationships (SAR), as shown in this hypothetical example.

developed from these related processes help medicinal chemists with optimizing activity and properties. Furthermore, through multivariate analysis of SAR, SPR, or both, the interactions of modifications at these sites help to further optimize series performance individually for activity and properties or together in parallel.[1]

21.4 Iterative Parallel Optimization

Activity and properties of new compounds can be tested simultaneously for both activity and properties.[2] The data are fed back to the team for improvement of structure redesign (Figure 21.3). The testing time should be comparable for activity and properties. This approach allows the project teams to most effectively use their resources, make rapid progress, and take a holistic view of optimization. It also ensures that properties do not deteriorate as structural modifications are made to improve activity, and vice versa. A set of automated high-throughput assays can be implemented using an automated parallel process to rapidly provide data for key properties and activity.

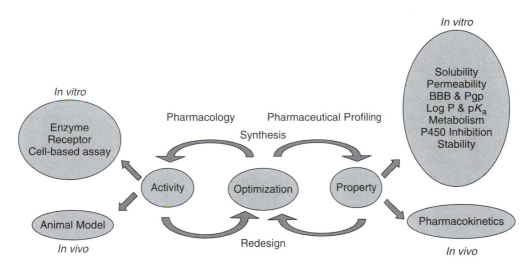

Figure 21.3 ▶ Iterative parallel optimization by simultaneous assessment of both activity and properties. (Reprinted with permission from [2].) (see Plate 4)

21.5 Obtain Property Data that Relates Directly to Structure

Structure is one of the most important elements of drug discovery. Therefore, property measurement should provide data that are directly related to structure. These data guide medicinal chemists in which structural modifications can be made to adjust the properties. For example, medicinal chemists can make modifications that change pK_a, lipophilicity, molecular weight (MW), and hydrogen bonding, block metabolically labile sites, and remove reactive groups. These modifications affect higher-level properties, such as solubility, permeability, metabolic stability, toxicity, and pharmacokinetic parameters. Rules, in vitro single variable assays, and in silico tools focus on individual properties for which medicinal chemists can readily envision specific structural modifications. Complex assays, such as pharmacokinetics

(PK), Caco-2 permeability, and hepatocyte metabolic stability, are affected by multiple properties and do not provide the specificity needed to make informed structural modification decisions. For example, Caco-2 permeability results are affected by passive, uptake, and efflux permeability as well as solubility and metabolism. Deciding which of these mechanisms is most determinative requires further experiments to obtain the specificity needed to plan specific structural modifications.

21.6 Apply Property Data to Improve Biological Experiments

The study of drug-like properties in discovery originally was implemented to select and optimize leads for in vivo PK. As data for properties have become widely available in discovery, awareness of the effects of these properties on discovery in vitro biological assays has been growing. Chemical stability and solubility in bioassay media, during dilutions, and stored compound solutions affect the concentration of test compound at the biological target (see Chapter 39). Membrane permeability affects compound access to intracellular targets in cell-based bioassays. Applying property data to optimize biological assays has proven to be very valuable in ensuring accurate SAR for discovery.

21.7 Utilize Customized Assays to Answer Specific Project Questions

An ensemble of high-throughput assays for key properties allows rapid access to data. However, the generic assay conditions and the limited number of properties screened with such assays only provide a general understanding of the compounds. It is advantageous to plan customized studies that provide data for specific project team questions that are more sophisticated than the generic assays provide data for. Examples of such studies are chemical stability or solubility in the specific bioassay media, phase II and extramicrosomal metabolic reactions, permeability of low-solubility compounds, and active transport. When an important research question goes beyond the appropriateness of a generic method to adequately answer, a customized assay can be developed and applied to obtain data that are reliable and more definitive for informed decisions.

21.8 Diagnose Inadequate Performance in Complex Systems Using Individual Properties

If inadequate in vivo performance is observed in PK studies, in vitro assays and structural rules can be used to diagnose the limiting property. Custom in vitro assays may be needed to test properties that have not been previously tested for the compound. If a limiting property is found, structure modifications can be made to improve this specific property. The new compound is retested using an in vitro assay to check for improvement of the property. If improvement of the property is observed, the compound can be retested in vivo to see if PK performance is improved. Testing first in vitro saves expensive in vivo resources. This diagnostic process is illustrated in Figure 21.4. This strategy works well as part of the lead optimization phase of discovery.

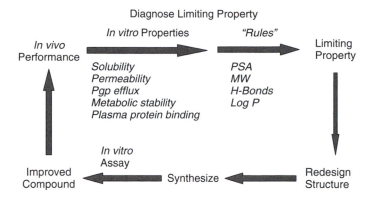

Figure 21.4 ▶ Diagnose inadequate in vivo performance using in vitro assay data and structural rules to identify the limiting property. Use these data to redesign the structure and test for property improvement.

Problems

(Answers can be found in Appendix I at the end of the book.)

1. Why is it inefficient to wait to optimize properties until after activity is optimized?

2. Having access to property data in 1 week, as opposed to 3 weeks, provides what benefit for project teams?

3. Why is single-property assay data more useful for medicinal chemists than data from assays that involve multiple properties?

References

1. Ellingboe, J. (2005). *The application of pharmaceutical profiling data to lead identification and optimization.* Abstracts, 37th Middle Atlantic Regional Meeting of the American Chemical Society, New Brunswick, NJ, United States, May 22–25, 2005, GENE-231.

2. Di, L., & Kerns, E. H. (2003). Profiling drug-like properties in discovery research. *Current Opinion in Chemical Biology, 7,* 402–408.

Part 4
Methods

Methods for Profiling Drug-like Properties: General Concepts

Overview

▶ *Property profiling should be rapid and use relevant assay conditions.*

▶ *Use a diverse set of assays that have high impact for the organization.*

▶ *Property assays consume resources and influence projects, so assays should be carefully implemented.*

▶ *Evaluate the cost versus benefit of assays.*

The assessment of compound properties has become an integral activity in drug discovery. It provides data for the selection of leads, the development of structure–property relationships (SPR), and the judging of whether project advancement criteria have been met. The data also guide property optimization by structural modification. Several aspects of the property-profiling program that should be carefully considered are discussed here.

22.1 Property Data Should be Rapidly Available

In order for property data to be relevant for discovery, they need to be reported to project teams rapidly.[1] This allows decisions to be made rapidly so that aggressive discovery time lines can be met. Faster data allow more iterations of trying new ideas to optimize the compounds, thus increasing the success rate. A general guide is to provide data in a few days to 2 weeks, the same time frame as biological data.

22.2 Use Relevant Assay Conditions

The conditions of property assays need to be relevant to the environment faced by the compound.[2] Variables, such as concentration, pH, matrix components in solution, and biological tissue extracts, must be controlled and designed to be reflective of the barriers faced by the compound.

22.3 Evaluate the Cost-to-Benefit Ratio for Assays

In any discovery organization there are always tradeoffs on the allocation of resources. Thus, it is important to decide which properties are of greatest importance to the organization.[3] The properties that have the greatest impact on the projects and goals of the organization should prevail. Assays should not be put in place because other organizations have them but should be decided based on what the organization considers to be the critical issues.

One approach to resource allocation is to use the "appropriate assay" for the particular stage of discovery. Higher-throughput and simpler methods can be used in earlier discovery stages, when the purpose is to select compounds that are in the right general portion of property space. In later discovery stages the purpose is to carefully optimize lead series; thus, more detailed methods are appropriate. This approach mirrors the strategy in discovery biology, where in vitro enzyme or receptor assays are used early, while functional cell-based assays and in vivo models are used in later stages. This approach balances data needs and resources so that there is not an "overkill" of detailed data from resource-intensive assays early and not overinterpretation of, or reliance on, undetailed data later in the process. Resources should be used wisely and evaluation of the cost-to-benefit ratio of assays should be constant.

An example of overkill is the use of Caco-2 for hits from high-throughput screening (HTS) or virtual screening, when structural rules on lipophilicity or polar surface area or high-throughput parallel artificial membrane permeability assay (PAMPA) methods would provide sufficient information for selecting the most favorable from among the fatally flawed compounds. An example of overinterpretation is the use of data for solubility in an aqueous buffer at pH 7.4 as the primary solubility value in late discovery, when solubility can be very different with the different pHs and concentrations of bile salts and lipids present throughout the length of the intestine.

Rules and in silico tools are useful for rapid and inexpensive screening of compounds to check if they meet certain criteria or to form a rapid opinion about whether the properties will be drug-like. *High-throughput in vitro assays* (HT) can process hundreds of compounds per day but tend to use generic conditions. HT assays are most appropriate for earlier drug discovery studies where a large number of compounds are studied in a short period of time and the data need not be highly detailed to be adequate for the decisions made in this stage. *Customized assays* usually are developed and used to answer specific questions for project teams. They can have quite an impact on team success. *In-depth assays* provide detailed information that is necessary for advancement to clinical development. They test specific issues and help to predict and plan for development studies that will be necessary to develop the candidate.

22.4   Choose an Ensemble of Key Properties to Evaluate

Each organization must select the properties they are most interested in monitoring. For example, the assays shown in Figure 22.1 are from one company's property profiling

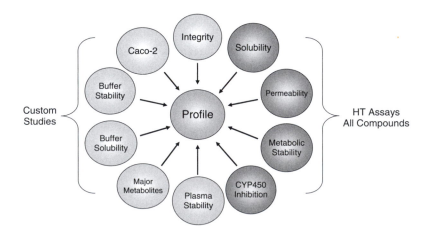

Figure 22.1 ▶   An ensemble of property assays can be available to support discovery, some as HT assays for all compounds and others as custom assays for selected compounds. (Reprinted with permission from [3].)

program.[4] Some assays can be run as standard assays for all compounds, while others are best used as custom assays to answer specific project team questions.

22.5 Use Well-Developed Assays

Data used for decision-making should be generated using assays that have been well developed. The assay conditions can greatly affect the results. For example, if the compound is not soluble in the assay, if a high compound concentration saturates the enzyme, if substrate turnover is too high, or if a co-solvent inhibits the enzyme, the data will be misleading. It is important that scientists generating the data fully validate the method against the known properties of well-characterized compounds, and that scientists using the data know how it was generated and discuss with the data generator how particular characteristics of the compounds in their series might affect the assay.

Problems

(Answers can be found in Appendix I at the end of the book.)

1. Is it useful to use data from a generic pH 7.4 solubility assay to predict solubility in the GI tract?

2. Is it useful to obtain Caco-2 A > B and B > A permeability data for 200 HTS hits for lead selection?

3. Is PAMPA data at pH 7.4 sufficient for inclusion in the data package for clinical candidate nomination?

4. Is microsomal stability data sufficient for assessing the potential of hydrolysis?

5. Is microsomal stability in multiple species necessary during lead optimization?

References

1. Di, L., & Kerns, E. H. (2005). Application of pharmaceutical profiling assays for optimization of drug-like properties. *Current Opinion in Drug Discovery & Development, 8*, 495–504.

2. Di, L., & Kerns, E. H. (2003). Profiling drug-like properties in discovery research. *Current Opinion in Chemical Biology, 7*, 402–408.

3. Kerns, E. H. (2001). High throughput physicochemical profiling for drug discovery. *Journal of Pharmaceutical Sciences, 90*, 1838–1858.

4. Kerns, E. H., & Di, L. (2003). Pharmaceutical profiling in drug discovery. *Drug Discovery Today, 8*, 316–323.

Lipophilicity Methods

Overview

▶ *In silico lipophilicity predictions are reliable and inexpensive.*

▶ *High-throughput methods include partitioning in a microtiter plate, high-performance liquid chromatography, and capillary electrophoresis.*

▶ *In-depth methods utilize equilibrium shake flask partitioning and titration.*

Lipophilicity is one of the most studied physicochemical properties. It also has one of the most developed sets of methods for its prediction and measurement. Lipophilicity values usually are measured and published for new drugs as part of the development process. Many laboratories have measured Log P and Log D values as part of their research. An extensive set of published lipophilicity values for drugs and other compounds has been compiled by Hansch and Leo.[1]

Although a lipophilicity experiment at first may seem simple and easy, it is important to remember the effect of conditions on lipophilicity (see Chapter 5). The following conditions should be carefully controlled and listed when reporting lipophilicity data: partitioning solvents, pH, ionic strength, temperature, buffer composition, co-solutes, co-solvents, and equilibration time. The data can vary significantly with the different conditions.

23.1 In Silico Lipophilicity Methods

Databases provide a rich source of lipophilicity data for historic compounds. The Med-Chem Database from BioByte Corp. is available through Daylight Chemical Information Systems (*www.daylight.com/products/Medchem.html*). It contains Log P values for 61,000 compounds.

Software tools for the prediction of properties typically are developed using a set of compounds for which the property has been measured. The quality of this "training" dataset is higher with greater structural diversity and the reliable measurement of the property data. Following development of the algorithm, the software should be tested using a separate "validation" set of compounds. The widespread and long-term use of octanol/water partitioning has produced a large published set of compounds that have been carefully measured. Thus, Log P and Log D prediction software can be some of the most reliable in silico tools.

Many commercial software packages for lipophilicity are available. A partial list is given in Table 23.1. Web sites are available for free calculations of Log P by the entry of a SMILES string (e.g., *www.vcclab.org*). Even structure drawing programs, such as ChemDraw, include Log P calculation software.

Hansch and Leo[1] developed the fragment method for calculating Log P. The Log P contributions of a large number of substructures have been determined. The Log P of a new

TABLE 23.1 ▶ Partial List of Commercial Software for Log P and Log D Calculation

Name	Company	Web site
CLog P	Daylight Chemical Information	*www.daylight.com*
PrologP, PrologD	CompuDrug	*www.compudrug.com*
Bio-Loom	BioByte	*www.biobyte.com*
KowWin	Syracuse Research Corporation	*www.syrres.com/esc/kowwin.htm*
Log*D and* Log*P* DB	Advanced Chemistry Development	*www.acdlabs.com/products/*
QikProp	Schrodinger	*www.schrodinger.com*
ADMET Predictor	Simulations Plus	*www.simulations-plus.com*
ADME Boxes	Pharma Algorithms	*www.ap-algorithms.com*
KnowItAll	Biorad	*www.biorad.com*
ALOGPS	Virtual Computational Chemistry Lab	*www.vcclab.org*
Molinspiration Prop. Calculator	Molinspiration	*www.molinspiration.com*
CSLogP, CSLogD	ChemSilico	*www.chemsilico.com*
SLIPPER-2001	ChemDB	*http://software.timtec.net*
DSMedChem Explorer	Accelrys	*www.accelrys.com*
SciLogP	Scivision	*www.amazon.com*
ChemDraw	CambridgeSoft	*www.cambridgesoft.com*

structure is predicted by breaking it up into these substructures and calculating the sum of the individual contributions. Other Log P calculation approaches use neural net algorithms to construct a model using structural descriptors that correlate to Log P.

The user should not assume that all software produces the same Log P or Log D values, or that the values will be accurate when compared to laboratory measurements. Data for comparison of data obtained from the software package Prolog D with literature values[1,2] are listed in Table 23.2. The average difference between the predicted and measured values is about 1.05 log units. This correlation is represented in the graph in Figure 23.1. $R^2 = 0.72$ indicates a typical correlation between experimental and calculated values of Log D. The same type of comparison for structurally diverse discovery compounds, which typically have less drug-like properties than commercial drugs, are shown in Figure 23.2. The experimental determinations of Log D were performed at Wyeth Research using the gold standard pH-metric assay (see Section 23.3.2). Figure 23.2 shows how discovery compounds follow the same trend as commercial compounds but often have a lower correlation between calculated and measured properties. In addition to having generally poorer drug-like properties, discovery compounds may contain substructures that are not covered by the software development training set. Therefore, the full contribution of each of the discovery compound's substructures or descriptors to Log D may not be fully calculated.

Another aspect of predictive software is its use within a series of analogs around the same core scaffold or template. Software usually is reliable when comparing the structure–lipophilicity relationships of compounds in a series. Thus, although the predicted Log D may differ from the experimental Log D by an average of 1.05 log units (see above), the comparison of compounds within a series involves small substructural differences, and the software should be more predictive for indicating the increasing or decreasing lipophilicity trends resulting from substructural modification.

TABLE 23.2 ▶ Comparison of Literature Values and Calculated Values of Log D (Using Prolog D) for 70 Commercial Drugs

Compound	Literature Value	pLogD 7.4	Difference	Compound	Literature Value	pLogD 7.4	Difference
Acetylsalicylic Acid	-1.14	-1.58	0.44	Oxprenolol	0.32	1.16	-0.84
Salicylic Acid	-2.11	-2.99	0.88	Labetalol	1.07	-1.2	2.27
Acetaminophen	0.51	0.61	-0.1	Flurbiprofen	0.91	0.33	0.58
Amoxicillin	-1.35	0.08	-1.43	Ibuprofen	1.37	0.85	0.52
Theophylline	-0.02	-2.03	2.01	Propranolol	1.26	1.69	-0.43
Ceftriaxone	-1.23	-5.61	4.38	Alprenolol	0.97	2.14	-1.17
Terbutaline	-1.35	-1.39	0.04	Dexamethasone	1.83	1.49	0.34
Metoprolol	-0.16	0.4	-0.56	Oxazepam	2.13	0.36	1.77
Atenolol	-1.38	-1.21	-0.17	Corticosterone	1.82	2.83	-1.01
Sulpiride	-1.15	-1.08	-0.07	Chloramphenicol	1.14	0.69	0.45
Cephalexin	-1.45	1.17	-2.62	Lorazepam	2.51	1.03	1.48
Pindolol	-0.21	0.58	-0.79	Desipramine	1.28	1.3	-0.02
Nadolol	-1.21	-0.62	-0.59	Promazine	2.52	2.16	0.36
Timolol	-0.047	-0.92	0.873	Imipramine	2.4	2.72	-0.32
Acebutolol	-0.29	0.18	-0.47	Diltiazem	2.06	2.93	-0.87
Warfarin	1.12	-0.42	1.54	Verapamil	1.99	4.98	-2.99
Ketoprofen	-0.13	-0.18	0.05	Triphenylene	5.49	6.71	-1.22
Sulfasalazine	0.08	0.91	-0.83	L-Dopa	-2.57	-2.13	-0.44

Compound				Compound			
3,4,5 - Trihydroxybenzoic Acid	−0.40	−2.22	1.82	Bifonazole	4.77	5.29	−0.52
3,5 - Dinitrobenzoic Acid	0.91	−1.9	2.81	Diethystilbestrol	5.07	6.25	−1.18
Pipemidic Acid	−1.52	−1.78	0.26	Clotrimazole	5.20	5.4	−0.2
2,4 - Dihydroxybenzoic Acid	2.06	−3.2	5.26	Ephedrine	−1.48	−0.82	−0.66
Furosemide	−1.02	−1.44	0.42	Sotalol	−1.35	−1.83	0.48
Sulfamerazin	−0.12	−0.73	0.61	Sumatriptaon	−1	−0.39	−0.61
Sulfathiazole	−0.43	0.55	−0.98	Disopyramide	−0.66	0.51	−1.17
Naproxen	0.3	0.18	0.12	Atropine	−0.25	0.57	−0.82
Allopurinol	−0.44	−3.74	3.3	Ranitidine	−0.29	1.14	−1.43
Thiamphenicol	−0.27	−0.4	0.13	Procaine	0.33	0.75	−0.42
Caffeine	−0.07	−1.8	1.73	Triflupromazine	3.61	3.31	0.3
Metronidazole	−0.02	−0.72	0.7	Clozapine	3.13	3.55	−0.42
Nitrofurazone	0.23	−0.98	1.21	Thioridazine	3.34	3.59	−0.25
Prednisone	1.41	0.96	0.45	Bupivacaine	2.65	4.37	−1.72
Carbamazepine	2.19	2.2	−0.01	Chlorpromazine	3.38	2.9	0.48
Testosterone	3.29	4.81	−1.52	Loratadine	4.4	5.61	−1.21
Estradiol	4.01	4.96	−0.95	Amiodarone	6.1	8.12	−2.02
				Average Difference:			1.05

The average difference is 10.5 log units

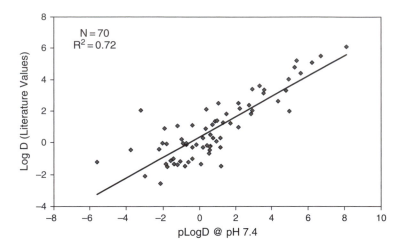

Figure 23.1 ▸ Correlation of literature values and calculated values of Log D (using Prolog D) for 70 commercial drugs, as listed in Table 23.2. $R^2 = 0.72$ indicates a typical correlation between experimental and calculated Log D.

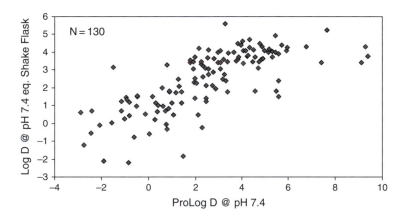

Figure 23.2 ▸ Correlation of experimental shake flask and calculated values of Log D for 130 discovery compounds at Wyeth Research. The general trends are comparable to the correlation for the commercial compounds.

Log P and Log D prediction from software has many advantages. Predictions are reasonably reliable. Software is fast and highly accessible, often via the company's computer network. The medicinal chemist does not need to have compound material to assay. There is no interference to the measurement by impurities, low solubility, or the need for co-solvents. Software covers a wide dynamic range. In silico predictions are much less expensive than experimental measurements, thus saving company resources for the measurement of other important properties.

23.2 Experimental Lipophilicity Methods

The three main in vitro high-throughput methods used for lipophilicity measurement are scaled-down shake flask, reversed-phase high-performance liquid chromatography (HPLC), and capillary electrophoresis (CE). Each method is meant to model the partitioning environment and can use internal standards or known compounds for calibration of the analytical response to Log P or Log D. Table 23.3 tabulates the methods and indicates characteristics such as throughput.

TABLE 23.3 ▶ Methods for Determination of Lipophilicity

Method	Type assay	Speed (min/cpd)[a]	Throughput (cpd/day/instrument)[b]
Scaled-down shake flask	High throughput	10	100
Reversed-phase HPLC	High throughput	5	200
Capillary electrophoresis	High throughput	10	100
Shake flask	In depth	60	20
pH-metric	In depth	60	20

[a] min/cpd: minutes per analysis of one compound
[b] cpd/day/instrument: number of compounds that can be assayed in one day using one instrument

23.2.1 Scaled-Down Shake Flask Method for Lipophilicity

The octanol/water partitioning experiment was traditionally performed in larger vials and flasks (see Section 23.3.1), but the procedure can be scaled down to the titer plate level for higher throughput.[3–6] As shown in Figure 23.3, a 1-mL deep, 96-well plate can be filled with 0.5 mL of aqueous buffer and 0.5 mL of octanol to form the shake flask setup. The test compound is dissolved in dimethylsulfoxide (DMSO), and a small volume is added to the well. DMSO can alter the experiment via interaction with the analyte and solvents, so the volume should be kept as low as possible (<1% of the aqueous volume). It is important to control the ionic strength (e.g., 0.15 M NaCl) and the buffer pH and molarity in order to achieve comparable results between compounds and on different days. The plate is sealed and then agitated to produce good mixing between the phases. An aliquot is obtained from each of the phases and is analyzed, typically by HPLC, to determine the concentration of compound in each phase. A complication of the method is that the concentration of compound in each phase can be quite different, depending on the difference of Log D from 0 (equal concentration in each phase). Also, injection of octanol onto the HPLC can disrupt the chromatography of the analyte, compared to injection in aqueous phase. Carryover from one HPLC injection to the next should be checked and reduced to a minimum. Analiza has commercialized the scaled-down shake flask method into an integrated instrument. Commercial instruments for various lipophilicity methods are listed in Table 23.4.

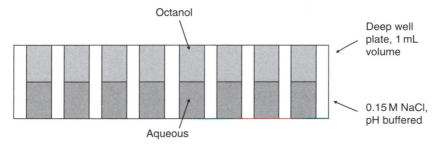

Figure 23.3 ▶ Scaled-down shake flask method for lipophilicity.

TABLE 23.4 ► Commercial Instruments for Measurement of Log D and Log P

Method	Product name	Company	Web site
Scaled-down shake flask	AlogPW	Analiza	*www.analiza.com*
Reversed-phase HPLC	PCA200	Sirius Analytical Instruments	*www.sirius-analytical.com*
Reversed-phase HPLC	Veloce	Nanostream	*www.nanostream.com*
Capillary electrophoresis	cePRO 9600	CombiSep	*www.combisep.com*
pH-metric	GLpKa	Sirius Analytical Instruments	*www.sirius-analytical.com*

23.2.2 Reversed-Phase HPLC Method for Lipophilicity

Reversed phase HPLC is fundamentally a separation scheme involving two immiscible media. It is technically related to a multistage partitioning experiment. Successive partitioning of the solute between the aqueous HPLC mobile phase and the stationary organic phase performs multiple partitions along the length of the HPLC column. The HPLC retention time is affected by these partitions. The retention time increases as the compound has higher affinity for the organic stationary phase than the aqueous phase. A common reversed-phase HPLC separation uses octadecane groups attached to the stationary support particles.

The lipophilicity of compounds is assessed by first injecting a series of standards for which Log D or Log P has already been determined using definitive analytical methods. The retention times of the standards are plotted against their respective previously measured Log D values (Figure 23.4). When the test compound is run, its retention time (t_R) is

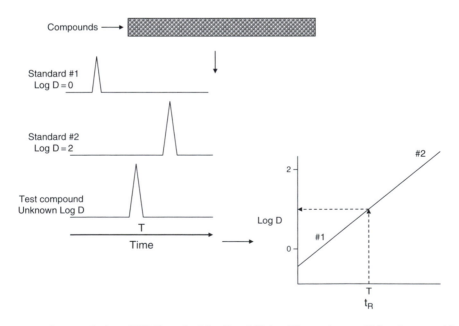

Figure 23.4 ► Reversed-phase HPLC method for lipophilicity. The stationary C18 column packing simulates the nonpolar phase and the mobile phase simulates water. t_R of the standards are plotted against their previously measured Log D. t_R of the test compound is compared to this calibration to determine the compound's Log D.

compared to the calibration curve to determine its Log D. Several articles have reported this type of method.[2,7−22] Any of the commercial HPLC instruments can be used. Nanostream has developed a 24-parallel HPLC system in one instrument for high-throughput analysis that accelerates lipophilicity determination. Sirius Analytical Instruments has developed an integrated instrument for Log D based on the HPLC approach. For HPLC-based methods, it is necessary to ensure that the same mobile phase and column product are always used because t_R varies with conditions.

23.2.3 Capillary Electrophoresis Method for Lipophilicity

As with the HPLC method, CE is another chromatographic method that, utilizing the microemulsion electrokinetic chromatographic (MEEKC) technique, has been used to determine lipophilicity.[21] The test compound partitions between the organic nonpolar microemulsion phase and the aqueous phase. When the compound is in the microemulsion phase, it moves more slowly through the column. Thus, the greater affinity it has for the lipophilic microemulsion phase, the longer will be t_R, which is calibrated to Log D in the same manner as in the HPLC method. CombiSep has developed an integrated 96-channel CE system for determination of lipophilicity.

23.3 In-Depth Lipophilicity Methods

The two primary in-depth methods for lipophilicity measurement are shake flask and pH-metric. Both methods provide reliable data for definitive lipophilicity determination of compounds. Both require significantly greater resources than the high-throughput methods, so they are applied to relatively few compounds. These methods are primarily used as gold standards for compounds entering development toward the end of discovery.

23.3.1 Shake Flask Method for Lipophilicity

The traditional shake flask method for lipophilicity is diagrammed in Figure 23.5. Solid test compound is placed into a flask or vial, and measured volumes of octanol and water are added. The flask is agitated for 24 to 72 hours, and the test compound in each phase is

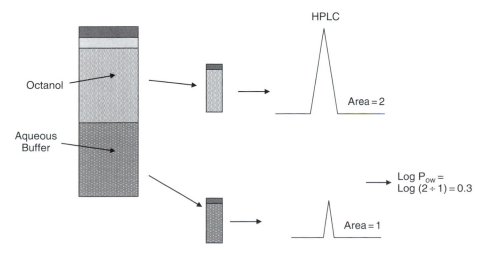

Figure 23.5 ▶ Traditional shake flask method for lipophilicity.

sampled and quantitated using HPLC. The areas under the peaks, or concentrations calculated from a standard curve, are divided and the log taken. The shake flask method is considered a gold standard for Log P (Log D) determination. It is necessary to control all of the conditions listed in the introduction to this chapter to obtain the highest quality data.

23.3.2 pH-Metric Method for Lipophilicity

Titration is a common method for determining pK_a. However, it also can be used to determine lipophilicity.[22] The test compound is first titrated by the addition of known equivalents of acid or base to produce a titration curve (Figure 23.6). Then the titration is repeated in the presence of octanol. A shift of the titration curve is obtained, owing to partitioning of the test compound into octanol. From the curve shift the lipophilicity is calculated. Sirius Analytical Instruments has developed an instrument that is in widespread use in the industry for pK_a and Log P.

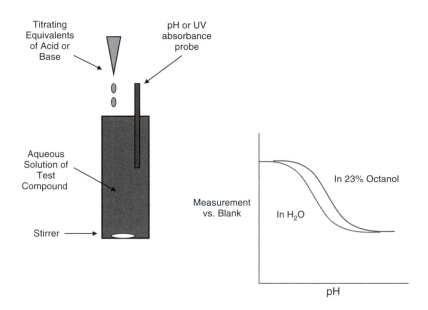

Figure 23.6 ► pH-metric method for lipophilicity.

▢ Problems

(Answers can be found in Appendix I at the end of the book.)

1. Why are in silico predictions often poorer for drug discovery compounds than for commercial drug compounds?

2. Are in silico predictions for lipophilicity any better than other property predictions?

3. Which of the following can affect measurement of Log D data?: (a) % DMSO, (b) the buffer components, (c) humidity, (d) temperature, (e) ionic strength, (f) pH, (g) time.

4. With chromatographic HPLC and CE methods for lipophilicity, how does the analyst correlate retention time to Log D?

5. In general, how close should one expect in silico Log D predictions to be to actual values?

6. Which of the following are common methods of measuring Log P?: (a) partition between octanol and aqueous buffer, (b) infrared, (c) NMR in the presence of octanol, (d) HPLC, (e) titration in the presence of octanol.

References

1. Hansch, C., Leo, A., & Hoekman, D. (1995). *Exploring QSAR. Fundamentals and applications in chemistry and biology, vol. 1. Hydrophobic, electronic and steric constants, vo. 2.* New York: Oxford University Press.

2. Lombardo, F., Shalaeva, M. Y., Tupper, K. A., Gao, F., & Abraham, M. H. (2000). ElogPoct: a tool for lipophilicity determination in drug discovery. *Journal of Medicinal Chemistry, 43,* 2922–2928.

3. Gulyaeva, N., Zaslavsky, A., Lechner, P., Chlenov, M., McConnell, O., Chait, A., et al. (2003). Relative hydrophobicity and lipophilicity of drugs measured by aqueous two-phase partitioning, octanol-buffer partitioning and HPLC. A simple model for predicting blood-brain distribution. *European Journal of Medicinal Chemistry, 38,* 391–396.

4. Andersson, J. T., & Schraeder, W. (1999). A method for measuring 1-octanol-water partition coefficients. *Analytical Chemistry, 71,* 3610–3614.

5. Wang, X., Xu, R., Wilson, D. M., Rourick, R. A., & Kassel, D. B. (2000). High throughput log D determination using parallel liquid chromatography/mass spectrometry. In *American Society for Mass Spectrometry: 48th Annual Conference on Mass Spectrometry and Allied Topics,* Long Beach, CA.

6. Zaslavsky, B., & Chait, A. (2000). Buffer effects on partitioning of organic compounds in octanol-buffer systems: lipophilicity in drug disposition. Log P 2000 Symposium: Lipophilicity in Drug Disposition, March 5–9, 2000, Lausanne, Switzerland.

7. Minick, D. J., Frenz, J. H., Patrick, M. A., & Brent, D. A. (1988). A comprehensive method for determining hydrophobicity constants by reversed-phase high-performance liquid chromatography. *Journal of Medicinal Chemistry, 31,* 1923–1933.

8. Lambert, W. J., Wright, L. A., & Stevens, J. K. (1990). Development of a preformulation lipophilicity screen utilizing a C-18-derivatized polystyrene-divinylbenzene high-performance liquid chromatographic (HPLC) column. *Pharmaceutical Research, 7,* 577–586.

9. Detroyer, A., Schoonjans, V., Questier, F., Vander Heyden, Y., Borosy, A. P., Guo, Q., et al. (2000). Exploratory chemometric analysis of the classification of pharmaceutical substances based on chromatographic data. *Journal of Chromatography, A, 897,* 23–36.

10. Yamagami, C., Araki, K., Ohnishi, K., Hanasato, K., Inaba, H., Aono, M., et al. (1999). Measurement and prediction of hydrophobicity parameters for highly lipophilic compounds: application of the HPLC column-switching technique to measurement of log P of diarylpyrazines. *Journal of Pharmaceutical Sciences, 88,* 1299–1304.

11. Abraham, M. H., Gola, J. M. R., Kumarsingh, R., Cometto-Muniz, J. E., & Cain, W. S. (2000). Connection between chromatographic data and biological data. *Journal of Chromatography, B: Biomedical Sciences and Applications, 745,* 103–115.

12. Valko, K., Bevan, C., & Reynolds, D. (1997). Chromatographic hydrophobicity index by fast-gradient RP-HPLC: a high-throughput alternative to log P/log D. *Analytical Chemistry, 69,* 2022–2029.

13. Valko, K., Plass, M., Bevan, C., Reynolds, D., & Abraham, M. H. (1998). Relationships between the chromatographic hydrophobicity indices and solute descriptors obtained by using several reversed-phase, diol, nitrile, cyclodextrin and immobilized artificial membrane-bonded high-performance liquid chromatography columns. *Journal of Chromatography, A, 797,* 41–55.

14. Du, C. M., Valko, K., Bevan, C., Reynolds, D., & Abraham, M. H. (1998). Rapid gradient RP-HPLC method for lipophilicity determination: a solvation equation based comparison with isocratic methods. *Analytical Chemistry, 70,* 4228–4234.

15. Valko, K., Du, C. M., Bevan, C., Reynolds, D. P., & Abraham, M. H. (2001). Rapid method for the estimation of octanol/water partition coefficient (Log Poct) from gradient RP-HPLC retention and a hydrogen bond acidity term (Sa2H). *Current Medicinal Chemistry, 8,* 1137–1146.

16. Abraham, M. H. (1993). Scales of solute hydrogen-bonding: their construction and application to physico-chemical and biochemical processes. *Chemical Society Reviews*, *22*, 73–83.

17. Abraham, M. H., Chadha, H. S., Leitao, R. A. E., Mitchell, R. C., Lambert, W. J., Kaliszan, R., et al. (1997). Determination of solute lipophilicity, as log P(octanol) and log P(alkane) using poly(styrene-divinylbenzene) and immobilized artificial membrane stationary phases in reversed-phase high-performance liquid chromatography. *Journal of Chromatography, A*, *766*, 35–47.

18. Pagliara, A., Khamis, E., Trinh, A., Carrupt, P.-A., Tsai, R.-S., & Testa, B. (1995). Structural properties governing retention mechanisms on RP-HPLC stationary phases used for lipophilicity measurements. *Journal of Liquid Chromatography*, *18*, 1721–1745.

19. Abraham, M. H., Chadha, H. S., & Leo, A. J. (1994). Hydrogen bonding. XXXV. Relationship between high-performance liquid chromatography capacity factors and water-octanol partition coefficients. *Journal of Chromatography, A*, *685*, 203–211.

20. Kerns, E. H., Di, L., Petusky, S., Kleintop, T., Huryn, D., McConnell, O., et al. (2003). Pharmaceutical profiling method for lipophilicity and integrity using liquid chromatography-mass spectrometry. *Journal of Chromatography, B: Analytical Technologies in the Biomedical and Life Sciences*, *791*, 381–388.

21. Poole, S. K., Durham, D., & Kibbey, C. (2000). Rapid method for estimating the octanol-water partition coefficient (log Pow) by microemulsion electrokinetic chromatography. *Journal of Chromatography, B: Biomedical Sciences and Applications*, *745*, 117–126.

22. Slater, B., McCormack, A., Avdeef, A., & Comer, J. E. A. (1994). pH-Metric log P. 4. Comparison of partition coefficients determined by HPLC and potentiometric methods to literature values. *Journal of Pharmaceutical Sciences*, *83*, 1280–1283.

Chapter 24

pK_a Methods

Overview

► *In silico* pK_a *predictions can be reliable, calculate tautomers, and indicate which ionizable center has which* pK_a.

► pK_a *can be predicted in higher throughput using spectral gradient analysis and capillary electrophoresis techniques.*

► *The in-depth method for* pK_a *is titration.*

Knowledge of the pK_a of substructures in the lead structure allows the "tuning" of ionizability in order to modify its permeability and solubility. Like lipophilicity, pK_a has been studied for many years, providing a solid body of research and quality measurements upon which reliable tools for pK_a prediction for compounds can be based for application in medicinal chemistry.

24.1 In Silico pK_a Methods

Databases can provide a useful source of comparison pK_a data for compounds that have been reported (Table 24.1). For example, the Daylight/BioByte product contains pK_a values

TABLE 24.1 ► Partial List of Commercial Software and Instruments for Determination of pK_a

Name	Company	Web site
Software		
MedChem Database	Daylight Chemical Information	*www.daylight.com*
Masterfile/CQSAR	BioByte	*www.biobyte.com*
pK_a DB	Advanced Chemistry Development	*www.acdlabs.com/products/*
pKalc (formerly PALLAS)	CompuDrug	*www.compudrug.com*
ADMET Predictor	Simulations Plus	*www.simulations-plus.com*
ADME Boxes	Pharma Algorithms	*www.ap-algorithms.com*
KnowItAll	Biorad	*www.biorad.com*
CSpKa	ChemSilico	*www.chemsilico.com*
SPARC	University of Georgia	*http://ibmlc2.chem.uga.edu/sparc*[a]
Instruments		
SGA	Sirius Analytical Instruments	*www.sirius-analytical.com*
GLpKa	Sirius Analytical Instruments	*www.sirius-analytical.com*
cePRO 9600	CombiSep	*www.combisep.com/pKa.html*

[a] Online calculator.

for 13,900 compounds, and ACD pKa/DB has 16,000 structures with over 31,000 experimental pK_a.

Software for the prediction of pK_a from structure has greatly progressed. A partial listing of available software is provided in Table 24.1. As with any software, the predictability should be validated by comparison to known compounds that have been measured using high-quality "gold standard" methods.

An example of the correlation of predictive software with measurements is shown in Figure 24.1. The compound set has 98 diverse real-world drug discovery compounds that were synthesized, measured, and predicted using ACD/pKa DB software at Wyeth Research. The R^2 for the correlation is 0.90, which is very good for in silico predictions.

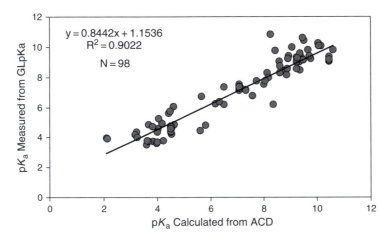

Figure 24.1 ▶ Correlation of pK_a measured using the GLpKa technique and calculated using software from ACDLabs.

pK_a software can provide more advanced information for chemists. Figure 24.2 shows how ACDLabs software provides structural assignment of which ionization center in the molecule has which pK_a. In addition, the different pK_a values of tautomers can be predicted.

Software can be applied when planning the synthesis of structural series analogs, when the goal is to modify permeability or solubility by the substitution of substructures with

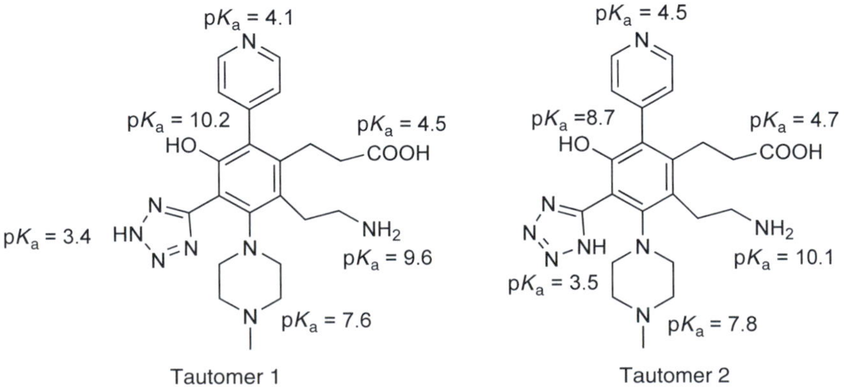

Tautomer 1   Tautomer 2

Figure 24.2 ▶ In silico products often indicate which ionization centers have which pK_a, as well as predicting pK_a for tautomers. These results are from ACDLabs software.

varying pK_a values in order to optimize absorption. The most promising of all the possible modifications then can be synthesized, or several analogs distributed over a pK_a range can be selected for synthesis.

pK_a has been measured accurately for many diverse compounds. Its relationship to structure has been studied in detail. Thus, in many cases, pK_a software is very predictive and is sufficient for many purposes of medicinal chemists.

24.2 Experimental pK_a Methods

Two higher-throughput methods have been introduced. Each can be implemented in drug discovery laboratories with the proper equipment, and each has been offered as an integrated commercial instrument. For both of these methods, only a small quantity of compound is needed (<0.1 mg), and throughput is about 5 minutes per compound.

24.2.1 Spectral Gradient Analysis Method for pK_a

Spectral gradient analysis (SGA) was invented at GSK by Bevan et al.[1,2] It is based on the concept of a gradient high-performance liquid chromatography pump, with the substitution of aqueous acidic and basic buffers for the two mixed liquid phases. The instrument is diagrammed in Figure 24.3. The test compound is dissolved in dimethylsulfoxide (DMSO) at 10 mM and placed in a 96-well plate. Each solution in the plate is diluted with aqueous buffer. The aqueous solution then is continuously mixed into the aqueous pH buffer. Throughout the 2-minute experiment, a gradient program is run, which starts with a high percentage of one buffer and progresses to a low percentage of that buffer over 2 minutes. Thus, the test compound is mixed with a continuously changing pH. As the pH changes, the fraction of the compound that is ionized changes. The absorption of an ultraviolet or visible chromophore, which is near the ionization center (within three to four bonds), changes with ionization. As the mixture flows into the diode array UV detector/spectrometer, the ultraviolet/visual (UV/VIS) absorption changes with pH (Figure 24.4). pK_a is the inflection point of this absorption curve. The instrument can produce a pK_a screen for test compounds about every 3 to 4 minutes for high throughput. A commercial instrument is available (Table 24.1).

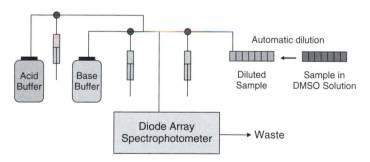

Figure 24.3 ► Spectral gradient analysis pK_a instrument diagram.

24.2.2 Capillary Electrophoresis Method for pK_a

Higher-throughput pK_a measurement is available using capillary electrophoresis (CE).[3-9] This is based on the different electrophoretic mobility of a compound in the neutral and ionized form. The test compound is diluted into aqueous buffer and injected into a CE column. The compound is run several times using CE mobile phase buffers at different

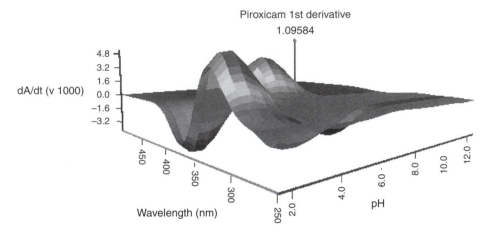

Figure 24.4 ▶ Spectral gradient analysis (SGA) pK_a system UV/VIS absorbance data output for piroxicam and its use in pK_a screening determination. (Reprinted with permission of John Comer.)

pHs. Ionized molecules move through the mobile phase faster because CE mobility is proportional to charge. Thus, as the fraction of ionized molecules increases, the effective mobility (retention time) is progressively shorter. By plotting the effective mobility versus mobile phase pH, pK_a is calculated as the inflection point.

The CE pK_a method can be automated using commercial CE instruments. A commercial instrument that runs 96 experiments in parallel for high throughput and has pK_a processing software is available (Table 24.1). The CE method separates components and, therefore, has low potential for interference from impurities in the sample.

24.3 In-Depth pK_a Method: pH-Metric

The definitive method for pK_a determination is potentiometric titration. A diagram of this approach is shown in Figure 24.5. The compound is dissolved in water and titrated with an acidic or basic buffer of known molarity. In the classic method, the pH of the test compound solution changes as the titrant is added. This change is monitored using a pH electrode. This change in pH with titrating equivalents is plotted versus solution pH. pK_a is the pH of the inflection point of the curve. A variation of this method is to measure the UV absorbance

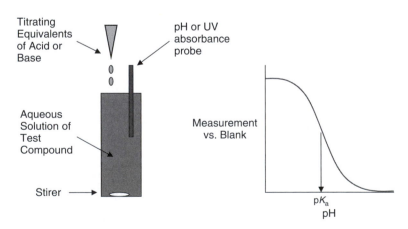

Figure 24.5 ▶ In-depth titration method for pK_a determination is the pH-metric method.

using a UV spectrophotometric probe in the solution of the test compound. As in the case of the SGA method, the absorbance of a chromophore near the ionization center changes with ionization and can be monitored to determine the extent of ionization.

This method has been studied in detail by Avdeef et al.[10] and termed the *pH-metric method*. It has been integrated as the GLpK_a instrument that is commercially available (Table 24.1). This method is considered the "gold standard" for pK_a and Log P analysis in drug development laboratories. If the compound has low solubility, a co-solvent is added to the test compound solution. Titrations at three co-solvent concentrations allow extrapolation to zero co-solvent aqueous concentration. The throughput for pK_a measurements is about 0.5 to 1 hour per soluble compound and up to 2 hours per compound requiring needs the co-solvent method.[11]

Problems

(Answers can be found in Appendix I at the end of the book.)

1. Detection methods for pK_a use the change in what three measurable parameters?

2. Rank the following pK_a methods in terms of throughput: CE, pH-metric, SGA, in silico.

3. What useful features can advanced software for pK_a provide?

4. Is low solubility a potential issue for pK_a measurement?

References

1. Bevan, C. D., Hill, A. P., & Reynolds, D. P. (1999). *Analytical method and apparatus therefor.* UK: Glaxo Group Limited.

2. Box, K., Bevan, C., Comer, J., Hill, A., Allen, R., & Reynolds, D. (2003). High-throughput measurement of pKa values in a mixed-buffer linear pH gradient system. *Analytical Chemistry, 75,* 883–892.

3. Cleveland, J. A., Jr., Benko, M. H., Gluck, S. J., & Walbroehl, Y. M. (1993). Automated pKa determination at low solute concentrations by capillary electrophoresis. *Journal of Chromatography, 652,* 301–308.

4. Gluck, S. J., & Cleveland, J. A., Jr. (1994). Capillary zone electrophoresis for the determination of dissociation constants. *Journal of Chromatography, A, 680,* 43–48.

5. Gluck, S. J., & Cleveland, J. A., Jr. (1994). Investigation of experimental approaches to the determination of pKa values by capillary electrophoresis. *Journal of Chromatography, A, 680,* 49–56.

6. Bartak, P., Bednar, P., Stransky, Z., Bocek, P., & Vespalec, R. (2000). Determination of dissociation constants of cytokinins by capillary zone electrophoresis. *Journal of Chromatography, A, 878,* 249–259.

7. Beckers, J. L., Everaerts, F. M., & Ackermans, M. T. (1991). Determination of absolute mobilities, pK values and separation numbers by capillary zone electrophoresis: effective mobility as a parameter for screening. *Journal of Chromatography, 537,* 407–428.

8. Sarmini, K., & Kenndler, E. (1998). Capillary zone electrophoresis in mixed aqueous-organic media: effect of organic solvents on actual ionic mobilities and acidity constants of substituted aromatic acids. III. 1-Propanol. *Journal of Chromatography, A, 818,* 209–215.

9. Mrestani, Y., Neubert, R., Munk, A., & Wiese, M. (1998). Determination of dissociation constants of cephalosporins by capillary zone electrophoresis. *Journal of Chromatography, A, 803,* 273–278.

10. Avdeef, A. (1993). pH-Metric log P. II. Refinement of partition coefficients and ionization constants of multiprotic substances. *Journal of Pharmaceutical Sciences, 82,* 183–190.

11. Box, K. J., Comer, J. E. A., Hosking, P., Tam, K. Y., & Trowbridge, L. (2000). Rapid physicochemical profiling as an aid to drug candidate selection. In G. K. Dixon, J. S. Major, & M. J. Rice (Eds.), *High throughput screening: the next generation* (pp. 67–74). Oxford, UK: BIOS Scientific Publishers.

Chapter 25

Solubility Methods

Overview

▶ *Solubility can be estimated using* pK_a, *pH, Log P, and melting point.*

▶ *Commercial software calculates equilibrium solubility for comparing series analogs.*

▶ *High-throughput (HT) kinetic solubility is most relevant in discovery because it mimics discovery conditions.*

▶ *HT kinetic methods include direct UV, nephelometry, and turbidimetry.*

▶ *Custom solubility methods are used to mimic a specific project issue.*

▶ *Equilibrium solubility is relevant for animal dosing and development studies.*

Earlier chapters have emphasized the importance of solubility in drug discovery. Solubility plays a crucial role in drug absorption in the gastrointestinal tract, accurate biological assays in vitro, and dosage form selection for in vivo administration. With this central role, measurement of solubility is a frequent activity of discovery scientists, and attention should be paid to accurate and relevant analyses. This chapter provides an overview of in silico and in vitro solubility methods. The fundamentals of kinetic and thermodynamic solubility are discussed in Chapter 7.

25.1 Literature Solubility Calculation Methods

The total solubility of an ionizable compound at a particular pH can be estimated using its intrinsic solubility (solubility of the neutral form) and pK_a, according to the Henderson-Hasselbalch equation.[1] The following equation is used for acids. Concentration is in molarity:

$$S_{tot} = S_{HA}(1 + 10^{(pH - pK_a)}),$$

where S_{tot} = total solubility, and S_{HA} = intrinsic solubility of the neutral acid. Solubility is in g/mL.

The intrinsic solubility can be estimated using the Yalkowsky equation.[2] It uses lipophilicity and melting point, as follows:

$$Log\ S = 0.8 - Log\ P - 0.01(MP - 25),$$

where Log S = log of the solubility, Log P = log of the octanol/water partition coefficient, and MP = melting point in degrees Centigrade. Solubility is in mol/L.

25.2 Commercial Software for Solubility

Several software products for solubility are commercially available (Table 25.1). No study has compared results from these products against each other. Results for one solubility software package have been compared to experimental measurements.[3] The training and validation sets for these algorithms use equilibrium (thermodynamic) solubility values from the literature. Therefore, they may not be predictive of kinetic solubility values.

TABLE 25.1 ► Commercial Software for Solubility

Name	Company	Web site
Aqueous Solubility		
QikProp	Schrodinger	www.schrodinger.com
Solubility DB, Solubility Batch	Advanced Chemistry Development	www.acdlabs.com
ADMET Predictor	Simulations Plus	www.simulations-plus.com
Volsurf	Tripos	www.tripos.com
ADME Boxes	Pharma Algorithms	www.ap-algorithms.com
DSMedChem Explorer	Accelrys	www.accelrys.com
WSKow	Syracuse Research Corporation	www.syrres.com
KnowItAll	Bio-Rad Laboratories	www.biorad.com
SLIPPER	ChemDB	www.chemdbsoft.com/SLIPPER
CSLogWS	ChemSilico	www.chemsilico.com
Quantum	Quantum Pharmaceuticals	www.q-pharm.com
DMSO Solubility		
Volsurf	Tripos	www.tripos.com
DMSO Box	Pharma Algorithms	www.ap-algorithms.com

One widely distributed software, QikProp, predicts several absorption, distribution, metabolism, excretion, and toxicity (ADME/Tox) properties based on regression models using molecular descriptors. William L. Jorgensen, the developer of QikProp, suggests three rules for compounds, based on 90% of 1,700 oral drugs. The three rules are as follows:

► QikProp Log S > −5.7 (M)

► QikProp Caco-2 Permeability >22 nm/s

► Number of primary metabolites predicted by QikProp <7

Thus, in addition to providing predictions of specific properties, software can rapidly provide a profile of the drug-like properties compared to a benchmark of success.

Dimethylsulfoxide (DMSO) solubility has recently become an interest of many organizations because of its effects on biological assays (see Chapter 40). Many discovery scientists have considered DMSO a universal solvent. However, investigations have shown that salts and polar compounds may not be fully soluble in DMSO. Two software products for DMSO solubility are listed in Table 25.1.

25.3   Kinetic Solubility Methods

Kinetic solubility methods first dissolve the solid compound in DMSO and then add an aliquot of DMSO solution to the aqueous buffer. Kinetic solubility is most appropriate for drug discovery for the following reasons:

▶ It mimics the conditions of discovery experiments (initial dissolution in DMSO followed by addition to aqueous buffer, incubation time of hours)

▶ Initially compounds are dissolved in DMSO, thus negating differences in the solid form among compounds (even batches of the same compound can vary in solid form: metastable crystal form, amorphous material)

▶ Only 1 to 2 mg of compound is consumed (consistent with the small amount of each compound initially synthesized in discovery)

▶ Only 1 day is necessary for the assay (consistent with the fast decision-making in discovery)

▶ In many companies, 10 to 20 mM DMSO solutions of compounds are already available in a company's screening laboratory or compound repository, which minimizes sample preparation work.

As more stable crystal forms are prepared in discovery and development, the equilibrium (thermodynamic) solubility drops. The solubility will be lower than the less stable crystal form that precipitates in kinetic solubility measurements. The focus on kinetic solubility in discovery is to provide an optimistic estimate of solubility values when only limited material with unknown crystal forms is available. The purpose of kinetic solubility measurements is to identify compounds that do not have good kinetic solubility even in aqueous buffer containing DMSO, to guide modification of structures to improve solubility, and to guide formulation selection for animal dosing. As compounds progress to development, typically the most stable crystal form is selected, and more detailed thermodynamic solubility studies are conducted in various solvents. Kinetic solubility is used during lead selection and optimization to:

▶ Rank order compounds

▶ Guide modification of structures to optimize solubility.

The final DMSO concentration should be kept as low as practical (typically 1%–5%) in solubility methods. This is because DMSO increases the aqueous solubility of lower solubility compounds, especially lipophilic compounds.

It is important to remember that kinetic solubility data vary with the conditions of the solution. Small changes in pH, organic solvent, ionic strength, ions in solution, co-solutes, incubation time, and temperature can result in large changes in the solubility of a compound. Methods for kinetic and thermodynamic solubility are listed in Table 25.2.

25.3.1   Direct UV Kinetic Solubility Method

The direct UV method, developed by Avdeef,[4] measures the concentration of a compound that is dissolved in solution. A diagram of the method is shown in Figure 25.1. A small volume of compound in DMSO solution is added to the well of a 96-well plate containing

TABLE 25.2 ▶ Methods for Determination of Solubility

Method	Type assay	Speed (min/cpd)	Throughput (cpd/day/instrument)
Kinetic Solubility			
Nephelometry	High throughput	4	300
Direct UV	High throughput	4	300
Turbidimetry	Moderate throughput	15	50
Thermodynamic Solubility			
Equilibrium shake flask	Low throughput	60	10
Potentiometric	Low throughput	60	10

Note: cpd = compound

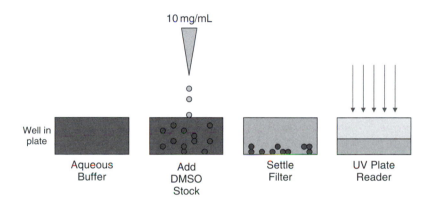

Figure 25.1 ▶ Direct UV kinetic solubility method diagram.

aqueous buffer and mixed. Typically the solution is covered and held at ambient temperature for a designated time (e.g., 1–18 hours). The maximum target concentration is set at about 100 μg/mL to minimize material consumption and volume of added DMSO. If a compound is not completely soluble, the extra portion will precipitate from solution. The precipitate is removed by filtration, and the UV absorbance of the supernatant is measured using a UV plate reader. The concentration of the compound is proportional to the UV absorbance, according to Beer's law. Solubility is determined against a single point standard, owing to the linearity of concentration versus UV absorbance. pION Inc. has implemented this method as a commercial instrument (Table 25.3). The pION software can compensate for the contribution of DMSO in enhancing compound aqueous solubility. Filters for the method can be obtained from Millipore, Corning, and other manufacturers. The filters must have a minimum of nonspecific absorption so that all of the dissolved compound stays in solution and is quantitated. Analiza has implemented this method in a commercial instrument using a nitrogen-specific high-performance liquid chromatography (HPLC) detector (Table 25.3).

The direct UV method is amenable to automation using laboratory robots. In addition to higher sample throughput, automation offers the opportunity of rapidly measuring solubility at multiple pHs. For example, it is useful to measure solubility at various pHs to simulate physiological and assay conditions: pH 1 (stomach), 6.5 (small intestine), 7.4 (neutral pH in blood, tissues, and bioassay buffers), and 8 (large intestine). One efficient strategy is to measure all compounds using generic conditions (e.g., pH 7.4, ambient temperature, 16 hours) for screening purposes and then to measure solubility using customized conditions for selected compounds of interest, as needed.

TABLE 25.3 ▶ Commercial Instruments and Products for Measurement of Solubility

Method	Product name	Company	Web site
Kinetic Solubility			
Direct UV	μSol Evolution, Explorer	pION Inc.	*www.pion-inc.com*
Direct UV	Multiscreen Solubility	Millipore	*www.millipore.com*
Direct UV	ASolW	Analiza Inc.	*www.analiza.com*
Nephelometry	NEPHELOstar	BMG LABTECH	*www.bmglabtech.com*
Nephelometry	Solubility Scanner	BG Gentest	*www.bdbiosciences.com*
Nephelometry	Npheloskan Accent	Thermo Electron	*www.thermo.com*
Thermodynamic Solubility			
Potentiometric	pSOL	pION Inc.	*www.pion-inc.com*

25.3.2 Nephelometric Kinetic Solubility Method

The nephelometric method, developed by Bevan and Lloyd,[5] measures the precipitation of a compound after it reaches its maximum concentration in solution. A diagram of the method is shown in Figure 25.2. A small volume of compound in DMSO solution is added to the first well of a row in a 96-well plate containing aqueous buffer and mixed using the pipetter. If the compound is not fully soluble, it will precipitate from solution and form particles. A small aliquot is removed from this solution, added to the next well in the row, and mixed. Subsequent dilutions result in lower concentrations of compound in subsequent wells. Serial dilutions are performed into 5% DMSO/95% phosphate-buffered saline (pH 7.4). Following a brief incubation, the plate is placed in a plate reader that sequentially impinges laser light on each well. Precipitate in the well from undissolved compound scatters the light. The "counts" of scattered light are plotted versus the well concentration, and the turning point is the maximum concentration, or solubility, of the compound. Bevan and Lloyd suggested the

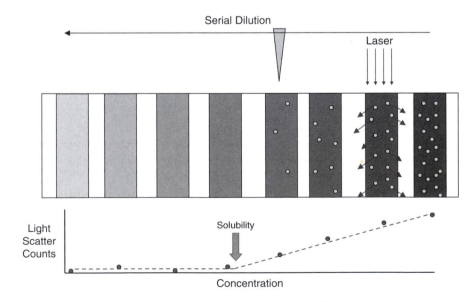

Figure 25.2 ▶ Nephelometric kinetic solubility method.

following solubility ranges: sparingly soluble (<10 µg/mL), partially soluble (10–100 µg/mL), and soluble (>100 µg/mL); these ranges are similar to those used by Lipinski et al.[6,7]

BMG LABTECH has implemented this method using its NEPHELOstar plate reader with a laser light source at 633 nm. A 96-well plate is scanned in 60 seconds. Thermo Electron has the Nepheloskan Accent reader (Table 25.3), which dispenses DMSO solutions into plate wells, shakes, scans the plate, and calculates the results. BD Gentest has implemented a variation of the method, which uses a flow cytometry detection system to detect the precipitate (Table 25.3).[8,9] The nephelometric method is the most commonly used method in industry[10,11]; however, users should be aware that the method measures precipitate, not actual compound concentration in solution, as in the direct UV method.

25.3.3 Turbidimetric In Vitro Solubility Method

The turbidimetric method also measures the precipitation of a compound from solution when it exceeds its solubility. The method is diagrammed in Figure 25.3. Small volumes (0.5 µL) of compound DMSO solution (20 mg/mL) are added stepwise at 1 min intervals to a cuvette containing stirred 2.5 mL of pH 7 phosphate buffer. When the concentration exceeds the compound solubility, precipitate forms and the turbidity scatters the light and reduces transmission (620–820 nm) through the cuvette. Subsequent additions of DMSO solution increase the precipitate and further decrease the light transmission. Fourteen additions of DMSO solution cover the solubility range of 5 to 65 µg/mL, with increasing DMSO concentration up to 0.375%. Light transmission is plotted versus compound, concentration and the turning point is the solubility of the compound. This method was developed by Lipinski et al.[6,7] and has not been commercialized.

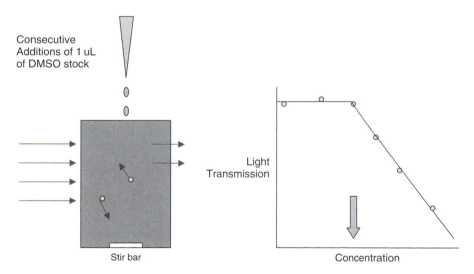

Figure 25.3 ► Turbidimetric kinetic solubility method.

Lipinski et al.[6] suggested that compound solubility <20 µg/mL is problematic. More than 85% of commercial drugs have solubilities >65 µg/mL by the turbidimetric method and have good bioavailability. Those commercial drugs having lower solubilities usually overcome this deficiency by having high potency, high permeability, or active transport across the intestinal epithelial cells.

25.3.4 Customized Kinetic Solubility Method

Often there are cases where generic solubility method conditions are not adequate to accurately assess the solubility of compounds under specific conditions. For example, the solubility of a compound can differ greatly among the following:

▶ Generic solubility assay buffer

▶ Enzyme or receptor assay buffer

▶ Cell-based assay buffer

▶ In vivo dosing vehicle

▶ Buffer with surfactant, protein, or DMSO

▶ pH 1–8

Solubility also varies with temperature, crystal form, and incubation time. In recent years, biological scientists have recognized that solubility differences between bioassay media can greatly affect IC_{50}. If the compound is not fully dissolved, IC_{50} is shifted to a higher concentration. This occurs to a greater extent for lower solubility and insoluble compounds in discovery.

When the solubilities of compounds are in question, a customized solubility experiment can be developed for a set of compounds. Such a "customized" solubility experiment can be performed using the same buffer composition, dilution procedure, incubation time, and temperature as the biological assay. Useful methodology for customized solubility assays was discussed by Di.[12,13] Compounds are added to the bioassay buffer under the conditions of the biological assay and then incubated according to bioassay time schedules. Precipitate is removed by filtration, and the supernatant is diluted accordingly and analyzed by LC/UV/mass spectrometry (MS) techniques. Results from such an assay are shown in Figure 25.4. In this example, the concentration of test compounds 1 to 3 were close to the target concentration of $10 \, \mu M$ in the receptor assay buffer, but in the cell-based assay buffer the concentrations were in the range of 1.4 to $4.8 \, \mu M$. This likely is due to the high amounts of bovine serum albumin (BSA) and DMSO in the receptor assay buffer compared to the cell-based assay buffer. BSA and DMSO enhance the solubility of lower-solubility compounds. It is advisable to perform such an analysis during the development of each biological assay so that the buffer and dilution protocols can be optimized to maximize

Compounds	Solubility in Receptor Binding Assay Buffer (uM)	Solubility in Cell-based Assay Buffer (uM)
1	11	2.4
2	10	4.8
3	10	1.4
Buffer	5% BSA, 2.5% DMSO	0.1% DMSO

* Target assay concentration $10 \, \mu M$

Figure 25.4 ▶ Customized solubility assays using different biological assay buffers indicate that the cell-based assay buffer does not keep the compounds fully solubilized at the target concentration of $10 \, \mu M$.

solubility of test compounds and produce reliable biological data. (See Chapter 40 for further discussion of solubility issues in biological assays.)

25.4 Thermodynamic Solubility Methods

Thermodynamic solubility methods add aqueous buffer directly to compound solid and mix the solution for an extended time until it reaches or approaches equilibrium. Thermodynamic solubility data are most appropriate for crystalline material that is studied in detail during late drug discovery and development.

Thermodynamic solubility is most often determined using the equilibrium shake flask method or the pH-metric method. A clinical candidate often will undergo thermodynamic solubility studies in a wide set of conditions, as discussed below.

25.4.1 Equilibrium Shake Flask Thermodynamic Solubility Method

Thermodynamic solubility is often measured by placing solid crystalline compound in a vial and adding the solvent (Figure 25.5). The vial then is shaken for 24 to 72 hours at a controlled temperature (25°C–37°C), and undissolved material is separated by filtration. The compound dissolved in the supernatant is diluted accordingly and measured using LC/UV/MS. This method is often considered the "gold standard" solubility method.

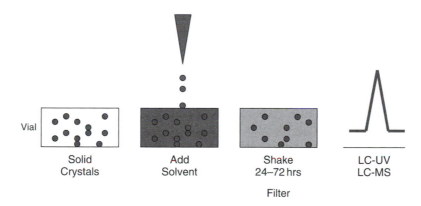

Figure 25.5 ▶ Equilibrium shake flask thermodynamic solubility method.

In a modification of this method,[14,15] 1 mg of solid compound is placed in a Whatman Mini-UniPrep filter chamber and 0.45 mL of aqueous buffer is added. After the filter cap is attached, the vials are placed in a HPLC autosampler vial block and shaken for 24 hours. The block is transferred to a Whatman filter processor device, and all the vials are simultaneously filtered. HPLC is used to measure the filtrate concentration. This approach automates portions of the thermodynamic solubility method.

25.4.2 Potentiometric In Vitro Thermodynamic Solubility Method

If the actual pK_a of a compound has been determined, potentiometric titration can be used to determine its intrinsic solubility (S_0; Figure 25.6). Known volumes of acid or base are added to a solution of the compound in the assay tube. The pH change during titration produces a titration curve. The apparent pK_a (pK_a^{app}) is shifted from the actual pK_a due to precipitation of the compound, and the S_0 is calculated from the equation shown in Figure 25.6. This method is implemented on the GLpK$_a$ instrument from Sirius Analytical, which is often used in development laboratories. The potentiometric solubility method is suitable only for compounds with ionization centers.

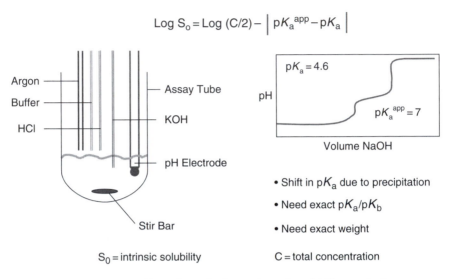

$$\text{Log } S_0 = \text{Log } (C/2) - \left| pK_a{}^{app} - pK_a \right|$$

Argon

Buffer

HCl

Assay Tube

KOH

pH Electrode

Stir Bar

$S_0 =$ intrinsic solubility

$pK_a = 4.6$

pH

$pK_a{}^{app} = 7$

Volume NaOH

• Shift in pK_a due to precipitation

• Need exact pK_a/pK_b

• Need exact weight

$C =$ total concentration

Figure 25.6 ▶ Potentiometric thermodynamic solubility method.

25.4.3 Thermodynamic Solubility in Various Solvents

Thermodynamic solubility studies are often performed to assist discovery and development formulation and in preparation for clinical development. Examples of solvents commonly tested are listed in Table 25.4. Such studies provide perspectives on compound solubility in various physiological fluids to aid absorption studies, in various formulary solvents to aid formulation, and in solvents that are used to measure lipophilicity.

TABLE 25.4 ▶ **Various Physiological and Formulary Solvents Used to Test Thermodynamic Solubility of Development Candidates**

Physiological buffer	Formulary solvent	Lipophilicity
pH 1	Tween 80	Octanol
pH 4.5	PEG 200	Cyclohexane
pH 6.6	PEG 400	
pH 7.4	Phosal 53 MCT	
pH 9	Phosal PG	
SGF	Benzyl Alcohol	
SIF	EtOH	
SIBLM	Corn Oil	
Plasma	2% Tween/0.5%MC	

Reprinted with permission from [16].
MC, Methylcellulose; SGF, simulated gastric fluid; SIF, simulated intestinal fluid; SIBLM, simulated intestinal bile lecithin media.

Problems

(Answers can be found in Appendix I at the end of the book.)

1. For the following compounds, estimate their equilibrium solubilities:

Compound	Melting point (°C)	Log P	Estimated solubility (mol/L)
1	125	0.8	
2	125	1.8	
3	225	1.8	
4	225	3.8	

2. For the following acids, estimate their equilibrium solubilities:

Compound	Intrinsic solubility	pK_a	pH	Total solubility (g/mL)
1	0.001	4.4	7.4	
2	0.001	4.4	4.4	
3	0.001	4.4	8.4	
4	0.00001	4.4	7.4	

3. What is/are measured in the nephelometric and direct UV methods?

4. Why is it useful to use customized solubility methods as opposed to generic high-throughput assays?

5. List some differences between kinetic and thermodynamic (equilibrium) solubility methods.

6. Which of the following are *true* about *thermodynamic* solubility measurements?: (a) aqueous buffer is added to solid compound and agitated for 2–3 days, (b) are unaffected by the form of the material (i.e., amorphous, crystalline, polymorphic), which varies with batch, (c) are used for discovery lead optimization, (d) are used for development and clinical batches, (e) assist with product development.

References

1. Lee, Y.-C., Zocharski, P. D., & Samas, B. (2003). An intravenous formulation decision tree for discovery compound formulation development. *International Journal of Pharmaceutics, 253,* 111–119.

2. Yalkowsky, S. H. (1999). Solubility and partial miscibility. In S. H. Yalkowsky (Ed.), *Solubility and solubilization in aqueous media* (pp. 49–80). Washington, DC: American Chemical Society.

3. Jorgensen, W. L., & Duffy, E. M. (2002). Prediction of drug solubility from structure. *Advanced Drug Delivery Reviews, 54,* 355–366.

4. Avdeef, A. (2001). High-throughput measurements of solubility profiles. *Pharmacokinetic optimization in drug research: biological, physicochemical, and computational strategies, [LogP2000, Lipophilicity Symposium],* 2nd, Lausanne, Switzerland, Mar. 5–9, 2000, pp. 305–325.

5. Bevan, C. D., & Lloyd, R. S. (2000). A high-throughput screening method for the determination of aqueous drug solubility using laser nephelometry in microtiter plates. *Analytical Chemistry, 72,* 1781–1787.

6. Lipinski, C. A., Lombardo, F., Dominy, B. W., & Feeney, P. J. (1997). Experimental and computational approaches to estimate solubility and permeability in drug discovery and development settings. *Advanced Drug Delivery Reviews, 23*, 3–25.

7. Lipinski, C. A. Computational and experimental approaches to avoiding solubility and oral absorption problems in early discovery. June 19–20, 2000, Princeton, NJ. Drew University Residential School on Medicinal Chemistry. Designing drugs with optimal in vivo activity after oral administration.

8. Goodwin, J. J. (2003). *Flow cell system for solubility testing.* Franklin Lakes, NJ: Becton Dickinson.

9. Goodwin, J. J. (2006). Rationale and benefit of using high throughput solubility screens in drug discovery. *Drug Discovery Today: Technologies, 3*, 67–71.

10. Fligge, T. A., & Schuler, A. (2006). Integration of a rapid automated solubility classification into early validation of hits obtained by high throughput screening. *Journal of Pharmaceutical and Biomedical Analysis, 42*, 449–454.

11. Dehring, K. A., Workman, H. L., Miller, K. D., Mandagere, A., & Poole, S. K. (2004). Automated robotic liquid handling/laser-based nephelometry system for high throughput measurement of kinetic aqueous solubility. *Journal of Pharmaceutical and Biomedical Analysis, 36*, 447–456.

12. Di, L., & Kerns, E. H. (2005). Application of pharmaceutical profiling assays for optimization of drug-like properties. *Current Opinion in Drug Discovery & Development, 8*, 495–504.

13. Di, L., & Kerns, E. H. (2006). Biological assay challenges from compound solubility: strategies for bioassay optimization. *Drug Discovery Today, 11*, 446–451.

14. Kerns, E. H. (2001). High throughput physicochemical profiling for drug discovery. *Journal of Pharmaceutical Sciences, 90*, 1838–1858.

15. Glomme, A., Maerz, J., & Dressman, J. B. (2005). Comparison of a miniaturized shake-flask solubility method with automated potentiometric acid/base titrations and calculated solubilities. *Journal of Pharmaceutical Sciences, 94*, 1–16.

16. Di, L., & Kerns, E. H. (2006). Application of physicochemical data to support lead optimization by discovery teams. In R. T. Borchardt, E. H. Kerns, M. J. Hageman, D. R. Thakker, & J. L. Stevens (Eds.), *Optimizing the drug-like properties of leads in drug discovery.* pp. 167–194. New York: Springer, AAPS Press.

Permeability Methods

Overview

▶ *In silico methods are available for calculating in vivo and in vitro permeability, such as intestinal absorption, Caco-2, and parallel artificial membrane permeability assay (PAMPA).*

▶ *High-throughput permeability methods utilize high-performance liquid chromatography (immobilized artificial membrane), artificial membranes (PAMPA), and cell layers (Caco-2).*

▶ *In-depth study of permeability uses portal vein cannulation and in situ perfusion.*

26.1 In Silico Permeability Methods

Computational models for intestinal drug absorption have been reported.[1] Several commercial software packages are available for the prediction of gastrointestinal (GI) absorption. A partial list of software is given in Table 26.1. As with all software, the user should evaluate the software using the compounds with which they are familiar and for which internal data are available before purchasing and implementation. Permeability is a more complex process than lipophilicity, and only a limited number of measurements[1–3] of intestinal absorption are available upon which to develop algorithms. Thus, the predictions should be taken as a guide but not overinterpreted. As with other software, predictions likely are best for comparing compounds in a series on a relative scale, as opposed to setting expectations for an exact value in vivo. One application of predictive permeability tools is for medicinal chemists to study the potential GI absorption effects of various substituents on a core scaffold when planning compounds to synthesize.

TABLE 26.1 ▶ Partial List of Commercial Software for Permeability

Name (prediction)	Company	Web site
GastroPlus (human intestinal)	Simulations Plus	*www.simulations-plus.com*
ADME Boxes (passive intestinal)	Pharma Algorithms	*www.ap-algorithms.com*
ADME Index (database of marketed drugs)	Lighthouse Data Sol.	*www.lighthousedatasolutions.com*
KnowItAll (human intestinal)	Biorad	*www.biorad.com*
CSHIA (human intestinal)	ChemSilico	*www.chemsilico.com*
Admensa (human intestinal)	Inpharmatica	*www.inpharmatica.com*
QikProp (Caco-2, MDCK)	Schrodinger	*www.schrodinger.com*
ADMET Predictor (MDCK, human jejunal)	Simulations Plus	*www.simulations-plus.com*

Egan et al.[4] demonstrated that Log P and polar surface area were good descriptors for a model of well-absorbed (>90%), moderately absorbed (30%–90%), and poorly absorbed compounds. Well-absorbed compounds group in an ellipse in the region of AlogP98 –1 to 5.9 and polar surface area of 0 to 132 $Å^2$.

Software for predicting permeability in cell layer methods is available, as for Caco-2 and the Madin Darby Canine Kidney cell line (MDCK), and are listed in Table 26.1. Quantitative structure–activity relationship methods for Caco-2, parallel artificial membrane permeability assay (PAMPA), PAMPA-BBB, and human intestinal absorption have been published.[5,6]

26.2 In Vitro Permeability Methods

Three types of permeability assays are most commonly used: (a) "immobilized artificial membrane" (IAM) high-performance liquid chromatography (HPLC), (b) cell layer, and (c) PAMPA. Each of these methods involves partitioning the test compound between aqueous and lipophilic phases. The IAM and PAMPA methods model only the passive diffusion mechanism. The cell layer method models passive diffusion, active uptake transport, efflux, and paracellular permeability mechanisms.

26.2.1 IAM HPLC

IAM is a convenient method because it uses the common HPLC format. Instead of octadecyl groups covalently bonded to the solid support, as in reversed-phase HPLC, the IAM technique uses phospholipids bonded to the solid support. These contain the polar head groups and aliphatic side chains of the lipids. Pidgeon developed this new HPLC phase concept.[7–16] Test compounds partition between the aqueous mobile phase and the phospholipid phase. The chromatographic capacity factor (k) increases with increasing affinity for the phospholipid phase. Compounds are rank ordered by k, which indicates a higher lipophilicity or phospholipids affinity. The parameters of this affinity correlate with permeation. Retention time is calibrated against permeability or absorption using standard compounds for which permeability or absorption has been measured using another technique.

The HPLC format is commonly available in discovery laboratories, is convenient to use, and is readily automated using autosamplers to save scientists' time. The correlation to permeability may be better than Log D because the lipids are close in structure to biological membrane lipids. IAM columns are commercially available (Table 26.2). Because IAM involves only physicochemical interaction with the phospholipids and not the polarity transitions and molecular volume space constraints of the lipid bilayer of a biological membrane, it is less predictive of permeability than some other methods. Traditionally, IAM uses an isocratic mobile phase and, therefore, has long retention times for highly lipophilic compounds. Valko et al.[16] developed a gradient IAM method that has higher throughput. Advantages of IAM are that the method requires very little material, and impurities do not interfere with the permeability prediction.

26.2.2 Cell Layer Method for Permeability

The first practical method for in vitro permeability assessment in discovery was the cell layer method. This assay models the epithelial cell layer permeability barrier that compounds encounter in the small intestine. Caco-2 is the best known cell line for this assay.[17–23] It is an immortal human colon carcinoma cell line that is readily available from the American Type Culture Collection (ATCC; Table 26.2). Desirable aspects of this cell line are its morphology

TABLE 26.2 ► Commercial Instruments and Products for the Measurement of Permeability

Method	Product Name	Company	Website
Caco-2	Transwell™	Corning	*www.corning.com/lifesciences/*
Caco-2	MultiScreen Caco-2™	Millipore	*www.millipore.com*
Caco-2	BioCoat™	BD Biosciences	*www.bdbiosciences.com*
Caco-2	BIOCOAT® HTS	BD Biosciences	*www.bdbiosciences.com*
Caco-2	Caco-2 Assay Kits	In Vitro Technologies	*www.invitrotech.com*
Caco-2	CAC-BD	Nichiryo	*www.nichiryo.co.jp/e/pdf/cac-bd.pdf*
Caco-2	EVOM, REMS TEER	World Precision Instr.	*www.wpiinc.com*
Caco-2	Biomek	Beckman Coulter	*www.beckman.com*
Caco-2	Caco-2 cells	ATCC	*www.atcc.org*
MDCK	MDCK cells	ATCC	*www.atcc.org*
MDCK	MDR1-MDCKII cells	Netherlands Cancer Ins.	*p.borst@nki.nl*
PAMPA	PAMPA Evolution™	pION Inc.	*www.pion-inc.com*
PAMPA	PAMPA Explorer™	pION Inc.	*www.pion-inc.com*
PAMPA	Gut-Box™	pION Inc.	*www.pion-inc.com*
PAMPA	AperW™	Analiza Inc.	*www.analiza.com*
PAMPA	MultiScreen PAMPA™	Millipore	*www.millipore.com*
IAM	IAM Columns	Regis Technologies	*www.registech.com*

and multiple permeability mechanisms. Caco-2 develops microvilli on its apical surface that resemble the morphology of GI epithelial cells that line the intestinal villi. Caco-2 cells also express cell membrane transporters on the apical surface, such as P-glycoprotein (Pgp), breast cancer resistance protein (BCRP), and multidrug resistance protein 2 (MRP2). This provides the opportunity to investigate various permeability mechanisms.

A variation of the cell layer method uses the MDCK (Madin Darby Canine Kidney) cell line or other cell lines such as LLC or A1/2/4. MDCK has been actively used in drug discovery for passive diffusion permeability prediction.[22]

A diagram of the cell layer experiment is shown in Figure 26.1. Cells are plated in the insert of a device that is commonly called the *cell culture insert*. The cells settle onto a porous filter support. Over approximately 21 days, the cells grow to confluence and cover the surface of the support. It is preferable to form a cell monolayer, but a mix of monolayers and multilayers might form. It is important that there be no gaps, otherwise the test compound will rapidly pass unimpeded through the porous filter. With time, the cells develop the microvilli morphology on the apical (upper) surface. They also increasingly express transporter proteins. The cell maintenance for media changes over this 21-day period makes Caco-2 a relatively expensive cell assay. A shorter 5-day culture technique[24] is commercially available (Table 26.2), but care must be taken to ensure full transporter functionality. By contrast, MDCK cells grow to confluence in 3 to 4 days and are ready

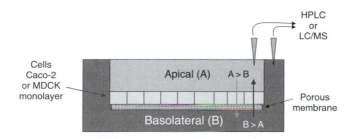

Figure 26.1 ► Cell layer permeability cell culture insert experiment setup.

for a permeability experiment. During the permeability experiment, the growth medium is replaced with buffered saline that contains glucose and the test compound.

Several different permeability experiments can be performed with this technique. The simplest experiment is to place the test compound in the buffer on the apical (A) side of the cell layer. Buffer with no test compound is placed on the basolateral (B) side. The test compound diffuses from the apical compartment through the cells and into the basolateral compartment. Aliquots from the two compartments are removed at specific time points over a 1- to 2-hour period. The concentration in each compartment is measured using HPLC or liquid chromatography/mass spectrometry (LC/MS) techniques, and the rate of permeation is calculated. This apical to basolateral (A > B) experiment provides a value for permeability in the absorptive direction, which models absorption in the GI. This permeability value (P) has been correlated by several research groups to absorption in the small intestine or to fraction of dose absorbed (FA). When such a correlation is established, it can assist medicinal chemists with predicting in vivo absorption of their compounds.

It is important to remember that Caco-2 data are not likely to completely agree between laboratories, even within the same company. This is due to differences caused by divergence of the characteristics of the cell lines, culturing conditions (e.g., serum source, frequency of media changes), and practice (e.g., experimental apparatus, percent dimethylsulfoxide [DMSO], media components). Thus, although trends are likely to agree, the actual permeability values can vary among laboratories.

Another experiment that can be performed is to place the test compound in the buffer on the basolateral side of the cell layer and buffer without compound on the apical side. This experiment is performed to study the permeability of test compounds by cell membrane transporters. If the permeability is the same in the A > B direction ($P_{A>B}$) and the B > A direction ($P_{B>A}$), then the compound primarily permeates by passive diffusion permeation. If the rates are significantly different, then a transporter may be involved. These experiments are discussed in greater detail in Section 27.2.1.

The cell layer permeability assay has been run under various conditions. Some of the important variables to consider are test compound concentration, pH of the aqueous buffers in the apical and basolateral compartments, use of materials that sequester the test compound in the basolateral compartment (sometimes called *sink compounds*), and use of solubilizers in the basolateral compartment. These are diagrammed in Figure 26.2.

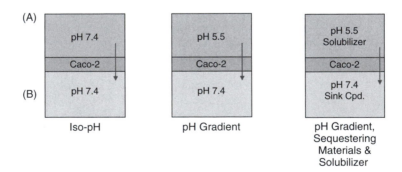

Figure 26.2 ▶ The cell layer assay has been run under various conditions, including the pH of the aqueous buffer, solubilizers, and sequestering materials. Shown are some of the variations in conditions in the apical (**A**) and basolateral (**B**) compartments.

Test compound concentration selection can affect the results. Many laboratories have used the concentration range of 5 to 10 μM. Alternatively, it has been argued that the concentration of compound in the GI lumen following an oral dose is closer to 50 to 100 μM, and some laboratories use this concentration range. At this concentration range, transporters are likely to be saturated. Thus, there may be differences in the measured permeability between low and high concentrations. If the test compound is an efflux transport substrate, the ratio of permeability by transporters relative to passive diffusion is lower as the concentration is increased, and transporters are saturated. Therefore, it is important to note the concentration of the experiment. Higher concentrations do not properly model other permeability barriers (e.g., blood–brain barrier [BBB]), where the concentration is much lower ($\sim$1 µM level) and the effects of transporters are much greater (e.g., Pgp efflux at the BBB).

It is important in cell layer assays to first test for the transepithelial electrical resistance (TEER). The TEER value indicates if the cell layer is forming a continuous network of tight junctions. Devices for measuring the TEER of individual wells in manual or automated format are commercially available (Table 26.2). If the TEER value is low, there will be much higher levels of paracellular permeation. Lucifer yellow can be added to each well to detect high paracellular permeability. Lucifer yellow is detected in the receiver well using a plate reader.

Two types of pH conditions are used in cell layer assays. The first is to use an acidic pH (e.g., pH 5–6.5) in the apical compartment and neutral pH in the basolateral compartment. This models the upper small intestine, where the intestinal lumen is at an acidic pH. The problem that occurs with using a pH gradient (as discussed in Section 8.1.1) is that passive diffusion of acids is enhanced in the direction of higher pH and of bases is enhanced in the direction of lower pH. Therefore, it may appear that acids are undergoing active uptake transport and bases are undergoing efflux transport. For data generated using a pH gradient, this effect should be kept in mind if the discovery project team is trying to elucidate the permeability mechanisms of a compound. For this reason, some laboratories use pH 7.4 in both chambers.

Another experimental variation is the use of solubilizers in the apical compartment. These help to solubilize low-solubility compounds during the experiment in order to produce a measurable and accurate value for the permeability. Materials used for this purpose include serum albumin. Sequestering materials can be used in the basolateral chamber. These compounds sequester the compound once it has passed through the permeability barrier and are meant to model the conditions of the intestine. Materials used for this purpose include bile salts. Such conditions are colloquially termed *sink conditions*. Alternatively, conditions could be chosen such that the concentration in the receiver compartment never exceeds 10% of the starting concentration.

Cell layer experiments are performed in several different cell culture insert plate formats. The number of wells varies from 12 to 96 per plate. The 96-well plates allow more compounds to be assayed per plate. The tradeoff is that as the number of wells per plate increase, the volume and surface area decrease, resulting in lower method sensitivity. If highly sensitive LC/MS/MS systems are available, the sensitivity issue may not be a problem, and higher-density plate formats may enable higher throughput. Most laboratories run the Caco-2 permeability assay in 24-well plates, although 96-well applications are becoming more common.

Caco-2 permeability values differ among laboratories; however, sometimes it is useful to have a benchmark for comparison. Here is one set of permeability ranges used for Caco-2:

▶ $P_{app} < 2 \times 10^{-6}$ cm/s   Low permeability

▶ $2 \times 10^{-6} < P_{app} < 20 \times 10^{-6}$ cm/s   Moderate permeability

▶ $P_{app} > 20 \times 10^{-6}$ cm/s   High permeability

26.2.3 Artificial Membrane Permeability Assay

PAMPA was invented by Kansy et al.[25] It reduces the cost and increases the throughput of permeability assays. Instead of a barrier made of living cells, the PAMPA barrier is made of phospholipids (e.g., phosphatidyl choline, egg lecithin) solubilized in a long-chain hydrocarbon (e.g., dodecane). A diagram of the PAMPA permeability experiment is shown in Figure 26.3. Other groups have also reported variations of the PAMPA method.[24,26–39]

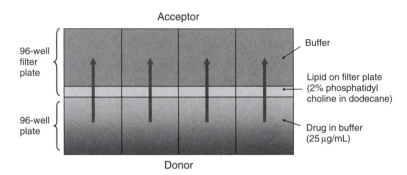

Figure 26.3 ► Parallel artificial membrane permeability assay (PAMPA) experiment.

The test compound is diluted in aqueous buffer ("donor solution") and is placed in the well of a 96-well plate. Each of these wells is termed a *donor*. A concentration of 25 µg/mL is commonly used. The liquid completely fills the wells. A 96-well filter plate, which has a porous filter on the bottom, is placed on top of this plate so that it is in contact with the aqueous buffer. A few microliters of a solution of the phospholipid is placed onto the top of the porous filter and soaks down into the holes of the filter to form the artificial barrier. Blank buffer is placed in the wells of the filter plate, on top of the artificial barrier. Each of these wells is termed an *acceptor*. This "sandwich" of 96-well plate and filter plate is maintained at a constant temperature and humidity for between 1 and 18 hours, depending on the laboratory's protocol and the permeability of the compounds. Samples are taken from the acceptor wells, the filter plate is removed, and samples are taken from the donor wells. The concentration of compound in the wells is quantitated using an LC/MS, LC/ultraviolet (UV), or a UV plate reader instrument. The unused "donor solution" that was not placed in the donor wells is used as a standard for quantitating the concentration of compound in the donor and acceptor wells, to calculate the permeability. The permeability often is termed *effective permeability* (P_e).

An advantage of PAMPA is that no cell culture maintenance is required. The artificial barrier is created at the time of the experiment. The capability of using the UV plate reader allows the method to be high throughput. The method measures only passive diffusion, but this provides a way to evaluate this important property independent of other permeability mechanisms.

PAMPA has been shown to correlate to human jejunal permeability with approximately the same reliability as Caco-2. An example of this correlation is shown in Figure 26.4.[28,40–42] Thus, if the project team wants to project far ahead to in vivo absorption, PAMPA provides a high-throughput approach with adequate predictability.

As with Caco-2, PAMPA has been run in many different ways in different laboratories. Therefore, it is not wise to compare data among laboratories unless the methods and quality control values are identical. Variations on PAMPA are similar to Caco-2. The pH of the buffers is sometimes run with the same pH on each side of the barrier and sometimes with

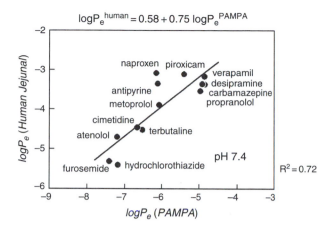

Figure 26.4 ▶ Prediction of human jejunal permeability using PAMPA.[28,40–42] Used with permission of A. Avdeef.

neutral pH on the acceptor side and lower pH on the donor side, to simulate the GI tract. The gradient pH condition has the same effect on acids. Compound flux (rate of flow through barrier) is increased in the direction of the higher pH. For bases, the flux is increased in the direction of the lower pH. Solubilizing components have been used on the donor side to enhance solubility, and sequestering materials have been used on the acceptor side to simulate conditions in the GI. The components in the artificial barrier also have been varied. Examples include 2% phosphatidylcholine in dodecane,[28] 20% egg lecithin in dodecane,[25] and hexadecane alone.[26] Stirring of the donor may be performed to reduce the gradient that forms due to diffusion rates and depletion at the barrier surface ("unstirred water layer"). Stirring can reduce the experiment time from 18 hours to 1 hour or less. Laboratories also have used various thicknesses of barriers by using different filter supports. All of these conditions will affect the data but allow modification of the experiment to model a specific set of conditions of interest to the project team.

26.2.4 Comparison of Caco-2 and PAMPA Methods

The widespread implementation of PAMPA in the pharmaceutical industry has prompted many questions regarding how the methods compare and which method produces "more reliable data." One way to compare the two methods is with relative characteristics, which reveal advantages and limitations.

Table 26.3 compares aspects of the two methods. PAMPA can be used over a wider pH range than Caco-2. This allows permeability at wider pHs to be studied. At the more

TABLE 26.3 ▶ **Comparison of PAMPA and Caco-2 Permeability assays.**

Comparison	PAMPA	Caco-2
Barrier	Phospholipids in Solvent	Cell Layer
pH	4–8	5.5–7.4
Mechanisms	Passive	Passive, Active, Efflux, Metabolism
Throughput	500/week	30/week
Cost	<$1/sample	~$30/sample
Manpower	0.35 FTE[1]	2 FTE[1]

*[1] FTE = full time equivalent (i.e., one scientist working full time)

extreme pH ranges, the pores between the cells in the Caco-2 cell layer increase in size and, thus, overpredict the paracellular mechanism. PAMPA correlates only with passive diffusion, whereas Caco-2 provides additional permeability mechanisms. The simplest Caco-2 permeability values ($P_{A>B}$) contain permeability components from each of these mechanisms. If the discovery project team is interested only in getting an estimate of permeability for absorption, then this may be sufficient. However, if a medicinal chemist wants to know the relative contributions of each mechanism, multiple Caco-2 experiments are necessary ($A>B$, $B>A$). PAMPA provides passive diffusion insights in only one experiment.

Throughput and cost are two clear advantages of PAMPA. In about one third to one half of a scientist's time and with one instrument, at least 500 compounds can be run by PAMPA in 1 week. Experimental supplies for PAMPA are less expensive than the cell culture materials used for Caco-2. For Caco-2, one to two scientists run 30 compounds in 1 week. This results in an approximately 30-fold lower throughput and 30-fold greater cost per analysis for Caco-2.

If data from Caco-2 and PAMPA on the same compounds are compared, the pattern in Figure 26.5 is obtained.[43] For compounds that are primarily permeable by passive diffusion, the data fall along the central correlation line. Compounds having strong Pgp efflux tend to plot in the section (lower right) of relatively higher PAMPA permeability than Caco-2 permeability. Compounds having strong paracellular or active uptake transport tend to plot in the section (upper left) of relatively higher Caco-2 than PAMPA permeability. This type of comparison can help to elucidate the predominating permeability mechanisms, if data are available from both methods.

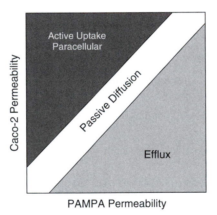

Figure 26.5 ▶ Comparison of Caco-2 and PAMPA permeability values from the same compound can provide insight on permeability mechanisms. (Reprinted with permission from [37].)

These comparisons suggest how Caco-2 and PAMPA can be used together synergistically (Figure 26.6). In early discovery, when large numbers of compounds are being considered by a project team, PAMPA is a rapid, low-cost, fast means for obtaining permeability insights. In mid-discovery, the two techniques can be used together to provide enhanced permeability mechanisms information. In late discovery, Caco-2 can be used for in-depth permeability mechanism studies.

26.3 In Depth Permeability Methods

The most detailed permeability studies are performed with living systems or organs from living systems. Investigators have cannulated the hepatic portal vein to measure the concentration of drug absorbed in the intestine, prior to first-pass liver metabolism. This provides a

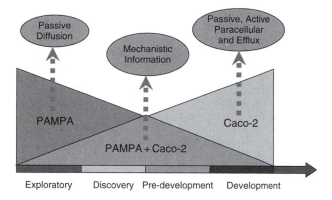

Figure 26.6 ► Strategy for combined use of Caco-2 and PAMPA in drug discovery. (Reprinted with permission from [37].)

concentration versus time profile of compound absorption by the intestine. Another in-depth approach, termed *in situ perfusion*, passes a drug solution through a section of the small intestine in a living animal and observes the rate of absorption.. This provides a profile of compound absorption in a specific portion of the intestine (e.g., jejunum). Such studies are performed rarely in discovery but are more common in development, when they provide data for biopharmatics classification system (BCS) studies that are submitted to the Food and Drug Administration (see Section 7.2.3.1).[44] A comparison of the throughput of these in depth methods and in vitro methods is provided in Table 26.4.

TABLE 26.4 ► **Methods for the Determination of Permeability**

Method	Type Assay	(min/cpd)	(cpd/day/instrument)
IAM	High throughput	10	120
PAMPA	High throughput	0.5	200
Cell layer-MDCK	Moderate throughput	40	12
Cell layer-Caco-2	Moderate throughput	80	6
In situ intestinal perfusion	Low throughput	250	2
Hepatic portal vein cannulation	Low throughput	250	2

Problems

(Answers can be found in Appendix I at the end of the book.)

1. For the following permeability methods, which permeation mechanisms contribute to the measured permeability value?

Method	Passive diffusion	Active Uptake	Efflux	Paracellular
IAM				
PAMPA				
Caco-2				

2. Why is IAM convenient for permeability estimation?

3. What factors add to the expense of Caco-2 compared to PAMPA?

4. What additional information can be obtained from Caco-2 compared to PAMPA?

5. How do IAM HPLC columns differ from other reversed-phase HPLC columns?

6. When the apical compartment in Caco-2 (donor in PAMPA) is more acidic than the basolateral compartment (acceptor in PAMPA), what artificial effects can occur for basic and acidic test compounds?

7. Compare Caco-2 and PAMPA permeability data for the following compounds:

Compound's el Permeability Mechanism(s)	PAMPA relatively higher than Caco-2	Caco-2 relatively higher than PAMPA	PAMPA and Caco-2 relatively the same
Passive diffusion only			
Passive diffusion, active uptake			
Passive diffusion, efflux			
Passive diffusion, paracellular			

References

1. Stenberg, P., Norinder, U.S., Luthman, K., & Artursson, P. (2001). Experimental and computational screening models for the prediction of intestinal drug absorption. *Journal of Medicinal Chemistry, 44*, 1927–1937.

2. Lennernaes, H. (1998). Human intestinal permeability. *Journal of Pharmaceutical Sciences, 87*, 403–410.

3. Winiwarter, S., Bonham, N. M., Ax, F., Hallberg, A., Lennernaes, H., & Karlen, A. (1998). Correlation of human jejunal permeability (in vivo) of drugs with experimentally and theoretically derived parameters. A multivariate data Analysis approach. *Journal of Medicinal Chemistry, 41*, 4939–4949.

4. Egan, W. J., Merz, K. M., & Baldwin, J. J. (2000). Prediction of drug absorption using multivariate statistics. *Journal of Medicinal Chemistry, 43*, 3867–3877.

5. Hansch, C., Leo, A., Mekapati, S. B., & Kurup, A. (2004). QSAR and ADME. *Bioorganic & Medicinal Chemistry, 12*, 3391–3400.

6. Verma, R. P., Hansch, C., & Selassie, C. D. (2007). Comparative QSAR studies on PAMPA/modified PAMPA for high throughput profiling of drug absorption potential with respect to Caco-2 cells and human intestinal absorption. *Journal of Computer-Aided Molecular Design, 21*, 3–22.

7. Pidgeon, C., Ong, S., Liu, H., Qiu, X., Pidgeon, M., Dantzig, A. H., et al. (1995). IAM chromatography: an in vitro screen for predicting drug membrane permeability. *Journal of Medicinal Chemistry, 38*, 590–594.

8. Ong, S., Liu, H., & Pidgeon, C. (1996). Immobilized-artificial-membrane chromatography: measurements of membrane partition coefficient and predicting drug membrane permeability. *Journal of Chromatography, A, 728*, 113–128.

9. Yang, C. Y., Cai, S. J., Liu, H., & Pidgeon, C. (1997). Immobilized artificial membranes: screens for drug-membrane interactions. *Advanced Drug Delivery Reviews, 23*, 229–256.

10. Ong, S., Liu, H., Qiu, X., Bhat, G., & Pidgeon, C. (1995). Membrane partition coefficients chromatographically measured using immobilized artificial membrane surfaces. *Analytical Chemistry, 67*, 755–762.

11. Liu, H., Ong, S., Glunz, L., & Pidgeon, C. (1995). Predicting drug-membrane interactions by HPLC: structural requirements of chromatographic surfaces. *Analytical Chemistry, 67*, 3550–3557.

12. Stewart, B. H., & Chan, O. H. (1998). Use of immobilized artificial membrane chromatography for drug transport applications. *Journal of Pharmaceutical Sciences, 87*, 1471–1478.

13. Masucci, J. A., Caldwell, G. W., & Foley, J. P. (1998). Comparison of the retention behavior of b-blockers using immobilized artificial membrane chromatography and lysophospholipid micellar electrokinetic chromatography. *Journal of Chromatography, A, 810*, 95–103.

14. Caldwell, G. W., Masucci, J. A., Evangelisto, M., & White, R. (1998). Evaluation of the immobilized artificial membrane phosphatidylcholine. Drug discovery column for high-performance liquid chromatographic screening of drug-membrane interactions. *Journal of Chromatography, A, 800*, 161–169.

15. Beigi, F., Gottschalk, I., Lagerquist Hagglund, C., Haneskog, L., Brekkan, E., Zhang, Y., et al. (1998). Immobilized liposome and biomembrane partitioning chromatography of drugs for prediction of drug transport. *International Journal of Pharmaceutics, 164*, 129–137.

16. Valko, K., Du, C. M., Bevan, C. D., Reynolds, D. P., & Abraham, M. H. (2000). Rapid-gradient HPLC method for measuring drug interactions with immobilized artificial membrane: comparison with other lipophilicity measures. *Journal of Pharmaceutical Sciences, 89*, 1085–1096.

17. Artursson, P., Palm, K., & Luthman, K. (2001). Caco-2 monolayers in experimental and theoretical predictions of drug transport. *Advanced Drug Delivery Reviews, 46*, 27–43.

18. Hidalgo, I. J. (2001). Assessing the absorption of new pharmaceuticals. *Current Topics in Medicinal Chemistry, 1*, 385–401.

19. Artursson, P., & Borchardt, R. T. (1997). Intestinal drug absorption and metabolism in cell cultures: Caco-2 and beyond. *Pharmaceutical Research, 14*, 1655–1658.

20. Balimane, P. V., Patel, K., Marino, A., & Chong, S. (2004). Utility of 96 well Caco-2 cell system for increased throughput of P-gp screening in drug discovery. *European Journal of Pharmaceutics and Biopharmaceutics, 58*, 99–105.

21. Irvine, J. D., Takahashi, L., Lockhart, K., Cheong, J., Tolan, J. W., Selick, H. E., et al. (1999). MDCK (Madin-Darby canine kidney) cells: a tool for membrane permeability screening. *Journal of Pharmaceutical Sciences, 88*, 28–33.

22. Chong, S., Dando, S. A., & Morrison, R. A. (1997). Evaluation of Biocoat intestinal epithelium differentiation environment (3-day cultured Caco-2 cells) as an absorption screening model with improved productivity. *Pharmaceutical Research, 14*, 1835–1837.

23. Gan, L.-S. L., & Thakker, D. R. (1997). Applications of the Caco-2 model in the design and development of orally active drugs: elucidation of biochemical and physical barriers posed by the intestinal epithelium. *Advanced Drug Delivery Reviews, 23*, 77–98.

24. Zhu, C., Jiang, L., Chen, T.-M., & Hwang, K.-K. (2002). A comparative study of artificial membrane permeability assay for high throughput profiling of drug absorption potential. *European Journal of Medicinal Chemistry, 37*, 399–407.

25. Kansy, M., Senner, F., & Gubernator, K. (1998). Physicochemical high throughput screening: parallel artificial membrane permeation assay in the description of passive absorption processes. *Journal of Medicinal Chemistry, 41*, 1007–1010.

26. Wohnsland, F., & Faller, B. (2001). High-throughput permeability pH profile and high-throughput alkane/water log P with artificial membranes. *Journal of Medicinal Chemistry, 44*, 923–930.

27. Avdeef, A. (2003). High-throughput measurement of permeability profiles. *Methods and Principles in Medicinal Chemistry, 18*, 46–71.

28. Avdeef, A. (2003). *Absorption and drug development: solubility, permeability, and charge state*. Hoboken, NJ: John Wiley & Sons.

29. Sugano, K., Nabuchi, Y., Machida, M., & Aso, Y. (2003). Prediction of human intestinal permeability using artificial membrane permeability. *International Journal of Pharmaceutics, 257*, 245–251.

30. Bermejo, M., Avdeef, A., Ruiz, A., Nalda, R., Ruell, J. A., Tsinman, O., et al. (2004). PAMPA-a drug absorption in vitro model 7. Comparing rat in situ, Caco-2, and PAMPA permeability of fluoroquinolones. *European Journal of Pharmaceutical Sciences, 21*, 429–441.

31. Kansy, M., Fischer, H., Bendels, S., Wagner, B., Senner, F., Parrilla, I., et al. (2004). Physicochemical methods for estimating permeability and related properties. *Biotechnology: Pharmaceutical Aspects, 1*, 197–216.

32. Kansy, M., Fischer, H., Bendels, S., Wagner, B., Senner, F., Parrilla, I., et al. (2004). Physicochemical methods for estimating permeability and related properties. In R. T. Borchardt, E. H. Kerns, C. A. Lipinski, D. R. Thakker, & B. Wang (Eds.), *Pharmaceutical profiling in drug discovery for lead selection* (pp. 197–216). AAPS Press. Springer Science, New York.

33. Avdeef, A. (2005). The rise of PAMPA. *Expert Opinion on Drug Metabolism & Toxicology*, *1*, 325–342.

34. Faller, B., Grimm, H. P., Loeuillet-Ritzler, F., Arnold, S., & Briand, X. (2005). High-throughput lipophilicity measurement with immobilized artificial membranes. *Journal of Medicinal Chemistry*, *48*, 2571–2576.

35. Obata, K., Sugano, K., Saitoh, R., Higashida, A., Nabuchi, Y., Machida, M., et al. (2005). Prediction of oral drug absorption in humans by theoretical passive absorption model. *International Journal of Pharmaceutics*, *293*, 183–192.

36. Sugano, K., Obata, K., Saitoh, R., Higashida, A., & Hamada, H. (2006). Processing of biopharmaceutical profiling data in drug discovery. In *Pharmacokinetic profiling in drug research: biological, physicochemical, and computational strategies*, [LogP2004, Lipophilicity Symposium], 3rd, Zurich, Switzerland, Feb. 29-Mar. 4, 2004, pp. 441–458.

37. Balimane, P. V., Pace, E., Chong, S., Zhu, M., Jemal, M., & Van Pelt, C. K. (2005). A novel high-throughput automated chip-based nanoelectrospray tandem mass spectrometric method for PAMPA sample analysis. *Journal of Pharmaceutical and Biomedical Analysis*, *39*, 8–16.

38. Loftsson, T., Konradsdottir, F., & Masson, M. (2006). Development and evaluation of an artificial membrane for determination of drug availability. *International Journal of Pharmaceutics*, *326*, 60–68.

39. Avdeef, A., Bendels, S., Di, L., Faller, B., Kansy, M., Sugano, K., & Yamauchi, Y. (2007). PAMPA—critical factors for better predictions of absorption. *Journal of Pharmaceutical Sciences*, *96*, 2893–2909.

40. Avdeef, A. Personal communication.

41. Avdeef, A. (2001). Physicochemical profiling (solubility, permeability and charge state). *Current Topics in Medicinal Chemistry*, *1*, 277–351.

42. Avdeef, A. (2003). High-throughput measurement of membrane permeability. In H. van de Waterbeemd, H. Lennernäs, & P. Artursson (Eds.), *Drug bioavailability/estimation of solubility, permeability, absorption and bioavailability* (pp. 46–71). Weinheim: Wiley-VCH.

43. Kerns, E. H., Di, L., Petusky, S., Farris, M., Ley, R., & Jupp, P. (2004). Combined application of parallel artificial membrane permeability assay and Caco-2 permeability assays in drug discovery. *Journal of Pharmaceutical Sciences*, *93*, 1440–1453.

44. Martin, A., Bustamante, P., & Chun, A. (1993). *Physical pharmacy: physical chemical principles in the pharmaceutical sciences*. Philadelphia: Lea & Febiger.

Chapter 27

Transporter Methods

Overview

▶ *In silico predictions are increasingly available for transporters.*

▶ *In vitro cell layer assays use inhibitors and cell lines with transfected transporters.*

▶ *High-throughput ATPase and calcein acetoxymethyl ester assays are used, but cell layer transport is most relevant.*

▶ *Transporters are knocked out genetically or chemically in vivo for in-depth study.*

Transporters are an emerging area in drug discovery. The extent of their impact on absorption, distribution, metabolism, excretion, and toxicity (ADME/Tox) is still being investigated. Most pharmaceutical companies have heeded the evidence that P-glycoprotein (Pgp) causes significant drug resistance and ADME/Tox effects. In response, methods have been implemented, and synthetic strategies that counteract Pgp efflux have gained wider understanding and application. The intensity of the effects of other transporters and their relative impact on the success of discovery projects in finding a quality clinical candidate remain under investigation. As with all properties, resources for transporters must be carefully allocated. If transporters are found to have minimal effects, in general, assaying all compounds for several transporters is unwise. Currently, assays for transporters are generally implemented to study selected compounds and diagnose poor or nonlinear pharmacokinetic performance or an unexplained observation. Various in vitro and in vivo assays for efflux transporters have been implemented.[1]

27.1 In Silico Transporter Methods

Initial approaches for developing in silico tools for Pgp transporter substrates have been reported. Crivori et al.[2] developed a model using VolSurf descriptors and partial least squares discriminant (PLSD) analysis. The model had 72% predictability for classifying Pgp substrates and inhibitors (not quantitative prediction of efflux ratio). Pharmacophore structural descriptors for substrates were determined using GRIND software.

Characterization of transporters currently is based on a multidisciplinary approach combining insights from chemistry, function, quantitative structure–activity relationships (QSAR), homology modeling, comparative modeling, and structural studies.[3,4] Comparative modeling using a low-resolution x-ray crystal structure of *Escherichia coli* lipid A transporter (MsbA) provided a model for Pgp, but this has been criticized.[5–7] If high-quality Pgp structural data become available, they will be very useful, however, at this time a multidisciplinary approach currently appears to be the strategy by which improved in silico predictions of transporter substrates will be achievable in the near term. These models will guide structural modifications to reduce efflux, such that it reduces drug exposure to the target, and enhance transport, when it is desirable to use transporters to enhance absorption, improve distribution, or reduce elimination.

Commercial software for transporters is limited to the classification of Pgp substrates (Table 27.1).

TABLE 27.1 ▶ Commercial Software for Transporters

Name	Company	Web site
ADME Boxes (Pgp)	Pharma Algorithms	*www.ap-algorithms.com*

27.2 In Vitro Transporter Methods

Several in vitro methods are available for assessing the susceptibility of a compound to transport. The indispensable method component is the presence of the transporter in a living cell or membrane system. Transporters have been expressed in various formats: cell lines transfected with a specific transporter gene, isolated primary cell cultures, immortalized cell cultures, microinjected oocytes, isolated membranes, and inverted vesicles. Each of these formats has its particular characteristics and applications. Commercial products for transporter studies are listed in Table 27.2.

TABLE 27.2 ▶ Commercial Suppliers of Products for Transporter Assays

Product name	Company	Web site
Transwell	Corning	*www.corning.com/lifesciences/*
MultiScreen Caco-2	Millipore	*www.millipore.com*
BioCoat	BD Biosciences	*www.bdbiosciences.com*
Caco-2 Assay Kits	In Vitro Technologies	*www.invitrotech.com*
EVOM, REMS TEER	World Precision Instr.	*www.wpiinc.com*
Biomek	Beckman Coulter	*www.beckman.com*
Transporter Vesicles	Solvo Biotechnology	*www.solvo.hu*
MultiDrugQuant Kits	Solvo Biotechnology	*www.solvo.hu*
PREDEASY ATPase Assay Kits	Solvo Biotechnology	*www.solvo.hu*
Transportocytes	BD Gentest	*www.gentest.com*
Human Vesicles	BD Gentest	*www.gentest.com*
Human Membranes	BD Gentest	*www.gentest.com*
Calcein AM Fluorescent Dye	BD Gentest	*www.gentest.com*
Caco-2 and MDCK Cell Lines	ATCC	*www.atcc.org*
Mdr1a/b Transgenic Mice	Taconic Laboratories	*www.taconic.com*

27.2.1 Cell Layer Permeability Methods for Transporters

The cell layer experiment for transporters is performed in the same manner as shown in Figure 26.1 and in 27.1 for Caco-2. In addition to the A>B experiment, a B>A experiment is performed for transporter permeability studies. This experiment is performed by placing buffer, which contains the test compound, in the basolateral compartment of the transwell apparatus and buffer without test compound in the apical compartment. The test compound passes through the porous membrane to reach the basolateral cell membrane and then permeates through the cells to reach the apical compartment. This basolateral to apical (B>A) experiment provides a value for permeation in the "secretory" direction. If the

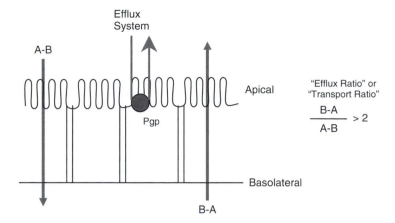

Figure 27.1 ► Cell layer efflux assays measure the permeability both the apical to basolateral (A-B) and basolateral to apical (B-A) directions and calculate an efflux or transport ratio. A ratio >2 indicates efflux.

compound permeates only by passive diffusion or paracellular permeation, then the $P_{A>B}$ and $P_{B>A}$ permeability values are approximately the same. However, if the compound is transported actively, then the values will differ. If $P_{A>B}$ is greater than $P_{B>A}$ and the *uptake ratio* ($P_{A>B}/P_{B>A}$) is ≥2, the compound likely will be actively transported for uptake. If $P_{B>A}$ is greater than $P_{A>B}$ and the *efflux ratio* ($P_{B>A}/P_{A>B}$) is ≥2, the compound likely will be actively transported for efflux. The actual permeability values from cell layer permeability experiments are known to vary among laboratories. Therefore, each laboratory should establish the values of these ratios that indicate convincing evidence for uptake or efflux transport. This validation is performed by assaying known transporter substrates and evaluating the resulting ratio values.

A complementary approach for assaying or confirming that a test compound is a transporter substrate is the use of an inhibitor. Certain compounds have been recognized as specific inhibitors for transporters (e.g., cyclosporin A [CsA] for Pgp). If the transport ratio (uptake or efflux) changes when the inhibitor is coincubated with the test compound, the test compound likely is a substrate for the transporter.

27.2.1.1 Caco-2 Permeability Method for Transporters

The Caco-2 method for permeability is discussed in Chapter 26. The most common application of Caco-2 is to estimate intestinal absorption. In this application, the A>B permeability has contributions from passive diffusion, paracellular permeation, and active transport. Caco-2 also is used to study some transporters because it expresses several transporters of pharmaceutical interest.

When Caco-2 is used for transporter studies, two points are important to understand and control. First, expression levels of transporters can vary. Caco-2 cells are not genetically identical, and, over time, the relative populations of different cell strains can vary. Therefore, the expression levels of different transporters can vary. In most companies, it is common to run quality control (QC) compounds with Caco-2 assays to monitor for Pgp efflux (e.g., digoxin), paracellular permeation (e.g., atenolol), and passive transcellular permeation (e.g., propranolol). However, it is uncommon to measure the permeability of control compounds for other transporters because the process is time consuming. If a specific transporter is being studied using Caco-2, one or more specific transporter QC compounds should be run to verify the expression level and activity of the transporter of interest.

A second point with Caco-2 use is that a test compound may be a substrate for more than one transporter. The multiple transporter expression of Caco-2 could confuse the results unless good control is built into the experiment with specific inhibitors.

Caco-2 has been reported to express several transporters of interest in drug discovery. These include Pgp (MDR1), breast cancer resistance protein (BCRP; ABCG2), PepT1, PepT2, multidrug resistance protein 2 (MRP2), and others. Culture conditions can affect the expression level of transporters. The functional activity of the transporters should be verified before a cell line is used for screening.

27.2.1.2 Transfected Cell Line Permeability Method for Transporters

Transfected cell lines are more commonly used for cell layer transporter assays. MDCK (a canine kidney line), LLC-PK1 (a porcine kidney epithelial line), 2008 (human ovarian carcinoma), and HEK 293 (human embryonic kidney line) cells have been transfected with transporter genes to produce stable immortal cell lines that express individual transporters. These cell lines have advantages over Caco-2 for the study of specific transporters:

▶ These cell lines naturally express only low levels of membrane transporters, so background signal is low.

▶ Transporter expression levels are high, so high signal-to-noise ratios are obtained. Larger transport ratios are produced. The assay can differentiate among project compounds for rank ordering.

▶ They require fewer resources than Caco-2 to maintain. For example, MDCK cells are ready for use in 3 days after plating, whereas Caco-2 requires 21 days of culture maintenance prior to use.

Various transfected cell lines have been prepared and discussed in the literature; examples are listed in Table 27.3.

TABLE 27.3 ▶ Various Cell Lines with Transfected Transporters Used in Cell Layer Permeability Assays

Transporter	Transporter species	Parental cell line	Cell line developer [a]	Reference
PepT1	Human	MDCK	Rutgers/BMS	[9,24]
BCRP	Human	MDCKII	Pfizer	[25]
OAT-K1	Rat	LLC-PK1	Kyoto	[26,27]
MDR1	Human	MDCKII	NCI	[28]
MDR1	Human	LLC-PK$_1$	NCI, Takeda	[29]
MDR1	Monkey	LLC-PK$_1$	Takeda	[29]
MDR1	Canine	LLC-PK$_1$	Takeda	[29]
mdr1a	Mouse	LLC-PK$_1$	NCI, Takeda	[29]
mdr1b	Mouse	LLC-PK$_1$	NCI, Takeda	[29]
MDR1a	Rat	LLC-PK$_1$	Takeda	[29]
MDR1b	Rat	LLC-PK$_1$	Takeda	[29]
MRP1	Human	MDCKII	NCI	[30]
MRP2	Human	MDCKII	NCI	[30]

[a] Rutgers University (Rutgers), Bristol-Myers Squibb Co. (BMS), Netherlands Cancer Institute (NCI), Takeda Pharmaceutical Company and Kanazawa University, Kyoto University Hospital.

Professor Dr. Piet Borst from Netherlands Cancer Institute has kindly shared transfected cell lines with researchers (Table 27.4). These cell lines have been provided without charge to academic laboratories and for a licensing fee to companies. Contact information: Professor Dr. P. Borst, Division of Molecular Biology, Netherlands Cancer Institute, Telephone: +31-20-512 2087, Fax: +31-20-669 1383, e-mail: p.borst@nki.nl.

TABLE 27.4 ▶ Cell Lines from the Netherlands Cancer Institute

Transfected cell line	Inserted transporter gene
2008 Parental	
2008 MRP1	Human MRP1 cDNA
2008 MRP2	Human MRP2 cDNA
2008 MRP3	Human MRP3 cDNA
HEK 293	
HEK 293 MRP5	Human MRP5 cDNA
LLC Parental	
LLC MRP1	Human MRP1 cDNA
LLC MDR1	Human MDR1 cDNA
LLC MDR3	Human MDR3 cDNA
LLC Mdr1a	Mouse Mdr1a cDNA
LLC Mdr1b	Mouse Mdr1b cDNA
LLC Bcrp1	Mouse Bcrp1 cDNA
MDCKII Parental	
MDCKII MDR1	Human MDR1 cDNA
MDCKII BCRP	Human BCRP cDNA
MDCKII Bcrp1	Mouse Bcrp1 cDNA
MDCKII MRP1	Human MRP1 cDNA
MDCKII MRP2	Human MRP2 cDNA
MDCKII MRP3	Human MRP3 cDNA
MDCKII MRP5	Human MRP5 cDNA

Such cell lines allow the development of an assay that is predictive of the effect of a specific transporter on test compounds. An example is a specific Pgp efflux assay using MDR1–MDCKII (Madin Darby Canine Kidney cell line II transfected with human MDR1 gene, which codes for Pgp protein product). In such an assay, the efflux ratio often is higher than that in Caco-2 for Pgp efflux substrates. Pgp is a major concern of drug discovery because it reduces GI absorption, reduces blood–brain barrier (BBB) penetration, and is responsible for drug resistance in cancer cells. LLC, 2008, and HEK cells also have been transfected with transporter genes for specific transporter assays.

A typical MDR1–MDCKII transport assay with cell layer is performed as follows. Cells are seeded in a transwell plate (3×10^6 cells/cm^2) and cultured for 3 days. The transepithelial electrical resistance (TEER) values are checked (e.g., >200 ohm/cm^2) to ensure minimal paracellular permeation. A paracellular marker, such as atenolol, is included in the assay to check layer integrity. The growth medium is removed and replaced with 37°C reduced-serum medium (to minimize protein binding) containing the test compound (5–20 μM). The transwell plates are maintained at 37°C, with gentle shaking. Small volumes from the apical and basolateral chambers are withdrawn at specific time points (e.g., 30, 60, 120 minutes) and quantitated using liquid chromatography/mass spectrometry (LC/MS) techniques.

27.2.2 Uptake Method for Transporters

The uptake assay is an alternative to the cell layer transport assay. Uptake assays measure the rate of test compound concentration increase inside cells that are living in standard solid-bottom culture plates, as opposed to the rate of test compound concentration increase in the basolateral chamber as a result of layer permeation. At an experiment time point, the medium is completely removed from a well, the cells are gently washed, and they are lysed using a detergent (e.g., TX-100, sodium dodecyl sulfate (SDS)) or organic solvent, combined with shaking or sonication. The concentration of test compound released into the lysate is measured, and the concentration in the cells is determined based on the total cell volume. The uptake experiment is a convenient format for higher-throughput assays because it does not require the complexity and high cost of the transwell system. The uptake experiment is a necessity when the cells do not form a confluent cell layer with tight junctions. Under these conditions, the test compound molecules in a transwell device would leak through by paracellular permeation. Uptake assays have been applied for Pgp efflux,[8] PepT1,[9] and BBB transporters.[10] A drawback of the uptake assay is that it does not guarantee the transcellular permeation of a test compound across the entire cell, as is necessary for absorption. Nonspecific binding to cell surface, rather than uptake, can complicate data interpretation.

Uptake assays have also proven beneficial for in vitro studies, in which the observed functional in vitro activity is correlated to the intracellular test compound concentration. This correlation is valuable to discovery project teams.

27.2.3 Oocyte Uptake Method for Transporters

BD Gentest has expressed several transporters in *Xenopus* oocytes (Table 27.2). The mRNA of a transporter is microinjected into each oocyte using a transcription vector.[11] Oocytes are useful because they are large, they are easy to handle, and they express in high abundance the transporters on the oocyte membrane. Oocytes are good for up to 1 week after injection. The assay is conducted by placing oocytes in suspension in the well of a plate and adding medium containing the test compound (Figure 27.2). Incubations occur for 30 to 120 minutes. At a specific time point, the oocytes are washed with ice-cold transport buffer and lysed with 10% SDS. The cell contents are assayed for the concentration of test compound, typically using LC/MS or scintillation counting techniques. Water-injected or uninjected

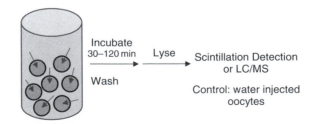

Figure 27.2 ▶ Schematic of the *Xenopus* oocyte transporter assay.

oocytes are used as controls. The extent and kinetics of uptake are calculated. Gentest Transportocytes are available in multiple species for OAT1, OAT2, OAT3, PEPT1, PEPT2, OATP1, OATP2, OATP4, OATP8, OATP1B3, OCT1, and NTCP. Oocyte transporter studies are not considered high throughput, but they could be beneficial for the study of selected compounds.

27.2.4 Inverted Vesicle Assay for Transporters

Transporter genes have been cloned into insect cells (*Spodoptera frugiperda*). From this expression system, vesicles are produced that contain the transporter in their membranes. With special treatment, the vesicle is inverted so that the normally extracellular face of the transporter is inside the vesicle. If the transporter is Pgp and the inverted vesicles are placed in a solution containing a Pgp efflux substrate, the compound will be taken up into the vesicle. At a specific time point, the vesicles are separated from the solution by filtration, washed, and lysed to release the compound. Quantitation uses scintillation counting or LC/MS. Solvo is a vendor of inverted vesicles (Table 27.2).

TABLE 27.5 ▶ Comparison of Three Assays Commonly Used In Drug Discovery for Pgp Effux[12]

Comparison	Pgp Assay		
	Cell Layer Efflux	ATPase	Calcein-AM
Activity Indicated for Test Compound	Pgp Efflux	ATPase Activation	Pgp Inhibition
Test Compound Diagnosis	Definitely a Pgp Efflux Substrate	May be a Substrate, Inhibitor or ATPase Activator	May be a Substrate or Inhibitor
Materials for Pgp Assay	MDR1-MDCKII Cell Line	MDR1-MDCKII Cell Line or Membrane Vesicles	MDR1 Membrane Vesicles from Sf9
Instrumentation Required	LC-MS-MS	UV Absorption Plate Reader	Fluorescence Plate Reader
Compounds Tested Per Week	20	200	200

27.2.5 ATPase Assay for ATP Binding Cassette Transporters

ATP binding cassette (ABC) transporters (e.g., Pgp) bind and hydrolyze ATP molecules as part of the transport process. ATP hydrolysis demonstrates that the test compound affects the ATPase activity of Pgp. The ATP hydrolysis reaction is shown in Figure 27.3. An ATPase assay measures ATP hydrolysis. The transporter material can be purchased from vendors in a convenient membrane-bound form that is produced from transduced insect cells. The inorganic phosphate released from ATP hydrolysis reacts with ammonium molybdate in the reaction solution to produce an intense color that is detected using an UV/Vis plate reader. Increasing color intensity is correlated with increasing ATPase activity. The concentration at half-maximal ATPase activation is frequently used to estimate the test compound binding affinity to the ABC transporter. The assay is readily automated for high throughput analysis. One drawback is that the assay only demonstrates that the test compound affects the ATPase activity of Pgp, not that the compound actually is effluxed by Pgp. This has been demonstrated by Polli et al.[12] and is summarized in Table 27.5.

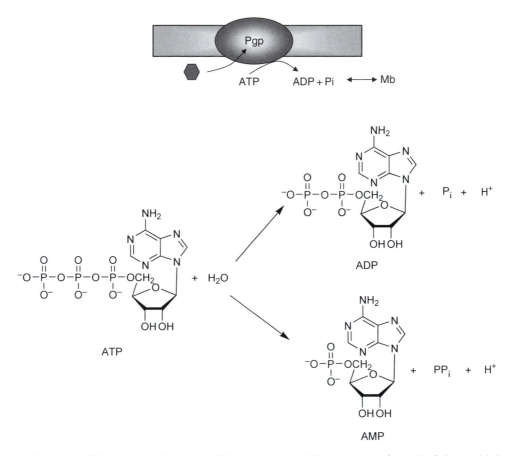

Figure 27.3 ▶ ATPase assay schematic. cDNA-expressed ABC transporter (e.g., Pgp) is provided on a membrane. Test compound, ATP, and ammonium molybdate are added. Binding a molecule of test compound to the ABC transporter results in ATP hydrolysis. The inorganic phosphate (Pi) reacts with molybdenum to produce a visible color.

27.2.6 Calcein AM Assay for Pgp Inhibitor

Calcein acetoxymethyl ester (calcein-AM; Figure 27.4) is a Pgp substrate, which limits its ability to enter the cells. Calcein-AM is rapidly hydrolyzed inside the cells to form calcein. Calcein is fluorescent and is readily detected using a fluorescence plate reader. If a coincubated test compound is a Pgp transport inhibitor, the test compound will inhibit Pgp efflux, and more calcein-AM will be able to enter the cells. This results in an increased intracellular concentration of calcein-AM, which subsequently is hydrolyzed to produce calcein and leads to increased fluorescence. Thus, a test compound that is a Pgp transport inhibitor causes a higher level of fluorescence than controls. A drawback of the method is that it determines inhibitors of Pgp transport, which may or may not be substrates of Pgp (see Table 27.5).

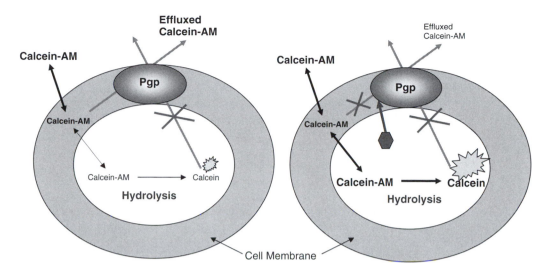

Figure 27.4 ► Calcein-AM fluorescence assay for Pgp efflux. Calcein-AM is a Pgp substrate and is effluxed from the cell (**left**), forming minimal fluorescent Calcein. In the presence of a Pgp inhibitor (hexagon), the efflux of calcein-AM is reduced, and more fluorescent calcein is produced (**right**).

27.3 In Vivo Methods for Transporters

It is valuable to follow up *in vitro* studies of key compounds with further study *in vivo*. Such studies confirm *in vitro* observations and provide greater confidence and understanding of the effect of the transporter in a dynamic living system. Two types of *in vivo* transporter experiments are commonly performed: genetic knockout and chemical knockout. Comparison of the compound performance in wild type animals to knockout animals provides strong evidence of the effect of a transporter and its extent.

27.3.1 Genetic Knockout Animal Experiments for Transporters

Mice strains have been developed in which the mdr1a, mdr1b, and mrp1 genes are knocked out, individually or in combinations.[13,14] These animals are commercially available. Examples of the use of this approach are as follows:

► For a neuroscience discovery project: Does the compound have a much higher BBB permeability in the knockout versus the wild-type?[15–18]

► For a compound that is poorly absorbed after oral dosing and has good solubility, passive diffusion permeability, and metabolic stability, does it have higher absorption in the knockout versus the wild-type?[19,20]

27.3.2 Chemical Knockout Experiments for Transporters

Coadministering or preadministering an inhibitor that is specific for the transporter can knock out the transporter. The inhibitor will specifically reduce the transporter activity. A difference in compound performance, pharmacokinetically or pharmacologically, with inhibitor coadministration versus no inhibitor is evidence that the transporter is having an effect on the compound's in vivo ADME/Tox properties.[16,21,22]

A variation of this experiment is to saturate the transporters with the test compound. If increasing permeability or absorption is observed with increasing doses, an efflux transporter likely is being saturated.[23]

Chemical knockout experiments can be performed with the animal efficacy/pharmacology model. The results will be better correlated to observations in regular in vivo biology experiments.

Problems

(Answers can be found in Appendix I at the end of the book.)

1. Define efflux ratio. What does it indicate?

2. What other variations on the cell monolayer assay can be used to detect or confirm efflux?

3. When is an uptake method useful, compared to a transport assay using transwell devices?

4. What does the ATPase assay measure?

5. What does the calcein-AM assay measure?

6. What are in vivo Pgp assays useful for?

7. How do the genetic and chemical knockout experiments differ?

References

1. Zhang, Y., Bachmeier, C., & Miller, D. W. (2003). In vitro and in vivo models for assessing drug efflux transporter activity. *Advanced Drug Delivery Reviews, 55,* 31–51.

2. Crivori, P., Reinach, B., Pezzetta, D., & Poggesi, I. (2006). Computational models for identifying potential P-glycoprotein substrates and inhibitors. *Molecular Pharmaceutics, 3,* 33–44.

3. Zhang, E. Y., Phelps, M. A., Cheng, C., Ekins, S., & Swaan, P. W. (2002). Modeling of active transport systems. *Advanced Drug Delivery Reviews, 54,* 329–354.

4. Chang, C., Ray, A., & Swaan, P. (2005). In silico strategies for modeling membrane transporter function. *Drug Discovery Today, 10,* 663–671.

5. Lee, J.-Y., Urbatsch Ina, L., Senior Alan, E., & Wilkens, S. (2002). Projection structure of P-glycoprotein by electron microscopy. Evidence for a closed conformation of the nucleotide binding domains. *Journal of Biological Chemistry, 277,* 40125–40131.

6. Chang, G., & Roth, C. B. (2001). Structure of Msb, A., from E. coli: a homolog of the multidrug resistance ATP binding cassette (ABC) transporters. *Science, 293,* 1793–1800.

7. Seigneuret, M., & Garnier-Suillerot, A. (2003). A structural model for the open conformation of the mdr1 P-glycoprotein based on the Msb, A., crystal structure. *Journal of Biological Chemistry, 278,* 30115–30124.

8. Kerns, E. H., Hill, S. E., Detlefsen, D. J., Volk, K. J., Long, B. H., Carboni, J., et al. (1998). Cellular uptake profile of paclitaxel using liquid chromatography tandem mass spectrometry. *Rapid Communications in Mass Spectrometry, 12,* 620–624.

9. Faria, T. N., Timoszyk, J. K., Stouch, T. R., Vig, B. S., Landowski, C. P., Amidon, G. L., et al. (2004). A novel high-throughput PepT1 transporter assay differentiates between substrates and antagonists. *Molecular Pharmaceutics, 1,* 67–76.

10. Terasaki, T., Ohtsuki, S., Hori, S., Takanaga, H., Nakashima, E., & Hosoya, K.-I. (2003). New approaches to in vitro models of blood-brain barrier drug transport. *Drug Discovery Today, 8,* 944–954.

11. Tamai, I., Tomizawa, N., Kadowaki, A., Terasaki, T., Nakayama, K., Higashida, H., et al. (1994). Functional expression of intestinal dipeptide/b-lactam antibiotic transporter in Xenopus laevis oocytes. *Biochemical Pharmacology*, *48*, 881–888.

12. Polli, J. W., Wring, S. A., Humphreys, J. E., Huang, L., Morgan, J. B., Webster, L. O., et al. (2001). Rational use of in vitro P-glycoprotein assays in drug discovery. *Journal of Pharmacology and Experimental Therapeutics*, *299*, 620–628.

13. Schinkel, A. H., Smit, J. J.M., van Tellingen, O., Beijnen, J. H., Wagenaar, E., van Deemter, L., et al. (1994). Disruption of the mouse mdr1a P-glycoprotein gene leads to a deficiency in the blood-brain barrier and to increased sensitivity to drugs. *Cell*, *77*, 491–502.

14. Schinkel, A. H., Mayer, U.S., Wagenaar, E., Mol, C. A.A.M., van Deemter, L., Smit, J. J.M., et al. (1997). Normal viability and altered pharmacokinetics in mice lacking MDR1-type (drug-transporting) P-glycoproteins. *Proceedings of the National Academy of Sciences of the United States of America*, *94*, 4028–4033.

15. Doran, A., Obach, R. S., Smith, B. J., Hosea, N. A., Becker, S., Callegari, E., et al. (2005). The impact of P-glycoprotein on the disposition of drugs targeted for indications of the central nervous system: evaluation using the MDR1A/1B knockout mouse model. *Drug Metabolism and Disposition*, *33*, 165–174.

16. Polli, J. W., Jarrett, J. L., Studenberg, S. D., Humphreys, J. E., Dennis, S. W., Brouwer, K. R., et al. (1999). Role of p-glycoprotein on the CNS disposition of amprenavir (141W94), an HIV protease inhibitor. *Pharmaceutical Research*, *16*, 1206–1212.

17. Kemper, E. M., van Zandbergen, A. E., Cleypool, C., Mos, H. A., Boogerd, W., Beijnen, J. H., et al. (2003). Increased penetration of paclitaxel into the brain by inhibition of P-glycoprotein. *Clinical Cancer Research*, *9*, 2849–2855.

18. Dagenais, C., Rousselle, C., Pollack, G. M., & Scherrmann, J.-M. (2000). Development of an in situ mouse brain perfusion model and its application to mdr1a P-glycoprotein-deficient mice. *Journal of Cerebral Blood Flow and Metabolism*, *20*, 381–386.

19. Schinkel, A. H., Wagenaar, E., van Deemter, L., Mol, C. A.A., & Borst, P. (1995). Absence of the mdr1a P-glycoprotein in mice affects tissue distribution and pharmacokinetics of dexamethasone, digoxin, and cyclosporin A. *Journal of Clinical Investigation*, *96*, 1698–1705.

20. Beaumont, K., Harper, A., Smith, D. A., & Bennett, J. (2000). The role of P-glycoprotein in determining the oral absorption and clearance of the NK2 antagonist, UK-224,671. *European Journal of Pharmaceutical Sciences*, *12*, 41–50.

21. Letrent, S. P., Pollack, G. M., Brouwer, K. R., & Brouwer, K. L. R. (1998). Effect of GF120918, a potent P-glycoprotein inhibitor, on morphine pharmacokinetics and pharmacodynamics in the rat. *Pharmaceutical Research*, *15*, 599–605.

22. Luker, G. D., Rao, V. V., Crankshaw, C. L., Dahlheimer, J., & Piwnica-Worms, D. (1997). Characterization of phosphine complexes of technetium(III) as transport substrates of the multidrug resistance P-glycoprotein and functional markers of P-glycoprotein at the blood-brain barrier. *Biochemistry*, *36*, 14218–14227.

23. Wetterich, U.S., Spahn-Langguth, H., Mutschler, E., Terhaag, B., Roesch, W., & Langguth, P. (1996). Evidence for intestinal secretion as an additional clearance pathway of talinolol enantiomers: concentration- and dose-dependent absorption in vitro and in vivo. *Pharmaceutical Research*, *13*, 514–522.

24. Herrera-Ruiz, D., Faria, T. N., Bhardwaj, R. K., Timoszyk, J., Gudmundsson, O. S., Moench, P., Wall, D. A., et al. (2004). A novel hPepT1 stably transfected cell line: establishing a correlation between expression and function. *Molecular Pharmaceutics*, *1*, 136–144.

25. Xiao, Y., Davidson, R., Smith, A., Pereira, D., Zhao, S., Soglia, J., et al. (2006). A 96-well efflux assay to identify ABCG2 substrates using a stably transfected MDCK II cell line. *Molecular Pharmaceutics*, *3*, 45–54.

26. Saito, H., Masuda, S., & Inui, K.-i. (1996). Cloning and functional characterization of a novel rat organic anion transporter mediating basolateral uptake of methotrexate in the kidney. *Journal of Biological Chemistry*, *271*, 20719–20725.

27. Masuda, S. (2003). Functional characteristics and pharmacokinetic significance of kidney-specific organic anion transporters, OAT-K1 and OAT-K2, in the urinary excretion of anionic drugs. *Drug Metabolism and Pharmacokinetics*, *18*, 91–103.

28. Hochman, J. H., Pudvah, N., Qiu, J., Yamazaki, M., Tang, C., Lin, J. H., et al. (2004). Interactions of human P-glycoprotein with simvastatin, simvastatin acid, and atorvastatin. *Pharmaceutical Research*, *21*, 1686–1691.

29. Takeuchi, T., Yoshitomi, S., Higuchi, T., Ikemoto, K., Niwa, S.-I., Ebihara, T., et al. (2006). Establishment and characterization of the transformants stably-expressing MDR1 derived from various animal species in LLC-PK1. *Pharmaceutical Research*, *23*, 1460–1472.

30. Baltes, S., Fedrowitz, M., Tortos, C. L., Potschka, H., & Loscher, W. (2007). Valproic acid is not a substrate for P-glycoprotein or multidrug resistance proteins 1 and 2 in a number of in vitro and in vivo transport assays. *Journal of Pharmacology and Experimental Therapeutics*, *320*, 331–343.

Blood–Brain Barrier Methods

Overview

▶ *Several in silico predictions for blood–brain barrier (BBB) have appeared as a result of the importance of brain penetration for drug research on neurological disorders.*

▶ *Various in vitro BBB methods are used: PAMPA-BBB, Log P, ΔLog P, brain-plasma dialysis, IAM HPLC, surface activity, and cell layer permeability (MDCK, Caco-2, BMEC).*

▶ *Brain penetration is studied in vivo using in situ perfusion, brain uptake, brain to plasma ratio (B/P), CSF sampling, and microdialysis.*

▶ *It is necessary to differentiate BBB permeation, brain/plasma partitioning, and free drug.*

Brain penetration is a major barrier for some compound series that otherwise would have efficacy for diseases of the brain. As discussed in Chapter 10, brain penetration is limited by (1) blood–brain barrier (BBB) permeation and (2) brain distribution. Each of these is determined by multiple underlying mechanisms. Understanding the difference between BBB permeation and brain distribution can lead to more effective data interpretation and successful planning for structure modification to enhanced brain penetration.

Methods for BBB permeation include the following:

▶ In silico models

▶ In vivo brain perfusion or uptake

▶ Passive diffusion

▶ P-glycoprotein (Pgp) efflux

▶ Uptake transport

▶ Physicochemical properties that are components of passive diffusion

BBB permeation is the velocity of compound flux from blood into brain tissue through the BBB endothelium. This velocity is determined by passive diffusion, efflux transport, and uptake transport. In vivo methods (e.g., in situ perfusion, brain uptake) accurately determine total BBB permeability or the impact of a transporter on total permeability (e.g., Pgp knockout mice). However, they are expensive. In vitro methods are useful for screening larger numbers of compounds and obtaining data on specific underlying mechanisms. In vitro methods for passive diffusion (e.g., parallel artificial membrane permeability assay) [PAMPA]-BBB, MDCK) and underlying structural properties that affect it (e.g., polar surface area [PSA], molecular weight [MW], Log P, H-bonding) assist the optimization of passive BBB permeability. Pgp efflux potential can be assessed using in vitro Pgp methods (e.g., MDR1-MDCKII, Caco-2). Screens expressing uptake transporters have been suggested as a means to support the enhancement of BBB permeation but have not been widely implemented.[1] Uptake transporter affinity usually is discovered by serendipity. Methods for uptake transporters are discussed in Chapter 27.

Methods for brain distribution include the following:

- ▶ In silico models

- ▶ In vivo B/P or Log BB

- ▶ Metabolic stability

- ▶ Plasma protein binding

- ▶ Equilibrium dialysis

In vivo pharmacokinetic (PK) methods for distribution of compound between plasma and brain (e.g., brain to plasma ratio [B/P]) are widely used. Chapter 10 discusses how the improvement of metabolic stability increases brain penetration. Methods for metabolic stability are discussed in Chapter 11.

Most companies performing neuroscience drug discovery research use a combination of methods to better determine the brain penetration characteristics of their leads. Combination of data from in vivo and in vitro methods can reveal and confirm predominant BBB mechanisms for a compound. General reviews of BBB methods are available.[2,3]

28.1   In Silico Methods for BBB

Considerable work has been invested in the development of in silico tools for predicting brain penetration.[4] As discussed in Chapter 10, brain penetration has many variables (e.g., Pgp efflux, passive diffusion, uptake, plasma protein binding, brain tissue binding, metabolic clearance), thus providing a complex challenge for in silico models.

28.1.1   Classification Models

Some in silico models seek simply to classify compounds as CNS+ (penetrates into the brain) or CNS- (does not appreciably penetrate into brain). Such methods have been reviewed.[5,6] At least five different computational classification methods have been used.[7–11] The classification of compounds in the training sets as CNS+ and CNS- is based on whether a compound showed in vivo CNS efficacy or had a B/P value above a certain level (e.g., B/P ≥ 1). Thus, the criteria are not quantitative. The average accuracy of the in silico models for validation set compounds for CNS+ classification was 90% and for CNS- classification was 74%. Such models are useful for an initial overview of structures, such as hits from high-throughput screening, compound library planning or assessment, and planning compound synthesis.

28.1.2   Quantitative Structure–Activity Relationship Methods

Quantitative structure–activity relationship (QSAR) models typically use brain distribution B/P or Log BB values for brain penetration with linear regression, partial least squares, and molecular descriptor mathematical methods.[11–17] The models developed using these approaches are consistent with experimental evidence from in vivo measurements of the importance of structural physicochemical properties in determining brain penetration, including hydrogen bonding (or PSA), lipophilicity, molecular size/weight, and basicity/acidity. Results for six models have been summarized as having an average predictive performance of within 0.4 log unit (2.5-fold in B/P).[5,6]

Newer QSAR models[6] have used BBB permeability–surface area coefficient (PS, permeability · surface area) data.[12,13] These models also have major descriptors for lipophilicity and hydrogen bonding. One opportunity for models based on passive diffusion is to compare the predictions with measurements to obtain an initial indication if the compound is primarily permeable by passive diffusion (good agreement of measurement with model), highly affected by efflux transport (measurement is much lower than model), or affected by uptake transport (measurement is much higher than model).

Other in silico models for brain distribution use MolSurf[14] and free energy calculations.[15] A model for BBB permeation used VSMP systematic variable selection.[16] The physiologically based pharmacokinetic modeling approach was used to model BBB permeability as the time to reach brain equilibrium.[17]

28.1.3 Commercial Software

Commercial software for predicting BBB penetration is available. Examples are listed in Table 28.1. Software is useful for an initial prediction of brain penetration potential. When evaluating any software for purchase, it is important to test its predictions for compounds that have been studied in-house. This gives a realistic demonstration of the accuracy of the software model on compounds that are well known to the company. The training set for the software may differ greatly from the compound series being studied in a discovery project,

TABLE 28.1 ▶ Partial List of Commercial Software and Products for Predicting BBB Penetration

Software name	Company	Web site
Software		
ADMET Predictor	Simulations Plus	*www.simulationsplus.com*
Volsurf	Molecular Discovery	*www.moldiscovery.com*
Volsurf	Tripos	*www.tripos.com*
MedChem Explorer	Accelrys	*www.accelrys.com*
KnowItAll	Bio-Rad	*www.biorad.com*
ADME Boxes™ (Pgp)	Pharma Algorithms	*www.ap-algorithms.com*
Absolv	Pharma Algorithms	*www.ap-algorithms.com/absolv.htm*
QikProp	Schrodinger	*www.schrodinger.com*
CSBBB	ChemSilico	*www.chemsilico.com*
PreADME	Bioinformatics and Molecular Design Research Center	*www.bmdrc.org/02_R&D/04_chem.asp*
Admensa	Inpharmatica Ltd.	*www.inpharmatica.co.uk*
Products		
HTD 96	HTDialysis	*www.htdialysis.com/*
Dialyzer 96-well	Harvard Apparatus	*www.harvardapparatus.com*
Dialyzer Multi-Equil	Harvard Apparatus	*www.harvardapparatus.com*
IAM.PC.DD2	Regis Technologies	*www.registech.com*
Delta-8	Kibron	*www.kibron.com*
CT Bovial	Cellial Technologies	*www.cellial.com*

so prediction should be verified using in-house data for a few compounds in the series. It is advantageous if the software allows using a training set of in-house data to develop a custom "local" model or to modify the model.

28.2   In Vitro Methods for BBB

28.2.1   Physicochemical Methods for BBB

28.2.1.1   PAMPA-BBB Method [BBB Permeability]

This method predicts a classification of passive BBB permeability in high throughput. PAMPA-BBB follows the PAMPA format (see Section 26.2.3) but uses porcine brain lipid in dodecane as the artificial permeability membrane (Figure 28.1).[18] The performance of the method with 30 commercial drugs is shown in Figure 28.2. Compounds in the figure are CNS+ or CNS− if they were reported in the scientific literature as penetrating into the brain or not penetrating into the brain, respectively. The PAMPA-BBB method

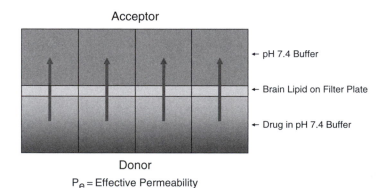

Figure 28.1 ▶   PAMPA-BBB assay diagram.

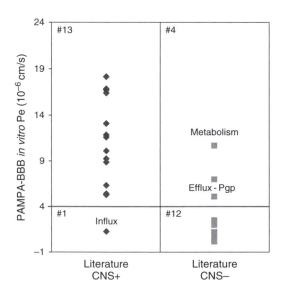

Figure 28.2 ▶   Application of PAMPA-BBB to 30 test compounds. (Reprinted with permission from [13].)

classified all of the CNS+ compounds correctly, except for one that in vivo penetrates into the brain by active uptake transport. It also classified all of the CNS– compounds correctly, except for two that in vivo are prevented from penetrating into the brain by efflux and one that in vivo is heavily metabolized and does not reach sufficient blood concentration to allow significant concentrations in the brain. PAMPA-BBB is a rapid, low-cost method for classifying compounds for predicted passive BBB permeability. Because PAMPA-BBB permeability indicates how fast a compound can permeate through the BBB, it does not necessary predict in vivo B/P because B/P relates to distribution properties of a compound in the static state rather than kinetic rate. PAMPA-BBB correlates well with in vivo perfusion data (Log PS or Log BB) and predicts well for acute in vivo clinical indications, which measures kinetic properties and not equilibrium static state properties of a given drug.

An example of the application of PAMPA-BBB is shown in Figure 28.3. In a discovery project, compounds from multiple compound series were evaluated in vitro for several features, including potency, structure diversity, BBB permeability, and metabolic stability. The data were used together in a holistic manner to prioritize compounds for in vivo dosing. In vivo results suggested where additional assays for other mechanisms should be performed to diagnose unexplained behavior. Finally, structures were modified to improve in vivo performance, as measured by efficacy and brain penetration. Several applications of the PAMPA-BBB assay have been reported in the literature.[19−21]

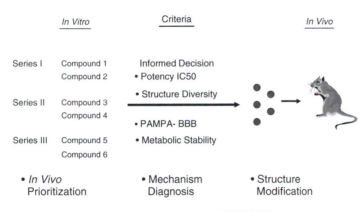

Figure 28.3 ▶ Applications of PAMPA-BBB assay.

28.2.1.2 Lipophilicity [BBB Permeability and Brain Distribution]

Lipophilicity affects both BBB permeation and brain distribution. Since the discovery of the BBB, it has been recognized that lipophilic molecules have greater access to the brain than hydrophilic molecules. The importance of lipophilicity, as measured by Log P or Log D, in predicting BBB permeability was discussed in Chapter 10. The optimal Log D for brain penetration is 1 to 3.[5,22−24] Lipophilicity affects BBB permeation by increasing lipid membrane permeability. It has been reported that in vivo brain permeability correlated to Log P and $MW^{-1/2}$.[25] Lipophilicity is also involved in partitioning between brain and blood, where it contributes to plasma protein binding and nonspecific binding to proteins and lipids in brain tissue. Lipophilicity could bring the compound into closer proximity with a brain target that is in a highly lipophilic environment. Log P and Log D can be calculated or measured, as discussed in Chapter 23. An example of the relationship of lipophilicity and BBB permeability is shown in Figure 28.4.[1]

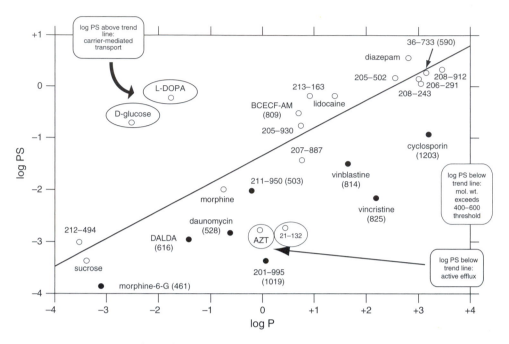

Figure 28.4 ▶ Lipophilicity (Log P) and the BBB permeability surface coefficient (Log PS) are correlated for compounds that permeate by passive diffusion. Compounds that are above the line tend to have enhanced uptake due to active transport. Compounds that are below the line tend to have efflux. (Reprinted with permission from [1].)

28.2.1.3 Equilibrium Dialysis [Brain Distribution]

This method predicts partitioning between plasma and brain homogenate by equilibrium dialysis.[26,27] Thus, it provides insights on brain distribution. Compound is dialyzed between plasma and buffer, then between brain homogenate and buffer. From this, the free drug in brain, free drug in plasma, and brain/plasma partitioning is calculated. By combining partitioning and Pgp efflux data, good correlations are provided to in vivo B/P. The assay is performed using 96-well format dialysis device from HTDialysis or Harvard Biosciences. Free drug in brain can be correlated to in vivo pharmacology.

28.2.1.4 ΔLog P [BBB Permeability]

ΔLog P is an assay for estimating the contribution of hydrogen bonds.[28] For this method, the Log P of a compound is measured by performing both octanol–aqueous buffer and cyclohexane–aqueous buffer partitioning. The aqueous buffer is at least 2 pH units away from the pK_a (i.e., below the pK_a of an acid and above the pK_a of a base). The basis of the method is that octanol forms hydrogen bonds with the test compound but cyclohexane does not; therefore, the ΔLog P value is mainly due to hydrogen bonding. The following calculation is used:

$$\Delta LogP = Log\ P_{octanol} - Log\ P_{cyclohexane}.$$

As ΔLog P increases, BBB permeability and brain distribution usually decrease. Optimal ΔLog P for brain uptake is < 2.[24,29] Young et al.[28] a reported that the brain to blood partition (Log BB) correlated to ΔLog P for 20 H_2 receptor histamine antagonists (see equation below):

$$Log\ BB = -0.485 \times \Delta LogP + 0.889.$$

28.2.1.5 Immobilized Artificial Membrane High-Performance Liquid Chromatography Column [BBB Permeability]

This method predicts permeability but not distribution. Immobilized artificial membrane (IAM) columns were described in Chapter 26. A lipid is bonded to the high-performance liquid chromatography (HPLC) stationary phase. The chromatographic retention time, using the IAM.PC.DD2 column, was shown to be a predictor of BBB permeability.[30,31] Estimation of B/P distribution values is not advised.

28.2.1.6 Surface Activity [BBB Permeability]

BBB permeability has been correlated to air/water interfacial tension partitioning coefficient K_{memb}.[32] This technique has been demonstrated using a high-throughput surface tension instrument from Kibron.

28.2.2 Cell-based In Vitro Methods [BBB Permeability]

Cell-based BBB assays have inherent appeal as in vitro models of BBB permeation because they are living and, presumably, closer to the BBB than physicochemical methods. However, they may not closely resemble the complex conditions at the BBB. They also consume considerably more resources. Most cell-based BBB assays use "transwell" procedures similar to Caco-2 (see Section 26.2.2). Compounds with high lipophilicity tend to depot in the cell monolayer and not move into the subsequent chamber (e.g., basolateral), so these should be tested in vivo. Cell-based permeability methods are not intended to correlate with in vivo B/P (distribution) values.

28.2.2.1 Microvessel Endothelial Cell Permeability [BBB Permeability]

The bovine microvessel endothelial cell (BMEC) method was developed as a model for BBB permeability. The microvessels of brain are isolated from fresh tissue and the endothelial cells are cultured.[33–35] When plated in trans-well culture plates they form monolayers with tight junctions. The correlation to the BBB is reasonable. A limitation of this assay is that the cells are used only as primary cultures. Fresh brain is obtained and the cells are immediately separated, cultured, and used soon in experiments. The expression level of transporters can vary from preparation to preparation. Porcine cultures also have been used.[29] An immortalized endothelial cell line has been produced.

Variations of this method are bovine brain endothelial cells (BBEC) and human primary brain endothelial cells (HPBEC).[2] Microvessels are isolated from brain and cultured. Endothelial cells grow out from the vessels and form microcolonies, which are harvested and plated. They have been cultured in proximity to rat astrocytes, which appear to release factors that enhance differentiation of BBB-like characteristics. The cultured endothelial cells are viable for a few passages.

Endothelial cell cultures have been used for detailed BBB studies but are not common for moderate- to high-throughput applications. It requires considerable resources to prepare endothelial cell cultures. The brain capillary endothelial monolayers in cell models typically are much leakier than BBB, with an electrical resistance more than 10-fold lower than BBB in vivo.[36] Furthermore, most BBB transporters or carriers are down-regulated, up to 100-fold in culture. For example, transport of L-dopa (a CNS drug for Parkinson's disease) was not detected in cell culture models because of the marked suppression of LAT1.[36]

Cellial Technologies supplies bovine brain capillary endothelial cell and glial cell kits, along with special culture media for cell layer BBB permeability studies (Table 28.1).

28.2.2.2 Caco-2 Method for BBB [BBB Permeability]

Caco-2 was developed as a permeation model for gastrointestinal absorption. Some companies have extended its use to screening BBB permeability. Caco-2 incorporates lipid bilayer membranes, Pgp efflux, and some other transporters. The limitations of using Caco-2 for BBB are that it does not form as tight junctions as brain endothelial cells, the lipid mixture of Caco-2 cells is not the same as BBB endothelial cells so it can exhibit different passive diffusion characteristics, and the profile of its transporters (which transporters and their expression levels) differs from those of the BBB.

A more limited, but powerful, application of Caco-2 and MDR1-MDCKII (below) is as a targeted Pgp efflux assay (see Chapter 27). Caco-2 expresses Pgp, and MDR1-MDCKII overexpresses Pgp. Therefore, they provide a means for predicting Pgp efflux, which has a crucial role in BBB permeability. The efflux ratio can be reliably used to evaluate the efflux potential of discovery compounds, diagnose BBB permeability limitations, and develop structure–Pgp efflux relationships to guide structural modification for the purpose of reducing Pgp efflux. Pgp efflux data also can be used in conjunction with other in vitro data to predict brain penetration. An example of this application is shown in Figure 28.5. PAMPA-BBB indicated that a compound should penetrate into the brain by passive diffusion (CNS+). However, an in vivo study showed poor brain penetration (CNS-). The compound was assayed for Pgp efflux using Caco-2. It was found to have an "efflux ratio" ($P_{B>A}/P_{A>B}$, see Section 27.2.1) of 7, which disappeared when cyclosporin A (CSA), a Pgp inhibitor, was coincubated with the compound in the Caco-2 assay media. Thus, the compound could enter the BBB by passive diffusion, but it was rapidly removed by Pgp efflux.

Pgp efflux data can be used in conjunction with equilibrium dialysis (see Section 28.1.3) to predict brain/plasma partitioning.[21,22]

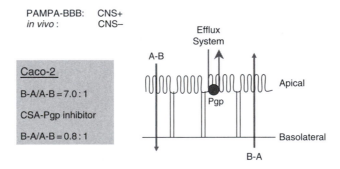

Figure 28.5 ► Diagnosis of BBB permeation mechanisms in a CNS project.

28.2.2.3 MDR1-MDCKII [BBB Permeability]

MDCK or MDR1-MDCKII cells are widely applied for BBB permeability prediction (but not brain distribution). MDCK cells have low transporter expression levels and have been used as a model of passive diffusion permeability.

MDR1-MDCKI and MDR1-MDCKII cell lines both have been transfected with human MDR1, the gene that codes for Pgp.[37] [This cell line is kindly provided by Professor Piet Borst, Netherlands Cancer Institute (NCI), Amsterdam, The Netherlands [see Section 27.2.1.2]. It is free to academic laboratories and can be licensed by companies for a fee. Another line of MDR1-MDCK cells is available from the National Institutes of Health, Bethesda, MD, USA.] Transfection with human MDR1 imparts overexpression of Pgp. A study of the (NCI) MDR1-MDCKII permeability of commercial drugs indicated that CNS drugs usually have a net apical to basolateral $P_{app} > 150$ nm/s.[37] Also, they have an efflux ratio $P_{app,B>A}/P_{app,A>B} < 2.5$. Interestingly, a few CNS drugs had significant efflux ratios and still were active in the CNS, indicating that some Pgp-effluxed compounds still may sufficiently penetrate the brain if their $P_{app,A>B}$ is sufficiently high. MDCK cells form tighter junctions than Caco-2 cells.[38]

A productive use of the MDR1-MDCKII cell line is as a specific assay for Pgp efflux. It overexpresses Pgp; therefore, it is a more sensitive than Caco-2 for detecting Pgp efflux and reliably differentiating among series analogs to indicate structure–Pgp efflux relationships to guide chemists in structural modification.

28.2.2.4 TR-BBB and TM-BBB Cell Lines [BBB Permeability]

The TR-BBB and TM-BBB cell lines have been immortalized from microvessel endothelial cells by Terasaki.[39,40] These cell lines express membrane transporters with good fidelity to endothelial cells. They do not form tight junctions; therefore, they are used only for uptake studies (see Section 27.2.2), not transcellular permeation. The experiment is run using 24-well culture plates. The test compound is added to the cell medium over the cells and incubated. Following incubation, the cell medium is washed off, the cells are lysed, and the concentration of drug inside the cells is measured to determine how much compound permeated into the cell.

28.3 In Vivo Methods for BBB

28.3.1 B/P Ratio or Log BB [Brain Distribution]

The B/P experiment uses standard in vivo PK approaches and is widely implemented in companies. B/P provides brain distribution data (not BBB permeability) over several hours. In this experiment, multiple animals are dosed (time zero), and at designated time points three animals are sacrificed. A sample of blood is retained and the brain is removed. The compound concentrations in the plasma and brain homogenate are measured. The concentrations are plotted versus time (Figure 28.6). B/P is calculated as brain area under the curve (AUC)/plasma AUC. Log BB is the log of B/P. The figure shows a case where the compound has good brain penetration (CNS+), and B/P = 1.23. In the second case, the compound has poor brain penetration (CNS−), and B/P = 0.02. B/P ≥ 0.3 are generally considered acceptable. Some lipophilic compounds produce B/P ≥ 10. An advantage of this approach is that the B/P values are easy to use for ranking compounds in a discovery project. Another advantage is that this experiment provides other PK insights, such as C_{max} and the length of time that the compound concentration remains above a certain value that is active in vitro (e.g., EC50), to help understand the relationship of pharmacodynamics (PD) and PK.

B/P has limitations. First, the experiment requires considerable resources. The seven time point studies shown in Figure 28.6 require 21 animals per compound. Each animal is used for only one time point. The experiment requires significant compound material, animal scientist's time, bioanalytical scientist's time, and instrumentation. Second, B/P is a brain distribution value that is largely determined by nonspecific binding of drug molecules by brain tissue (i.e., partitioning between plasma and brain lipid) and much less by brain permeability.[36]

Third, the values from in vivo studies are total concentration. The experiment does not provide data for the free (unbound) drug concentration in the extracellular fluid (ECF) of the brain. Only free drug interacts with the receptor or enzyme to produce the pharmacological action.[36,41] (See Section 28.2.1.3 for a method for predicting free drug concentration in brain.)

It has been suggested that B/P measurements should be replaced by Log PS (Log BBB permeability-surface area coefficient) in drug discovery because BBB permeation is the greatest limitation to brain exposure.[36,41]

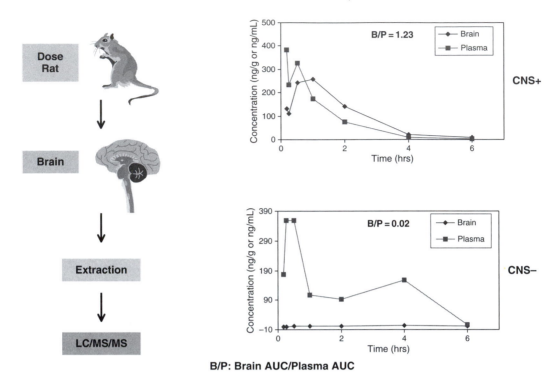

B/P: Brain AUC/Plasma AUC

Figure 28.6 ▶   In vivo brain uptake experiment for B/P and Log BB determination.

Abbreviated in vivo brain penetration studies are often performed as part of an in vivo pharmacology study. During the study, a couple of dosed animals are sacrificed at one or two time point(s) to measure the concentration of the dosed compound in brain and plasma. From these samples, the total brain concentration and a single time point B/P value are obtained. The data allow a simple correlation of brain concentration (or B/P) data to pharmacological data. The information is valuable in early dosing studies. If pharmacological activity is observed, the study shows that a particular brain concentration is efficacious. If no activity is observed, the study confirms whether any compound reached the brain tissue or at what total brain concentration the compound is not efficacious. The limitation of such an experiment is that there can be differences in t_{max}, $t_{1/2}$, and C_{max} between brain and blood, so single time points do not give a complete assessment of brain penetration or PK/PD relationships.

It is useful to perform an initial brain PK study prior to full pharmacology dosing experiments. This study is conducted with different dose levels or vehicles to determine what dose will achieve a desired brain concentration and duration for the subsequent pharmacology experiment.

28.3.2   Brain Uptake Index [BBB Permeability]

The brain uptake index method was one of the earliest techniques for studying in vivo BBB permeability. A quantity of radiolabeled drug is injected into the common carotid artery of an animal along with tritiated water. After 15 seconds, the experiment is stopped and the brain removed. The amount of compound in the brain is calculated relative to the amount of tritiated water (used as an internal standard). This method was used for many years but has been replaced by in situ perfusion, which allows greater control of the fluid composition for research purposes.

28.3.3   In Situ Perfusion [BBB Permeability, Log PS, μL/min/g]

The in situ perfusion method provides high-quality in vivo BBB permeability data.[42-49] The experimental setup is shown in Figure 28.7. A catheter is placed in the common carotid artery of an anesthetized animal, and the external carotid artery is ligated. The pterygopalatine artery remains open. The blood flow is stopped, and the syringe pump is switched in line. The perfusate, which contains dissolved drug, physiological electrolytes, oxygen, and nutrients, provides the fluid flow to the brain. The perfusate rapidly flows through the brain arteries and throughout the brain capillaries. The perfusate replaces the blood. During the perfusion, circulation to half of the brain is totally taken over. BBB integrity has been shown to remain high throughout the experiment. The perfusion is conducted for periods ranging from 5 seconds to several minutes, after which compound-free perfusate is used to purge the brain vessels. The brain is removed, and the compound concentration in the half of the brain that was perfused is measured. This experiment provides high-quality data for studying BBB permeation and specific mechanisms. The short time periods allow minimization of nonspecific binding. From this experiment, PS is determined.[50] Other methods for PS determination include carotid arterial injection and/or infusion methods, and quantitative intravenous injection methods.[12,36]

Coadministration of transporter inhibitor or application of the techniques to transgenic animals that lack a transporter allows study of the extent of transporter contribution to the penetration of a particular compound. For example, in situ perfusion was used with Pgp-deficient and wild-type mice to demonstrate that brain uptake of colchicine, a Pgp substrate, was reduced several-fold by Pgp.[43] The flexibility of selecting constituents of the perfusion fluid allows a wide range of studies of the BBB permeability effects of factors in the blood (e.g., proteins, electrolytes).

This method provides high-quality data, but it consumes significant resources for the surgery, experiment, and quantitation. It is useful for advanced BBB studies. When used by

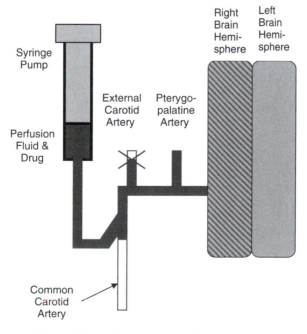

Figure 28.7 ▶   In situ brain perfusion experiment.

a pharmaceutical company, it typically is implemented for detailed studies that are critical to a discovery project.[51,52]

28.3.4   Mouse Brain Uptake Assay [BBB Permeability and Brain Distribution]

The in vivo mouse brain uptake assay (MBUA) is a short duration in vivo experiment.[2,53] The MBUA experiment is initiated with a 6.5 μmol/kg injection (50 μL) into the tail vein of a mouse. At 5 minutes, the animal is anesthetized, and a blood sample is collected and the brain removed. It is assumed that at 5 minutes there is minimal nonspecific brain tissue binding, back flow of test compound out of the brain, and brain metabolism. The concentration of compound in each sample is analyzed using LC/MS/MS. PS is calculated, and from this value BBB Papp is derived (based on a surface area of 240 cm^2 per gram brain in mouse). Running the same experiment for 60 minutes indicates whether the compound is cleared from the brain at the same rate as from the plasma, suggesting fast equilibration across the BBB, or more slowly from the brain, indicating that the compound accumulates in the brain. This method is rapid and uses a minimum of animals and laboratory resources. MBUA Papp data can be plotted versus calculated Log D at pH 7.4 (Clog D7.4). As shown in Figure 28.8., and Table 28.2, this provides insights on the predominant mechanism(s) of

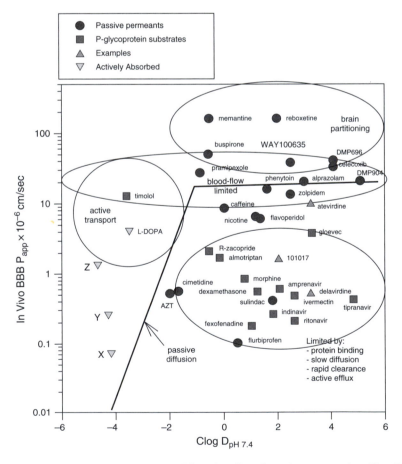

Figure 28.8 ▶   Mouse brain uptake BBB permeability data (Papp) can be compared to Clog $D_{pH7.4}$ to gain insights on potential mechanisms of BBB penetration. (Reprinted with permission from [49].)

TABLE 28.2 ▶ BBB Penetration Mechanisms Suggested by Plotting Location of Data in Figure 28.8

Predominant mechanism	Diagram location
Passive diffusion	Ascending diagonal line
High passive diffusion limited by blood flow	Horizontal line
High distribution (nonspecific binding) into brain	Above horizontal line
Hydrophobic compounds with high plasma protein binding, slow passive diffusion, rapid clearance, or active efflux	Below horizontal line
Hydrophilic compounds with active uptake	Plots to left of passive diffusion diagonal

a compound's BBB penetration. Raub et al.[53] have used this data, in combination with in vitro assays (i.e., plasma protein binding, MDR1-MDCKII) to discern and estimate multiple permeability mechanisms.

28.3.5 Microdialysis Method for BBB

The components in brain ECF can be sampled directly using microdialysis methodology.[54−57] This technique is common for measuring neurotransmitters and has been applied for measuring drug candidate concentrations in the ECF. A small probe having concentric tubes and tipped with a dialysis membrane is implanted into brain tissue (Figure 28.9) and allowed to equilibrate. Artificial ECF fluid is pumped through the probe at a low flow rate (e.g., 1 μL/min). Compounds below the MW cutoff freely diffuse across the membrane and are trapped in the flow of fluid out of the probe. The compound concentration in the probe ("recovery") is 5% to 20% of the concentration in the ECF, depending on the compound and flow rate. The probe is calibrated by placing it in solutions of known concentrations and measuring the dialysate.

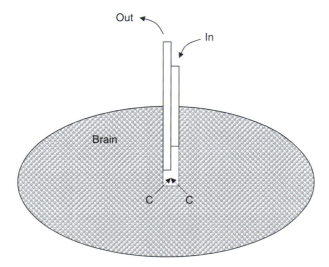

Figure 28.9 ▶ A microdialysis probe is implanted in brain. After equilibration, the ECF is sampled by perfusion with artificial ECF. Compound (C) and other components (protein is excluded) from the ECF are collected into the microdialysate and quantitated using LC/MS/MS.

Data from this method are useful and reliable because only free compound in the ECF is sampled. The limitations of the method are that it is relatively time-consuming to perform, it can disrupt the BBB by causing leakage from capillaries, and very lipophilic compounds can adsorb to the dialysis membrane, resulting in low recovery.

28.3.6   Cerebrospinal Fluid Method for BBB

Cerebrospinal fluid (CSF) is relatively easy to sample in vivo and thus is attractive as a means of assessing brain penetration. Use of CSF is controversial.[36] Compound gets into the CSF by crossing the blood–cerebrospinal fluid barrier (see Chapter 10), which has about 1/5000th the surface area of the BBB, or by being cleared from the ECF, which is slow. CSF is completely refreshed every 5 hours, so there is a rapid turnover. Doran et al.[58] sampled CSF for 31 commercial drugs as part of a major in vivo brain penetration study. Compound CSF concentrations have been predictive of pharmacological activity in some studies.[36,41]

28.4   Assessment Strategy for Brain Penetration

Several methods are available for assessing brain penetration. As the complexity of the data increases, the cost increases and the throughput decreases (Figure 28.10). Therefore, methods should be selected with care.

It is advisable to first assess discovery compounds for brain penetration using higher-throughput methods, such as in silico, physicochemical properties (especially PSA; see Section 10.3 for useful structural property rules), and PAMPA-BBB (Figure 28.11). These provide an initial prediction of passive BBB permeability, which is useful for early prediction of in vivo BBB permeability. This provides discovery scientists insights on whether their

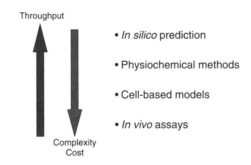

Figure 28.10 ▶   Several methods are available for BBB penetration.

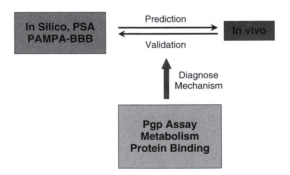

Figure 28.11 ▶   Screening strategy for BBB penetration.

project compounds are expected to have any problems with penetrating to the therapeutic target in the brain. Selected series examples then can be studied using in vivo studies. These results provide more in-depth assessment of penetration into the brain as well as feedback (validation) on whether the early assessment of passive brain penetration was correct. If the in vivo and in vitro brain penetration data differ significantly, the contributions of other mechanisms, such as Pgp efflux, hepatic clearance, plasma protein binding, and nonspecific brain tissue binding, can be assessed using other in vitro assays. The limiting mechanisms can be discerned, and structure modifications can be undertaken to improve the brain penetration properties of the compound series.

Problems

(Answers can be found in Appendix I at the end of the book.)

1. What is the difference between BBB permeation and brain distribution?

2. What structural properties affect brain penetration?

3. PAMPA-BBB provides what predictions of brain penetration?

4. In the equilibrium dialysis method, what are the fluids between which the compounds are dialyzed? Is this a BBB permeability or brain distribution method? What predictions are obtained from this experiment?

5. What does ΔLog P mainly indicate?

6. What does IAM predict?

7. Why is the microvessel endothelial cell method challenging to perform? What are the limitations of this method?

8. What BBB permeability insights can be gained using MDR1-MDCKII cells?

9. What is the most commonly used in vivo measurement for brain penetration? What are its limitations?

10. The brain uptake and in situ perfusion methods at short time points are good measurements of what?

References

1. Pardridge, W. M. (1998). CNS drug design based on principles of blood-brain barrier transport. *Journal of Neurochemistry, 70*, 1781–1792.

2. Garberg, P., Ball, M., Borg, N., Cecchelli, R., Fenart, L., Hurst, R. D., et al. (2005). In vitro models for the blood-brain barrier. *Toxicology in Vitro, 19*, 299–334.

3. Abbott, N. J. (2004). Prediction of blood-brain barrier permeation in drug discovery from in vivo, in vitro and in silico models. *Drug Discovery Today: Technologies, 1*, 407–416.

4. Goodwin, J. T., & Clark, D. E. (2005). In silico predictions of blood-brain barrier penetration: considerations to "keep in mind." *Journal of Pharmacology and Experimental Therapeutics, 315*, 477–483.

5. Clark, D. E. (2003). In silico prediction of blood-brain barrier permeation. *Drug Discovery Today, 8*, 927–933.

6. Clark, D. E. (2005). Computational prediction of blood-brain barrier permeation. *Annual Reports in Medicinal Chemistry, 40*, 403–415.

7. Ajay, Bemis, G. W., & Murcko, M. A. (1999). Designing libraries with CNS activity. *Journal of Medicinal Chemistry, 42*, 4942–4951.

8. Crivori, P., Cruciani, G., Carrupt, P.-A., & Testa, B. (2000). Predicting blood-brain barrier permeation from three-dimensional molecular structure. *Journal of Medicinal Chemistry, 43,* 2204–2216.

9. Keseru, G. M., Molnar, L., & Greiner, I. (2000). A neural network based virtual high throughput screening test for the prediction of CNS activity. *Combinatorial Chemistry and High Throughput Screening, 3,* 535–540.

10. Doniger, S., Hofmann, T., & Yeh, J. (2002). Predicting CNS permeability of drug molecules: comparison of neural network and support vector machine algorithms. *Journal of Computational Biology. 9,* 849–864.

11. Engkvist, O., Wrede, P., & Rester, U. (2003). Prediction of CNS activity of compound libraries using substructure analysis. *Journal of Chemical Information and Computer Sciences, 43,* 155–160.

12. Abraham, M. H. (2004). The factors that influence permeation across the blood-brain barrier. *European Journal of Medicinal Chemistry, 39,* 235–240.

13. Liu, X., Tu, M., Kelly, R. S., Chen, C., & Smith, B. J. (2004). Development of a computational approach to predict blood-brain barrier permeability. *Drug Metabolism and Disposition, 32,* 132–139.

14. Norinder, U., Sjoeberg, P., & Oesterberg, T. (1998). Theoretical calculation and prediction of brain-blood partitioning of organic solutes using MolSurf parametrization and PLS statistics. *Journal of Pharmaceutical Sciences, 87,* 952–959.

15. Lombardo, F., Blake, J. F., & Curatolo, W. J. (1996). Computation of brain-blood partitioning of organic solutes via free energy calculations. *Journal of Medicinal Chemistry, 39,* 4750–4755.

16. Narayanan, R., & Gunturi, S. B. (2005). In silico ADME modelling: prediction models for blood-brain barrier permeation using a systematic variable selection method. *Bioorganic & Medicinal Chemistry, 13,* 3017–3028.

17. Liu, X., Smith, B. J., Chen, C., Callegari, E., Becker, S. L., Chen, X., et al. (2005). Use of a physiologically based pharmacokinetic model to study the time to reach brain equilibrium: an experimental analysis of the role of blood-brain barrier permeability, plasma protein binding, and brain tissue binding. *Journal of Pharmacology and Experimental Therapeutics, 313,* 1254–1262.

18. Di, L., Kerns, E. H., Fan, K., McConnell, O. J., & Carter, G. T. (2003). High throughput artificial membrane permeability assay for blood-brain barrier. *European Journal of Medicinal Chemistry, 38,* 223–232.

19. Wexler, D. S., Gao, L., Anderson, F., Ow, A., Nadasdi, L., McAlorum, A., et al. (2005). Linking solubility and permeability assays for maximum throughput and reproducibility. *Journal of Biomolecular Screening, 10,* 383–390.

20. Rodriguez-Franco, M. I., Fernandez-Bachiller, M. I., Perez, C., Hernandez-Ledesma, B., & Bartolome, B. (2006). Novel tacrine-melatonin hybrids as dual-acting drugs for Alzheimer disease, with improved acetyl-cholinesterase inhibitory and antioxidant properties. *Journal of Medicinal Chemistry, 49,* 459–462.

21. Pavon, F. J., Bilbao, A., Hernandez-Folgado, L., Cippitelli, A., Jagerovic, N., Abellan, G., et al. (2006). Antiobesity effects of the novel in vivo neutral cannabinoid receptor antagonist 5-(4-chlorophenyl)-1-(2,4-dichlorophenyl)-3-hexyl-1H-1,2,4-triazole-LH 21. *Neuropharmacology, 51,* 358–366.

22. Hartmann, T., & Schmitt, J. (2004). Lipophilicity—beyond octanol/water: a short comparison of modern technologies. *Drug Discovery Today: Technologies, 1,* 431–439.

23. Comer, J. E. A. (2004). High-throughput measurement of log D and pKa. In H. van de Waterbeemd, H. Lennernäs, & P. Artursson (Eds.), *Drug bioavailability,* Weinheim, Germany: Wiley-VCH.

24. ter Laak, A. M., Tsai, R. S., Donne-Op den Kelder, G. M., Carrupt, P.-A., Testa, B., & Timmerman, H. (1994). Lipophilicity and hydrogen-bonding capacity of H1-antihistaminic agents in relation to their central sedative side-effects. *European Journal of Pharmaceutical Sciences, 2,* 373–384.

25. Levin, V. A. (1980). Relationship of octanol/water partition coefficient and molecular weight to rat brain capillary permeability. Journal of Medicinal Chemistry, *23,* 682–684.

26. Maurer, T. S., DeBartolo, D. B., Tess, D. A., & Scott, D. O. (2005). Relationship between exposure and nonspecific binding of thirty-three central nervous system drugs in mice. *Drug Metabolism and Disposition, 33,* 175–181.

27. Summerfield, S. G., Stevens, A. J., Cutler, L., Osuna, M. d. C., Hammond, B., Tang, S.-P., et al. (2006). Improving the in vitro prediction of in vivo central nervous system penetration: integrating permeability, P-glycoprotein efflux, and free fractions in blood and brain. *Journal of Pharmacology and Experimental Therapeutics, 316,* 1282–1290.

28. Young, R. C., Mitchell, R. C., Brown, T. H., Ganellin, C. R., Griffiths, R., Jones, M., et al. (1988). Development of a new physicochemical model for brain penetration and its application to the design of centrally acting H2 receptor histamine antagonists. *Journal of Medicinal Chemistry, 31*, 656–671.

29. Kramer, S. D. (1999). Absorption prediction from physicochemical parameters. *Pharmaceutical Science & Technology Today, 2*, 373–380.

30. Salminen, T., Pulli, A., & Taskinen, J. (1997). Relationship between immobilised artificial membrane chromatographic retention and the brain penetration of structurally diverse drugs. *Journal of Pharmaceutical and Biomedical Analysis, 15*, 469–477.

31. Lazaro, E., Rafols, C., Abraham, M. H., & Roses, M. (2006). Chromatographic estimation of drug disposition properties by means of immobilized artificial membranes (IAM) and C18 columns. *Journal of Medicinal Chemistry, 49*, 4861–4870.

32. Suomalainen, P., Johans, C., Soederlund, T., & Kinnunen, P. K. J. (2004). Surface activity profiling of drugs applied to the prediction of blood-brain barrier permeability. *Journal of Medicinal Chemistry, 47*, 1783–1788.

33. Audus, K. L., & Borchardt, R. T. (1987). Bovine brain microvessel endothelial cell monolayers as a model system for the blood-brain barrier. *Annals of the New York Academy of Sciences, 507*, 9–18.

34. Gumbleton, M., & Audus, K. L. (2001). Progress and limitations in the use of in vitro cell cultures to serve as a permeability screen for the blood-brain barrier. *Journal of Pharmaceutical Sciences, 90*, 1681–1698.

35. Audus, K. L., Ng, L., Wang, W., & Borchardt, R. T. (1996). Brain microvessel endothelial cell culture systems. *Pharmaceutical Biotechnology, 8*, 239–258.

36. Pardridge, W. M. (2004). Log(BB), PS products and in silico models of drug brain penetration. *Drug Discovery Today, 9*, 392–393.

37. Doan, K. M. M., Humphreys, J. E., Webster, L. O., Wring, S. A., Shampine, L. J., Serabjit-Singh, C. J., et al. (2002). Passive permeability and P-glycoprotein-mediated efflux differentiate central nervous system (CNS) and non-CNS marketed drugs. *Journal of Pharmacology and Experimental Therapeutics, 303*, 1029–1037.

38. Wang, Q., Rager, J. D., Weinstein, K., Kardos, P. S., Dobson, G. L., Li, J., et al. (2005). Evaluation of the MDR-MDCK cell line as a permeability screen for the blood-brain barrier. *International Journal of Pharmaceutics, 288*, 349–359.

39. Terasaki, T., Ohtsuki, S., Hori, S., Takanaga, H., Nakashima, E., & Hosoya, K.-I. (2003). New approaches to in vitro models of blood-brain barrier drug transport. *Drug Discovery Today, 8*, 944–954.

40. Deguchi, Y., Naito, Y., Ohtsuki, S., Miyakawa, Y., Morimoto, K., Hosoya, K.-I., et al. (2004). Blood-brain barrier permeability of novel [D-Arg2] Dermorphin (1–4) analogs: transport property is related to the slow onset of antinociceptive activity in the central nervous system. *Journal of Pharmacology and Experimental Therapeutics, 310*, 177–184.

41. Martin, I. (2004). Prediction of blood-brain barrier penetration: are we missing the point? *Drug Discovery Today, 9*, 161–162.

42. Smith, Q. R. (1996). Brain perfusion systems for studies of drug uptake and metabolism in the central nervous system. *Pharmaceutical Biotechnology, 8*, 285–307.

43. Dagenais, C., Rousselle, C., Pollack, G. M., & Scherrmann, J.-M. (2000). Development of an in situ mouse brain perfusion model and its application to mdr1a P-glycoprotein-deficient mice. *Journal of Cerebral Blood Flow and Metabolism, 20*, 381–386.

44. Pardridge, W. M., Triguero, D., Yang, J., & Cancilla, P. A. (1990). Comparison of in vitro and in vivo models of drug transcytosis through the blood-brain barrier. *Journal of Pharmacology and Experimental Therapeutics, 253*, 884–891.

45. Rapoport, S. I., Ohno, K., & Pettigrew, K. D. (1979). Drug entry into the brain. *Brain Research, 172*, 354–359.

46. Ohno, K., Pettigrew, K. D., & Rapoport, S. I. (1978). Lower limits of cerebrovascular permeability to nonelectrolytes in the conscious rat. *American Journal of Physiology, 235*, H299–H307.

47. Takasato, Y., Rapoport, S. I., & Smith, Q. R. (1984). An in situ brain perfusion technique to study cerebrovascular transport in the rat. *American Journal of Physiology, 247*, H484–H493.

48. Deane, R., & Bradbury, M. W. B. (1990). Transport of lead-203 at the blood-brain barrier during short cerebrovascular perfusion with saline in the rat. *Journal of Neurochemistry, 54*, 905–914.

49. Gratton, J. A., Lightman, S. L., & Bradbury, M. W. (1993). Transport into retina measured by short vascular perfusion in the rat. *Journal of Physiology, 470*, 651–663.

50. Tanaka, H., & Mizojiri, K. (1999). Drug-protein binding and blood-brain barrier permeability. *Journal of Pharmacology and Experimental Therapeutics, 288*, 912–918.

51. Killian, D. M., Gharat, L., & Chikhale, P. J. (2000). Modulating blood-brain barrier interactions of amino acid-based anticancer agents. *Drug Delivery, 7*, 21–25.

52. Murakami, H., Takanaga, H., Matsuo, H., Ohtani, H., & Sawada, Y. (2000). Comparison of blood-brain barrier permeability in mice and rats using in situ brain perfusion technique. *American Journal of Physiology, 279*, H1022–H1028.

53. Raub, T., Lutzke, B., Andrus, P., Sawada, G., & Staton, B. (2006). Early preclinical evaluation of brain exposure in support of hit identification and lead optimization. In R. T. Borchardt, E. H. Kerns, M. J. Hageman, D. R. Thakker,l & J. L. Stevens (Eds.), *Optimizing the "drug-like" properties of leads in drug discovery,* New York: Springer.

54. Deguchi, Y. (2002). Application of in vivo brain microdialysis to the study of blood-brain barrier transport of drugs. *Drug Metabolism and Pharmacokinetics, 17*, 395–407.

55. Masucci, J. A., Ortegon, M. E., Jones, W. J., Shank, R. P., & Caldwell, G. W. (1998). In vivo microdialysis and liquid chromatography/thermospray mass spectrometry of the novel anticonvulsant 2,3:4,5-bis-O-(1-methylethylidene)-b-D-fructopyranose sulfamate (topiramate) in rat brain fluid. *Journal of Mass Spectrometry, 33*, 85–88.

56. Chen, Y.-F., Chang, C.-H., Wang, S.-C., & Tsai, T.-H. (2005). Measurement of unbound cocaine in blood, brain, and bile of anesthetized rats using microdialysis coupled with liquid chromatography and verified by tandem mass spectrometry. *Biomedical Chromatography, 19*, 402–408.

57. Qiao, J.-P., Abliz, Z., Chu, F.-M., Hou, P.-L., Zhao, L.-Y., Xia, M., et al. (2004). Microdialysis combined with liquid chromatography-tandem mass spectrometry for the determination of 6-aminobutylphthalide and its main metabolite in the brains of awake freely-moving rats. *Journal of Chromatography, B: Analytical Technologies in the Biomedical and Life Sciences, 805*, 93–99.

58. Doran, A., Obach, R. S., Smith, B. J., Hosea, N. A., Becker, S., Callegari, E., et al. (2005). The impact of P-glycoprotein on the disposition of drugs targeted for indications of the central nervous system: evaluation using the MDR1A/1B knockout mouse model. *Drug Metabolism and Disposition, 33*, 165–174.

Chapter 29

Metabolic Stability Methods

Overview

▶ *It is difficult to predict metabolic rate using software, but predicting the labile sites is more successful.*

▶ *Various combinations of cell fractions, hepatocytes, and cofactors are used to study metabolism in vitro. Each studies a different profile of metabolic enzyme activities.*

▶ *Structure elucidation of major metabolites guides structure modification to increase stability.*

Metabolic stability is one of the most important properties in drug discovery. Stability often is a major liability for a lead series and needs improvement. Quantitative metabolic stability data are used to assess the extent of metabolic conversion, prioritize compounds for in vivo studies, and set priorities for discovery activities. Qualitative metabolite structure information indicates the sites of metabolic instability ("hot spots") that medicinal chemists use to plan structural modifications for improved stability. Reviews of metabolic stability[1–3] and hepatobiliary clearance[4] methods have been published.

Metabolic Stability Methods

Five types of methods are commonly used in discovery (Table 29.1):

▶ *Microsomal Stability Assay:* Uses liver microsomes and NADPH cofactor to assess phase I oxidations by CYP and flavin monooxygenases (FMO); applied to all discovery compounds as soon as they are synthesized.

▶ *Phase II Stability Assay:* Uses liver microsomes, S9, or hepatocytes with appropriate cofactors (e.g., uridine diphosphate glucuronic acid [UDPGA] to assess glucuronidation); applied to selected compounds with moieties susceptible to conjugation (e.g., phenols, aliphatic hydroxyls).

▶ *S9 or Hepatocyte Stability Assay:* Encompasses a broader range of metabolizing enzymes for assessment of microsomal and extramicrosomal (e.g., cytosolic, mitochondrial) metabolic reactions; applied to selected compounds, especially in late discovery.

▶ *Metabolite Structure Elucidation:* Uses spectroscopy to determine metabolite structures; applied to selected series examples during optimization or late discovery.

▶ *Metabolic Phenotyping:* Examines which enzymes or isozymes metabolize the compound; applied to selected compounds during optimization or late discovery.

TABLE 29.1 ▶   In Vitro Methods for Determination of Metabolic Stability

Method	Metabolic material	Cofactor	Throughput	Purpose
Microsomal	Microsomes[a]	NADPH	500 cpd/wk	Phase I stability ($t_{1/2}$)
Phase II	Microsomes, S9, or hepatocytes	UDPGA, PAPS	500 cpd/wk	Phase II conjugation stability ($t_{1/2}$)
S9	S9	NADPH, UDPGA, SAPS	500 cpd/wk	Phase I and II stability ($t_{1/2}$)
Hepatocyte	Fresh or cryopreserved hepatocytes	None	100 cpd/wk	Broader stability from phase I and II and permeability ($t_{1/2}$)
Metabolite structure elucidation (LC/MS or NMR)	Microsomes, S9, hepatocytes	NADPH, UDPGA, SAPS	5 cpd/wk	Sites of metabolic liability to modify structure; synthesize to test activity and safety
Metabolic phenotyping	rhCYP, individual enzymes	NADPH	100 cpd/wk	Identify isozymes(s) of metabolism to modify structure; plan clinical DDI studies

[a]Microsomes refer to liver microsomes.

A common strategy for using these assays is illustrated in Figure 29.1.

It is beneficial to measure the in vitro metabolic stability in species that are relevant for the project. Metabolic stability can be quite different in various species. It is efficient to initially obtain stability data in a single generic species (e.g., rat). This provides a common

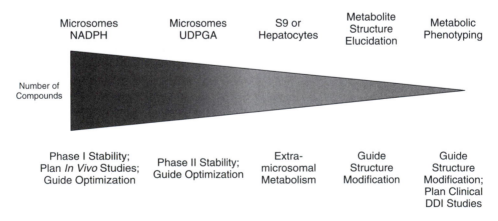

Figure 29.1 ▶   Metabolic assays and their corresponding applications in drug discovery.

point of comparison for all projects. Rat is also commonly used for toxicity screening, so data from this species are useful for planning and interpreting safety studies. In addition to rat, project teams evaluate metabolic stability in their efficacy species (e.g., transgenic mouse, cynomolgus monkey) in vitro to assist with achieving sufficient dosing levels for pharmacology proof of concept studies. In vitro metabolic stability in human is useful for projecting ahead to clinical studies. Metabolic stability data are useful for prospectively selecting compounds for in vivo dosing studies and for retrospectively diagnosing the causes of poor pharmacokinetic (PK) performance or lack of in vivo activity.

29.1 In Silico Metabolic Stability Methods

Three software companies offer tools for predicting metabolite structures and metabolic stability (Table 29.2). Metasite does a very good job at predicting sites of metabolic instability in compound structures.[5] Such tools can be used during structure design to determine if planned substructure modifications will be liable for metabolism or to suggest sites on the molecule to block (see Section 11.3) if the structures of major metabolites have not been spectroscopically elucidated.

TABLE 29.2 ▶ Commercial Software or Metabolic Stability

Name	Company	Purpose	Web site
Metasite	Molecular Discovery	Metabolite structures	*www.moldiscovery.com*
KnowItAll	Biorad	Metabolic stability	*www.biorad.com*
ADMENSA	Inpharmatica	Metabolic stability, metabolite structures	*www.inpharmatica.com*
Meteor	Lhasa	Metabolite structures	*www.lhasalimited.org*

29.2 In Vitro Metabolic Stability Methods

Metabolic stability assays incubate the test compound with a selected metabolic stability material. Detection methods are used to quantitate the remaining (unmetabolized) compound or to spectroscopically study the structures of metabolites.

29.2.1 General Aspects of Metabolic Stability Methods

29.2.1.1 Metabolic Stability Materials

Materials for in vitro metabolic stability methods are readily available because of extensive use of the assay. The activity levels of each batch of material should be prechecked using appropriate quality control compounds because activity can vary with isolation procedure, vendor, and liver source and among batches from the same laboratory.

Metabolic enzymes for metabolic assays are available from several sources. These materials are selected based on the data required by the discovery project team. It is important to know the differences between the materials for proper interpretation of the data. A partial list of vendors of materials is given in Table 29.3. Human-derived materials are generally preferred for in-depth studies in later discovery because they appear to more accurately predict human in vivo stability and are useful for projection to human clinical studies. In earlier stages of discovery, metabolic materials derived from animal species are used in running thousands of assays, for making generic species comparisons, and for planning and

TABLE 29.3 ▶ Partial List of Commercial Suppliers of Materials and Instruments for Metabolic Stability Studies

Company	Product	Web site

BD Gentest	Superzomes (rhCYP isozymes)
	Supermix (mix of rhCYP isozymes)
	Individual metabolic enzymes
	  (other than CYP)
	Liver microsomes (various species and
	  human)
	S9, cytosol
	NADPH regenerating system
	Fresh and cryopreserved hepatocytes
		www.gentest.com

XenoTech	Bactosomes (rhCYP isozymes) Liver microsomes (various species and human) S9, cytosol, mitochondria Fresh and cryopreserved hepatocytes	*www.xenotech.com*
In Vitro Technologies	InVitroSomes (rhCYP isozymes) Liver microsomes (various species and human) Fresh and cryopreserved hepatocytes S9	*www.invitrotech.com*
ADMET Technologies	Human liver microsomes S9, cytosol Fresh and cryopreserved hepatocytes	*www.admettechnologies.com*
Invitrogen	Baculosomes (rhCYP isozymes)	*www.invitrogen.com*
Sigma-Aldrich	CYP isozymes Individual metabolic enzymes (other than CYP)	*www.sigmaaldrich.com*
Qualyst	B-Clear sandwich hepatocyte culture	*www.qualyst.com*
Instruments		
Perkin Elmer	Packard Multiprobe Robot	*www.perkinelmer.com*
Tecan	EVO Platform	*www.tecan.com*
Beckman Coulter	Biomek Laboratory Automation Workstation	*www.beckman.com*

interpreting experiments with efficacy species. Liver microsomes from rats are one third the cost of liver microsomes from humans.

▶ *rhCYP isozymes:* Individual recombinant human CYP (rhCYP) isozymes are available in bulk. They are produced from cDNA cloning of human CYP genes and transfection using baculovirus into insect cells in culture or cloning into bacteria. The isozymes are used individually or in mixtures.

▶ *Liver Microsomes:* These are the most frequently used metabolizing material in drug discovery. They contain the metabolizing enzymes that are bound to the endoplasmic reticulum (ER) membrane in the cells. Liver microsomes contain CYP and other metabolizing enzymes that have a major role in drug clearance. They are a convenient reagent for in vitro metabolic studies. They represent a major portion of the

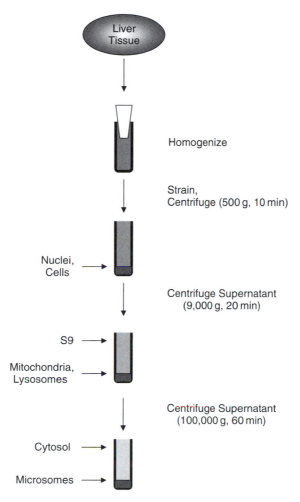

Figure 29.2 ▶ General scheme for preparation of subcellular fractions for metabolic stability studies.

metabolizing enzymes in living systems. Structural optimization guided by liver microsome metabolism has proven to be a very successful strategy in drug discovery. Liver microsomes are prepared by differential centrifugation of liver tissue homogenate (Figure 29.2). The S9 (supernatant) is centrifuged at 100,000*g* to obtain microsomes (pellet) and cytosol. Many of the important metabolizing enzyme classes in liver microsomes are listed in Table 29.4. Cytosol and mitochondria fractions can be separately purchased from vendors.

▶ *S9:* This material contains a broader set of metabolizing enzymes than liver microsomes (Table 29.4). S9 is the supernatant obtained from differential centrifugation at 9,000*g* of strained liver homogenate. This fraction contains both the cellular cytosol and the ER membranes with their respective metabolic enzymes. S9 has about 20% to 25% of the CYP activity of liver microsomes because the liver microsomes make up about one fifth of the S9.

▶ *Hepatocytes:* These are liver cells and contain an ensemble of all the metabolizing enzymes. They are prepared from fresh livers by treatment with calcium-free chelating agent, followed by collagenase to dissociate the cells from the liver matrix. Hepatocytes

TABLE 29.4 ▶ Major Metabolizing Enzyme Activity Classes and Subcellular Liver Fractions in Which Class Members are Commonly Found

Metabolizing activity[a]	Liver microsomes	Cytosol	S9	Mitochondria
Cytochrome P450	X		X	
Alcohol dehydrogenase		X	X	
Aldehyde dehydrogenase		X		X
Monoamine oxidase				X
Diamine oxidase		X		X
Flavin-monooxygenases	X		X	
Reductases	X	X	X	
Esterases	X	X	X	
UDP-Glucuronosyltransferases	X		X	
Sulfotransferase		X	X	
N-Acetyltransferase		X	X	X
Amino acid conjugations	X		X	X
Glutathione-S-transferases	X	X	X	
Methyltransferases	X	X	X	

[a]These activities represent classes of enzymes. The specific enzyme that metabolizes a particular compound may or may not be present in this fraction. The degree of an enzyme's expression varies with fraction, tissue, species, and individual.

can be used freshly prepared (primary culture) or frozen in liquid nitrogen ("cryopreserved").

▶ *Liver slices:* "Precision-cut" liver slices are sections of whole liver tissue. They represent all of the natural liver metabolizing system, including transporters, enzymes, and cofactors.[6] They are used for in-depth studies of selected compounds.

29.2.1.2 Detection Methods for Metabolic Stability

Detection of test compounds after incubation usually is performed using liquid chromatography/mass spectrometry (LC/MS) techniques following the incubation and sample preparation.[7–11] The sensitivity, selectivity, and cost of LC/MS are necessary because the assays are commonly performed at a low compound concentration (e.g., 1 μM). The measurement of 1% of test compound remaining after incubation requires a lower limit of quantitation of 10 ng/mL. The selectivity of high-performance liquid chromatography (HPLC) separation and MS or MS/MS analysis is required because the sample matrix is quite complex and can interfere with detection of the test compound. The matrix contains microsomal components, incubation cofactors, and compound metabolites that must be resolved to obtain a reliable signal for the trace levels of test compound. Recent LC/MS methods for use in metabolic stability assays include trap and inject (no HPLC),[7,8] fast chromatography (1–2 minutes per injection) using fast ("ballistic") gradients,[9,10] or "ultraperformance liquid chromatography" (UPLC). Multichannel HPLC with a switching ion source (MUX) has been used,[11] but it can be challenging because each channel (usually four) analyzes a different compound and must use unique MS/MS conditions, so the constant switching creates a lot of overhead time and requires additional instrument maintenance. MS/MS analysis, using a unique precursor–product ion pair, provides high levels of selectivity and sensitivity. All MS/MS instrument vendors now offer automated procedures for the unattended selection of optimum MS/MS conditions (i.e., precursor and product ions, ion source voltages, collision energy), which saves considerable time for the laboratory scientist and works well. Electrospray ionization (ESI) is used for most compounds. Atmospheric pressure chemical ionization (APCI) is useful for compounds that do not ionize well using ESI.

29.2.2 In Vitro Microsomal Assay for Metabolic Stability

An overview of the metabolic stability assay incubation is shown in Figure 29.3. The test compound is dissolved in dimethylsulfoxide (DMSO) and a small volume is diluted into the solution containing liver microsomes. This solution is added to buffer (5 mM EDTA, 100 mM potassium phosphate buffer, pH 7.4) containing NADPH regenerating system (glucose-6-phosphate, NADP$^+$, and glucose-6-phosphate dehydrogenase). The final microsomal protein concentration typically is 0.5 mg/mL. The mixture is incubated at 37°C, and aliquots are removed at specific time points and quenched. For a single time point assay, the entire incubation solution is quenched. Quenching is accomplished by adding at least one volume of cold acetonitrile, which inactivates the enzymes and precipitates the microsomal material (e.g., protein, lipid). The incubate is centrifuged, and the precipitate pellets to the bottom. The supernatant is injected into an LC/MS/MS system to quantitate the test compound remaining after incubation. Specific LC/MS/MS conditions are used for each test compound. The results are reported as percent remaining and $t_{1/2}$ (half-life). Half-life is calculated from first-order kinetics. A fully automated assay method has been described.[12,13]

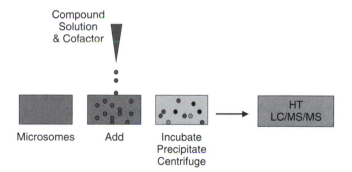

Figure 29.3 ▶ Microsomal stability assay overview. (Reprinted with permission from [28].)

The concentration of the test compound used in the assay differs among laboratories from 0.5 to 15 μM[10–13]; however, the concentration has a dramatic effect on the results.[12] For example, propranolol (Figure 29.4) in rat liver microsomes had 76% remaining after 5 minutes at 10 μM compared to 1% remaining at 1 μM. Apparently, the enzymes are saturated at higher concentrations. The effect is compound dependent. Thus, when comparing (rank ordering) compounds for metabolic stability, it is important that all the data derive from the same assay conditions. The test compound concentration should be low and physiologically relevant. A good concentration for the assay is 1 μM or lower.

Figure 29.4 also illustrates that fact that metabolic rates differ among species. For propranolol, the metabolic rate followed the order: rat > mouse > human. Therefore, it is important to select a species that is relevant to the discovery project and to rank order compounds using data from the same species.

Several microsomal stability assays have been reported.[9–12] With time, improvements to the quality and throughput of the method have been achieved. A 96-well plate assay using the high-throughput aspects outlined has been described.[12] High throughput can be reliably achieved using a "single time point" approach,[13] as opposed to the multiple time

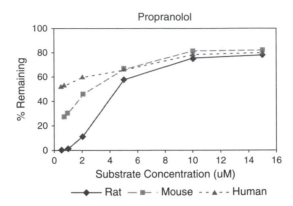

Figure 29.4 ► Species and substrate concentration dependence of percent propranolol remaining after 5 minutes of incubation with liver microsomes. (Reprinted with permission from [12].)

points of many methods. Microsomal stability reactions follow first-order kinetics, in which there is a linear relationship between a plot of log % remaining and linear time. Thus, $t_{1/2}$ can be determined using a linear fit of two points: the first is log (100% remaining) at t_0 and the second is log (% remaining) at the time the reaction is quenched (e.g., 15 minutes). Two points define the line. It is common practice in many laboratories to measure percent remaining at multiple time points. This is not necessary for determination of $t_{1/2}$. Multiple time points only add data points for increased precision and extend the maximum predictive half-life. If the method is reliable, duplicate measurements at t_0 and t_{15min} provide sufficient precision for discovery stability assessment. An example of the agreement of multiple time point data with single time point data (15 minutes) is shown in Figure 29.5 and Table 29.5. A single-point assay of 15 minutes allows highly unstable compounds with $t_{1/2}$ values in the range from 1 to 5 minutes to be measured and provides an upper limit of 30 min $t_{1/2}$. This is a good working range for discovery, which is primarily concerned with being alerted to and stabilizing the highly unstable compounds. Concerns decrease at $t_{1/2}$ greater than 30 minutes.

It is important to know the limitations of the assay in reporting $t_{1/2}$. It has been shown that the logarithmic relationship of $t_{1/2}$ to percent remaining and the error of the method set limits on the maximum and minimum limits of reliable predictive half-life values.[13] For an assay having a single time point at 15 minutes, it is reliable to report $t_{1/2}$ values from 0 to

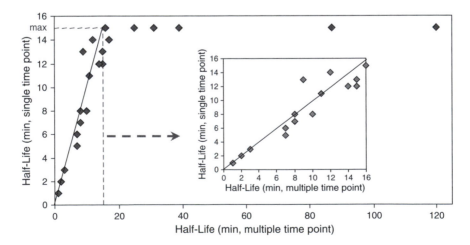

Figure 29.5 ► Correlation of single time point and multiple time point data for drug discovery compounds.

TABLE 29.5 ▶ Agreement of Single Time Point HT Microsomal Assay Results with Multiple Time Point Literature Results[3]

Compound	$T_{1/2}$ (min) Wyeth	$t_{1/2}$ (min) Literature
Midazolam	3	4
Verapamil	6	10
Diltiazem	15	21
Zolpidem	<30	44
Tenoxicam	>30	38

30 minutes. Above 30 minutes, the logarithmic amplification of the error makes differences in the values insignificant. If an assay runs for 30 minutes, the maximum predictive half-life should be no more than 60 minutes.

It is important to optimize microsomal stability assays in order to produce reliable data for low-solubility compounds. Many drug discovery programs have a lead series that is highly lipophilic and, therefore, has low solubility. These types of compounds are not properly assayed by many in vitro assays. Many scientists use well-behaved drug compounds during method development and validation. These do not have the property limitations of many drug discovery compounds (e.g., low solubility, high lipophilicity). Low-solubility compounds precipitate or adhere to the walls of the well. Precipitation eliminates interaction of the test compound with the enzymes, so the compound is not metabolized. When the quenching organic solvent (e.g., acetonitrile) is added to the reaction, the precipitated compound is redissolved and is quantitated as a high percentage of original compound material remaining. Therefore, the compound falsely appears to be stable. It has been shown that metabolic stability methods can be improved for low-solubility compounds by the following critical method conditions and steps[14]:

1. Diluting test compound DMSO stock solutions into organic solvent, not aqueous buffer, to reduce precipitation at higher concentrations (10–20 µM)

2. Adding the diluted test compound into diluted liver microsomes, not into aqueous buffer, to utilize the microsomal lipid and protein for solubilization and to bring the compound into close proximity to the membrane-bound metabolic enzymes

3. Maintaining the final incubation solution at the highest organic solvent and type of solvent that can be tolerated by the assay without affecting the results (e.g., 0.8% acetonitrile/0.2% DMSO rather than 0.2% DMSO alone)

These precautions in method conditions and protocol will keep insoluble test compounds in solution and provide a more accurate measurement of metabolic stability.

Higher throughput in metabolic stability assays is achieved using automated instruments. Robotics using a multichannel robot allows unattended liquid handling with reproducible pipette volumes, timing, and protocol. A diagram of the automation scheme is shown in Figure 29.6. The robot takes a 96-well plate containing stock solutions of each test compound in DMSO (10 mg/mL) and dilutes it to an "optimization plate," which is used for automated MS/MS method optimization, as well as an "incubation plate," which is used for the stability assay. The robot configuration for the microsomal stability assay is shown in Figure 29.7. The assay produces a "sample plate," which is taken to the LC/MS/MS for quantitation. The MS/MS method, developed using the optimization plate and the automated method optimization procedure, is used to quantitate the test compound percent remaining after

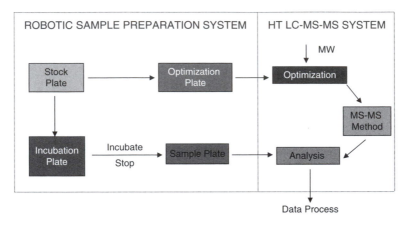

Figure 29.6 ▶ Automation scheme for microsomal stability. (Reprinted with permission from [12].)

Figure 29.7 ▶ Microsomal stability assay setup using a multichannel robot.

incubation. Figure 29.8 shows a high throughput LC/MS/MS quantitation system that uses an HPLC with a short analysis time, an autosampler that can manage up to twelve 96-well plates, and an MS/MS instrument for quantitation.

It is important to qualify each batch of liver microsomes used in the assay. For example, results from liver microsomes of three vendors are shown in Figure 29.9. The enzyme activity in each vendor's material differed greatly. Results can vary among batches from the same vendor or laboratory (Figure 29.10). It is good practice to measure the activity of each new batch using consistent quality control standards and to adjust the method in order to obtain consistent data.

29.2.3 In Vitro S9 Assay for Metabolic Stability

The S9 stability assay conditions are nearly identical to the microsomal assay (e.g., NADPH, buffer, EDTA, Mg).[15,16] S9 contains all of the metabolizing enzymes in liver microsomes plus some additional enzymes found in the cytosol (Table 29.4). However, in S9 the microsomal enzymes are present at 20% to 25% of their concentrations in liver microsomes, so there

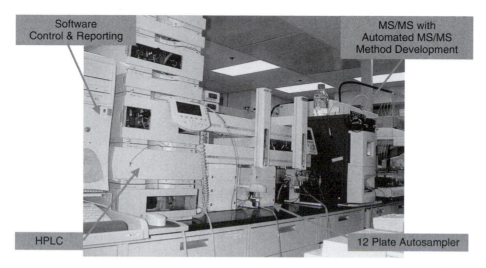

Figure 29.8 ► High-throughput LC/MS/MS system for quantitation.

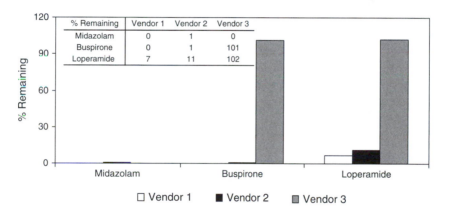

% Remaining	Vendor 1	Vendor 2	Vendor 3
Midazolam	0	1	0
Buspirone	0	1	101
Loperamide	7	11	102

☐ Vendor 1 ■ Vendor 2 ▨ Vendor 3

Figure 29.9 ► Vendor-to-vendor variation in findings from rat liver microsomes. (Reprinted with permission from [12].)

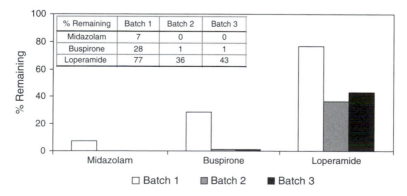

% Remaining	Batch 1	Batch 2	Batch 3
Midazolam	7	0	0
Buspirone	28	1	1
Loperamide	77	36	43

☐ Batch 1 ▨ Batch 2 ■ Batch 3

Figure 29.10 ► Batch-to-batch variation in findings from rat liver microsomes from the same vendor. (Reprinted with permission from [12].)

is a lower rate of metabolism for ER-bound enzymes. One application of S9 used 2.5 mg/mL protein compared to the normal 0.5 mg/mL protein for microsomal stability assays.[15]

An advantage of S9 is that it accesses extramicrosomal enzymes, such as sulfotransferase, alcohol dehydrogenase, and N-acetyltransferase. It is especially useful for compounds having moieties that are prone to metabolic reactions catalyzed by extramicrosomal enzymes. (An alternative is to mix microsomes and cytosol.) As shown in Figure 29.1, S9 typically is used for more detailed study of a selected set of compounds. S9 also is often used for producing metabolites for structure elucidation to ensure a broad coverage of possible metabolic reactions.

29.2.4 In Vitro Hepatocytes Assay for Metabolic Stability

Metabolic stability studies can be conducted using living hepatocytes.[16–19] Isolated hepatocytes closely parallel the conditions of in vivo liver cells,[19] having a complete ensemble of metabolizing enzymes of all isoforms, cofactors, cellular components, and membrane permeation mechanisms (e.g., passive, transporter). Hepatocytes are used immediately as primary cell cultures, are directly prepared from livers, and can be cryopreserved for later use.[17] They also can be used as "sandwich cultured hepatocytes" for uptake and biliary excretion studies.[4] Hepatocytes are available from several suppliers (Table 29.3). They are considerably more expensive than microsomes. Cryopreserved hepatocytes are convenient because they do not depend on the immediate availability of fresh livers. They can be stored for later use, thawed just before the assay time, and used in suspension. It is prudent to check hepatocytes for activity levels of key metabolizing enzymes because expression of metabolizing enzyme can vary as a result of isolation and with the characteristics of the donor.[19] Pooled hepatocytes from several donors are available. Hepatocytes should be used within a few hours before activity diminishes.

Hepatocytes contain all of the enzyme cofactors and physiological concentrations, so they do not require supplementation with NADPH and other cofactors. Some discovery scientists believe that hepatocytes are better correlated to expected in vivo metabolism than other metabolic stability materials. A much lower hepatocyte $t_{1/2}$ than microsome $t_{1/2}$ may indicate significant extramicrosomal or phase II metabolism for the compound.

Fresh cells are more apt to adhere to the cell culture surface. Otherwise, the cells are used in suspension. As with S9 and phase II assays, hepatocytes are used for selected compounds for more detailed studies, although high-throughput approaches using 96-well plates are being used.[18] A commercial product (Qualyst, Table 29.3) is a sandwich-cultured hepatocyte system that predicts in vivo hepatic uptake, metabolism, and biliary clearance in one assay.[4]

29.2.5 In Vitro Phase II Assay for Metabolic Stability

Phase II metabolic stability assays are performed using liver microsomes, hepatocytes, or S9.[20,21] Use of this assay is suggested when a compound has a moiety that is susceptible to phase II conjugation (e.g., phenol, aliphatic hydroxyls) to evaluate the stability of a compound against glucuronidation. Some drugs are primarily cleared by glucuronidation.[19,21] Sulfation usually is studied using S9 because the sulfotransferase enzyme is located in the cytosol. Sulfation is rapid compared to glucuronidation but has low capacity and is saturable.

Phase II metabolic reactions in microsomes, cytosol, and S9 require cofactors. UDPGA is required for glucuronidation, and 3'-phosphoadenosine-5'-phosphosulfate (PAPS) is necessary for sulfation.

The UDP-glucuronosyltransferases (UGTs) have many isozymes. UGTs are found on the inside of the ER membranes and remain inside the liver microsomes prepared from ER membranes. The membrane may reduce access of some compounds to the UGTs. Use of the pore-forming peptide alamethicin at 50 μg/mL of protein in the assay medium is thought by some researchers to increase the rate of in vitro glucuronidation without affecting the enzyme activity. Other researchers find little difference use of Mg^{2+} at 1 mM also increases the rate of glucuronidation.[20]

Glucuronidation is a rapid reaction relative to CYP oxidation. Therefore, if a compound must be hydroxylated by CYP prior to glucuronidation, the CYP reaction is the rate-limiting step and glucuronidation follows immediately. A glucuronidation assay with liver microsomes uses the same conditions and detection as a microsomal assay, except UDPGA is added to the medium at 1.9 mg/mL. Another method uses [14C]UDP-glucuronic acid in the incubation, separates the metabolite(s) from the unreacted UDPGA by solid-phase extraction, and measures the 14C-labeled metabolites using a scintillation detector.[22]

29.2.6 Metabolic Phenotyping

This assay examines which isozymes metabolize the compound (i.e., differentiates among CYP isozymes).[23] This experiment is performed by incubating the test compound with individual isozymes under the same conditions as the microsomal stability assay. The percent remaining of test compound after incubation is quantitated using LC/MS techniques. The relative $t_{1/2}$ for each isozyme (Figure 29.11) indicates whether the compound is a substrate and the rate of the reaction. Metabolic phenotyping has also been performed with UGTs.[22]

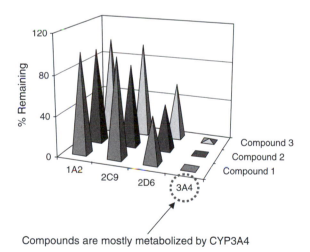

Figure 29.11 ▶ Metabolic phenotyping. Incubation of three compounds with individual rhCYP isozymes indicates which isozyme(s) metabolizes the compound and to what extent.

Metabolic phenotyping has been performed using the microsomal assay and adding an inhibitor that is specific for an isozyme. An increase in $t_{1/2}$ in the presence of the inhibitor is an indication that the isozyme is important in the metabolism of the compound. However, once this isozyme has been inhibited, it is possible for one or more other isozymes to metabolize an increased portion of the compound (colloquially termed *metabolic switching*) and result in a similar $t_{1/2}$.

Metabolic phenotyping has been used together with structural knowledge of the isozyme's active site or substrate specificity (see Chapter 11) to guide structure modifications intended to make the molecule less prone to reaction at the particular isozymes. Metabolic phenotyping data also are necessary for early human clinical studies because they indicate which CYP inhibitors may interact with the new drug (see Chapter 15). Possible drug–drug interactions must be studied using co-dosing experiments during clinical development. It is advantageous if a new clinical candidate is metabolized by three isozymes or more so that if one metabolic route is blocked by a coadministered inhibitor drug, then clearance can occur by "metabolic switching" to another uninhibited isozymes.

29.2.7   In Vitro Metabolite Structure Identification

During the active chemical synthetic phase of discovery optimization, it is useful to determine the structures of the major metabolites of a lead so that they can be synthesized. This can enable testing for activity and toxicity. Metabolite structures also can enable synthetic modifications to block metabolism and increase stability. A selected series lead is incubated with liver microsomes or another metabolic material, and the structures of the metabolites are elucidated using spectroscopy.

A rapid profile of the major metabolites can be obtained using microsomal incubation, followed by LC/MS/MS analysis (Figure 29.12).[24–27] First, the parent compound is analyzed to obtain the HPLC retention time, molecular ion (e.g., MH+), and MS/MS product ion spectrum (Figures 29.12 and 29.13). Modern instruments allow these data to be obtained from a single HPLC injection using automated "data-dependent" analysis. The MS/MS product ion spectrum is interpreted using the scientist's experience and software (e.g., Mass Frontier from Thermo Fisher) to assign specific product ions to specific substructures of the molecule. A microsomal extract control containing the compound and incubation components but quenched at time zero is analyzed. The actual microsomal incubation extract is injected, and components that appear in the incubated sample and not in the microsomal extract control likely are metabolites (Figure 29.14). Their molecular weights determine the mass difference from the parent compound. This can be interpreted in terms of common metabolic reactions (e.g., +16 for hydroxylation or N-oxidation, +32 for dihydroxylation, −14 for demethylation). Interpretation of the MS/MS product spectra of each metabolite (Figure 29.15) demonstrates the mass shifts in specific substructures and indicates the substructure that was metabolized. In some cases, the metabolite is unambiguously elucidated (e.g., dealkylation).

In other cases, such as that shown in Figure 29.15, the specific position of hydroxylation is not determined and NMR is required.[24–27] In these cases, the microsomal incubation

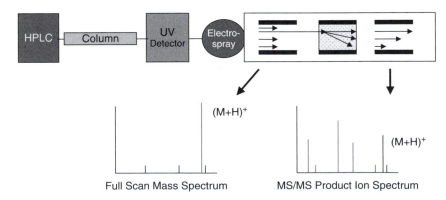

Full Scan Mass Spectrum   MS/MS Product Ion Spectrum

Figure 29.12 ▶ Diagram of LC/MS/MS system and its use in determining the molecular weight of metabolites and MS/MS product ion spectra.

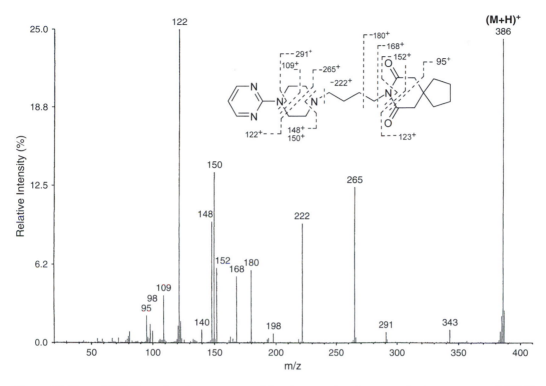

Figure 29.13 ► MS/MS product ion spectrum of buspirone and assignment of substructures to MS/MS product ions. (Reprinted with permission from [24].)

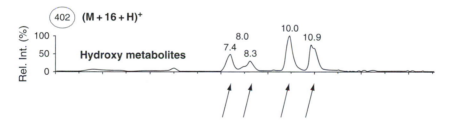

Figure 29.14 ► Selected ion chromatogram of m/z 402 of microsomal incubation of buspirone. Four (M+16) hydroxyl buspirone metabolites are indicated. (Reprinted with permission from [24].)

is repeated at a larger scale. The specific metabolite is isolated using HPLC. The HPLC peak can be monitored using UV detection or MS detection of the specific metabolite molecule ion. Fractions from multiple injections are collected in order to obtain 10 to 50 µg of the metabolite. The mobile phase is removed using low heat and vacuum or inert gas impingement.

Often ^{1}H-NMR is sufficient for determining the site of metabolism. The spectrum of the compound is first obtained and the resonances assigned. Then the spectrum of the metabolite is obtained and changes in the resonances examined. Figure 29.16 shows an example where the hydroxylation was determined to be at C_{17} based on the lack of the C_{17} proton in the ^{1}H-NMR spectrum of the metabolite.

An iterative strategy for optimizing the metabolic stability of a lead series is shown in Figure 29.17. Metabolic assays provide a quantitative stability assessment of the compound. If its stability is low, the specificity for the metabolizing isozyme(s) can be determined

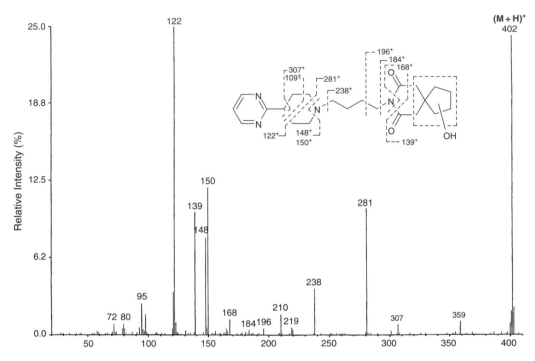

Figure 29.15 ► MS/MS product ion spectrum of a (M+16) buspirone metabolite. (Reprinted with permission from [24].)

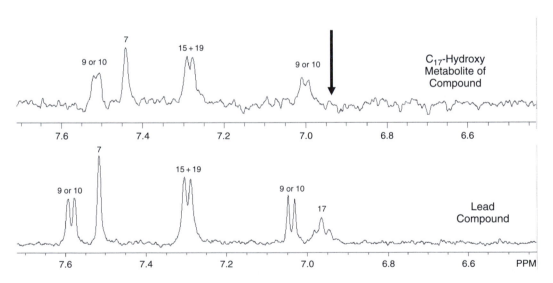

Figure 29.16 ► H-NMR spectra of the aromatic region for a discovery lead and 10 µg of isolated hydroxyl metabolite. The proton signal at C_{17} in the spectrum for the compound was not present in the spectrum for the metabolite, indicating the hydroxylation had occurred at C_{17}.

and the metabolite structures elucidated. The structure can be modified to block metabolic hotspots, and the new compound can be assayed for its stability. The metabolites themselves also can be synthesized to test them for pharmacological activity, for activity at other targets where side effects may be produced, and for safety testing.

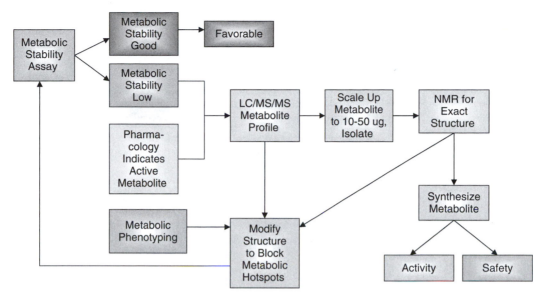

Figure 29.17 ▶ Strategy for metabolic stability assessment and optimization. Metabolites also can be synthesized to test their activity and safety.

Problems

(Answers can be found in Appendix I at the end of the book.)

1. What does S9 or hepatocyte incubation provide compared to microsome incubation?

2. When is it useful to screen for phase II metabolic stability?

3. Which metabolic stability materials include CYP enzymes? Which contain sulfotransferase enzymes?

4. What cofactor provides the energy for the CYP reaction?

5. Why is it best to screen metabolic stability at low concentration ($\sim 1\,\mu M$)?

6. Why is it useful to measure metabolic stability in multiple species?

7. Why is it possible to screen metabolic stability using only time zero and one additional concentration time point?

8. Why is it important to check the activity of each batch of metabolic stability material (e.g., microsomes)?

9. What are the cofactors for glucuronidation and sulfation?

10. What are the cofactors for the hepatocyte metabolic stability assay?

11. What is metabolic phenotyping useful for?

12. Why is structure elucidation of major metabolites useful?

13. When are LC/MS/MS and NMR used for structural studies of metabolites?

☐ References

1. Lin, J. H., & Rodrigues, A. D. (2001). In vitro models for early studies of drug metabolism. In *Pharmacokinetic Optimization in Drug Research: Biological, Physicochemical, and Computational Strategies*, [LogP2000, Lipophilicity Symposium], 2nd, Lausanne, Switzerland, Mar. 5–9, 2000, pp. 217–243.

2. Ansede, J. H., & Thakker, D. R. (2004). High-throughput screening for stability and inhibitory activity of compounds toward cytochrome P450-mediated metabolism. *Journal of Pharmaceutical Sciences*, *93*, 239–255.

3. Obach, R. S. (1999). Prediction of human clearance of twenty-nine drugs from hepatic microsomal intrinsic clearance data: an examination of in vitro half-life approach and nonspecific binding to microsomes. *Drug Metabolism and Disposition*, *27*, 1350–1359.

4. Zamek-Gliszczynski, M. J., & Brouwer, K. L. R. (2004). In vitro models for estimating hepatobiliary clearance. *Biotechnology: Pharmaceutical Aspects*, *1*, 259–292.

5. Cruciani, G., Carosati, E., DeBoeck, B., Ethirajulu, K., Mackie, C., Howe, T., et al. (2005). MetaSite: understanding metabolism in human cytochromes from the perspective of the chemist. *Journal of Medicinal Chemistry*, *48*, 6970–6979.

6. Gebhardt, R., Hengstler, J. G., Mueller, D., Gloeckner, R., Buenning, P., Laube, B., et al. (2003). New hepatocyte in vitro systems for drug metabolism: metabolic capacity and recommendations for application in basic research and drug development, standard operation procedures. *Drug Metabolism Reviews*, *35*, 145–213.

7. Janiszewski, J. S., Rogers, K. J., Whalen, K. M., Cole, M. J., Liston, T. E., Duchoslav, E., et al. (2001). A high-capacity LC/MS system for the bioanalysis of samples generated from plate-based metabolic screening. *Analytical Chemistry*, *73*, 1495–1501.

8. Kerns, E. H., Kleintop, T., Little, D., Tobien, T., Mallis, L., Di, L., et al. (2004). Integrated high capacity solid phase extraction-MS/MS system for pharmaceutical profiling in drug discovery. *Journal of Pharmaceutical and Biomedical Analysis*, *34*, 1–9.

9. Korfmacher, W. A., Palmer, C. A., Nardo, C., Dunn-Meynell, K., Grotz, D., Cox, K., et al. (1999). Development of an automated mass spectrometry system for the quantitative analysis of liver microsomal incubation samples: a tool for rapid screening of new compounds for metabolic stability. *Rapid Communications in Mass Spectrometry*, *13*, 901–907.

10. Caldwell, G. W., Masucci, J. A., & Chacon, E. (1999). High throughput liquid chromatography-mass spectrometry assessment of the metabolic activity of commercially available hepatocytes from 96-well plates. *Combinatorial Chemistry and High Throughput Screening*, *2*, 39–51.

11. Xu, R., Nemes, C., Jenkins, K. M., Rourick, R. A., Kassel, D. B., & Liu, C. Z. C. (2002). Application of parallel liquid chromatography/mass spectrometry for high throughput microsomal stability screening of compound libraries. *Journal of the American Society for Mass Spectrometry*, *13*, 155–165.

12. Di, L., Kerns, E. H., Hong, Y., Kleintop, T. A., McConnell, O. J., & Huryn, D. M. (2003). Optimization of a higher throughput microsomal stability screening assay for profiling drug discovery candidates. *Journal of Biomolecular Screening*, *8*, 453–462.

13. Di, L., Kerns, E. H., Gao, N., Li, S. Q., Huang, Y., Bourassa, J. L., & Huryn, D. M. (2004). Experimental design on single-time-point high-throughput microsomal stability assay. *Journal of Pharmaceutical Sciences*, *93*, 1537–1544.

14. Di, L., Kerns, E. H., Li, S. Q., & Petusky, S. L. (2006). High throughput microsomal stability assay for insoluble compounds. *International Journal of Pharmaceutics*, *317*, 54–60.

15. Rajanikanth, M., Madhusudanan, K. P., & Gupta, R. C. (2003). Simultaneous quantitative analysis of three drugs by high-performance liquid chromatography/electrospray ionization mass spectrometry and its application to cassette in vitro metabolic stability studies. *Rapid Communications in Mass Spectrometry*, *17*, 2063–2070.

16. Li, A. P. (2001). Screening for human ADME/Tox drug properties in drug discovery. *Drug Discovery Today*, *6*, 357–366.

17. Cross, D. M., & Bayliss, M. K. (2000). A commentary on the use of hepatocytes in drug metabolism studies during drug discovery and development. *Drug Metabolism Reviews*, *32*, 219–240.

18. Jouin, D., Blanchard, N., Alexandre, E., Delobel, F., David-Pierson, P., Lave, T., et al. (2006). Cryopreserved human hepatocytes in suspension are a convenient high throughput tool for the prediction of metabolic clearance. *European Journal of Pharmaceutics and Biopharmaceutics*, *63*, 347–355.

19. Gomez-Lechon, M. J., Donato, M. T., Castell, J. V., & Jover, R. (2004). Human hepatocytes in primary culture: the choice to investigate drug metabolism in man. *Current Drug Metabolism*, *5*, 443–462.

20. Fisher, M. B., Campanale, K., Ackermann, B. L., Vandenbranden, M., & Wrighton, S. A. (2000). In vitro glucuronidation using human liver microsomes and the pore-forming peptide alamethicin. *Drug Metabolism and Disposition 28*, 560–566.

21. Soars, M. G., Burchell, B., & Riley, R. J. (2002). In vitro analysis of human drug glucuronidation and prediction of in vivo metabolic clearance. *Journal of Pharmacology and Experimental Therapeutics*, *301*, 382–390.

22. Di Marco, A., D'Antoni, M., Attaccalite, S., Carotenuto, P., & Laufer, R. (2005). Determination of drug glucuronidation and UDP-glucuronosyltransferase selectivity using a 96-well radiometric assay. *Drug Metabolism and Disposition*, *33*, 812–819.

23. Lu, C., Miwa, G. T., Prakash, S. R., Gan, L.-S., & Balani, S. K. (2006). A novel model for thr prediction of drug-drug interactions in humans based on in vitro phenotypic data. *Drug Metabolism and Disposition*, 106.011346.

24. Kerns, E. H., Rourick, R. A., Volk, K. J., & Lee, M. S. (1997). Buspirone metabolite structure profile using a standard liquid chromatographic-mass spectrometric protocol. *Journal of Chromatography, B: Biomedical Sciences and Applications*, *698*, 133–145.

25. Zhang, M.-Y., Pace, N., Kerns, E. H., Kleintop, T., Kagan, N., & Sakuma, T. (2005). Hybrid triple quadrupole-linear ion trap mass spectrometry in fragmentation mechanism studies: application to structure elucidation of buspirone and one of its metabolites. *Journal of Mass Spectrometry*, *40*, 1017–1029.

26. Nassar, A.-E. F., Kamel, A. M., & Clarimont, C. (2004). Improving the decision-making process in the structural modification of drug candidates: enhancing metabolic stability. *Drug Discovery Today*, *9*, 1020–1028.

27. Keating, K. A., McConnell, O., Zhang, Y., Shen, L., Demaio, W., Mallis, L., et al. (2006). NMR characterization of an S-linked glucuronide metabolites of the potent, novel, nonsteroidal progesterone agonist tanaproget. *Drug Metabolism and Disposition*, *34*, 1283–1287.

28. Di, L., & Kerns, E. H. (2007). High throughput screening of metabolic stability in drug discovery. *American Drug Discovery*, 28–32.

Chapter 30

Plasma Stability Methods

Overview

▶ *Compounds are incubated with plasma in vitro to predict in vivo stability.*

▶ *Identification of decomposition products indicates the hydrolytically labile sites.*

Hydrolysis in plasma can be a major route of clearance for compounds having certain hydrolytically labile functional groups. Incubation with plasma in vitro can quickly determine if a compound is susceptible to plasma degradation and to what extent. It also supports structure modification studies to reduce plasma degradation. Identification of the plasma decomposition products indicates which group in the molecule is being attacked. It also is important for pharmacokinetic (PK) studies to determine if a compound is stable in plasma after the sample is taken and stored.

Plasma is a fraction of blood that is prepared as follows. Blood is collected into a vessel containing an anticoagulant (e.g., heparin) and mixed immediately. The liquid is centrifuged to remove cells, and the clear fluid is the plasma. It can be used immediately or stored frozen at −80°C for later use. It can be transported frozen on dry ice.

Several in vitro methods for plasma stability have been published. The emphasis of these articles has often been on high-throughput instrumentation using direct plasma injection liquid chromatography/mass spectrometry/mass spectrometry (LC/MS/MS) with restricted access high-performance liquid chromatography (HPLC) columns,[1,2] automated HPLC column switching,[3] and robotic sample preparation.[4] Conditions for plasma stability assays are quite diverse with regard to compound concentration (3 μM to 6 mM), percent organic solvent (0%–5%), and dilution of plasma with buffer before use (0- to 4-fold).[5–11]

Di et al.[12] investigated the effects of method conditions. They found no significant difference in percent remaining for most compounds over the concentration range from 1 to 20 μM, indicating that plasma stability is not sensitive to drug concentration up to 20 μM. Saturation of plasma enzymes has been reported at 100 μg/mL.[9] A test compound concentration of 1 μM is suggested because it reflects an average plasma concentration after dosing. Percent dimethylsulfoxide (DMSO) up to 2.5% had little effect on the assay. At higher percentages of DMSO, the solvent may denature the enzyme or interfere with protein binding, causing a decrease in activity. Dilution of plasma with buffer does not reduce the hydrolytic rate until it is diluted to 20% to 40% plasma in buffer. Plasma appears to have high catalytic capacity. Plasma pH increases to pH 8 to 9 during long-term storage have been reported; therefore, it is advisable to dilute plasma with pH 7.4 buffer so that stability results are not complicated by decomposition due to a basic pH. Plasma enzyme activity does not appear to decrease for at least 22 hours at 37°C. A 3-hour incubation time is suggested as consistent with in vivo dosing experiments that often are conducted out to 6 hours in discovery. Plasma should be centrifuged (e.g., 3,000 rpm, 10 minutes, 10°C) to remove particulates prior to the assay. Overall, plasma stability assays are relatively tolerant of varying experimental conditions compared to microsomal stability assays[12]; however, consistent use of optimized conditions is prudent.

Plasma for in vitro stability studies is readily available from many suppliers (Table 30.1). Activity can vary greatly with vendor and batch (Figure 30.1) and is compound dependent.[12] A 2- to 20-fold difference can occur. This could alter the compound ranking in a discovery project if assays are performed with different plasma batches. Careful quality control (QC) and adjustment of incubation time to produce the same extent of degradation are suggested for each new plasma batch that enters the laboratory. Alternatively, some compounds can be rerun to "bridge" datasets from different plasma batches. The best solution is to purchase a large batch of plasma to use for an extended time.

TABLE 30.1 ▶   Partial List of Commercial Suppliers of Plasma for Stability Studies

Company	Product	Web site
Bioreclamation	Animal plasma	*www.bioreclamation.com*
Innovative Research	Animal and human plasma	*www.innov-research.com*
Taconic	Animal plasma	*www.taconic.com*

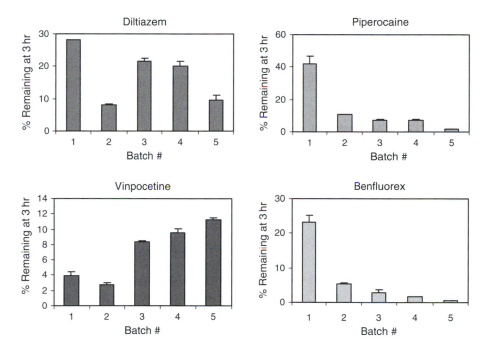

Figure 30.1 ▶   Batch-to-batch variation in plasma from the same vendor.

Detection and quantitation of test compounds after plasma stability incubation are performed using LC/MS techniques (see Chapter 29). Plasma contains many components that may interfere with quantitation; thus, a detection instrument with good selectivity is important.

Few companies assay plasma stability for all new compounds because it is only a problem for compounds having functional groups that are susceptible to hydrolysis. The most common plasma for use in a generic assay is Sprague-Dawley male rat plasma (heparinized and filtered). Because plasma stability varies with species, many discovery project teams may prefer to use a relevant animal species for their particular project. This often is the efficacy species used for the project's pharmacology experiments. Human plasma stability is important to obtain when planning for human clinical studies. Proper safety precautions should be followed when using human plasma to avoid disease.

A diagram of the plasma stability assay is shown in Figure 30.2.[12] To each well is added 195 μL of plasma that has been diluted 1:1 with phosphate buffer (pH 7.4). Five microliters of DMSO stock compound solution 40 μM is added to obtain a final test compound concentration of 1 μM in the plasma and final DMSO content of 2.5%. The plate is sealed with a plate mat, vortexed, and placed on a 37°C shaker (gentle) for 3 hours. The reaction is quenched with 600 μL of cold acetonitrile, followed immediately by mixing and centrifugation. The time zero samples are quenched immediately after the sample is added to plasma. The supernatant (400 μL) is transferred to a 96-well plate for LC/MS analysis. As discussed for metabolic stability assays, it is sufficient to use a single time point concentration plus the initial (time zero) concentration for the calculation of $t_{1/2}$. No problems are anticipated for low-solubility compounds in this assay because of the high protein and DMSO concentrations, which help solubilize compounds. Use of a laboratory robot for the assay is efficient if sufficient samples are to be assayed.

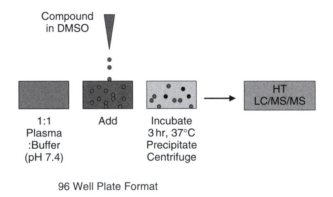

Compound
in DMSO

1:1   Add   Incubate
Plasma   3 hr, 37°C
:Buffer   Precipitate
(pH 7.4)   Centrifuge

HT
LC/MS/MS

96 Well Plate Format

Figure 30.2 ► Plasma stability assay overview.

It is important to co-assay QC controls with each set of test compounds. These ensure that the assay is operating properly and that the activity of the plasma is consistent with established criteria. Four suggested QC compounds are shown in Figure 30.3, and others have been discussed.[12] The assay described here produces consistent QC results over a range of experimental dates.

As with metabolites, it is useful to obtain structural information about plasma degradation products. This provides useful guidance during the active chemical synthetic phase of discovery optimization. Structural modifications can be made to reduce or eliminate plasma hydrolysis reactions. When more than one site in the molecule might be hydrolyzed, these data indicate which site(s) is labile. Often, LC/MS (single-stage) analysis with molecular weight determination of the degradants is sufficient because plasma hydrolysis reactions occur at predictable sites in the molecule, and the molecular weights of the putative products are easily calculated. Because hydrolysis can produce acidic, basic, or neutral products, it is useful to operate the mass spectrometer in alternating positive and negative ion polarities to ensure detection of all of the degradation products. For more detailed structural studies, the LC/MS/MS and NMR techniques, as described for metabolite structure elucidation (see Section 29.2.6), can be used.

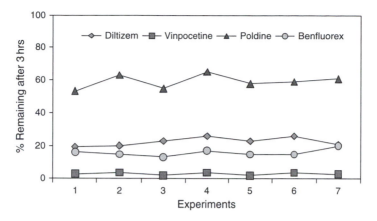

Figure 30.3 ▶ QC standards and results over multiple experiments are reproducible. (Reprinted with permission from [12].)

Problems

(Answers can be found in Appendix I at the end of the book.)

1. What is in plasma that makes compounds unstable?

2. Before using plasma from a new preparation or vendor batch, it is good to do what?

3. Are the following statements true or false?: (a) plasma stability assays are highly sensitive to % DMSO, buffer dilution, and substrate concentration, (b) plasma stability varies with species, (c) degradation in plasma will increase clearance in vivo, (d) stability in plasma is important for accurate PK study results, (e) QC controls are not important for plasma stability assays because the results are so consistent.

References

1. Wang, G., Hsieh, Y., Lau, Y., Cheng, K.-C., Ng, K., Korfmacher, W. A., et al. (2002). Semi-automated determination of plasma stability of drug discovery compounds using liquid chromatography-tandem mass spectrometry. *Journal of Chromatography B, 780,* 451–457.

2. Wang, G., & Hsieh, Y. (2002). Utilization of direct HPLC-MS-MS for drug stability measurement. *American Laboratory, 12,* 24–27.

3. Peng, S. X., Strojnowski, M. J., & Bornes, D. M. (1999). Direct determination of stability of protease inhibitors in plasma by HPLC with automated column-switching. *Journal of Pharmaceutical and Biomedical Analysis, 19,* 343–349.

4. Linget, J.-M., & du Vignaud, P. (1999). Automation of metabolic stability studies in microsomes, cytosol and plasma using a 215 Gilson liquid handler. *Journal of Pharmaceutical and Biomedical Analysis, 19,* 893–901.

5. Pop, E., Rachwal, S., Vlasak, J., Biegon, A., Zharikova, A., & Prokai, L. (1999). In vitro and in vivo study of water-soluble prodrugs of dexanabinol. *Journal of Pharmaceutical Sciences, 88,* 1156–1160.

6. Rautio, J., Taipale, H., Gynther, J., Vepsalainen, J., Nevalainen, T., & Jarvinen, T. (1998). In vitro evaluation of acyloxyalkyl esters as dermal prodrugs of ketoprofen and naproxen. *Journal of Pharmaceutical Sciences, 87,* 1622–1628.

7. Kim, D.-K., Ryu, D. H., Lee, J. Y., Lee, N., Kim, Y.-W., Kim, J.-S., et al. (2001). Synthesis and biological evaluation of novel A-ring modified hexacyclic camptothecin analogues. *Journal of Medicinal Chemistry, 44,* 1594–1602.

8. Udata, C., Tirucherai, G., & Mitra, A. K. (1999). Synthesis, stereoselective enzymatic hydrolysis, and skin permeation of diastereomeric propranolol ester prodrugs. *Journal of Pharmaceutical Sciences, 88,* 544–550.

9. Nomeir, A. A., McComish, M. F., Ferrala, N. F., Silveira, D., Covey, J. M., & Chadwick, M. (1998). Liquid chromatographic analysis in mouse, dog and human plasma; stability, absorption, metabolism and pharmacokinetics of the anti-HIV agent 2-chloro-5-(2-methyl-5,6-dihydro-1,4-oxathiin-3-yl carboxamido) isopropylbenzoate (NSC 615985, UC84). *Journal of Pharmaceutical and Biomedical Analysis, 17,* 27–38.

10. Greenwald, R. B., Zhao, H., Yang, K., Reddy, P., & Martinez, A. (2004). A new aliphatic amino prodrug system for the delivery of small molecules and proteins utilizing novel PEG derivatives. *Journal of Medicinal Chemistry, 47,* 726–734.

11. Geraldine, C., & Jordan, M. (1998). How an increase in the carbon chain length of the ester moiety affects the stability of a homologous series of oxprenolol esters in the presence of biological enzymes. *Journal of Pharmaceutical Sciences, 87,* 880–885.

12. Di, L., Kerns, E. H., Hong, Y., & Chen, H. (2005). Development and application of high throughput plasma stability assay for drug discovery. *International Journal of Pharmaceutics, 297,* 110–119.

Solution Stability Methods

Overview

▶ *Decomposition in various in vitro and in vivo solutions can confuse research if it is not recognized.*

▶ *Reactions can occur in bioassay matrix, dosing solution, or gastrointestinal fluids.*

▶ *Compounds can be incubated in relevant solutions in well plates and analyzed by liquid chromatography/mass spectrometry.*

Solution stability encompasses a broad range of drug barriers. Chapter 13 discussed the most frequently encountered solution stability issues: biological assay buffers, pH, and in vivo gastrointestinal fluids. (Plasma stability was discussed in Chapters 12 and 30.) Because of the diversity of solution decomposition situations and conditions, the testing of all new company compounds using generic high-throughput solution stability assays is not common. Instead, assays often are performed to prospectively check compounds that contain groups that might be unstable or to retrospectively diagnose unexpectedly poor performance in an in vitro biological assay, pharmacokinetics, or in vivo pharmacology. Solution stability assays are customized to the specific physicochemical conditions (e.g., pH, temperature, time, ionic strength, light), solution components (e.g., enzymes, buffer, modifiers, excipients, oxygen), and protocol that are relevant to the team's solution stability questions. In-depth methods are performed on late discovery compounds that are being considered for clinical development, in order to meet established criteria for advancement. In vitro methods for chemical stability have been reviewed.[1,2]

 ## 31.1 General Method for Solution Stability Assays

Solution stability assays have a unique problem with quenching the reaction. If the solution is injected into a high-performance liquid chromatograph (HPLC) column immediately at a particular incubation time point, the concentration of the test compound is accurately measured. However, if a period of time passes before the solution is injected, then the compound continues to be exposed to the solution conditions and can continue to react. In the past, stability studies were performed individually for a few compounds, and they could be immediately manually injected. However, with larger numbers of discovery compounds and more frequent study of solution stability, assays must be automated and performed in 96-well plates. Injection of the solution in the last well in the plate can occur 7 hours after injection of the solution in the first well (with a 5-minute analysis cycle). Thus, the incubation time incrementally lengthens for later samples in the plate, and different wells have different incubation times. Short-duration stability studies that simulate intestinal transit time or in vitro assays are particularly affected. For a pH stability study, one might assume that the reaction can be quenched by titration to pH 7.4, but compounds still may be unstable at neutral pH, and they continue to be exposed to solution components (e.g., enzymes, excipients). For most solutions (e.g., bioassay buffer), there is no way to effectively quench the reaction.

One answer for this problem is to use a multichannel HPLCs that simultaneously analyze many samples.[3] Nanostream (*www.nanostream.com*) and Eksigent (*www.eksigent.com*) offer instruments that can simultaneously analyze 8 or 24 samples by HPLC with UV detection. This allows studies of 24 solution conditions or compounds at the same time. These systems are not directly coupled to a mass spectrometer (MS) for online liquid chromatography (LC)/MS analysis.

Another approach is to start the incubation of each sample at a different time preceding the time zero injection. The samples are spaced according to their HPLC run time so that they will have the same incubation time at the time they are injected. This process is complicated and time consuming for 96-well plates.

Others groups have frozen samples at –80°C to stop the reaction. Individual samples are thawed just prior to HPLC injection. In this way, the samples have the proper incubation time.[4] This process requires manual manipulation and is time consuming.

An efficient and widely applicable method that solves the quenching problem is use of a conventional HPLC instrument that has an autosampler (e.g., Agilent 1100) capable of programmed liquid handling.[5] Into the autosampler are placed two 96-well plates, one containing dimethylsulfoxide (DMSO) stock solutions of the test compounds and the other containing wells filled (e.g., 180 µL) with the stability solution (Figure 31.1). The autosampler is programmed using the system software (e.g., Chemstation) to perform the following steps:

1. Transfer test compound (e.g., 20 µL) from the DMSO stock to the stability solution

2. Thoroughly mix the combined solution

3. Incubate the samples at a programmed temperature (e.g., 37°C)

4. Remove solution at programmed time points

5. Inject solution into the HPLC

6. Manage the other samples in the same manner by interlacing them

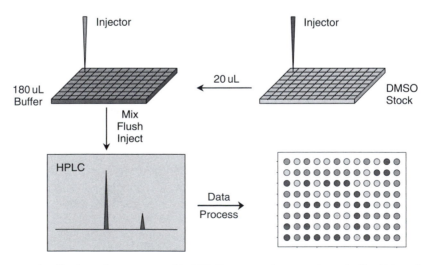

Figure 31.1 ► Application of a programmable HPLC autosampler to perform the liquid handling, incubation, and injection steps of the solution stability assay and obtain reliable time point data. (Reprinted with permission from [15].)

Programmed injections provide the exact time point for each data point, which is useful for accurate kinetic analysis and avoids the need to quench the reaction. This system is readily interfaced to a MS, which is valuable for high sensitivity and spectroscopic structural analysis. Structural data for the reaction products facilitate rapid identification of the labile sites in the structure and complement the kinetic stability results. Figure 31.2 shows a typical instrument used for this assay.

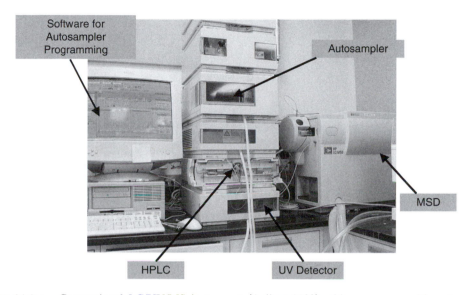

Figure 31.2 ▶ Conventional LC/UV/MS instrument (Agilent 1100) with a programmable autosampler that can be used for solution stability assays as shown in Figure 31.1.

This method has been demonstrated to be robust for stability studies in biological assay buffers, simulated gastric fluids, dosing vehicle, and pH stability studies.[5] Minimal analyst time and intervention are required. The assay can run unattended overnight, on weekends, or on holidays. It can be flexibly applied for custom assay design, such as simultaneous testing of 1 to 96 compounds, one to multiple time points, and different solution conditions on the same plate. Examples of applications of this method include single time point pH solution stability screening (Figure 13.7), high-throughput single time point study of the stability of 96 compounds in bioassay buffer (Figure 13.8), and kinetic analysis of one compound at multiple pHs (Figure 13.6). The last application is an example of the use of high-throughput methods to provide enhanced detail for one compound, as opposed to the more conventional use of high-throughput methods to provide single time point data for a large number of compounds. Kinetic analysis of the data can be performed using Scientist software (Micromath Research, St. Louis, MO, USA).

31.2 Method for Solution Stability in Biological Assay Media

If erratic results are obtained from bioassays, the cause may be low compound solubility or instability in the bioassay buffer. A set of project compounds can be rapidly screened, using the method discussed in Section 31.1, to check compound stability. The assay should mimic as closely as possible the conditions and protocol of the biological assay to ensure meaningful results. The assay conditions should ensure compounds are completely soluble in stability assays, either using co-solvent or by lowering the screening concentration. If compounds precipitate during the assay, the area counts for the test compound will decrease and lead to inconclusive results. Alternatively, percent purity can be used to monitor formation of degradation products, which is less sensitive to solubility and precipitation. An example is shown in Figure 13.8.

31.3 Methods for pH Solution Stability

Compounds in drug discovery encounter a wide range of pHs. Many assay buffers and physiological fluids are at pH 7.4. Oral dosing exposes compounds to pH 1 to 2 in the stomach, pH 4.5 at the beginning of the small intestine, pH 6.6 as an average pH for the small intestine, and pH 5 to 9 in the colon. These are useful pHs for profiling discovery compounds.

An integrated generic method profiles solution stability at multiple pHs (e.g., 2, 7, 12) and oxidation with 3% hydrogen peroxide in a 96-well plate using a Gilson 215 robot.[6] Compounds were incubated at 100 μM, and low-solubility compounds were kept in solution using 1:1 acetonitrile to buffer. (The reaction rate at this high organic content may be low compared to 100% aqueous because of a low dielectric constant. An alternative is to reduce the test compound concentration, allowing a lower organic content.) Samples were immediately injected into the HPLC at their time points.

The decomposition of compounds under acidic conditions in the stomach have been screened in solution at pH 2 for 75 minutes to simulate the time spent in the stomach.[7] These data were combined with solubility data for a "liberation ranking" as part of prospectively predicting oral bioavailability classification prior to in vivo dosing.

pH stability studies have been used for in vitro evaluation of prodrugs.[8] Solution stability studies used buffers at 0.02 M concentration with acetate (pH 5.0), phosphate (pH 3.0, 6.9, 7.4), borate (pH 8.5 and 9.75), and hydrochloric acid (pH 1). Constant ionic strength was maintained because of the effect of ionic strength on reaction rate. The experiment was run at 100 μM in an HPLC vial, with samples withdrawn at selected time points and immediately injected into the HPLC column.

The programmable autosampler technique described in Section 31.1 is very useful for automated custom pH solution stability studies.[5] Buffers of pHs that are relevant to the pH stability issues of the particular project team are placed into wells of the 96-well plate. The autosampler adds test compound solution to the wells, mixes, incubates at 37° C, and injects into the LC/MS system at programmed time points. The data from such an experiment are shown for a β-lactam compound in Figure 13.6. Such an assay provides reliable kinetics that can be used during discovery and development.

31.4 Methods for Solution Stability in Simulated Gastrointestinal Fluids

Important insights can be obtained from incubating discovery compounds with simulated gastrointestinal fluids. These include simulated gastric fluid (SGF) and simulated intestinal fluid (SIF). These materials are specified in the United States Pharmacopeia (USP).[9]

The components of SGF are as follows:

▶ Pepsin (an acidic protease)

▶ pH 1.2 adjusted using HCl

▶ NaCl

SGF simulates stomach fluid and incorporates acidic and enzymatic hydrolysis conditions. The components of SIF are as follows:

► Pancreatin (mixture of amylase, lipase, and protease from hog pancreas)

► pH 6.8 using monobasic phosphate buffer adjusted with NaOH

SIF mimics the pH and hydrolytic enzymes in the intestine.

The main purpose of these assays is the prediction or diagnosis of stability after oral dosing. The data guide medicinal chemists in structural modifications to improve GI stability for optimizing bioavailability and prioritization of compounds for in vivo studies.

The stability of compounds with these materials is readily determined using the solution stability assay system described in Section 31.1. Incubation in these fluids is a rapid way to determine if a compound will be stable under conditions found in the GI. The compounds shown in Figure 13.9 have a range of stabilities in SIF and SGF, with compounds 1 and 3 having favorable profiles.

31.5 Identification of Degradation Products from Solution Stability Assays

The structure elucidation of decomposition products from solution stability experiments follows the same methodology as discussed for plasma stability (see Chapter 30) and metabolic stability (see Section 29.2.6). Molecular weight data from a single-stage MS attached to the HPLC provides valuable information for the reaction products. The hydrolytically labile sites in a molecule usually are obvious; thus, the molecular weight of putative decomposition products can be rapidly calculated and compared to the results. Such information is superior to quantitative data alone because it can guide medicinal chemists in making structure modifications to improve stability.

31.6 In-Depth Solution Stability Methods for Late Stages of Drug Discovery

In late stages of drug discovery, in-depth assessment usually is undertaken of the solution stability of just a couple of potential development candidates from a project. Standard assays are used for the scrutiny of all compounds so that the data are comparable across all projects and are applicable to established advancement requirements. The assays often are similar to those suggested by the Food and Drug Administration for regulatory filings in new drug applications[10,11] but may not be as rigorous. The following are examples of stability experiments that often are performed in aqueous solutions to model drug barriers:

► pH: Aqueous buffers (37°C, pH 1–12)

► Oxidation: 3% hydrogen peroxide in pH 7.4 buffer for 10 minutes

► GI: Simulated intestinal fluid (USP, 37°C, 1–24 hours)

► GI: Simulated gastric fluid (USP, 37°C, 1–24 hours)

► GI: Simulated bile/lecithin mixture (USP, 37°C, 1–24 hours)

► Plasma: Human plasma (37°C, 1–24 hours)

► Light: High-intensity cool white fluorescent light (200 watt h/m^2, 1.2 million lux hour, room temperature, 1–7 days)

► Temperature: Heat (30–75°C, 1–7 days)

For studies over a long time, vials can be prepared for each condition and placed into an HPLC autosampler. Samples of each reaction condition are injected at time points. This is convenient because of the low number of compounds being evaluated. Kinetics can be calculated and used to predict long-term stability. Co-solvent can be added to help dissolve insoluble compounds.

Problems

(Answers can be found in Appendix I at the end of the book.)

1. List some of the solution conditions under which compounds may be unstable or which may accelerate decomposition.

2. What is a difficulty of solution stability assays?

3. What approach can be used to effectively and efficiently run solution stability assays?

4. What conditions should be used for solution stability studies?

5. In addition to percent remaining at a certain time point, what additional data can be obtained from solution stability studies and how can they be used?

6. For what purposes can solution stability data be used?

References

1. Kerns, E. H., & Di, L. (2006). Accelerated stability profiling in drug discovery. In B. Testa, S. D. Kramer, H. Wunderli-Allenspach, & G. Folkers (Eds.), *Pharmacokinetic profiling in drug research: biological, physicochemical and computational strategies* (pp. 281–306), Zurich: Wiley.

2. Kerns, E. H., & Di, L. (2007). Chemical stability. In: John B. Taylor & David J. Triggle (Eds.), *Comprehensive medicinal chemistry II*, Chapter 5.20, (pp. 489–508), Oxford: Elsevier.

3. Patel, P., Osechinskiy, S., Koehler, J., Zhang, L., Vajjhala, S., Philips, C., et al. (2004). Micro parallel liquid chromatography for high-throughput compound purity analysis and early ADMET profiling. *Journal of the Association for Laboratory Automation, 9*, 185–191.

4. Dias, C. S., Anand, B. S., & Mitra, A. K. (2002). Effect of mono- and di-acylation on the ocular disposition of ganciclovir: physicochemical properties, ocular bioreversion, and antiviral activity of short chain ester prodrugs. *Journal of Pharmaceutical Sciences, 91*, 660–668.

5. Di, L., Kerns, E. H., Chen, H., & Petusky, S. L. (2006). Development and application of an automated solution stability assay for drug discovery. *Journal of Biomolecular Screening, 11*, 40–47.

6. Kibbey, C. E., Poole, S. K., Robinson, B., Jackson, J. D., & Durham, D. (2001). An integrated process for measuring the physicochemical properties of drug candidates in a preclinical discovery environment. *Journal of Pharmaceutical Sciences, 90*, 1164–1175.

7. Caldwell, G. W. (2000). Compound optimization in early- and late-phase drug discovery: acceptable pharmacokinetic properties utilizing combined physicochemical, in vitro and in vivo screens. *Current Opinion in Drug Discovery & Development, 3*, 30–41.

8. Nielsen, A. B., Buur, A., & Larsen, C. (2005). Bioreversible quaternary N-acyloxymethyl derivatives of the tertiary amines bupivacaine and lidocaine: synthesis, aqueous solubility and stability in buffer, human plasma and simulated intestinal fluid. *European Journal of Pharmaceutical Sciences, 24*, 433–440.

9. United States Pharmacopeia. (2005). Test solution. In *United States Pharmacopeia (USP)*, *vol. 28* (p. 2858). Rockville, MD: United States Pharmacopeia (USP).

10. U.S. Department of Health and Human Services, Food and Drug Administration, Center for Drug Evaluation and Research (CDER), Center for Biologics Evaluation and Research (CBER), ICH. (1996, November). *Guidance for industry, Q1B photostability testing of new drug substances and products.* www.fda.gov/cder/guidance/1318.htm.

11. Department of Health and Human Services, Food and Drug Administration, Center for Drug Evaluation and Research (CDER), Center for Biologics Evaluation and Research (CBER), ICH. (2001, August). *Guidance for industry, Q1A stability testing of new drug substances and products, revision 1.* www.fda.gov/cder/guidance/4282fnl.htm.

CYP Inhibition Methods

Overview

▶ *CYP inhibition is assessed by coincubating the compound, substrate, and isozyme.*

▶ *Reduction in substrate metabolic rate indicates inhibition by the compound.*

▶ *Substrate concentration should be at the K_m for the isozyme. Less than 10% of substrate and compound should be metabolized for accurate measurement.*

▶ *Double cocktail method uses a mixture of substrates and of recombinant CYP isozymes with liquid chromatography/mass spectrometry.*

Cytochrome P450 (CYP) inhibition is a major cause of drug–drug interactions (DDI). Because DDI can lead to toxicity, considerable attention has been given to methods for measuring CYP inhibition. Most pharmaceutical companies have implemented CYP inhibition testing during the discovery process. A working group of experts from Pharmaceutical Research and Manufacturers of America (PhRMA) companies has published a consensus paper on DDI testing.[1] Measurement of CYP inhibition during discovery provides early warning of potential safety issues. It is used with other data to select leads and helps develop structure–CYP inhibition relationships that are used by medicinal chemists to overcome CYP inhibition by structural modification.[2–9]

In vitro CYP inhibition data are important in late discovery for planning human clinical studies.[10–12] In vitro CYP inhibition data indicating minimal inhibition are allowed by the Food and Drug Administration (FDA) for concluding that the clinical candidate lacks CYP inhibition potential and does not require clinical DDI studies. Approaches have been developed to predict in vivo DDI using in vitro data,[13] but other researchers have shown this to be a difficult task.[14]

Leaders in the field continue to work on optimum standardized methods for CYP inhibition testing.[15] For efficient discovery research and for planning of clinical DDI studies, high-quality in vitro data are crucial.

32.1 In Silico CYP Inhibition Methods

Two software companies offer tools for predicting CYP inhibition (Table 32.1).

Several groups have studied the in silico prediction of CYP inhibition.[16–24] The greatest current use of in silico CYP inhibition methods appears to be screening compound libraries for compounds with potential problems and guidance of medicinal chemists during the structure optimization stage on modifications that may yield reduced CYP inhibition.[25] Computational docking to individual CYP isozymes structures using information from contemporary reports[26] should enhance the in silico predictions of the future.

TABLE 32.1 ▶ Commercial Software for CYP Inhibition

Name	Company	Web site
KnowItAll	Biorad	*www.biorad.com*
ADMENSA	Inpharmatica	*www.inpharmatica.com*

32.2 In Vitro CYP Inhibition Methods

CYP inhibition methodology is an active area of research, and several methods have been reported.[2] Assays can be distinguished by four method elements: (1) CYP material, (2) substrate compounds, (3) detection method, and (4) strategy of analysis.

Commonly used *CYP materials* are recombinant human CYP isozymes (rhCYP) and human liver microsomes (HLM). rhCYPs come from individually cloned human CYP isozyme genes that are transfected using baculovirus into insect cells and produced in bulk. They can be purchased as individual CYP isozyme reagents from suppliers (Table 32.2) and in kits with substrates, buffers, and NADPH. HLM also are commercially available. They are prepared from human livers that are pooled from many individuals, so they represent a broad human population. The livers are tested for human disease to ensure safety. HLM contain all of the natural human CYP isozymes at the natural concentrations.

TABLE 32.2 ▶ Some Commercial Suppliers of Materials and Equipment for CYP Inhibition Studies

Company	Product	Web site	Detection
BD Gentest	P450 inhibition kits Superzomes Human liver microsomes	*www.gentest.com*	Fluorescence
Invitrogen	VIVID Screening Kits Baculosomes	*www.invitrogen.com*	Fluorescence
Promega	P450-Glo Screening Systems	*www.promega.com*	Luminescence
In Vitro Technologies	Human liver microsomes	*www.invitrotech.com*	
XenoTech	Human liver microsomes	*www.xenotech.com*	
BMG Labtech	Fluorescence plate readers	*www.bmglabtech.com*	Fluorescence

Differences in IC_{50} values have been reported in the literature between data generated using rhCYP with fluorescent probes (lower IC_{50}) and drug probes with HLM. This difference can be attributed to the effects of HLM, including protein binding,[27] substrate turnover,[28] and test compound metabolism[28] (see Section 32.2.3).

The *substrate compounds* for CYP inhibition assays typically consist of two types: fluorogenic substrates and drug substrates. *Luminogenic* and radioactive substrates also have been used. Fluorogenic substrates have very low fluorescence, but their metabolites are fluorescent. Examples of fluorogenic substrates are listed in Table 32.3 and shown in Figure 32.1. When the fluorogenic substrate is metabolized by the CYP isozyme, fluorescence intensity increases and is quantitated using a fluorescence plate reader. If the test compound reduces the fluorescence signal compared to control, it is an inhibitor. Some fluorescent substrates can be metabolized by more than one CYP isozyme; therefore, they must be used with only one isozyme at a time. Drug substrates used for CYP inhibition assays are selected because

TABLE 32.3 ►   Examples of Fluorogenic Substrates

Isozyme	Abbreviation	Substrate/metabolite
3A4	BFC	7-benzyloxy-4-trifluoromethylcoumarin 7-hydroxy-4-trifluoromethylcoumarin
2D6	AMMC	3-[2-(N,N-diethyl-N-methylamino) ethyl]-7-methoxy-4-methylcoumarin (3-[2-(N,N-diethyl-N-methylammonium) ethyl]-7-hydroxy-4-methylcoumarin)
2C9	MFC	7-methoxy-4-trifluoromethylcoumarin 7-hydroxy-4-trifluoromethylcoumarin
1A2	CEC	7-ethoxy-3-cyanocoumarin 7-hydroxy-3-cyanocoumarin

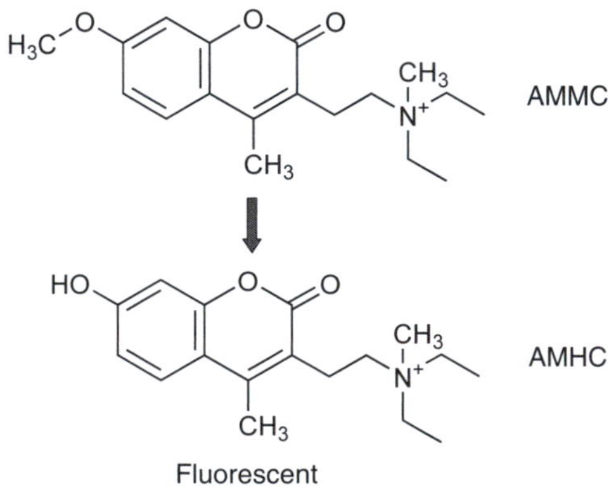

Figure 32.1 ► Structures of fluorogenic substrate AMMC and its metabolite AMHC used for the fluorescent inhibition assay for CYP2D6. (Reprinted with permission from [45].)

they are specifically metabolized by one CYP isozyme. As the drug substrate is metabolized by the CYP isozyme, the concentration of its metabolite increases and is quantitated using liquid chromatography/mass spectrometry/mass spectrometry (LC/MS/MS). If the test compound reduces the concentration of drug substrate metabolite compared to control, it is an inhibitor. The luminogenic substrates (Figure 32.2) are derivatives of luciferin. The metabolites react with a proprietary luciferin detection reagent and produce light. Reduction in the light produced when test compound is coincubated compared to control is an indication that the test compound is an inhibitor.[29,30] Luminogenic substrates can have interference from test compounds that affect luciferase enzymatic activity. Radioactive substrates are more expensive to purchase and produce radioactive waste.

Detection methods include fluorescence, LC/MS/MS, radioactivity, and luminescence.[31] Fluorescence and LC/MS/MS are the two most commonly used methods. Fluorescence detection can be performed using a plate reader in 96- or 384-well formats and therefore is much higher throughput than LC/MS/MS. Fluorescence-based methods are commonly used during the lead selection and optimization stages, when thousands of compounds require testing. LC/MS/MS detection is medium throughput and typically is used during late discovery and development to provide the most definitive data. An advantage of LC/MS/MS is that a cocktail of substrates can be used in a single well for measurement of inhibition at multiple CYP isozymes. The selectivity of LC/MS/MS distinguishes between the metabolites of each substrate. This parallel processing increases the throughput by the number of separate substrates analyzed. The luminescence method is reported to have high sensitivity, no

Figure 32.2 ▶ Structures of general luminescent substrate and its metabolite that are used for the luminescent inhibition CYP inhibition assay. Different R_1 and R_2 groups are used for different CYP isozymes.

fluorescence interference, and low false-positive rate. With radioactive assays, the radioactive metabolite must be separated from the radioactive substrate using high-performance liquid chromatography (HPLC) and quantitated with a radiometric detector, or a scintillation proximity assay is used.[32]

The CYP materials, substrate materials, and detection methods have been combined using four major assay strategies:

1. *Fluorescent Assay:* Individual rhCYP isozymes, individual fluorescent substrates, fluorescent detection

2. *Single Substrate HLM Assay:* HLM, individual specific drug substrates, LC/MS/MS detection

3. *Cocktail Substrate HLM Assay:* HLM, cocktail of specific drug substrates, LC/MS/MS detection

4. *Double Cocktail Assay:* Cocktail of rhCYP isozymes, cocktail of drug substrates, LC/MS/MS detection

Each of these methods is summarized in Table 32.4 and discussed in the following. Many companies use different assays at different stages of discovery and development. Therefore, it is important that medicinal chemists verify that data from the different methods are consistent.[31,33–37] Some studies show good correlation of methods and some show weak correlation, depending on the assay conditions and materials used. Differences in data between methods are primarily affected by differences in the concentration of CYP material, which affects substrate turnover and test compound metabolism[28] (see Section 32.2.3).

A general diagram of CYP inhibition methods is shown in Figure 32.3. The CYP material is added to buffer, NADPH cofactor, and substrate in the well of a 96- or 384-well plate. A small volume of test compound dissolved in dimethylsulfoxide (DMSO) is added, and the solution is incubated at 37°C for a specific time (e.g., 20 minutes). The reaction is quenched by the addition of acetonitrile, and the substrate metabolite concentration is measured. A reduction in metabolite production compared to control indicates inhibition by the test compound. If a single concentration of test compound is used (e.g., 3 µM), then a percent inhibition value is produced.[38] Testing at multiple concentrations of test compound

TABLE 32.4 ▶ Methods for Determination of CYP Inhibition

Method	CYP material	Substrate material	Detection method	Throughput for four isozymes
Fluorescent	Individual rhCYP isozyme	Individual fluorescent substrate	Fluorescence	500 cpd/wk
Single substrate HLM	HLM	Isozyme-specific drug	LC/MS/MS	50 cpd/wk
Cocktail substrate HLM	HLM	Cocktail of isozyme-specific drugs	LC/MS/MS	200 cpd/wk
Double cocktail	Cocktail of rhCYP isozymes	Cocktail of isozyme-specific drugs	LC/MS/MS	200 cpd/wk

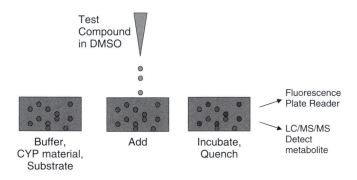

Figure 32.3 ▶ General diagram of an in vitro CYP inhibition assay.

provides an IC_{50} value. K_i values are produced from assays at multiple concentrations of test compounds and substrates.

32.2.1 Fluorescent Assay for CYP Inhibition

The fluorescent assay uses individual rhCYP isozymes, individual fluorescent substrates, and fluorescent detection. It is one of the most common techniques for higher-throughput CYP inhibition assessment.[33,35,36,39–46] The method scheme is shown in Figure 32.4. The fluorescent "probe" substrate forms its fluorescent metabolite at a predictable rate. When an inhibitory test compound is coincubated, it reduces the rate of production of the fluorescent metabolite. The CYP material for this assay is individual rhCYP isozymes. Following the incubation, the reaction is quenched, and the plate is placed in a fluorescence plate reader for quantitation.

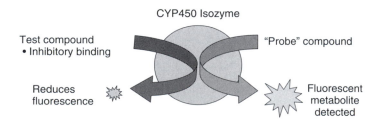

Figure 32.4 ▶ Scheme of the in vitro fluorescent CYP inhibition method using individual rhCYP isozymes, individual fluorescent substrates, and fluorescent detection.

Cloning and expression of individual human CYP isozymes has provided access to commercial supplies of each rhCYP isozyme, which can be prepared at a specific concentration with cofactors, substrate, and inhibitor for the generation of reliable enzymatic data under initial rate conditions. Furthermore, the design and development of substrates that produce fluorescent metabolites have enabled rapid plate reader quantitation of the reaction for high throughput.

A shortcoming of the fluorescent method is revealed if the test compound is fluorescent. The data for such compounds indicate negative inhibition values. It is advisable to assay these test compounds using a nonfluorescent detection method, such as LC/MS/MS.

32.2.2   Single Substrate HLM Assay for CYP Inhibition

The single substrate HLM assay uses HLM, individual specific drug substrates, and LC/MS/MS detection.[47] The scheme of the assay is shown in Figure 32.5. A reduction of substrate metabolite (e.g., hydroxyl midazolam) signal, relative to control, indicates inhibition (Figure 32.5). This method does not have the problem of fluorescent interference (see Section 32.2.1). The concentrations of substrate and enzyme can be set up so that this method does not have the problems of substrate depletion and test compound observed in the cocktail substrate HLM assay (see Section 32.2.3).

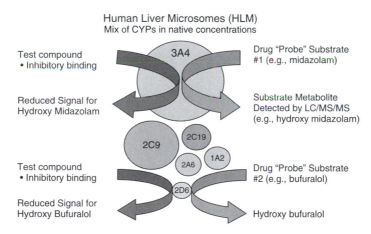

Figure 32.5 ▶   Scheme of the in vitro CYP inhibition methods using HLM, drug substrates, and LC/MS/MS detection. For the single substrate HLM method, only one drug substrate is used. For the cocktail substrate HLM method, a mixture of drug substrates is used.

The specificity of LC/MS/MS quantitation provides confidence of accuracy and reduction in interference. However, LC/MS/MS is more expensive and slower than a plate reader. The determination of definitive K_i values, late in discovery, often is performed using this method by varying both inhibitor and substrate concentrations. Higher throughput can be achieved by pooling the supernatants from incubations of single substrates and then analyzing the single pool in one LC/MS/MS analysis.[48]

32.2.3   Cocktail Substrate HLM Assay for CYP Inhibition

The cocktail substrate HLM assay uses HLM, a cocktail of specific drug substrates, and LC/MS/MS detection. The assay has higher throughput than the single substrate HLM assay. Each substrate in the cocktail is specifically metabolized by a single CYP isozyme and detected with specific LC/MS/MS conditions (i.e., retention time, precursor ion, product ion). Therefore, the inhibition at each isozyme is independently assessed. For these reasons, the assay design is efficient and specific.[49,50]

There are three concerns with this method. First, the native abundance of each of the CYP isozymes varies widely in HLM (Table 15.2). For example, CYP3A4 makes up 28% of the total CYP material, whereas CYP2D6 is only 2%. Because this assay typically is run with similar initial drug substrate concentrations, the substrates for the high abundance CYP3A4 and CYP2C9 isozymes are depleted to a much greater extent (35% and 40% depletion, respectively) than those for low abundance CYP2D6 (1%–4% depletion). This is shown in Table 32.5.[51] The high substrate depletion for CYP3A4 and CYP 2C9 exceed ideal initial rate conditions for enzyme kinetics.

TABLE 32.5 ▶ Examples of High Substrate Depletion and Inhibitor Metabolism by the Cocktail Substrate HLM Assay at CYP3A4

Assay	Cocktail substrate HLM assay	Cocktail substrate HLM assay	Fluorescent assay	Double cocktail assay	Human liver microsome stability
Total protein concentration	0.5 mg/mL	0.1 mg/mL	NA	NA	0.5 mg/mL
3A4 concentration	42 pmol/mL	8.4 pmol/mL	5.0 pmol/mL	0.78 pmol/mL	42 pmol/mL
Substrate	Midazolam	Midazolam	BFC	Midazolam	NA
Test compound	IC_{50} (µM)	IC_{50} (µM)	IC_{50} (µM)	IC_{50} (µM)	Percent remaining of test compound at 20 min
Clotrimazole	0.46	0.041	<0.01	0.005	25
Ethynylestradiol	>90	>100	1.2	7.6	0
Miconazole	5.4	0.49	0.21	0.23	43
Nicardipine	>20	2.1	0.24	0.39	3
Fluconazole	>100	19	24	13	97
Terfenadine	>20	18	1.2	1.9	58
Verapamil	>40	26	3.9	5.9	21
Erythromycin	>50	28	15	23	65
Nifedipine	>100	24	11	12	10
Clomipramine	>100	28	6.9	14	87
Ketoconazole	0.60	0.14	0.05	0.05	89
Percent conversion of substrate	35%	16%	NA	1%	NA

Reprinted with permission from [28].
NA, Not applicable.

The second concern with this method is that many test compounds are highly metabolized by the HLM. This is shown in Table 32.5, with percent remaining of some test compounds after the 20-minute incubation in the range from 0% to 50%. Thus, the actual concentration of inhibitor is changing and unknown during the incubation.

A third concern with this method is nonspecific protein binding to the HLM material, which sequesters test compound away from the enzymes.

The three aspects of concern regarding this method (high substrate depletion, high inhibitor metabolism, nonspecific protein binding) can produce falsely high IC_{50} values for some discovery test compounds by underestimating their CYP inhibition potentials.[51] Table 32.5 shows that IC_{50} values from the cocktail substrate HLM assay approach those of the fluorescent method if the total HLM protein concentration is reduced. However, operating the method at lower HLM concentration can reduce the substrate metabolite below the detection limit of LC/MS/MS.

It has been suggested that HLM is a better model of living systems. However, CYP inhibition is often used in discovery like a selectivity filter, and the concerns discussed can confuse interpretation of the data. No other selectivity experiment (e.g., nearest target neighbors) includes the decomposition of compound during the assay or the high nonspecific binding of test compound that restricts it from interacting with the enzyme.

32.2.4 Double Cocktail Assay for CYP Inhibition

The double cocktail assay uses a cocktail of rhCYP isozymes, cocktail of drug substrates and LC/MS/MS detection.[37,51] The method has high throughput, high specificity, and high sensitivity for CYP inhibition, and it overcomes problems of the other methods. The "double cocktail" name comes from the cocktail of rhCYP isozymes (first cocktail) and the cocktail of specific drug substrates (second cocktail).[51] This method is very cost effective because it uses small amounts of rhCYP isozymes. Substrate metabolites are quantitated using LC/MS/MS (Table 32.6).

TABLE 32.6 ▶ Commonly Used Drug Substrates for CYP Inhibition Assays and LC/MS/MS Conditions for Quantitation of the Substrate Metabolite[49]

Drug substrate	Drug metabolite	CYP isozyme	Precursor ion (m/z)	Product ion (m/z)
Midazolam	1'-Hydroxymidazolam	3A4	342	203
Bufuralol	1'-Hydroxybufuralol	2D6	278	186
Diclofenac	4'-Hydroxydiclofenac	2C9	312	231
Ethoxyresorufin	Resorufin	1A2	214	214
S-Mephenytoin	4'-Hydroxymephenytoin	2C19	235	150
Coumarin	7'-Hydroxycoumarin	2A6	163	107
Paclitaxel	6α-Hydroxypaclitaxel	2C8	870	286

The assay has the advantages of parallel incubation and parallel LC/MS/MS analysis of multiple substrate metabolites. Moreover, the rhCYP isozymes can be mixed at low concentrations that are optimized for enzyme kinetics. Therefore, the reaction conditions each can be controlled to keep the assay under initial rate conditions, with low substrate depletion, low test compound metabolism, and low protein binding. High specificity is provided by the LC/MS/MS analysis.

32.3 CYP Inhibition Assessment Strategy

An efficient strategy for CYP inhibition assessment in drug discovery is shown in Figure 32.6:

▶ The fluorescent assay at a single test compound concentration (3 μM) is used during "hit-to-lead" and optimization phases. Thousands of compounds are tested initially using the high-throughput characteristics of the fluorescent assay.

▶ Inhibition >50% triggers a fluorescent assay IC_{50} study for more definitive data on lead compounds of interest.

▶ Negative inhibition, caused by test compound fluorescence, triggers retesting using the double cocktail method.

▶ Compounds considered for development are tested for IC_{50} using the double cocktail method. A K_i value should be produced and compared to C_{max} (see Chapter 15).

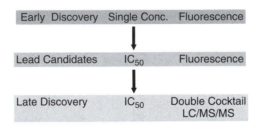

Figure 32.6 ▶ Strategy for CYP inhibition at various discovery stages.

Problems

(Answers can be found in Appendix I at the end of the book.)

1. What two types of materials are commonly used for CYP inhibition assays?

2. What are some negative effects of using microsomes in CYP inhibition assays that can affect the data?

3. What four types of substrates (probes) have been used for CYP inhibition assays? Which are most common?

4. CYP inhibition assays measure which of the following?: (a) increased production of substrate metabolite, (b) decreased production of substrate metabolite, (c) increased production of inhibitor metabolite.

5. What is a drawback of the fluorescent assay?

6. What is an advantage of the cocktail substrate HLM assay?

7. What is a drawback of the cocktail substrate HLM assay?

8. How does the double cocktail assay improve on the drawbacks of the cocktail substrate HLM assay?

References

1. Bjornsson, T. D., Callaghan, J. T., Einolf, H. J., Fischer, V., Gan, L., Grimm, S., et al. (2003). The conduct of in vitro and in vivo drug-drug interaction studies: a Pharmaceutical and Research Manufacturers of America (PhRMA) perspective. *Drug Metabolism and Disposition, 31*, 815–832.

2. Rodrigues, A. D., & Lin, J. H. (2001). Screening of drug candidates for their drug-drug interaction potential. *Current Opinion in Chemical Biology, 5*, 396–401.

3. Riley, R. J., & Grime, K. (2004). Metabolic screening in vitro: metabolic stability, CYP inhibition and induction. *Drug Discovery Today: Technologies, 1*, 365–372.

4. Jenkins, K. M., Angeles, R., Quintos, M. T., Xu, R., Kassel, D. B., & Rourick, R. A. (2004). Automated high throughput ADME assays for metabolic stability and cytochrome P450 inhibition profiling of combinatorial libraries. *Journal of Pharmaceutical and Biomedical Analysis, 34*, 989–1004.

5. Saunders, K. C. (2004). Automation and robotics in ADME screening. *Drug Discovery Today: Technologies*, *1*, 373–380.

6. Kassel, D. B. (2004). Applications of high-throughput ADME in drug discovery. *Current Opinion in Chemical Biology*, *8*, 339–345.

7. Kerns, E. H., & Di, L. (2003). Pharmaceutical profiling in drug discovery. *Drug Discovery Today*, *8*, 316–323.

8. Li, A. P. (2001). Screening for human ADME/Tox drug properties in drug discovery. *Drug Discovery Today*, *6*, 357–366.

9. Di, L., & Kerns, E. H. (2003). Profiling drug-like properties in discovery research. *Current Opinion in Chemical Biology*, *7*, 402–408.

10. Obach, R. S., Walsky, R. L., Venkatakrishnan, K., Houston, J. B., & Tremaine, L. M. (2005). In vitro cytochrome P450 inhibition data and the prediction of drug-drug interactions: qualitative relationships, quantitative predictions, and the rank-order approach. *Clinical Pharmacology & Therapeutics*, *78*, 582.

11. Obach, R. S., Walsky, R. L., Venkatakrishnan, K., Gaman, E. A., Houston, J. B., & Tremaine, L. M. (2006). The utility of in vitro cytochrome P450 inhibition data in the prediction of drug-drug interactions. *Journal of Pharmacology and Experimental Therapeutics*, *316*, 336–348.

12. Huang, S., Lesko, L., & Williams, R. (1999). Assessment of the quality and quantity of drug-drug interaction studies in recent NDA submissions: study design and data analysis issues. *Journal of Clinical Pharmacology*, *39*, 1006–1014.

13. Shou, M. (2005). Prediction of pharmacokinetics and drug-drug interactions from in vitro metabolism data. *Current Opinion in Drug Discovery and Development*, *8*, 66–77.

14. Andersson, T. B., Bredberg, E., Ericsson, H., & Sjoberg, H. (2004). An evaluation of the in vitro metabolism data for predicting the clearance and drug-drug interaction potential of CYP2C9 Substrates. *Drug Metabolism and Disposition*, *32*, 715–721.

15. Walsky, R. L., & Obach, R. S. (2004). Validated assays for human cytochrome P450 activities. *Drug Metabolism and Disposition*, *32*, 647–660.

16. Ekins, S., Bravi, G., Binkley, S., Gillespie, J. S., Ring, B. J., Wikel, J. H., & Wrighton, S. A. (1999). Three- and four-dimensional quantitative structure activity relationship analyses of cytochrome P-450 3A4 inhibitors. *Journal of Pharmacology and Experimental Therapeutics*, *290*, 429–438.

17. Ekins, S., Berbaum, J., & Harrison, R. K. (2003). Generation and validation of rapid computational filters for CYP2D6 and CYP3A4. *Drug Metabolism and Dispositions*, *31*, 1077–1080.

18. Ekins, S. (2004). Predicting undesirable drug interactions with promiscuous proteins in silico. *Drug Discovery Today*, *9*, 276–285.

19. Molnar, L., & Keseru, G. M. (2002). A neural network based virtual screening of cytochrome P450 3A4 inhibitors. *Bioorganic & Medicinal Chemistry Letters*, *12*, 419–421.

20. O'Brien, S. E., & De Groot, M. J. (2005). Greater than the sum of its parts: combining models for useful ADMET prediction. *Journal of Medicinal Chemistry*, *48*, 1287–1291.

21. Balakin, K. V., Ekins, S., Bugrim, A., Ivanenkov, Y. A., Korolev, D., Nikolsky, Y. V., et al. (2004). Kohonen maps for prediction of binding to human cytochrome P450 3A4. *Drug Metabolism and Disposition*, *32*, 1183–1189.

22. Kemp, C. A., Flanagan, J. U., van Eldik, A. J., Marechal, J.-D., Wolf, C. R., Roberts, G. C. K., et al. (2004). Validation of model of cytochrome P450 2D6: an in silico tool for predicting metabolism and inhibition. *Journal of Medicinal Chemistry*, *47*, 5340–5346.

23. Kriegl, J. M., Arnhold, T., Beck, B., & Fox, T. (2005). A support vector machine approach to classify human cytochrome P450 3A4 inhibitors. *Journal of Computer-Aided Molecular Design*, *19*, 189–201.

24. Egan, W. J., Zlokarnik, G., & Grootenhuis, P. D. J. (2004). In silico prediction of drug safety: despite progress there is abundant room for improvement. *Drug Discovery Today: Technologies*, *1*, 381–387.

25. Le Bourdonnec, B., Ajello, C. W., Seida, P. R., Susnow, R. G., Cassel, J. A., Belanger, S., et al. (2005). Arylacetamide [kappa] opioid receptor agonists with reduced cytochrome P450 2D6 inhibitory activity. *Bioorganic & Medicinal Chemistry Letters*, *15*, 2647–2652.

26. Williams, P. A., Cosme, J., Vinkovic, D. M., Ward, A., Angove, H. C., Day, P. J., et al. (2004). Crystal structures of human cytochrome P450 3A4 bound to metyrapone and progesterone. *Science*, *305*, 683–686.

27. Tran, T. H., von Moltke, L. L., Venkatakrishnan, K., Granda, B. W., Gibbs, M. A., Obach, R. S., et al. (2002). Microsomal protein concentration modifies the apparent inhibitory potency of CYP3A inhibitors. *Drug Metabolism and Disposition, 30*, 1441–1445.

28. Di, L., Kerns, E. H., Li, S. Q., & Carter, G. T. (2007). Comparison of cytochrome P450 inhibition assays for drug discovery using human liver microsomes with LC-MS, rhCYP450 isozymes with fluorescence, and double cocktail with LC-MS. *International Journal of Pharmaceutics, 335*, 1–11.

29. *P450-Glo^{TM} screening systems technical bulletin, TB340*, Promega Corporation. www.promega.com

30. Cali, J. (2003). Screen for CYP450 inhibitors using P450-GLO^{TM} luminescent cytochrome P450 assays. *Cell Notes, 7*, 2–4.

31. Zlokarnik, G., Grootenhuis, P. D. J., & Watson, J. B. (2005). High throughput P450 inhibition screens in early drug discovery. *Drug Discovery Today, 10*, 1443–1450.

32. Delaporte, E., Slaughter, D. E., Egan, M. A., Gatto, G. J., Santos, A., Shelley, J., et al. (2001). The potential for CYP2D6 inhibition screening using a novel scintillation proximity assay-based approach. *Journal of Biomolecular Screening, 6*, 225–231.

33. Favreau, L. V., Palamanda, J. R., Lin, C.-C., & Nomeir, A. A. (1999). Improved reliability of the rapid microtiter plate assay using recombinant enzyme in predicting CYP2D6 inhibition in human liver microsomes. *Drug Metabolism and Disposition, 27*, 436–439.

34. Bapiro, T. E., Egnell, A.-C., Hasler, J. A., & Masimirembwa, C. M. (2001). Application of higher throughput screening (HTS) inhibition assays to evaluate the interaction of antiparasitic drugs with cytochrome P450s. *Drug Metabolism and Disposition, 29*, 30–35.

35. Nomeir, A. A., Ruegg, C., Shoemaker, M., Favreau, L. V., Palamanda, J. R., Silber, P., et al. (2001). Inhibition of CYP3A4 in a rapid microtiter plate assay using recombinant enzyme and in human liver microsomes using conventional substrates. *Drug Metabolism and Disposition, 29*, 748–753.

36. Cohen, L. H., Remley, M. J., Raunig, D., & Vaz, A. D. N. (2003). In vitro drug interactions of cytochrome P450: an evaluation of fluorogenic to conventional substrates. *Drug Metabolism and Disposition, 31*, 1005–1015.

37. Weaver, R., Graham, K. S., Beattie, I. G., & Riley, R. J. (2003). Cytochrome P450 inhibition using recombinant proteins and mass spectrometry/multiple reaction monitoring technology in a cassette incubation. *Drug Metabolism and Disposition, 31*, 955–966.

38. Gao, F., Johnson, D. L., Ekins, S., Janiszewski, J., Kelly, K. G., Meyer, R. D., et al. (2002). Optimizing higher throughput methods to assess drug-drug interactions for CYP1A2, CYP2C9, CYP2C19, CYP2D6, rCYP2D6, and CYP3A4 in vitro using a single point IC50. *Journal of Biomolecular Screening, 7*, 373–382.

39. Crespi, C. L., Miller, V. P., & Penman, B. W. (1997). Microtiter plate assays for inhibition of human, drug-metabolizing cytochromes P450. *Analytical Biochemistry, 248*, 188–190.

40. Crespi, C. L., Miller, V. P., & Penman, B. W. (1998). High throughput screening for inhibition of cytochrome P450 metabolism. *Medicinal Chemistry Reviews, 8*, 457–471.

41. Stresser, D. M., Blanchard, A. P., Turner, S. D., Erve, J. C. L., Dandeneau, A. A., Miller, V. P., et al. (2000). Substrate-dependent modulation of CYP3A4 catalytic activity: analysis of 27 test compounds with four fluorometric substrates. *Drug Metabolism and Disposition, 28*, 1440–1448.

42. Crespi, C. L., & Stresser, D. M. (2000). Fluorometric screening for metabolism-based drug–drug interactions. *Journal of Pharmacological and Toxicological Methods, 44*, 325–331.

43. Stresser, D. M., Turner, S. D., Blanchard, A. P., Miller, V. P., & Crespi, C. L. (2002). Cytochrome P450 fluorometric substrates: identification of isoform-selective probes for rat CYP2D2 and human CYP3A4. *Drug Metabolism and Disposition, 30*, 845–852.

44. Chauret, N., Tremblay, N., Lackman, R. L., Gauthier, J.-Y., Silva, J. M., Marois, J., et al. (1999). Description of a 96-well plate assay to measure cytochrome P4503A inhibition in human liver microsomes using a selective fluorescent probe. *Analytical Biochemistry, 276*, 215–226.

45. Chauret, N., Dobbs, B., Lackman, R. L., Bateman, K., Nicoll-Griffith, D. A., Stresser, D. M., et al. (2001). The use of 3-[2-(N,N-diethyl-N-methylammonium)ethyl]-7-methoxy-4-methylcoumarin (AMMC) as a specific CYP2D6 probe in human liver microsomes. *Drug Metabolism and Disposition, 29*, 1196–1200.

46. Onderwater, R. C. A., Venhorst, J., Commandeur, J. N. M., & Vermeulen, N. P. E. (1999). Design, synthesis, and characterization of 7-methoxy-4-(aminomethyl)coumarin as a novel and selective cytochrome

P450 2D6 substrate suitable for high-throughput screening. *Chemical Research in Toxicology*, *12*, 555–559.

47. Chu, I., Favreau, L. V., Soares, T., Lin, C.-C., & Nomeir, A. A. (2000). Validation of higher-throughput high-performance liquid chromatography/atmospheric pressure chemical ionization tandem mass spectrometry assays to conduct cytochrome P450s CYP2D6 and CYP3A4 enzyme inhibition studies in human liver microsomes. *Rapid Communications in Mass Spectrometry*, *14*, 207–214.

48. Yin, H., Racha, J., Li, S. Y., Olejnik, N., Satoh, H., & Moore, D. (2000). Automated high throughput human CYP isoform activity assay using SPE-LC/MS method: application in CYP inhibition evaluation. *Xenobiotica*, *30*, 141–154.

49. Dierks, E. A., Stams, K. R., Lim, H.-K., Cornelius, G., Zhang, H., & Ball, S. E. (2001). A method for the simultaneous evaluation of the activities of seven major human drug-metabolizing cytochrome P450s using an in vitro cocktail of probe substrates and fast gradient liquid chromatography tandem mass spectrometry. *Drug Metabolism and Disposition*, *29*, 23–29.

50. Bu, H.-Z., Magis, L., Knuth, K., & Teitelbaum, P. (2001). High-throughput cytochrome P450 (CYP) inhibition screening via a cassette probe-dosing strategy. VI. Simultaneous evaluation of inhibition potential of drugs on human hepatic isozymes CYP2A6, 3A4, 2C9, 2D6 and 2E1. *Rapid Communications in Mass Spectrometry*, *15*, 741–748.

51. Di, L., Kerns, E. H., Li, S. Q., & Petusky, S. L. (2006). High throughput microsomal stability assay for insoluble compounds. *International Journal of Pharmaceutics*, *317*, 54–60.

Chapter 33

Plasma Protein Binding Methods

Overview

▶ *Literature and commercial in silico predictors of plasma protein binding are available.*

▶ *Equilibrium dialysis is the gold standard method. Ultrafiltration, ultracentrifugation, microdialysis, and plasmon resonance methods also are used.*

▶ *Red blood cell binding is important to measure for some compounds.*

The importance of free (unbound) drug for distribution into disease tissues and clearance organs has prompted many drug companies to implement plasma protein binding (PPB) measurement methods. Several of these methods have been developed in higher-throughput 96-well format.

33.1 In Silico PPB Methods

33.1.1 Literature In Silico PPB Methods

In silico models for PPB have been reported.[1–4] PPB was found to increase with increasing lipophilicity, increasing acidity, and increasing number of acid moieties.[1] PPB decreases with increasing basicity and number of basic substructures. One group used the chemical structure to calculate the ionization state and lipophilicity (Log P) for predicting PPB.[2]

33.1.2 Commercial In Silico PPB Methods

Several commercial software products for predicting PPB are available. Many of these products are listed in Table 33.1.

TABLE 33.1 ▶ Partial List of Commercial Software for Plasma Protein Binding

Name	Company	Web site
QikProp	Schrodinger	*www.schrodinger.com*
ADMET Predictor	Simulations Plus	*www.simulations-plus.com*
ADME Boxes Distribution	Pharma Algorithms	*www.ap-algorithms.com*
KnowItAll	Biorad	*www.biorad.com*
CSPB	ChemiSilico	*www.chemsilico.com*
DSMedChem Explorer	Accelrys	*www.accelrys.com*
Wombat database	Sunset Molecular Discovery	*www.sunsetmolecular.com*

33.2 In Vitro PPB Methods

In vitro PPB methods have been reviewed.[4,5] Several methods have been developed and described. Concomitant studies of PPB using more than one method have indicated the comparability of results from several of these methods,[6–8] with equilibrium dialysis generally considered the leading method and reference for other methods.

33.2.1 Equilibrium Dialysis Method

Equilibrium dialysis is considered the "gold standard" method for PPB.[4,9,10] It has been implemented in many formats, from individual chambers to 96-well plates. In this method, two chambers are separated by a dialysis membrane (Figure 33.1). The membrane has a certain molecular weight (MW) cutoff (e.g., 30 kDa). In one chamber is placed plasma with added test compound; in the other chamber is placed the same volume of blank buffer. The plasma protein molecules and bound compounds cannot pass through the membrane, but the free drug molecules can pass through. The chamber is incubated for 24 hours at a constant temperature. The concentration of free drug reaches equilibrium on each side of the membrane. The fraction unbound in plasma ($f_{u, plasma}$) is determined by measuring the concentration of drug in the buffer chamber and dividing by the total drug concentration in the plasma chamber.

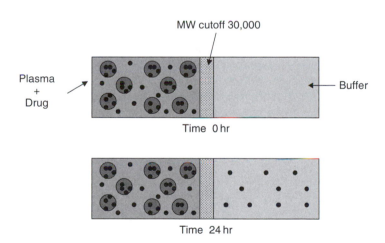

Figure 33.1 ▶ Equilibrium dialysis method for plasma protein binding.

A 24-well apparatus for measuring PPB by equilibrium dialysis has been described.[4] Throughput can be enhanced using a mixture of up to five compounds per dialysis cell (each at 10-μM concentration) in a cassette-type experiment. At this concentration, the albumin is at only a fraction of saturation. Two different 96-well apparatuses that work well have been described.[9,10] Several devices, from single-chamber to 96-well format, are available from commercial vendors (Table 33.2). An example of application of the method was measurement of the PPB of docetaxel.[11]

TABLE 33.2 ► **Commercial Instruments and Supplies for Plasma Protein Binding**

Method	Product name	Company	Web site
Equilibrium dialysis	HTD 96	HTDialysis	*www.htdialysis.com/*
Equilibrium dialysis	Dialyzer 96 well	Harvard Apparatus	*www.harvardapparatus.com*
Equilibrium dialysis	Dialyzer Multi-Equilib.	Harvard Apparatus	*www.harvardapparatus.com*
Equilibrium dialysis	Rapid Equilib. Dialysis (RED)	Pierce	*www.piercenet.com*
Equilibrium dialysis	Serum Binding System	BD Gentest	*www.bdgentest.com*
Ultrafiltration	Centrifree (single sample)	Millipore	*www.millipore.com*
Ultrafiltration	MultiScreen Ultracel-PPB	Millipore	*www.millipore.com*
Immobilized HPLC	Chiral-AGP, -HSA	Chrom Tech	*www.chromtech.com*
Microdialysis	Microdialysis probe	CMA/Microdialysis	*www.microdialysis.com*
Microdialysis	Microdialysis probe	BASi	*www.bioanalytical.com*

33.2.2 Ultrafiltration Method

In the ultrafiltration method, the test compound is added to plasma in a container and mixed. An aliquot of this solution is loaded into the upper chamber of an ultrafiltration apparatus (Figure 33.2) that has a membrane with a certain MW cutoff (e.g., 30 kDa). Individual sample vials and a 96-well ultrafiltration device are available (Table 33.2). The device is centrifuged (e.g., 2,000g, 45 minutes). The solution is ultrafiltered through the membrane by the force of the centrifugation. Unbound test compound moves with the liquid through the membrane into the receiver chamber, whereas compound bound to plasma protein remains in the loading chamber. Typically less than one fifth of the filtrate is collected. The concentration of test compound in the receiver is quantitated, and the fraction unbound (f_u) is calculated as this ultrafiltrate concentration divided by the total initial concentration. It is useful to perform mass balance studies to calculate the recovery to ensure lack of significant nonspecific binding to the apparatus.

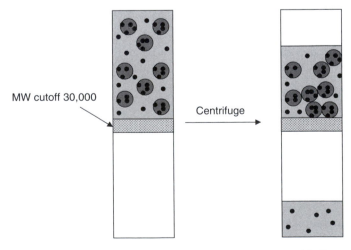

Figure 33.2 ► Ultrafiltration method for plasma protein binding.

33.2.3 Ultracentrifugation Method

In the Ultracentrifugation method test compound is added to plasma and incubated as above. A sample is transferred to an ultracentrifuge tube and centrifuged at high sedimentation rate (e.g., 6 hours at 160,000g, 15 hours at 100,000g). The concentration of test compound is measured in the original plasma and in the second layer of the supernatant.[8,12] Some researchers are concerned that the long period and conditions of the sedimentation may upset the equilibrium.

33.2.4 Immobilized Protein High-Performance Liquid Chromatography Column Method

High-performance liquid chromatography columns that have human serum albumin or α_1-acid glycoprotein bonded to the silica stationary support particles are available for purchase.[13,14] The test compound is injected onto the column and is moved along the column by the mobile phase. The test compound binds to the immobilized protein molecules to a degree that depends on its binding affinity. Binding slows the compound's progress through the column; thus, the retention time correlates to the affinity of the compound for the protein. Some researchers are concerned that the method does not provide good resolution for binding greater than 95%. Binding kinetics (association and dissociation rate constants) also can be determined by measuring the peak width and position.[15] A comparison of the most commonly used plasma protein binding methods is provided in Table 33.3.

TABLE 33.3 ► **Comparison of Plasma Protein Binding Methods**

Method	Advantage	Disadvantage
Equilibrium dialysis	Temperature controlled Widely used	Time to reach equilibrium Stability of protein and drug
Ultrafiltration	Technically simple, rapid, and inexpensive Widely used	Uncontrolled temperature Absorption of drugs to filter Dissociation of the bound drug
Ultracentrifugation	No absorption to membrane	Time: long (12–15 hr) Sedimentation, back diffusion Expensive equipment
Immobilized protein columns	Technically simple and inexpensive	Binding limited to albumin Limited physiological relevance
Circular dichroism	Technically rapid	Limited to specific albumin binding Limited physiological relevance

33.2.5 Microdialysis Method

Microdialysis is discussed in Chapter 28. Briefly, an isotonic solution bathes the inner surface of a dialysis membrane, which is placed into a fluid or tissue. The membrane is incorporated in a small probe and, after flowing past the membrane (e.g., 1 µL/min), the fluid is collected and analyzed. By placing this probe into plasma containing the test compound, the free drug is trapped in the dialysis fluid while the plasma proteins and drug binding to plasma protein stays in the sample.[16,17] Good success has been reported in determining percent binding compared to other methods, but the method is considered time consuming.

33.2.6 Other PPB Methods

Many other methods for analysis of drug–protein complexes have been reported.[5] Some methods measure specific properties of the drug–protein complex:

► Fluorescence spectroscopy

► Circular dichroism/optical rotatory dispersion (CD/ORD)

► NMR

► Electron spin resonance (ESR)

► Microcalorimetry

► Surface plasmon resonance (Biacore)[18]

Surface plasmon resonance can be used to measure the rate of association (k_a) and dissociation (k_a), in addition to the percent binding.

Other methods are chromatography based:

► Size exclusion chromatography

► Frontal analysis chromatography

► Capillary electrophoresis

33.3 Red Blood Cell Binding

Some compounds bind to red blood cells (RBCs). It is important to account for this binding because it can be a significant percentage of the compound in blood. RBC-bound drug is not measured in pharmacokinetics (PK) experiments because the RBCs are removed prior to storage of the sample for later analysis. A study of red cell binding can be performed retrospectively if there is difficulty in accounting for the mass balance of a compound in a PK study. It also can be performed prospectively as a screening method if the lead series is known to have an RBC binding issue.

The test compound is added to whole blood and incubated (e.g., 2 hours at 37°C). The blood samples are transferred to measurement of sedimentation (MESED™) devices, and the RBCs are separated from plasma. Each fraction then is analyzed using liquid chromatography/mass spectrometry techniques.[19]

Problems

(Answers can be found in Appendix I at the end of the book.)

1. Which PPB method is generally considered to be most reliable?

2. How is the unbound fraction determined using the equilibrium dialysis method?

3. If a compound is highly bound to red blood cells, what will be the PK consequences?

4. Why is a membrane filter with a high MW cutoff necessary in some PPB methods?

⬛ References

1. Gleeson, M. P. (2007). Plasma protein binding affinity and its relationship to molecular structure: an in-silico analysis. *Journal of Medicinal Chemistry, 50,* 101–112.

2. Lobell, M., & Sivarajah, V. (2003). In silico prediction of aqueous solubility, human plasma protein binding and volume of distribution of compounds from calculated pKa and AlogP98 values. *Molecular Diversity, 7,* 69–87.

3. Gunturi, S. B., Narayanan, R., & Khandelwal, A. (2006). In silico ADME modelling: computational models to predict human serum albumin binding affinity using ant colony systems. *Bioorganic & Medicinal Chemistry, 14,* 4118–4129.

4. Fessey, R. E., Austin, R. P., Barton, P., Davis, A. M., & Wenlock, M. C. (2006). The role of plasma protein binding in drug discovery. In *Pharmacokinetic Profiling in Drug Research: Biological, Physicochemical, and Computational Strategies,* [LogP2004, Lipophilicity Symposium], 3rd, Zurich, Switzerland, Feb. 29-Mar. 4, 2004, pp. 119–141.

5. Oravcova, J., Boehs, B., & Lindner, W. (1996). Drug-protein binding studies. New trends in analytical and experimental methodology. *Journal of Chromatography, B: Biomedical Applications, 677,* 1–28.

6. Melten, J. W., Wittebrood, A. J., Willems, H. J. J., Faber, G. H., Wemer, J., & Faber, D. B. (1985). Comparison of equilibrium dialysis, ultrafiltration, and gel permeation chromatography for the determination of free fractions of phenobarbital and phenytoin. *Journal of Pharmaceutical Sciences, 74,* 692–694.

7. Barre, J., Chamouard, J. M., Houin, G., & Tillement, J. P. (1985). Equilibrium dialysis, ultrafiltration, and ultracentrifugation compared for determining the plasma-protein-binding characteristics of valproic acid. *Clinical Chemistry, 31,* 60–64.

8. Kurz, H., Trunk, H., & Weitz, B. (1977). Evaluation of methods to determine protein-binding of drugs. Equilibrium dialysis, ultrafiltration, ultracentrifugation, gel filtration. *Arzneimittel-Forschung, 27,* 1373–1380.

9. Kariv, I., Cao, H., & Oldenburg, K. R. (2001). Development of a high throughput equilibrium dialysis method. *Journal of Pharmaceutical Sciences, 90,* 580–587.

10. Banker, M. J., Clark, T. H., & Williams, J. A. (2003). Development and validation of a 96-well equilibrium dialysis apparatus for measuring plasma protein binding. *Journal of Pharmaceutical Sciences, 92,* 967–974.

11. Acharya, M. R., Baker, S. D., Verweij, J., Figg, W. D., & Sparreboom, A. (2004). Determination of fraction unbound docetaxel using microequilibrium dialysis. *Analytical Biochemistry, 331,* 192–194.

12. Piekoszewski, W., & Jusko, W. J. (1993). Plasma protein binding to tacrolimus in humans. *Journal of Pharmaceutical Sciences, 82,* 340–341.

13. Valko, K., Nunhuck, S., Bevan, C., Abraham, M. H., & Reynolds, D. P. (2003). Fast gradient HPLC method to determine compounds binding to human serum albumin. Relationships with octanol/water and immobilized artificial membrane lipophilicity. *Journal of Pharmaceutical Sciences, 92,* 2236–2248.

14. Noctor, T. A. G., Diaz-Perez, M. J., & Wainer, I. W. (1993). Use of a human serum albumin-based stationary phase for high-performance liquid chromatography as a tool for the rapid determination of drug-plasma protein binding. *Journal of Pharmaceutical Sciences, 82,* 675–676.

15. Talbert, A. M., Tranter, G. E., Holmes, E., & Francis, P. L. (2002). Determination of drug-plasma protein binding kinetics and equilibria by chromatographic profiling: exemplification of the method using L-tryptophan and albumin. *Analytical Chemistry, 74,* 446–452.

16. Ekblom, M., Hammarlund-Udenaes, M., Lundqvist, T., & Sjoeberg, P. (1992). Potential use of microdialysis in pharmacokinetics: a protein binding study. *Pharmaceutical Research, 9,* 155–158.

17. Herrera, A. M., Scott, D. O., & Lunte, C. E. (1990). Microdialysis sampling for determination of plasma protein binding of drugs. *Pharmaceutical Research, 7,* 1077–1081.

18. Rich, R. L., Day, Y. S., Morton, T. A., & Myszka, D. G. (2001). High-resolution and high-throughput protocols for measuring drug/human serum albumin interactions using BIACORE. *Analytical Biochemistry, 296,* 197–207.

19. Dumez, H., Guetens, G., De Boeck, G., Highley, M. S., de Bruijn, E. A., van Oosterom, A. T., & Maes, R. A. A. (2005). In vitro partition of docetaxel and gemcitabine in human volunteer blood: the influence of concentration and gender. *Anti-Cancer Drugs, 16,* 885–891.

Chapter 34

hERG Methods

Overview

▶ *hERG blocking in silico models are being developed within companies and are increasingly commercially available.*

▶ *Patch clamp methods are most successful for in vitro predictions of hERG blocking*

▶ *In vitro assays should be followed up with in vivo ECG studies of long QT.*

Regulatory agencies insist on having experimental hERG data for all compounds moving into clinical development because of the risk of death from arrhythmia in a small portion of the population. Pharmaceutical companies need to develop a strategy for hERG block assessment during drug discovery to avoid investment in a chemical series that has significant risk.

For this reason, several methods have been developed to profile potential hERG channel blocking by discovery compounds. As with other properties, there is an emerging trend to begin screening for potential hERG blocking by applying in silico tools to hits from high-throughput screening. This initiates awareness of potential issues for the project team. However, because in silico predictions currently have low correlation to in vivo electrophysiology, they should not be used to eliminate potentially promising pharmacophores. The hERG screening strategy continues throughout discovery using in vitro methods, as part of the optimization process, which allows high throughput and fast turnaround of data.[1] Manual patch clamp is the most definitive in vitro assay. It is used for detailed study of the effects of drugs on the K$^+$ ion channel. Patch clamp often is referred to as the "gold standard" method. The method has been accelerated to higher throughput by instrument companies and is the most definitive of the high-throughput in vitro assays. Ultimately, in vivo electrophysiology methods using electrocardiography (ECG) studies must be performed for selected compounds moving toward development. Whereas in vitro studies indicate the likelihood of hERG block, the in vivo ECG will indicate the most important element, the production of long QT, and initiation of torsades de pointes arrhythmia. In vivo hERG methods are the most expensive and time consuming, but they produce the most definitive data. The Food and Drug Administration (FDA) requires that human ECG studies be conducted in the clinic for many drug candidates. As a general strategy, it is useful for drug research companies to have tools for in silico, in vitro, and in vivo hERG testing to meet the needs at each stage of discovery. Various methods for hERG are listed in Table 34.1.

TABLE 34.1 ▶ Methods of Screening for hERG Blocking

Method	Type assay	Comments
Membrane potential dye	HT in vitro	HT, LCP
Binding	HT in vitro	
Rubidium flux	HT in vitro	
HT patch clamp	HT in vitro	
LT patch clamp	In-depth in vitro	LT, expensive
Purkinje Fibers	In-depth in vitro	LT, expensive
Electrocardiogram	In-depth in vivo	LT, expensive

HT, High throughput; LCP, low correlation to patch clamp; LT, low throughput.

34.1 In Silico hERG Methods

Diverse compounds block the hERG K$^+$ channel. Evidence is emerging about the binding site(s) of drugs in the hERG channel and should eventually enable inexpensive and useful predictions.

Structural features associated with hERG blocking are discussed in Chapter 16. Various organizations have developed proprietary in silico methods for early prediction of hERG blocking potential. There is currently no commercial software for hERG blocking prediction based on compound structure. However, commercial software is likely in the near future as databases grow.

34.2 In Vitro hERG Methods

In vitro hERG methods can be classified into two types. Indirect methods (membrane potential-sensitive dye, ligand binding, rubidium efflux) measure effects associated with K$^+$ ion channel activity. These methods are favorable for high-throughput analysis but have low correlations to patch clamp and in vivo electrophysiology. The other type of method, high-throughput patch clamp, directly measures ion channel activity and has emerged as the in vitro method of choice. All in vitro hERG methods are cell based and use cell lines, such as CHO and HEK293, which have been transfected with the hERG gene and express K$^+$ channels on their membranes. The implementation of hERG screening in drug discovery has been reviewed.[1]

34.2.1 Membrane Potential–Sensitive Dye Method for hERG

Membrane potential–sensitive dyes are used in many ion channel experiments and have been adopted for hERG screening.[2] Cells transfected with the hERG gene have a more negative membrane potential than do wild-type cells as a result of the potassium channel activity, which lowers the internal K$^+$ ion concentration. If a compound blocks the hERG potassium channel, the membrane potential of the transfected cells increases, which can be monitored using a membrane potential–dye.[3]

The structure of one dye, DiBAC$_4$(3), is shown in Figure 34.1. This dye interacts with cytoplasmic components to produce fluorescence, but no fluorescence is produced in the extracelluar solution. The cells are preincubated with dye in the extracellular solution, then the test compound is added and the fluorescence is monitored using a fluorometric plate reader (e.g., FLIPR [Molecular Devices Corp.] or POLARstar [BMG Labtech]) for 3 to 15 minutes.

The dye permeates in and out of the cell in response to the membrane potential. DiBAC$_4$(3) is negatively charged, so there is a higher concentration in the cell when there are more K^+ ions in the cell. If the test compound blocks the K^+ channel, the fluorescence increases. The method is diagrammed in Figure 34.2, A.

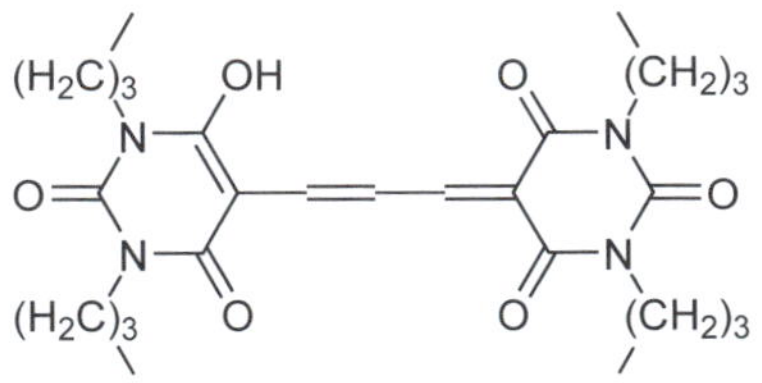

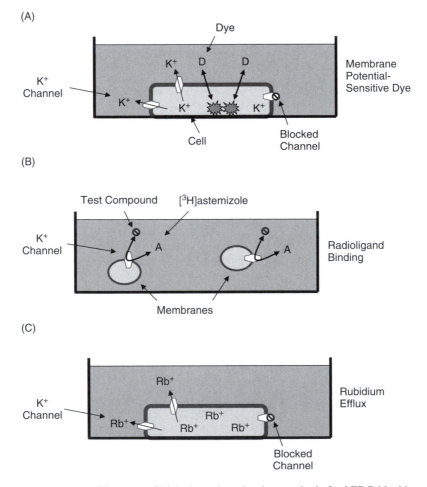

Figure 34.1 ► DiBAC$_4$(3) membrane potential–sensitive dye for FLIPR assay.

Figure 34.2 ► Diagrams of high-throughput in vitro methods for hERG blocking.

The drawback of DiBAC$_4$(3) is slow response time.[3,4] A new dye, FLIPR membrane potential dye (FMP, Molecular Devices Corp.), responds 14-fold faster than DiBAC$_4$(3), is less sensitive to temperature, and does not require a wash step; thus, it is more useful for high-throughput automation and screening. Dye methods are susceptible to fluorescence quenching and test compound interference, which can be compensated for by running the assay also on wild-type cells. The IC_{50} values average fivefold higher than by the patch-clamp method and do not have a high correlation.[3] It is not a functional assay, so other ion channels and cellular factors can compromise the results. The method can detect potent inhibitors and does not produce false–positive results, but it does produce false–negative results for less potent inhibitors.[5] An advantage of the method is high throughput using conventional high-throughput fluorescence plate readers, speed, and low cost.

34.2.2 Ligand Binding Method for hERG

Binding of test compounds to hERG channel protein can be screened using a radioligand binding assay.[6,7] In this assay, membranes are prepared from HEK293 cells transfected with the hERG gene (Figure 34.2, *B*). The membranes are incubated with a radiolabled compound that is known to bind to the hERG channel (e.g., [^{3}H]dofetilide, [^{3}H]astemizole) and the test compound. The radioligand and test compound compete for binding to the hERG channel. After incubation at 37°C for 30 to 60 minutes, the membranes are filtered and washed multiple times with cold buffer. Bound radioligand is detected by scintillation counting, and the binding compared to control is plotted against test compound concentration. The stronger the affinity of the test compound for the hERG channel, the lower will be the radioligand response in the filtered membranes.

There is good correlation between receptor binding K_i values and patch-clamp IC_{50} values, with correlation $r^2 = 0.91$. In some laboratories, binding results are many-fold less than obtained by the patch-clamp method.[6,7] In other laboratories, the results are consistent and robust and the method is considered the best indirect in vitro hERG blocking method in side-by-side tests.[5] It is not a functional assay.

34.2.3 Rubidium Efflux Method for hERG

The rubidium (Rb) efflux method (Figure 34.2, *C*) is a functional assay based on hERG channel opening. Rb is found at only trace levels in cells and media, is the same size and charge as K^+, and is permeable through the K^+ channels. CHO cells transfected with hERG gene are preincubated with Rb-containing media to establish an equilibrium of Rb between, inside, and outside the cells. Then the media is removed and the cells are washed multiple times with buffer to remove Rb from the extracellular solution. Buffer containing K^+ (to manipulate the membrane potential and open the channels) and test compound is added. Following incubation, the media is removed. The cells are washed and lysed. The media and lysate are separately measured for Rb concentration using either scintillation counting of ^{86} Rb[8] or atomic absorption spectroscopy of Rb.[9] If the K^+ channel is blocked by the test compound, the Rb^+ stays in the cells; if the K^+ channel is not blocked, the Rb^+ permeates through the channel into the medium. From the measured values, the percentage of Rb permeation is calculated.

As with the other high-throughput methods, IC_{50} values often are higher in the Rb efflux method than in the patch-clamp method. The underprediction by the Rb efflux method is greater for more potent inhibitors.[9] This appears to result from the presence of Rb, which reduces the affinity of the compound for the hERG channel and reduces activation of the channel. The method appears to be useful for screening for stronger hERG inhibitors in early discovery.[5]

34.2.4 Patch-Clamp Method for hERG

The "gold standard" of in vitro assays for hERG blocking is the patch-clamp technique. This method is widely used for the study of all ion channels. A diagram of the patch-clamp

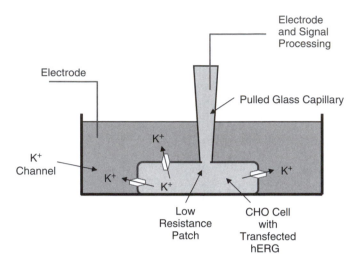

Figure 34.3 ▶ Diagram of patch-clamp method for hERG blocking.

method is shown in Figure 34.3. A glass capillary is melted, pulled to about 1-μm diameter, and cut off. The glass capillary is filled with buffer that is compatible with the cell cytoplasm, and an electrode is inserted. The capillary is attached to a micromanipulator and, while observing with a microscope, the tip is placed in contact with the membrane of a CHO or HEK293 cell that expresses the hERG K^+ ion channel. A negative pressure (vacuum) is applied, and the cell membrane is "patched" or held tightly to the capillary and forms a high resistance seal. Further application of vacuum causes the small portion of membrane at the tip of the capillary to be pulled into the capillary tip and ruptured, forming a "whole-cell patch." Most of the cell remains outside the capillary. The cell cytoplasm and the capillary fluid mix, establishing an electrical connection. The electrodes are used to set ("clamp") the membrane potential at a specified voltage. In response to a positive voltage applied by the investigator, the K^+ ion channel proteins undergo conformational change and open. Ions move out of the cell through the ion channels, creating a current. The membrane potential is held at a constant level by the electronics, and the current needed to maintain this voltage is recorded. This current is directly related to the total ion current of all the K^+ ion channels in the cell.

If a test compound blocks the hERG K^+ ion channel, then the current is reduced relative to control. An example of a patch-clamp profile for a hERG blocking experiment is shown in Figure 34.4. For the control, changing the membrane potential from −80 to +40 mV opens the channels and K^+ ion current flows. When the membrane potential then is reduced to −50 mV, ion current has a momentary spike due to the mechanics of the channel and then reduces to the unopen state. When hERG blocker thioridazine is added to the assay medium, the channel is blocked and the ion current is greatly reduced compared to control.

The detail provided by such data allows in-depth study of the function of the ion channel under hERG block. Patch clamp provides complete control of the crucial experimental conditions, such as membrane potential, which allows full ion channel function. Other in vitro methods cannot control this, except crudely with the K^+ ion concentration. Patch clamp also is able to monitor ion channel events on the millisecond level, whereas other techniques monitor events that are indirectly related to the ion channel. The method is more sensitive to hERG blocking than are other methods and consistently provides reliable data, even for compounds of higher IC_{50}.

The major problem with the patch-clamp method for drug discovery is the considerable time and skill needed for conducting each experiment. Therefore, the method is not

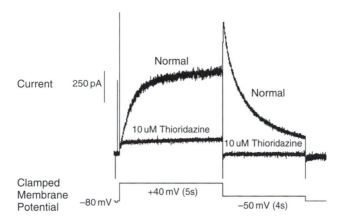

Figure 34.4 ▶ Examples of ion current profiles from patch clamp for K^+ ion channels.

amenable to high throughput in this configuration. This experiment is normally performed for detailed study of a few selected compounds.

34.2.5 High-Throughput Patch-Clamp Method for hERG

High-throughput patch clamp is achieved by modifying the patch-clamp experiment (Figure 34.5). Instead of pressing a glass capillary against the membrane, cells are seeded into a well that contains one or more small pores on the bottom. A vacuum is applied to draw a cell to the pore and hold it there. The electrical connection is established by further applying vacuum to burst the membrane. Alternatively, a modifier (e.g., amphotericin B, nystatin) is added to the fluid below the pore, which produces small openings in the cell membrane ("perforated patch"), allowing good electrical conductance. Membrane potential is controlled across the cell membrane. Several experiments can be performed in parallel using multiple wells for greatly enhanced throughput. This technique has become the method of choice for hERG block screening. Several commercial instruments now are available (Table 34.2). These instruments achieve both high throughput and highly informative data. High-throughput patch-clamp instruments have been reviewed.[10]

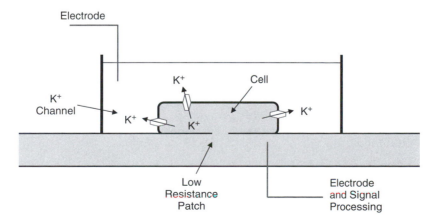

Figure 34.5 ▶ Diagram of planar patch-clamp method used for high-throughput hERG block screening.

TABLE 34.2 ▶   Commercial Instruments for Measurement of hERG Blocking

Method	Product name	Company	Web site
Fluorescent dye	FLIPR	Molecular Devices Corp.	*www.moleculardevices.com*
Rb efflux assay	ICR 12000	Aurora Biomed	*www.aurorabiomed.com*
Patch clamp	PatchXpress 700A	Molecular Devices	*www.moleculardevices.com*
Patch clamp	Flyscreen 8500	Flyion	*www.flyion.com*
MT patch clamp	Qpatch HT	Sophion Bioscience	*www.sophion.com*
MT patch clamp	CytoPatch	Cytocentrics	*www.cytocentrics.com*
MT patch clamp	NPC-16 Patchliner	Nanion	*www.nanion.de*
HT patch clamp	IonWorks Quatro	Molecular Devices	*www.moleculardevices.com*
HT patch clamp	PatchXpress 7000A	Molecular Devices	*www.moleculardevices.com*

Note: Medium throughput (MT)

34.3   In Vivo hERG Methods

The ultimate method for assessing and studying hERG block by discovery compounds is ECG using animal models (e.g., dog, monkey). The method produces data like the ECG traces shown in Figure 16.3. Electrodes are placed on the surface of the heart. The effects of the test compound of lengthening the QT interval and inducing torsades de pointes are directly observed. The method is definitive, but it is labor intensive, slow, low throughput, and expensive. Other in vivo methods involve Purkinje fiber action potential, isolated perfused heart ECG, perfused ventricular wedge, and repolarization reserve. Electrophysiologists conduct these studies.

Problems

(Answers can be found in Appendix I at the end of the book.)

1. In silico hERG blocking tools are best used for what?

2. Which in vitro hERG blocking method is generally considered to be most predictive?

3. Which hERG blocking method does the FDA require in humans, if there is a potential problem?

4. Dye methods for hERG rely on which of the following?: (a) membrane potential, (b) K^+ transporters, (c) interaction with cytoplasmic components to produce fluorescence, (d) interaction with extracellular components to produce fluorescence.

5. What is a drawback of the hERG membrane potential dye and Rb methods?

6. Which are true about the ligand binding hERG method?: (a) radiolabeled hERG protein is used, (b) the radiolabeled ligand is first incubated with the hERG protein and then test compound is added, (c) higher detected radioactivity correlates to lower test compound hERG binding, (d) radiolabeled hERG is used.

7. In the "efflux method," Rb is a surrogate for what? Why is Rb used?

8. In the patch-clamp method, rupturing the cell membrane provides which of the following?: (a) a delivery route for introducing the test compound into the cell, (b) a way to introduce the hERG gene, (c) an electrical connection between the inside and outside of the cell, (d) a means of controlling the transmembrane potential, (e) a way to introduce hERG inhibitor into the cell.

9. What is measured in patch clamp?

10. What is a drawback of the standard patch-clamp method for use in drug discovery?

11. In high-throughput patch-clamp methods, which of the following is true?: (a) the through-put is much higher than manual patch clamp, (b) a known competitive hERG blocker standard is added to the media, (c) the patch is created with a pore on the bottom of the well and negative pressure is applied, (d) the automated commercial instruments require less operator time and skill.

References

1. Dunlop, J., & Bowlby, M. (2006). Implementing hERG screening early in the preclinical drug discovery process. *American Drug Discovery*, *1*, 8–13.

2. Tang, W., Kang, J., Rampe, D., & Dunnington, D. (2000). Development of a FLIPR based HERG channel assay. *Molecular Devices Corp. 4th International Drug Discovery Conference.* http://www.moleculardevices .com/pdfs2/2000_abstracts/Tang.pdf

3. Dorn, A., Hermann, F., Ebneth, A., Bothmann, H., Trube, G., Christensen, K., et al. (2005). Evaluation of a high-throughput fluorescence assay method for HERG potassium channel inhibition. *Journal of Biomolecular Screening*, *10*, 339–347.

4. Taylor, B. (2002). Progress toward a high-throughput hERG channel assay. *Molecular Devices Corp. 6th International Drug Discovery Conference.* http://www.moleculardevices.com/pdfs2/2002_abstract/ taylor_ presentation.pdf

5. Murphy, S. M., Palmer, M., Poole, M. F., Padegimas, L., Hunady, K., Danzig, J., et al. (2005). Evaluation of functional and binding assays in cells expressing either recombinant or endogenous hERG channel. *Journal of Pharmacological and Toxicological Methods*, *54*, 42–55.

6. Finlayson, K., Turnbull, L., January, C. T., Sharkey, J., & Kelly, J. S. (2001). [^{3}H]dofetilide binding to HERG transfected membranes: a potential high throughput preclinical screen. *European Journal of Pharmacology*, *430*, 147–148.

7. Chiu, P. J., Marcoe, K. F., Bounds, S. E., Lin, C. H., Feng, J. J., Lin, A., et al. (2004). Validation of a [^{3}H]astemizole binding assay in HEK293 cells expressing HERG K$^+$ channels. *Journal of Pharmacological Sciences*, *95*, 311–319.

8. Cheng, C. S., Alderman, D., Kwash, J., Dessaint, J., Patel, R., Lescoe, M. K., et al. (2002). A high-throughput HERG potassium channel function assay: an old assay with a new look. *Drug Development and Industrial Pharmacy*, *28*, 177–191.

9. Rezazadeh, S., Hesketh, J. C., & Fedida, D. (2004). Rb$^+$ flux through hERG channels affects the potency of channel blocking drugs: correlation with data obtained using a high-throughput Rb$^+$ efflux assay. *Journal of Biomolecular Screening*, *9*, 588–597.

10. Wood, C., Williams, C., & Waldron, G. J. (2004). Patch clamping by numbers. *Drug Discovery Today*, *9*, 434–441.

Chapter 35

Toxicity Methods

Overview

▶ *In silico tools predict potentially toxic substructures and metabolites.*

▶ *In vitro assays test many toxic mechanisms, including metabolite reactivity, induction, mutagenicity, and cytotoxicity. These findings alert teams to potential problems.*

▶ *In vivo acute and chronic dosing studies, in combination with study of physiological function and histology, are necessary for advanced candidates. The complexity of studies increases as candidates progress through development.*

▶ *Toxicometabonomics, toxicoproteomics, and toxicogenomics provide higher-throughput approaches that can recognize in vivo toxic responses soon after dosing and indicate underlying toxic mechanisms.*

As with other absorption, distribution, metabolism, excretion, and toxicity (ADME/Tox) properties, toxicity uses a combination of in vitro and in vivo assays to obtain data for key variables. At the in vitro level, a large number of compounds are examined for key indicators of toxicity. At the in vivo level, detailed studies are performed for medically defined end points that are clear indicators of health. Toxicity is so critical to the success of developing a new drug that tremendous attention and priority are accorded to it. The predictability of each method is thoroughly examined. New in vitro methods should produce a minimum of false-negative results so that unexpected toxicities do not appear at a later stage. False-positive results also are problematic because they place a stigma on the compound series. Indications of toxicity are a significant deterrent to further work on a compound or lead series but might be overcome during discovery if the mechanism can be determined. This chapter introduces discovery scientists to some of the terms and methods used in discovery and preclinical toxicity studies. For greater depth of understanding, detailed reviews, books, and experts should be consulted.

Toxicity studies during discovery focus on key mechanisms of toxicity. As in other areas, there is a hierarchy of assays and resources applied to compounds:

▶ Earlier studies use higher-throughput in silico and in vitro methods for indications of toxicity.

▶ Toxicity indications are followed up with sophisticated diagnostic testing of selected compounds.

▶ Lead or preclinical compounds are subjected to standard advanced toxicity procedures:

    ▶ Off-target selectivity screening

    ▶ Animal dosing experiments

Some organizations have found that shifting toxicology resources into discovery allows the early identification of toxicity issues.[1] These groups suggest the following advantages of early toxicity testing:

▶ More informed decisions during discovery for deprioritization of toxic leads or modification of leads to eliminate toxicity.

▶ Any issues are more robustly understood going into the investigational new drug stage.

▶ Saving resources on development of candidates that would fail because of toxicity.

▶ Dosing studies in phase I development can be better planned by having these data.

 ## 35.1   In Silico Toxicity Methods

The expense of in vivo toxicity testing and the large number of compounds to be assessed in drug discovery have encouraged the development of in silico and in vitro techniques. The ability of certain chemical moieties and structural templates to predictably induce certain types of toxic responses has led to quantitative structure–activity relationship (QSAR) studies for the purpose of predicting toxic responses using in silico algorithms.[2,3] Several limitations to the development of in silico models include the quality and quantity of the toxicological data, the intricacy of each type of toxic mechanism, and the multiplicity of toxic mechanisms. A listing of commonly used in silico tools is provided in Table 35.1.

TABLE 35.1 ▶ Commercial Products for Toxicity Methods

Product	Type	Company	Web site
DEREK	in silico	LHASA Ltd.	*www.lhasalimited.org*
OncoLogic	in silico	LogiChem Inc.	*woo.yintak@epa.gov*
Hazard Expert	in silico	CompuDrug	*www.compudrug.com*
MCASE	in silico	MultiCASE Inc.	*www.multicase.com*
TOPKAT	in silico	Accelrys	*www.accelrys.com*
LDH kit	in vitro	Roche Applied Science	*www.rocheappliedscience.com*
GreenScreen GC	in vitro	Gentronix	*www.gentronix.co.uk*

35.1.1   Knowledge-Based Expert System In Silico Toxicity Methods

General classifications and expert opinions have been combined into knowledge-based methods. These provide rules for the evaluation of new chemical entities based on structural features as well as their associated toxicity probabilities. Thousands of compounds have been studied, and thousands of rules have been developed.

DEREK is an expert system. Rules are agreed upon by a committee and incorporated into the model. Predictions, rule(s), and literature references are shown on the screen, so the user can follow up. DEREK can be linked to METEOR for inclusion of metabolite toxicity predictions. The DEREK interface reportedly is easy to use, and the system can be operated in batch mode. DEREK predicts many types of toxicity.

OncoLogic, which was developed by the U.S. Environmental Protection Agency, uses a large array of rules. The user interacts with the software to optimize the assessment. Carcinogenicity predictions and mechanisms are provided.

HazardExpert evaluates compounds by toxicity associated with structural fragments. The software makes bioavailability and bioaccumulation estimations for inclusion in the toxicity predictions. It links to MetabolExpert to make predictions for compound metabolites.

35.1.2   Statistically Based In Silico Toxicity Methods

In these programs, parameters are calculated for structures and substructure connectivity. Computational models are derived using statistical methods.

TOPKAT uses a QSAR-based model. Structures are evaluated using electro-topological structural descriptors and fit with statistical linear regression and linear free energy relationships to produce the models.

MCASE divides the structure into active and inactive moieties for pattern recognition. It provides a text output of predictions for each moiety. It interacts with META to make predictions for compound metabolites.

In a comparison study, four in silico packages were used by the National Toxicity Program to predict the rodent carcinogenicity of 44 compounds. The overall accuracies were as DEREK: 59%; TOPKAT: 57%; COMPACT: 53%; MCASE: 49%.[3,4] In a subsequent carcinogenicity study of 30 compounds, the overall accuracies were OncoLogic: 65%; COMPACT: 43%; DEREK: 43% and MCASE: 25%.[2,3] By comparison, the in vitro Ames Assay is 85% accurate for carcinogenicity. In general, the best predictions are provided by expert systems. Significant uncertainty still exists for in silico predictions.

35.2   In Vitro Toxicity Assays

As in other areas of drug discovery, in vitro assays are being increasingly applied for toxicity screening.[5] This strategy allows study of a greater number of discovery compounds, specific toxicity mechanisms,[6] and low compound expenditure, and reduces animal usage. In vitro methods also are applied at later stages to investigate specific toxicity questions. The following sections discuss many of the in vitro toxicity methods encountered by discovery scientists. Toxicity studies during earlier phases of discovery have been termed *predictive toxicology* or *exploratory toxicology*. The International Committee on Harmonisation (ICH) has provided extensive guidance for many of the in vitro toxicity assays.[7] Ultimately, all in vitro assays are judged by their predictivity of in vivo toxicity results.

35.2.1   Drug–Drug Interaction

Coadministered drugs can interact in many ways and result in toxic effects. The usual outcome of this interaction is a change in the pharmacokinetics of the effected drug. The effected drug increases in the bloodstream to a toxic concentration.

35.2.1.1   CYP and Other Metabolizing Enzyme Inhibition

A drug may inhibit the metabolism of a second coadministered drug. Much of the focus in this area has been on drug–drug interactions at cytochrome P450 (CYP) isozymes. Methods for CYP inhibition are discussed in Chapter 32. Other metabolizing enzymes (e.g., monoamine oxidase (MAO)) can be similarly inhibited.

35.2.1.2 CYP Induction (Pregnane X Receptor and Isozyme Expression)

Some drugs are CYP isozyme inducers. This triggers an increase in the expression of a CYP isozyme. If a second drug is primarily metabolized by this isozyme, its metabolism will occur at a higher rate than normal, resulting in an unexpectedly low concentration of the effected drug in the bloodstream.

A higher-throughput method for detecting compounds that are CYP 3A4 isozyme inducers is the pregnane X receptor (PXR) assay. The inducer drug activates PXR, which binds to PXR response element (PXRE) in the 3A4 gene. A PXRE-luciferase reporter gene as been engineered in a special cell line to assay this mechanism.[8] When the inducer binds to PXR, which binds to PXRE, luciferase is expressed and detected.

Induction of CYP isozymes by the test compound can be directly measured.[9,10] After test compound treatment of hepatocytes, enzyme substrates that are specifically metabolized by one isozyme are incubated with the hepatocytes (or microsomes obtained from the treated cells), and the activity of each isozyme is measured using the specific substrates. An increase in the metabolic rate of a substrate indicates that the isozyme was induced by the test compound. Other induction methods biochemically quantitate the actual amount of each isozyme or its mRNA.

35.2.1.3 Aryl Hydrocarbon Receptor Assay

The aryl hydrocarbon receptor (AhR) assay examines AhR-mediated induction of gene expression, which can lead to carcinogenicity.[11] AhR activation is an initial event in adverse effects of dioxins and related compounds that bind to AhR and trigger several toxic biochemical responses, such as induction of CYP1A1 and CYP1A2. These are important in the metabolic activation of many promutagens. CYP1A1 metabolizes polynuclear aromatic hydrocarbons (PAHs) to carcinogenic arene oxides.

One in vitro assay for detecting this induction, called chemical-activated luciferase expression (CALUX), quantifies AhR-mediated reporter gene expression. A rat hepatoma cell line is stably transfected with a luciferase reporter gene, which is under control of dioxin-responsive enhancers. The cells are grown in 96-well plates for 24 hours and then treated with test compound. After 24 hours, the medium is removed, the cells are washed, luciferase is extracted by lysis, and luciferase activity is measured using a luminometer. Increasing luciferase activity indicates increasing AhR induction.

35.2.2 hERG Block Assays

These assays test the blockage of the hERG K^+ ion channel. hERG methods are discussed in Chapter 34.

35.2.3 Mutagenicity/Genotoxicity

Mutagens (genotoxic compounds) change the DNA sequence of genes, or they damage chromosomes. This leads to changes in the gene products (proteins), which usually makes them either less functional or nonfunctional. Most importantly, mutation can cause cancer. Several in vitro and in vivo assays for detection of DNA damage caused by test compounds have been developed. Use of two or more tests in parallel provides the highest sensitivity for detecting the mutagenicity of a compound.[12] These assays detect many, but not all, carcinogens, especially those that work through mechanisms other than DNA damage. Many of the following assays metabolically activate the test compound as an early step in the assay by adding liver S9 and cofactors to the assay matrix with the cells.

35.2.3.1 Micronucleus Assay

This mutagenicity assay detects compounds that damage chromosomes or the cell division apparatus, producing abnormal DNA fragments. During mitosis, the chromosomes do not migrate properly because the centromere is damaged or lacking. The resulting DNA pieces, called *micronuclei*, adhere to membranes and can be observed by microscope.

The assay in single point mode is conducted by incubating cells in culture with test compound at 10 mM, with and without metabolic activation. The treated cells are cultured after treatment. Chromosome damage causes formation of micronuclei. The cells are stained and examined by microscope for the number of cells that contain micronuclei and how many micronuclei are present per cell. If the test compound has a dose dependency in producing micronuclei or if the production is reproducible, the test compound is classified as "positive."[13–15] Advances in cellular imaging technologies have allowed more rapid unattended examination of micronuclei.[16]

35.2.3.2 Chromosomal Aberration Assay

Chromosome aberrations, caused by compounds, can lead to mutations, induce cancer, and cause other genetic diseases. The test compound is added to a mammalian cell culture and incubated for 3 hours, with and without metabolic activation. The medium then is replaced with compound-free medium, and the cells are cultured for 15 hours. Colcemid is added to arrest the cells in metaphase. The cells are fixed and examined by microscope for aberrant chromosomes.

35.2.3.3 Comet Assay

Cells in culture are treated with test compound, with and without metabolic activation, at concentrations up to 10 mM for 6 hours. The cells then are embedded in agarose on a microscope slide and lysed under mildly basic conditions. The cells are subjected to gel electrophoresis. DNA fragments from single- and double-stranded breaks and relaxed chromatin migrate faster in the electric field than does unchanged DNA, making them appear as a comet tail. The DNA is visualized using fluorescence microscopy after staining. The relative fluorescence intensity in the tail (DNA fragments) indicates the frequency of DNA breaks. The method is also called single-cell gel electrophoresis (SCGE) assay. The method is very sensitive to DNA breakage caused by genotoxic compounds.[17–19]

35.2.3.4 Ames Assay

The Ames assay examines the ability of a compound or its metabolite to cause DNA mutations. Both frameshift and base-pair substitutions are detected. Special mutant strains of *Salmonella typhimurium* or *Escherichia coli* bacteria are grown overnight in culture medium. Then the test compound (5 mg per plate), liver S9 fraction, and metabolic cofactors are added and incubated for 1 hour. This solution is mixed with agar and plated. After incubation for 72 hours, the number of colonies is counted. Greater numbers of colonies indicate a greater mutation rate.

The bacterial strains contain base-pair deletions that cause them to be able to grow only in the presence of supplemental histidine (*Salmonella* strains) or tryptophan (*E. coli* strains). If a mutation occurs at the location of the histidine or tryptophan base-pair deletions such that a functional enzyme can be made, the subsequent bacterial generations can grow without the supplemental amino acids. The colonies that are counted are called *revertants*. A strain has been engineered that lacks the genes for DNA excision repair, and another in which the cell wall is more permeable to mutagens. These strains produce a larger number of reversions.[20–25] Similar methods are the SOS Chromotest[26] and the RNR3-lacZ yeast test.[27]

35.2.3.5 Thymidine Kinase Mouse Lymphoma Cell Assay

Like the Ames assay, the mouse lymphoma assay determines test compounds that cause gene mutations by base-pair substitutions and frameshifts, but in mammalian cells. The detection system for this assay is mutation of the thymidine kinase (TK) gene to a nonfunctional form. The normal function of TK is phosphorylation of thymidine to produce thymidine monophosphate (TMP). The TMP concentration controls the rate of DNA synthesis in cells. A thymidine analog trifluorothymidine (TFT) is introduced to cells after treatment with the test compound. Cells that have normal TK die, but cells that have TK mutations are unaffected by the TFT.

Mouse lymphoma cells suspended in culture medium are treated for 4 hours with the test compound (over a range of concentrations), liver S9, and cofactors. After this treatment, cells are centrifuged and washed to remove the test compound, resuspended, and incubated for 2 days. Cells then are seeded into 96-well plates containing TFT. After 14 days, the cell colonies are counted. The number of living colonies is indicative of the mutagenicity of the test compound.[28–30]

35.2.3.6 HPRT Chinese Hamster Ovary Cell Assay

This assay is another mammalian cell mutagenicity assay. The assay is sensitive to mutation of the HPRT gene to a nonfunctional form. HPRT is involved in DNA synthesis. As part of the detection process for this assay, the nucleoside analog 6-thioguanine (6-TG) is introduced to cells after they have been treated with the test compound. Cells that have normal HPRT die, but cells that have HPRT mutations are unaffected by the 6-TG.

Chinese hamster ovary (CHO) cells suspended in culture medium are treated for 4 hours with the test compound (over a range of concentrations), liver S9, and cofactors. After this treatment, cells are centrifuged and washed to remove the test compound, resuspended, and incubated for 3 days. Cells then are seeded into 96-well plates containing 6-TG. After 10 days, the cell colonies are counted. The number of living colonies is indicative of the mutagenicity of the test compound.

35.2.3.7 GADD45a-GFP Genotoxicity Assay

The GADD45a (growth Arrest and DNA damage) gene is a biomarker for genomic stress. It is involved in regulating DNA repair, mitosis delay, and apoptosis. The promoter for GADD45a has been linked to green fluorescent protein (GFP) and transfected into human p53-proficient TK6 cells.[31] This stable cell line (GenM-T01) is used in 96-well format for detection of genotoxins. An assay kit (GreenScreen GC) is available from Gentronix. In initial tests, the performance of the GreenScreen assay exceeded that of other assays in sensitivity (success in detecting carcinogens) and selectivity (success in identifying noncarcinogens). In addition to mutagens, the assay detects aneugens (interfere with chromosome segregation in mitosis), which usually are detected using the micronucleus assay. The presence of p53 in the cell line appears to be important in accurate induction of the cell's response to genotoxins, and p53 is missing in cell lines used for some other in vitro genotoxicity assays. This high-throughput assay could precede current genotoxicity assays to increase efficiency and early detection.

35.2.4 Cytotoxicity

Cytotoxicity is the killing of viable cells by the test compound.[32] This approach surveys many diverse mechanisms by which the compound can impede the normal function of the cell and trigger cell death. Hepatocytes are advantageous for cytotoxicity assays because both the test compound and its metabolites can cause toxicity.

35.2.4.1 MTT Human Hepatotoxicity Assay

This assay detects compounds that are toxic to human hepatocytes in culture. The indicator of healthy function is the reduction of yellow-colored 3-[4,5-dimethylthiazol-2-yl]-2,5-diphenyltetrazolium bromide (MTT) to purple-colored formazan by mitochondria. The concentration of formazan is measured by absorption at 570 nm.

Human hepatocytes are plated and cultured for 2 days. The test compound is added and incubated at concentrations of 10 to 1,000 µM for 24 hours. The medium containing the test compound is removed and replaced with medium containing MTT. After 3 hours of incubation, the cells are lysed, formazan is extracted by organic solvent, and the absorption at 550 nm is measured. Absorption below that of the control indicates that the test compound is toxic to the cells. Other dyes used for cytotoxicity assays are 3-(4,5-dimethylthiazol-2-yl)-5-3(3-carboxymethoxyphenyl)-2-(4-sulfophenyl)-2H-tetrazolium (MTS) and (2,3)-bis-(2-methoxy-4-nitro-5-sulphenyl)-(2H)-tetrazolium-5-carboxanilide (XTT). Chinese hamster cells also have been used for the MTT assay. These cells lack metabolizing enzymes, so they often are coincubated with liver S9 and cofactors.

35.2.4.2 Lactate Dehydrogenase Assay

The plasma membrane of a dead cell will lyse and release the cell contents into the medium. Lactate dehydrogenase (LDH) is an abundant enzyme that is released. The concentration of LDH is measured using a colorimetric reaction. Cells in culture are treated with the test compound at a range of concentrations for 24 hours. The medium is harvested and tested for LDH activity using a test kit (Roche) that quantitatively produces formazan, which is measured by UV absorption. Assays for other enzymes that are released by membrane lysis include alanine aminotransferase and aspartate aminotransferase.

35.2.4.3 Neutral Red Assay

Neutral red is a dye that is absorbed by hepatocytes and concentrated in lysosomes. It is a marker for healthy cells. Hepatocytes are treated with the test compound, as in other hepatotoxicity assays. Increasing uptake of neutral red is associated with increasing cell survival. This assay is claimed to be simpler and more sensitive than the LDH leakage test.[33]

35.2.5 Teratogenicity: Zebrafish Model

Developmental toxicology has been studied using rodent embryo development. An emerging model is zebrafish development.[34] Zebrafish are small (adults are 1–1.5 inches) and easily maintained. They require only a small amount of test compound for dosing. Embryos are easily handled in 384-well plates. Their size and transparency make them easily examined for specific developmental end points of teratogenicity studies. Abnormalities are readily observed.

35.2.6 Selectivity Screens

In order to avoid disadvantageous activity at another enzyme, ion channel, or receptor in the body, a screen of diverse targets usually is performed on selected advanced leads. This screening usually is performed by a contract laboratory (e.g., MDS Panlabs, Ambit Biosciences, NovaScreen).

35.2.7 Reactivity Screens

These assays examine the reactivity of compounds and metabolites with cellular components.[35]

35.2.7.1 Glutathione Trapping

Glutathione is a ubiquitous compound that traps reactive compounds and prevents damage to vital proteins and nucleic acids. Reactive metabolites are detected by adding glutathione to the in vitro metabolic stability incubation (see Chapter 29) along with microsomes or hepatocytes. The glutathione and *N*-acetylcysteine adducts are extracted and analyzed using liquid chromatography/mass spectrometry/mass spectrometry (LC/MS/MS) techniques to detect adducts by the added molecular weight and then elucidate the structures of the adducts to determine the reactive site of the molecule.[36,37] This approach is useful in discovery, when no radiolabeled compound is available.

35.2.7.2 Covalent Protein Binding

A radiolabeled compound is used in the in vitro metabolic stability incubation (see Chapter 29) along with microsomes or hepatocytes for 1 hour. The protein is separated and analyzed for incorporated (covalently bound) radioactivity. A guideline for assessment is whether >50 pmol equivalents per milligram of the compound has been incorporated in the protein. The assay can be performed in vivo.[38]

35.3 In Vivo Toxicity

In vivo studies are of great importance because they are the ultimate indicator of toxicity. They are especially useful during studies leading up to first-in-human (FIH) phase I dosing and those conducted during phase II to III.

35.3.1 Discovery In Vivo Toxicity

Toxicity studies are performed during discovery for selected lead compounds or as specific issues arise. In vivo toxicity studies supporting drug discovery are intended to examine critical safety issues. A limited number of animals is used, and studies are short duration (e.g., 2 weeks). Many of the same parameters studied in preclinical toxicity studies are examined during discovery, but in less detail and under non-Good Laboratory Practices (GLP) conditions. Preclinical and clinical toxicity studies are discussed in Section 35.3.2. Discovery toxicity studies are sometimes conducted in coordination with in vivo pharmacology dosing studies, so fewer resources are needed. For discovery lead compounds, the purposes of in vivo toxicity studies are as follows:

► Check for any signs of toxicity

► Obtain data for lead prioritization

► Follow up with toxicological examination of other leads if toxicity is observed

35.3.2 Preclinical and Clinical In Vivo Toxicity

In vivo toxicity studies are described here in general terms to provide discovery scientists with an introduction to many of the tests to which clinical candidates are subjected once the discovery project team advances them to development. In vivo toxicity tests are required prior to FIH dosing and are performed under GLP conditions using highly detailed protocols.[39] A general scheme used in industry for preclinical studies between candidate advancement and phase I trials is as follows:

▶ Acute (single-dose) toxicity

▶ Chronic toxicity: 2 to 14 weeks, daily dosing, rodent and nonrodent

▶ Carcinogenicity: 2 weeks of chronic dosing

▶ Genotoxicity: In vitro Ames test, in vivo mouse micronucleus test, chromosome aberration

▶ Safety pharmacology: Monitoring of normal health, behavior, and function using medical examinations and tests for central nervous system, cardiovascular (including radiotelemetry), respiratory, gastrointestinal systems, and kidney. Tests include physical appearance, body weight, food consumption, eye function, electrocardiography, blood chemistry, urine, and organ weight. Many of these studies are defined by the ICH guidelines.[7]

A full microscope histology examination is performed on as many as 50 tissues from dosed animals.

Pharmacokinetic studies that are performed and correlated with toxicology studies are called *toxicokinetics*. Toxicokinetic studies allow determination of the following:

▶ No effect level (NOEL): Highest dose or exposure that produces no toxicity

▶ No adverse effect level (NOAEL): Highest dose or exposure that produces manageable toxicity

▶ Therapeutic index (margin of safety): NOEL or NOAEL/efficacious dose or exposure

A candidate is terminated if severe toxicity is observed or if the therapeutic index is too narrow.

The purposes of preclinical studies are to predict patient hazards, ensure a wide therapeutic index, plan clinical phase I studies and dosing regimens, determine what organs are affected, determine which toxicity markers to look for in humans, determine any toxic metabolites, and examine drug responses that cannot be studied in humans.[40]

More detailed animal toxicity tests are performed during the phase I to III time period and include the following:

▶ Toxicity: 3 to 12 months, rodent and nonrodent (e.g., dog, monkey)

▶ Reproductive health: Mating behavior, estrous cycles, sperm, fertility

▶ Embryonic development: Survival, normal fetus and offspring growth, health, and responses (rodent, nonrodent)

▶ Oncology: 2 years, rat and mouse

▶ Immunotoxicity (immunosuppression or enhancement)[41]

▶ Toxicokinetics

35.3.3 Biomarkers of In Vivo Toxic Responses

In vivo toxicity studies have relied heavily on phenotypic response using histological examinations and microscopic examination of tissues prepared from dosed animals. New technologies have demonstrated the ability to observe toxic responses at the biochemical level. These techniques profile small-molecule biochemical intermediates, proteins, and mRNA as biomarkers of toxicity. This is an emerging and developing field. If successful, this approach will increase the number of compounds that can be evaluated and reduce the time required for toxicity studies. These emerging technologies are expected to have a major impact on toxicity studies in the future.

35.3.3.1 Toxicometabonomics

It has been shown that toxicity can, in some cases, be detected earlier using spectroscopic analysis of body fluids.[42,43] Toxic activity, such as inhibition of an enzyme by the drug or a drug metabolite, will cause an imbalance in the normal biochemical intermediates in the organism. The concentrations of intermediates in the pathway of the inhibited enzyme increase or decrease. The study of endogenous biochemical intermediates is called *metabonomics*.

Animals are dosed with the test compound once or daily for several weeks. Urine or blood samples are collected and analyzed. The change in endogenous compound concentration is detected using LC/MS or NMR techniques. Hundreds of components are present in these samples, and sophisticated analysis methods are required. Detection of the affected endogenous compounds compares the spectra and chromatograms of samples from treated individuals to samples obtained before treatment. One challenge is to determine if changes are due to normal biological fluctuations or due to effects of the test compound. In many cases, biochemical changes are detectable in this manner before behavioral or morphological signs of toxic response are observed. The actual intermediates that change with dosing indicate which pathway is affected.

35.3.3.2 Toxicoproteomics

In a similar manner to metabonomics, the balance of proteins in a biological system can change in response to the administration of a drug. Some of these changes are advantageous and consistent with the pharmacological goals of affecting the disease. However, other protein pattern changes, are indicative of a toxic response. The study of the protein ensemble of a cell or an organism is called *proteomics*.[44] These studies use two-dimensional polyacrylamide gel electrophoresis (2D-PAGE) and matrix assisted laser desorption/ionization time of flight (MALDI-TOF) mass spectrometric analysis of protein mixtures from the sample.

35.3.3.3 Toxicogenomics

Genomics is another indicator of toxic response. The mRNA in the cells is profiled to monitor gene expression in response to compound dosing.[45–48] This technique is also called *transcriptional profiling*, and application of this technique has been termed *toxicogenomics*. cDNA and oligonucleotide microarrays are used to profile the thousands of mRNAs that might be modified by drug administration. Strong correlations among histopathology, clinical chemistry, and gene expression profiles have been reported.[45]

Problems

(Answers can be found in Appendix I at the end of the book.)

1. On what are the currently most successful in silico tools for toxicity prediction based?

2. What are in silico toxicity tools useful for?

3. What does the PXR assay test?

4. What method other than PXR can be used for this?

5. Which of the following compounds damage DNA?: (a) enzyme inducers, (b) cytotoxic, (c) mutagenic, (d) potentially carcinogenic.

6. In the following table, link the assay with the observed end point:

Assay	DNA fragments move faster in gel electrophoresis than normal DNA	Abnormally divided DNA is observed by microscope	Reversion mutations allow colonies to grow without histidine	Unusually shaped chromosomes	Normal mammalian cells mutate so TMP does not kill them
Micronucleus					
Chromosomal aberration					
Comet					
Ames					
TK mouse lymphoma					

7. The LDH assay works by which of the following mechanisms?: (a) uptake by healthy cells is detected colorimetrically, (b) LDH is taken up into the cells and reacts in the mitochondria to form a detectable product, (c) lysis of unhealthy cells releases enzymes that are detected with a biochemical assay.

8. Teratogenicity can be determined using what?

9. What compound is used to trap reactive metabolites?

10. What toxicity studies are always performed prior to human phase I dosing?

11. What newer "-omics" methods have much potential for future toxicity studies? What do these assays measure?

References

1. Car, B. D. (2007). Discovery approaches to screening toxicities of drug candidates. AAPS Conference: Critical Issues in Discovering Quality Clinical Candidates, Philadelphia, PA, April 24–26, 2006.

2. Greene, N. (2002). Computer systems for the prediction of toxicity: an update. *Advanced Drug Delivery Reviews*, *54*, 417–431.

3. Benigni, R. (2004). Computational prediction of drug toxicity: the case of mutagenicity and carcinogenicity. *Drug Discovery Today: Technologies*, *1*, 457–463.

4. Benigni, R. (1997). The first US National Toxicology Program exercise on the prediction of rodent carcinogenicity: definitive results. *Mutation Research*, *387*, 35–45.

5. Li, A. P. (2001). Screening for human ADME/Tox drug properties in drug discovery. *Drug Discovery Today*, *6*, 357–366.

6. Ulrich, R. G., Bacon, J. A., Cramer, C. T., Peng, G. W., Petrella, D. K., Stryd, R. P., et al. (1995). Cultured hepatocytes as investigational models for hepatic toxicity: practical applications in drug discovery and development. *Toxicology Letters*, *82/83*, 107–115.

7. ICH. (2005). *ICH Steering Committee Guidelines*. www.ich.org/cache/compo/502–272–1.html.

8. Moore, J. T., & Kliewer, S. A. (2000). Use of the nuclear receptor PXR to predict drug interactions. *Toxicology*, *153*, 1–10.

9. Li, A. P. (1997). Primary hepatocyte cultures as an in vitro experimental model for the evaluation of pharmacokinetic drug-drug interactions. *Advances in Pharmacology (San Diego)*, *43*, 103–130.

10. Mudra, D. R., & Parkinson, A. (2004). In vitro CYP induction in human hepatocytes. In: Yan, Zhengyin & Caldwell, Gary W. (Eds.), *Optimization in Drug Discovery: In Vitro Methods*, Totowa; Humana Press, pp. 203–214.

11. Machala, M., Vondracek, J., Blaha, L., Ciganek, M., & Neca, J. V. (2001). Aryl hydrocarbon receptor-mediated activity of mutagenic polycyclic aromatic hydrocarbons determined using in vitro reporter gene assay. *Mutation Research*, *497*, 49–62.

12. Kirkland, D., Aardema, M., Henderson, L., & Mueller, L. (2005). Evaluation of the ability of a battery of three in vitro genotoxicity tests to discriminate rodent carcinogens and non-carcinogens. I. Sensitivity, specificity and relative predictivity. *Mutation Research*, *584*, 1–256.

13. Fenech, M. (2000). The in vitro micronucleus technique. *Mutation Research*, *455*, 81–95.

14. Fenech, M. (1993). The cytokinesis-block micronucleus technique and its application to genotoxicity studies in human populations. *Environmental Health Perspectives*, *101*, 101–107.

15. Kirsch-Volders, M., Elhajouji, A., Cundari, E., & Van Hummelen, P. (1997). The in vitro micronucleus test: a multi-endpoint assay to detect simultaneously mitotic delay, apoptosis, chromosome breakage, chromosome loss and non-disjunction. *Mutation Research*, *392*, 19–30.

16. Lang, P., Yeow, K., Nichols, A., & Scheer, A. (2006). Cellular imaging in drug discovery. *Nature Reviews Drug Discovery*, *5*, 343–356.

17. Tice, R. R., Agurell, E., Anderson, D., Burlinson, B., Hartmann, A., Kobayashi, H., et al. (2000). Single cell gel/Comet assay: guidelines for in vitro and in vivo genetic toxicology testing. *Environmental and Molecular Mutagenesis*, *35*, 206–221.

18. Olive, P. L., Banath, J. P., & Durand, R. E. (1990). Detection of etoposide resistance by measuring DNA damage in individual Chinese hamster cells. *Journal of the National Cancer Institute*, *82*, 779–782.

19. Zegura, B., & Filipic, M. (2004). Application of in vitro comet assay for genotoxicity testing. *Optimization in Drug Discovery*, 301–313.

20. Ames, B. N., McCann, J., & Yamasaki, E. (1975). Methods for detecting carcinogens and mutagens with the salmonella/mammalian-microsome mutagenicity test. *Mutation Research*, *31*, 347–363.

21. Maron, D. M., & Ames, B. N. (1983). Revised methods for the Salmonella mutagenicity test. *Mutation Research*, *113*, 173–215.

22. Brusick, D. J., Simmon, V. F., Rosenkranz, H. S., Ray, V. A., & Stafford, R. S. (1980). An evaluation of the Escherichia coli WP2 and WP2 uvrA reverse mutation assay. *Mutation Research*, *76*, 169–190.

23. Mortelmans, K., & Zeiger, E. (2000). The Ames salmonella/microsome mutagenicity assay. *Mutation Research*, *455*, 29–60.

24. Hakura, A., Suzuki, S., & Satoh, T. (2004). Improvement of the Ames test using human liver S9 preparation. In: Yan, Zhengyin & Caldwell, Gary W. (Eds.), *Optimization in Drug Discovery: In Vitro Methods*, Totowa; Humana Press, pp. 325–336.

25. Kirkland, D. J. (1989). Statistical evaluation of mutagenicity test data. In *UKEMS subcommittee on guidelines for mutagenicity testing*, *Report Part III*. Cambridge, UK: Cambridge University Press.

26. Kevekordes, S., Mersch-Sundermann, V., Burghaus, C. M., Spielberger, J., Schmeiser, H. H., Arlt, V. M., et al. (1999). SOS induction of selected naturally occurring substances in Escherichia coli (SOS chromotest). *Mutation Research*, *445*, 81–91.

27. Jia, X., & Xiao, W. (2004). Assessing DNA damage using a reporter gene system. In: Yan, Zhengyin & Caldwell, Gary W. (Eds.), *Optimization in Drug Discovery: In Vitro Methods*, Totowa; Humana Press, pp. 315–323.

28. Clive, D., & Spector, J. A. F. S. (1975). Laboratory procedure for assessing specific locus mutations at the TK locus in cultured L5178Y mouse lymphoma cell. *Mutation Research, 31*, 17–29.

29. Oberly, T. J., Yount, D. L., & Garriott, M. L. (1997). A comparison of the soft agar and microtitre methodologies for the L5178Y tk+- mouse lymphoma assay. *Mutation Research, 388*, 59–66.

30. Chen, T., & Moore, M. M. (2004). Screening for chemical mutagens using the mouse lymphoma assay. In: Yan, Zhengyin & Caldwell, Gary W. (Eds.), *Optimization in Drug Discovery: In Vitro Methods*, Totowa; Humana Press, pp. 337–352.

31. Hastwell, P. W., Chai, L.-L., Roberts, K. J., Webster, T. W., Harvey, J. S., Rees, R. W., et al. (2006). High-specificity and high-sensitivity genotoxicity assessment in a human cell line: validation of the GreenScreen HC GADD45a-GFP genotoxicity assay. *Mutation Research, Genetic Toxicology and Environmental Mutagenesis, 607*, 160–175.

32. Crouch, S. P.M., & Slater, K. J. (2001). High-throughput cytotoxicity screening: hit and miss. *Drug Discovery Today, 6*, S48–S53.

33. Zhang, S. Z., Lipsky, M. M., Trump, B. F., & Hsu, I. C. (1990). Neutral Red (NR) assay for cell viability and xenobiotic-induced cytotoxicity in primary cultures of human and rat hepatocytes. *Cell Biology and Toxicology, 6*, 219–234.

34. Hill, A. J., Teraoka, H., Heideman, W., & Peterson, R. E. (2005). Zebrafish as a model vertebrate for investigating chemical toxicity. *Toxicological Sciences, 86*, 6–19.

35. Caldwell, G. W., & Yan, Z. (2006). Screening for reactive intermediates and toxicity assessment in drug discovery. *Current Opinion in Drug Discovery & Development, 9*, 47–60.

36. Tang, W., & Miller, R. R. (2004). In vitro drug metabolism: thiol conjugation. In: Yan, Zhengyin & Caldwell, Gary W. (Eds.), *Optimization in Drug Discovery: In Vitro Methods*, Totowa; Humana Press, pp. 369–383.

37. Chen, W. G., Zhang, C., Avery, M. J., & Fouda, H. G. (2001). Reactive metabolite screen for reducing candidate attrition in drug discovery. *Advances in Experimental Medicine and Biology, 500*, 521–524.

38. Evans, D. C., Watt, A. P., Nicoll-Griffith, D. A., & Baillie, T. A. (2004). Drug-protein adducts: an industry perspective on minimizing the potential for drug bioactivation in drug discovery and development. *Chemical Research in Toxicology, 17*, 3–16.

39. Van Zwieten, M. (2006). Preclinical toxicology. The Residential School on Medicinal Chemistry, Drew Chemistry, Madison, NJ, June 12–26, 2006.

40. Jones, T. W. (2006). Pre-clinical safety assessment: it's no longer just a development activity. *Drug Discovery Technology (R) and Development World Conference*, Boston, MA, August 8–10, 2006.

41. Dean, J. H., Cornacoff, J. B., Haley, P. J., & Hincks, J. R. (1994). The integration of immunotoxicology in drug discovery and development: investigative and in vitro possibilities. *Toxicology in Vitro, 8*, 939–944.

42. Beckonert, O., Bollard, M. E., Ebbels, T. M.D., Keun, H. C., Antti, H., Holmes, E., et al. (2003). NMR-based metabonomic toxicity classification: hierarchical cluster analysis and k-nearest-neighbour approaches. *Analytica Chimica Acta, 490*, 3–15.

43. Robertson, D. G. (2005). Metabonomics in toxicology: a review. *Toxicological Sciences, 85*, 809–822.

44. Bandara, L. R., & Kennedy, S. (2002). Toxicoproteomics—a new preclinical tool. *Drug Discovery Today, 7*, 411–418.

45. Waring, J. F., Jolly, R. A., Ciurlionis, R., Lum, P. Y., Praestgaard, J. T., Morfitt, D. C., et al. (2001). Clustering of hepatotoxins based on mechanism of toxicity using gene expression profiles. *Toxicology and Applied Pharmacology, 175*, 28–42.

46. Fielden, M. R., & Zacharewski, T. R. (2001). Challenges and limitations of gene expression profiling in mechanistic and predictive toxicology. *Toxicological Sciences, 60*, 6–10.

47. Waters, M. D., & Fostel, J. M. (2004). Toxicogenomics and systems toxicology: aims and prospects. *Nature Reviews Genetics, 5*, 936–948.

48. Maggioli, J., Hoover, A., & Weng, L. (2006). Toxicogenomic analysis methods for predictive toxicology. *Journal of Pharmacological and Toxicological Methods, 53*, 31–37.

Chapter 36

Integrity and Purity Methods

Overview

▶ *High-throughput integrity and purity methods commonly use high-performance liquid chromatography/ultraviolet detection/mass spectrometry (LC/UV/MS).*

▶ *UV response indicates the relative percentage of each sample component.*

▶ *MS indicates consistency with the compound's molecular weight or potential identity of impurities.*

▶ *Follow-up of inconsistent samples that are active can provide additional project leads.*

36.1 Criteria for Integrity and Purity Assays

Compounds that were synthesized within the last few years and were subjected to independent analytical review typically have high confidence with regard to their identity and purity. In contrast, compounds that are older may not have been rigorously confirmed or may have undergone improper handling or storage. In this case, a check of structural identity and purity will provide greater assurance for structure–activity relationships (SAR).[1–3]

The common practice in synthetic chemistry is to follow reaction steps for new compounds using structure-specific analytical techniques, such as mass spectrometry (MS) for molecular weight (MW) confirmation and NMR for more detailed characterization. High-performance liquid chromatography (HPLC) is used for purity estimation. Often these resources are available in a walk-up open-access format for convenience and flexibility that fits the workflow of chemists. Many organizations have a central analytical department that independently verifies that the spectra are consistent with the putative synthetic product. This is a wise strategy. Unbiased confirmation of the identity and purity of a compound that will be registered in the company's valuable compound collection is worthwhile.

The techniques used by medicinal chemists and analytical departments for this process are relatively detailed and time consuming. When projects are dealing with the thousands of high-throughput screening (HTS) hits that come from a screen, thousands of compounds that might be purchased from a vendor, hundreds of compounds that might come from an alliance collaborator, or hundreds of compounds that might come from a library similarity search, a faster and less resource-consuming assay method is necessary. Tradeoffs must be made to efficiently assay these samples. Turnaround time and resources expended per compound usually must be kept as low as possible. This dictates the need for faster methods and results in reduced resolution. For example, a typical high-resolution HPLC analysis of purity may use an analysis time of 30 to 60 minutes, whereas a higher-throughput integrity and purity assay may use an analysis time of 3 to 5 minutes. Techniques are combined for efficiency, such as integrating HPLC and MS to obtain purity and integrity data from one analytical run. Such compromises reduce the HPLC resolution for separation of impurities and the spectroscopic identity check is reduced to molecular weight from the molecule ion. These are still acceptable levels of data for the discovery project team to proceed with increased confidence for decision making.

These tradeoffs are a common part of a strategy of "*appropriate methods.*" In appropriate methods, the analytical method is streamlined such that it meets the needs of the discovery project team at the current stage of their work and does not consume unnecessary resources and time delays in obtaining too much detail for the question being asked. Integrity and purity profiling addresses the question of whether the compound appears to have the identity that is shown in the corporate database and if its purity is high enough to ensure that the observed activity derives from the putative compound. Resources are not available for more in-depth analysis, and the answer must come quickly to stay in synchronization with the discovery time line. The key activity is prioritization of many compounds for the next level of research experiments. Analytical detail can be assured at later discovery stages using lower-throughput techniques with higher analytical figures of merit for the fewer compounds that are studied in great detail (Table 36.1). Drug discovery currently is strongly guided by business strategies, such as *risk management*. Integrity and purity assays help to reduce the risk at earlier drug discovery stages with appropriate investment of resources.

TABLE 36.1 ▶ Comparison of High-Throughput and Low-Throughput Integrity and Purity Methods

	High throughput	Low throughput
Purpose	Rapid screen to avoid identity and purity mistakes	Detailed data to provide assurance at later stages
Samples/wk	100–1,000	10–100
HPLC column	2 × 20 mm, 3- to 5-μM particles	4.6 × 150 mm, 3- to 5-μM particles
HPLC run time (min); function	1–3; screen	30–120; resolve all components
Detector	UV relative area percent	UV, ELSD, CLND high-accuracy quantitation
Mass spectrometer	Single-stage quadrupole, ion trap, or time-of-flight MS	Single-stage MS to MS/MS; accurate mass analysis
Other spectroscopy	None	NMR

36.2   Samples for Integrity and Purity Profiling

When considering which samples to profile, it is important to remember the question to be answered. If the question is: "What is the quality of the solid sample from the compound repository?", then it is appropriate to obtain the sample in a vial and analyze it. However, if the question is: "What is the quality of the HTS hit?", then it must be remembered that HTS is performed with plates of compounds in solution that have been stored for a period of time. For this question, it is appropriate to obtain a small aliquot (2–5 μL) of the solution used for the HTS run. It is this solution that is linked to the HTS activity hit. This sample may differ from the solid that is in the compound repository because of handling and storage.

A high-throughput integrity and purity screen also allows the rapid profiling of compounds that come into laboratories from outside sources. Thus, at low expense potentially costly and time-consuming mistakes from the use of an inaccurate compound can be avoided.

36.3 Requirements of Integrity and Purity Profiling Methods

A wide array of analytical technologies potentially could be incorporated into integrity and purity methods. Typically, these exhibit the common analytical tradeoffs of selectivity, sensitivity, speed, and cost. Thus, different techniques can be chosen to match the needs of the assay to answer the particular research question at a specific place in the discovery time line. In late discovery, there is a need for *in-depth profiling* of a few selected compounds. With these few compounds and the importance of specific information for decision making, it is appropriate to use techniques with slower speed, such as extended HPLC separations with longer columns, slower gradients, and smaller particle size for high-resolution separation of sample components. Other examples are MS/MS product ion spectra for detailed fragmentation analysis, multiple NMR experiments (proton, carbon, two-dimensional), x-ray crystallography, and elemental analysis. These techniques provide unambiguous confirmation of identity, regiochemistry, stereochemistry, and quantitative measurement of purity on which critical late-stage experiments can be based for go/no-go decisions. Unfortunately, this approach is not appropriate for early discovery HTS integrity profiling. It requires considerable scientist's time, materials, instrumentation, and time line to accomplish. A *high-throughput profiling* approach is necessary for earlier studies.

In early discovery, only a small amount of each compound is available for property profiling. Only submilligram levels of HTS hits are available from HTS plates, and only milligrams are available as solids from the compound repository. This necessitates sensitive methods, such as HPLC, UV detection, and MS.

Integrity profiling methods also must provide data that are correlated to structure. NMR is highly correlated to detailed structural moieties, but NMR interpretation is too time consuming for integrity screening. MS provides a confirmation of MW, which is closely associated with structure. MS also provides initial structural information for impurities. Degradants and reaction by-products usually differ greatly in MW from the primary component.

Selectivity can be provided for a method by using a separation technique, such as HPLC. MS adds selectivity by distinguishing compounds by MW, even if they co-elute. Fast analytical methods are required to provide rapid turnaround. This allows discovery teams to make major decisions rapidly. Biological assay data usually are available within days or weeks, so integrity and purity data also should be provided in this time frame to make a meaningful contribution.

Integrity and purity methods for early discovery need speed so that they can handle a large number of samples, on the order of 1,000 samples per week. The large number of samples also demands that costs be kept low. The figure of $250,000 per scientist per year is commonly used in the industry. An analyst must process about 125 compounds per day in order to keep the analysis cost down to $10 per sample.

36.4 Integrity and Purity Method Advice

The integrity and purity assay may seem to be a simple assay; however, several aspects can ensure a reliable and efficient analysis. These are discussed in terms of the assay's steps: sample preparation, component separation, detection, quantitation, and confirmation. It is useful to begin by considering all of the available techniques that might be used for this method (Table 36.2). The appropriateness of these techniques is discussed below (Section 36.4.1–36.4.5).

TABLE 36.2 ▶   Various Techniques that may be Incorporated into Integrity and Purity Assays

Technique	Throughput (samples/h)	Analytical detail	Appropriate for HT profile	Appropriate for in-depth profile
Flow injection	60	L	M	L
Fast HPLC	20	M	H	L
High-resolution HPLC	2	H	L	H
NMR	5	H	L	H
MS	20	M	H	L
MS/MS	5	H	L	H
IR	5	M	L	M
UV	20	M	H	M
ELSD	20	H	M	H
CLND	20	H	M	H

Used with permission from [13].
The characteristics of the techniques make them appropriate for methods to address the needs of different parts of the discovery timeline.
H, High; L, low; M, moderate.

36.4.1   Sample Preparation

Accurately weighed and labeled samples can be readily handled for high throughput in well plates. For low-throughout analysis, vials are adequate. Sample handling and solvent addition using laboratory robots are very efficient.

It is important to completely dissolve the sample in order to assure accurate purity quantitation and to observe all of the sample components. The solubilities of different sample components can differ, and the main component may not, at first, be completely soluble. Dimethylsulfoxide (DMSO) is often used as a dilution solvent because of its "universality," but not all compounds are soluble. Freeze–thaw cycles tend to cause a compound to recrystallize to a stable polymorph, and this polymorph may not readily redissolve.[4] Some polar compounds (salts) have low DMSO solubility. In these cases, addition of a small volume of a second miscible solvent of differing polarity or use of a solvent mixture for the original dilution can assist dissolution. Precipitates cannot always be seen by visual inspection. A careful examination, under conditions where fine particulates can be observed, is warranted. Another issue with using DMSO is its strong UV absorption, which produces an intense peak at the void volume. This peak may obscure compounds that are not well retained on the HPLC column.

36.4.2   Component Separation

Some laboratories use flow injection analysis (FIA), in which the sample is injected directly into the detector. This is commonly done for a quick check at open-access facilities. FIA is very fast, requiring less than 1 minute per sample. However, there is no component separation, and the method is not adequate for purity estimation. In addition, the dilution solvent causes a strong signal, termed *solvent front*, which can interfere with component detection. In most samples, multiple components are present, even if they are at trace levels. The putative compound usually is the main component; however, many times the impurities are the main components. Integrity and purity assays are best accomplished using HPLC separation.

HPLC can separate the sample components from the solvent front that would interfere in FIA. High-resolution HPLC assays require up to 1 hour per sample. This would be appropriate for late discovery release of batches for toxicology studies. For early discovery studies in high throughput, this would consume too many resources. One issue for selection of a "generic" method for all samples is that compounds vary in their chromatographic characteristics. HPLC conditions that work for a wide range of compound polarities must be selected. Typically, reversed-phase HPLC is used with a wide mobile phase polarity gradient. For example, the gradient might start with 100% aqueous buffer and proceed to 100% acetonitrile. In recent years the technique of "fast HPLC" has been widely implemented, where the gradient is completed in a short time (1–2 minutes), the small-particle stationary phase is used (3–5 μm), and the mobile phase is at high flow rate (e.g., 1 mL/min).[5] These conditions may not provide sufficient chromatographic resolution and an acceptable tradeoff of resolution and speed must be found.

Recently, instrument manufacturers have introduced higher-resolution HPLC systems that use particle sizes of around 1.7 μm. If higher flow rates of 1 mL/min are used, very high pressures are generated that are beyond the capability of standard HPLC instruments. One company has termed the new technique *ultraperformance liquid chromatography*. Very high resolutions can be achieved with gradient times of about 1 to 1.5 minutes for high throughput, plus 0.5 to 1 min for re-equilibration. Supercritical fluid chromatography has been used to achieve higher resolution and increased throughput in some laboratories,[6] but it is not a common technique.

There are pitfalls to any HPLC separation method for which the analyst should be vigilant. A sample component may co-elute with another component and remain undetected. These components can be deconvoluted using MS if they have different MWs. Polar components may be buried under the solvent front. Very lipophilic components may not elute during the gradient and require an extended hold of the solvent ratio at the top of the gradient to assure elution. Enantiomers require chiral methods to be independently quantitated.

If MS is used for identity checking, the mobile phase must be compatible with the MS. Commonly, water, acetonitrile, and methanol are used as solvents. Common modifiers are ammonium acetate and formic acid. If a chemiluminsecent nitrogen detector (CLND) is used, acetonitrile cannot be used as a solvent. The mobile phase must be completely volatile if an evaporative light scattering detector (ELSD) or CLND is used.

36.4.3 Quantitation

A technique for detecting and measuring each component is necessary for purity estimation. A "universal" detector would be useful, but no detector responds to every compound with the same response on a molar basis. The most common detector is UV, which is sensitive and cost efficient. Most compounds studied in discovery absorb in the UV range. It is important to consider the UV wavelength that is used. The wavelength 254 nm typically is associated with aromatic groups and is commonly used. Some compounds do not contain an aromatic group; thus, the 214- to 220-nm region is often used for a broader compound response. Unfortunately, DMSO also absorbs in this region and makes an intense solvent front peak. Diode array UV detectors can be set to scan over a broader portion of the UV spectrum for more universal compound detection. The analyst should keep in mind that the diversity of UV spectra and molar absorptivities at a given UV wavelength can cause great differences in the quantitation of purity, depending on the wavelength chosen. Minor impurities at one wavelength may appear to be much greater at another wavelength. Despite these limitations, UV detection usually is sufficient for most integrity and purity profiling needs. It helps provide information that reduces the risk of wasted time and resources on inaccurate or impure compounds.

The ELSD and CLND appear to have somewhat greater universality than the UV. The ELSD nebulizes the HPLC effluent and evaporates the volatile solvents to condensed particles that are detected by light scattering.[7] The CLND vaporizes the HPLC effluent.[8,9] Components are pyrolized at high temperature, and nitrogen oxides produce light, which is detected. CLND requires relatively higher maintenance than other detectors, and its response is proportional to the number of nitrogen atoms in the molecule. This poses a problem with quantitation of unknown impurities.

Purity estimation typically is done using the relative response of each component. It is assumed that each component has an equal molar response. This approach results in some inaccuracy compared to quantitation using standards for each component. However, it is approved by the Food and Drug Administration for the purpose of quantitating unknown impurities in clinical batch release for development and manufacturing. Quantitation can be more accurate using ELSD with internal standards and CLND for known compounds. This is most useful for later discovery work. Despite these precautions, some components, such as trifluoroacetic acid, inorganic salts, silica, plastic extracts, and volatile solvents, can remain undetected.[9]

36.4.4   Identity Characterization

Another technique needed in the analysis is one that produces data that are related to structure and can be used to confirm the compound identity. For high-throughput profiling, the technique must produce signals at the nanogram compound level and be easily interpreted. Newer probes have allowed NMR to work with flowing aqueous streams. Eventually, the throughput will be consistent with higher throughput and sensitivity needs. However, NMR spectra continue to be time consuming for interpretation.

Single-stage MS is often used for this application. It is sensitive and interfaces readily to HPLC. The mass-to-charge ratio (m/z) of the molecule ion is rapidly interpreted in terms of the MW of the compound. Electrospray ionization (ESI) is often used. Both positive and negative ions are readily provided by ESI. Amines typically produce positive ions, and acids typically form negative ions. Occasionally, atmospheric pressure chemical ionization (APCI) and atmospheric pressure photo ionization (APPI) are used to produce ions for compounds that are not sensitive by ESI. Ion sources that combine two ionization methods, such as ESI and APCI, have begun to be commercially available and broaden the opportunities for producing ions from sample components in a single HPLC run.[10] All of these techniques usually produce ions that are easy to interpret as $(M + H)^+$ or $(M - H)^-$ ions. MS instruments can alternate between positive and negative ion analysis in a single HPLC run, thus efficiently providing a broad detection of diverse sample components.

In examining the mass spectra, less experienced scientists should be aware of potential interpretation mistakes. A compound may form adduct ions, which are produced when an ion from the mobile phase attaches to the analyte molecule and produces ions such as $(M + NH_4)^+$, $(M + Na)^+$, $(M + H + CH_3CN)^+$, $(M + HCOO)^-$, or a dimmer molecule ion $(2M + H)^+$. A labile ion may fragment to lose a molecule of water $(M + H - H_2O)^+$. The presence of a chlorine or bromine atom in the molecule can confuse interpretation because of the high abundance of $M + 2$ ions from naturally abundant stable isotopes. It is not sufficient to plot an ion chromatogram for the $(M + H)^+$ or $(M - H)^-$ ions of a compound to confirm its presence or absence. The spectra must be examined to determine that they are consistent with the putative structure and to assign the MW of an unknown impurity.

Several types of mass spectrometer are equally useful for integrity profiling: quadrupole, ion trap, and time of flight (TOF). TOF provides the possibility of highly accurate mass analysis when it is used with a mass calibration standard. Some groups prefer the greater confidence provided by accurate mass, but it may be more time-consuming to maintain. Quadrupole analyzers are the least expensive systems. Two MS manufacturers offer options for interfacing more than one HPLC column stream (2–8) to a single MS. The interface sequentially shifts each stream for analysis every fraction of 1 second. With one HPLC stream and a 5-minute analysis, 240 samples can be analyzed in 20 hours; with 4 HPLC streams, 960 samples can be analyzed in 20 hours.[11] Although some MS vendors provide software for an automated check of spectra, it is advisable for an experienced analyst to review the assignments, which can be time consuming.

NMR remains the best approach for high confidence in identity assignment. Proton NMR now can be coupled to HPLC for analysis of submilligram amounts of samples. However, this requires examination by a trained analyst for confident assignment and is more expensive than MS. Late-stage studies require the detail provided by one-dimensional and two-dimensional NMR analysis.

36.5 Follow-up on Negative Identity Results

In certain circumstances, it may be beneficial to follow up and try to identify compounds that were not confirmed using a standard method. One useful case is when an HTS hit is found not to be the expected compound but is present in high purity. Despite not being what it was supposed to be, it still is an active compound and may be a unique and valuable pharmacophore. It may be worth some additional effort to use more detailed analytical techniques, such as MS/MS and NMR, to elucidate the structure. When multiple components are present, they are sometimes isolated for individual activity testing. Modern fraction collection systems are used in combinatorial chemistry groups and are an effective approach. The structure of the active isolate can be elucidated to provide an active lead. However, this approach may be too time consuming for some companies.

Another follow-up opportunity presents when the compound's characteristics do not allow separation or detection using the standard method. Follow-up can use different chromatographic conditions or detection methods. Examining the structure can indicate a useful alternate method.

36.6 Example Method

Five microliters of a 10 mg/mL stock solution of test compound in DMSO is transferred to a well in a 96-well plate. The sample is diluted further with DMSO (or 50% acetonitrile/50% isopropanol) to a concentration of 500 ng/µL using a laboratory multiprobe robot. Each well is examined for any undissolved material. A small volume of polar solvent is added, or the plate is sonicated if some material remains undissolved. A 2- to 5 µL aliquot is injected onto the HPLC column. Separation utilizes an Aquasil column (2 ×50 mm, 5-µM particles) at 40°C and a flow rate of 0.8 mL/min. This column has a mixed stationary phase consisting of bonded C18, which retains lipophilic compounds, and bonded ethanol, which retains polar compounds, for use with a broad diversity of compound polarities. The mobile phase gradient is as follows: 100% mobile phase A (95% 10 mM ammonium acetate/5% acetonitrile)/0% mobile phase B (5% 10 mM ammonium acetate/95% acetonitrile) to 0% A/100% B in 2.5 minutes, hold for 1.5 minutes, and re-equilibrate for 1.5 minute. The mobile phase eluent flows into a UV diode array detector that is scanned from 190 to 600 nm to detect diverse components. Purity is estimated from the relative area under the

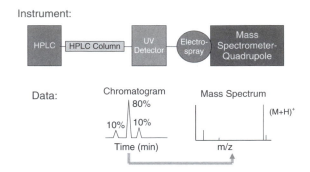

Figure 36.1 ▶ Schematic diagram of LC/UV/MS instrumentation in integrity and purity profiling.

chromatographic peaks of each component at 214 nm. A single-stage quadrupole MS with ESI obtains both positive and negative ion spectra each 1.0 second, while scanning from m/z 100 to 1,000. Analysis automation and postacquisition data processing are performed using the mass spectrometer's quantitation software. A schematic of the instrumentation is shown in Figure 36.1.

36.7 Method Case Studies

Kyranos et al.[12] profiled the identity and purity of compounds from high-throughput synthetic chemistry libraries. They found LC/MS to be more successful than FIA/MS or FIA/NMR for samples containing multiple components. APCI was found to be more universal than ESI but produced more fragment ions. "Fast HPLC" conditions were very reliable and had high throughput. The ELSD detector had consistent absolute molar purity quantitation if the compounds were known, and CLND had good absolute molar purity accuracy (±5%) for compounds containing nitrogen. Multiplexed parallel HPLC with a single MS provided very high throughput but was limited to a minimum HPLC peak width for accurate quantitation.

Hsu et al.[7] studied methods for monitoring drug discovery compounds. They found that not all compounds have UV chromophores and that molar absorptivities can vary widely. They found that the ELSD responded better to the weight percent of each component and thus is better for absolute quantitation, especially if an internal standard is used. ELSD did not produce a solvent front that can interfere with quantitation of early eluting components in UV. ELSD only produced signals for compounds that did not solidify in the drift tube. They found that UV was better for generic relative purity estimation and ELSD was better for absolute purity quantitation.

For cases needing absolute purity and yield quantitation, Taylor et al.[8] demonstrated that CLND is a useful tool for universal quantitation. A very consistent molar response for nitrogen-containing compounds, the majority of drug-like compounds, was observed. CLND response was independent of the mobile phase. An internal standard is recommended for accurate quantitation.

Yan et al.[9] found that many "invisible" impurities occur in combinatorial chemistry-derived library compounds, and these are not observed with common UV or MS detection schemes. Such compounds include trifluoroacetic acid, plastic extracts, inorganic salts, catalysts, silica, and resin washout.

Fang et al.[11] described the application of a high-capacity autosampler with eight parallel reversed-phase HPLC columns that are interfaced to eight UV detectors and a single TOF mass spectrometer using a multiplexed ion source. A cycle time of 3.5 minutes produced analysis throughputs of 3,200 samples per day.

Gallagher et al.[10] applied a combined ESI/APCI ion source and alternated both ion polarity $(+/-)$ and ionization methods within a single HPLC run. They found that ESI provided useful spectra for about 80% of discovery compounds. APCI was useful for an additional 10% of the compounds. Combination of all these modes increased throughput at a low cost.

Ventura et al.[6] applied supercritical fluid chromatography with MS and reduced the analytical cycle time to one third that of LC/MS. SFC also exhibited enhanced resolution compared to HPLC. By using an APCI interface with SFC, throughputs of 400 samples per 24 hours were obtained.

Kerns et al.[13] combined integrity and purity analysis with Log D estimation in the same run. Log D was estimated from the same HPLC retention time data (see Section 23.2.2) that was used for integrity and purity profiling. Thus, multiple data types were obtained from one analysis.

Problems

(Answers can be found in Appendix I at the end of the book.)

1. Why is it useful to use a short (1–5 minutes) HPLC analysis as part of an integrity and purity assay?

2. Why is NMR not used for higher-throughput integrity and purity assays?

3. Why is it necessary to completely dissolve the sample for integrity and purity analysis?

4. How can newer HPLC columns with small particles enhance the assay?

5. What is a drawback of HPLC detectors for purity analysis?

6. What is a drawback of mass spectrometer interfaces for identity analysis?

References

1. Kerns, E. H. (2001). High throughput physicochemical profiling for drug discovery. *Journal of Pharmaceutical Sciences*, *90*, 1838–1858.

2. Di, L., & Kerns, E. H. (2003). Profiling drug-like properties in discovery research. *Current Opinion in Chemical Biology*, *7*, 402–408.

3. Kerns, E. H., & Di, L. (2003). Pharmaceutical profiling in drug discovery. *Drug Discovery Today*, *8*, 316–323.

4. Lipinksi, C. A. (2003). Solubility in water and DMSO: issues and potential solutions. In R. T. Borchardt, E. H. Kerns, C. A. Lipinski, D. R. Thakker, & B. Wang (Eds.), *Pharmaceutical profiling in drug discovery for lead selection*. Arlington, VA: AAPS Press, pp. 93–125.

5. Romanyshyn, L., Tiller, P. R., & Hop, C. E.C.A. (2000). Bioanalytical applications of "fast chromatography" to high-throughput liquid chromatography/tandem mass spectrometric quantitation. *Rapid Communications in Mass Spectrometry*, *14*, 1662–1668.

6. Ventura, M. C., Farrell, W. P., Aurigemma, C. M., & Greig, M. J. (1999). Packed column supercritical fluid chromatography/mass spectrometry for high-throughput analysis. *Analytical Chemistry*, *71*, 2410–2416.

7. Hsu, B. H., Orton, E., Tang, S.-Y., & Carlton, R. A. (1999). Application of evaporative light scattering detection to the characterization of combinatorial and parallel synthesis libraries for pharmaceutical drug discovery. *Journal of Chromatography, B: Biomedical Sciences and Applications*, *725*, 103–112.

8. Taylor, E. W., Qian, M. G., & Dollinger, G. D. (1998). Simultaneous online characterization of small organic molecules derived from combinatorial libraries for identity, quantity, and purity by reversed-phase HPLC with chemiluminescent nitrogen, UV, and mass spectrometric detection. *Analytical Chemistry*, *70*, 3339–3347.

9. Yan, B., Fang, L., Irving, M., Zhang, S., Boldi, A. M., Woolard, F., et al. (2003). Quality control in combinatorial chemistry: determination of the quantity, purity, and quantitative purity of compounds in combinatorial libraries. *Journal of Combinatorial Chemistry*, *5*, 547–559.

10. Gallagher, R. T., Balogh, M. P., Davey, P., Jackson, M. R., Sinclair, I., & Southern, L. J. (2003). Combined electrospray ionization-atmospheric pressure chemical ionization source for use in high-throughput LC-MS applications. *Analytical Chemistry*, *75*, 973–977.

11. Fang, L., Cournoyer, J., Demee, M., Zhao, J., Tokushige, D., & Yan, B. (2002). High-throughput liquid chromatography ultraviolet/mass spectrometric analysis of combinatorial libraries using an eight-channel multiplexed electrospray time-of-flight mass spectrometer. *Rapid Communications in Mass Spectrometry*, *16*, 1440–1447.

12. Kyranos, J. N., Cai, H., Zhang, B., & Goetzinger, W. K. (2001). High-throughput techniques for compound characterization and purification. *Current Opinion in Drug Discovery & Development*, *4*, 719–728.

13. Kerns, E. H., Di, L., Petusky, S., Kleintop, T., Huryn, D., McConnell, O., et al. (2003). Pharmaceutical profiling method for lipophilicity and integrity using liquid chromatography-mass spectrometry. *Journal of Chromatography, B: Analytical Technologies in the Biomedical and Life Sciences*, *791*, 381–388.

Chapter 37

Pharmacokinetic Methods

Overview

▶ *In vivo pharmacokinetic (PK) parameter assessment is highly important for project teams.*

▶ *PK experiments have been accelerated using cassette dosing.*

▶ *PK quantitation ("bioanalysis") is accelerated by sample pooling and liquid chromatography/mass spectometry/mass spectrometry.*

▶ *Uptake into tissues can be very useful, as for brain or tumor targets.*

Considerable resources are dedicated to pharmacokinetics (PK) studies in drug discovery. Advancement criteria in drug discovery and development usually include guidelines for the major PK parameters. PK behavior represents a composite of the underlying physicochemical and biochemical properties of the compound in the dynamic living system. Individual PK parameters can be used as guides for diagnosing the properties of the compound (see Chapter 38) in order to improve performance by structural modification.

Discovery methods for animal PK studies utilize all of the techniques that are applied for human clinical PK study samples.[1] Good Laboratory Practice (GLP) and Good Manufacturing Practice (GMP) are not required for discovery PK studies.

37.1 PK Dosing

37.1.1 Single-Compound Dosing

Pharmacokinetic studies in drug discovery typically dose two to four animals with a test compound in a generic formulation (see Chapter 41). Doses vary, depending on the project, but a common dose level is 10 mg/kg oral (PO) and 1 mg/kg intravenous (IV). PO dosing typically is performed gavage administration of a compound solution or suspension directly into the stomach. Intraperitoneal (IP) dosing and subcutaneous (SC) dosing (see Chapter 41) also are used. Dosing one compound into an animal is sometimes called *discrete dosing*.

37.1.2 Cassette Dosing

The large number of compounds studied in drug discovery has stimulated interest in accelerated methods that increase the throughput of pharmacokinetic analysis. In one approach, called *cassette dosing* or *N-in-one*, several compounds (often 4-10) are mixed together in the same dosing solution ("cocktail") and coadministered.[2] Each compound is subjected to the barriers of the living system and exhibits concentrations in plasma and tissues concomitant with its properties. The concentration of each compound is independently measured in the plasma sample by means of unique liquid chromatography/mass spectometry/mass spectrometry (LC/MS/MS) signals (unique molecular weight and product ions). With this strategy, if five compounds are co-dosed and coanalyzed in one LC/MS/MS analysis, the throughput is increased by about five-fold. The major concern with this strategy is that compounds may interact, that is, they compete for the same proteins involved in membrane transport, metabolism, and elimination (see Chapter 15). Thus, their PK parameters can be changed by inhibition from another compound in the same dose. Interaction can be minimized by lowering the dose of each compound, which lowers the concentration of individual compounds at the proteins at which interaction may occur, thus reducing the risk of interaction. Another approach for reducing interaction is to include a compound of known PK parameters (previously measured) in the cassette mixture. If its PK parameters are changed in the cassette, the investigators know that at least one compound in the cassette causes drug–drug interactions. In the case of demonstrated interaction, the compounds in that set can be restudied individually to assure accurate data. In an industry-wide benchmarking study, the majority of the respondents (64%) believe that cassette dosing can be used to rank order compounds, provided proper controls are used, but <10% use cassette dosing as their primary PK screening strategy.[3] The technique remains controversial, and shortcomings have been discussed.[4]

Cassette dosing has several advantages. Data can be generated for more compounds using the same resources (scientist's time, instrumentation). Fewer animals are used for more compassionate use. Data can be provided to teams faster than if they had to wait for each discrete dosing study to be performed in sequence.

37.2 PK Sampling and Sample Preparation

Blood samples are collected manually at specific time points after dosing (0.03 to 24 hours), treated with anticoagulant, and centrifuged to remove the blood cells and produce plasma. Automated sampling instruments are commercially available to collect samples unattended at specified time points (Table 37.1). Such systems can collect blood, bile, urine, and feces samples from awake and freely moving animals.

TABLE 37.1 ► Commercial Suppliers of Products for PK Studies

Product name	Company	Web site
Culex	BASi	*www.culex.net*
DR-II Automated Sampling System	Protech International	*www.protechinternational.com*
AccuSampler	DiLab	*www.dilab.com*
WinNonLin	Pharsight	*www.pharsight.com*

Samples are prepared for instrumental analysis by addition of two or more volumes of organic solvent (e.g., acetonitrile) and agitation. This treatment precipitates much of the plasma protein material so that instrumental analysis interference is greatly reduced. This technique is often called *acetonitrile crash*. The solution is centrifuged and the supernatant analyzed. Alternatively, plasma samples are extracted using solid-phase extraction. A measured volume of plasma sample is applied to a porous solid phase cartridge and washed with aqueous phase. The cartridge is eluted with organic phase, and the eluent is instrumentally analyzed. Solid-phase extraction cartridges are commercially available in 96-well formats for automated processing.

Another analytical strategy to accelerate analysis is the cassette-accelerated rapid rat screen (CARRS) approach.[5] Two animals are dosed with a compound, and samples are collected from 0.5 to 6 hours. The samples from the two animals from the same time point are pooled into one mixture. This allows analysis of only a single sample per time point per compound. The method provides all of the PK parameters but requires only half the sample preparation and LC/MS/MS analyses of standard approaches. A disadvantage of the approach is that there is no information on the variability in PK parameters between the animals, which can vary greatly.

Samples from discrete dosing experiments are sometimes mixed for processing, commonly termed *pooling*. This allows for less sample preparation and LC/MS/MS analyses. It is necessary to pool only compounds of different molecular weight.

37.3 Instrumental Analysis

The adoption of LC/MS/MS quantitation for PK studies in the early 1990s greatly accelerated the speed and sensitivity of the analyses. A low-energy ion source (atmospheric pressure chemical ionization or electrospray) produces molecule ions for the analyte (e.g., MH^+, $(M-H)^-$). The first MS stage is set to selectively pass only the mass-to-charge ratio (m/z) value for the molecule ion. The second MS stage contains an inert gas (e.g., argon) with which the analyte molecules collide and become vibrationally excited. This energy is dissipated by fragmentation of the molecule ion into specific product (fragment) ions. The third MS stage is set to selectively pass only the m/z value of one of these specific product ions. Thus, the LC/MS/MS system provides three stages of separation (high-performance liquid chromatography [HPLC], MS for molecule ions, and MS for the product ion) for a highly specific technique that has a high signal-to-noise ratio (S/N).

When a cassette of compounds in the same sample is analyzed (from either cassette dosing [see Section 37.1.2] or pooling [see Section 37.2]), different molecule and product ions are used for specific analysis of each compound in the mixture. Compounds also usually are partially or fully separated by the HPLC system prior to entering the mass spectrometer. The instrument can be set to sequence between multiple parent–product ion pairs for selective quantitation of each compound.

Usually, rapid HPLC chromatography is used for PK. An analytical cycle time (injection to injection) of 2.5 to 3 minutes is common. Cycle times have been reduced to as low as 1.25 minutes.[6] New HPLC techniques, such as ultraperformance liquid chromatography, have accelerated the chromatographic separation further by use of small particle size, which allows reduction of cycle time to about 1.5 min.[7] Automation and software calculations assist in data handling and analysis.

Other instrumental techniques allow accelerated LC/MS/MS analysis. In the LC/MS/MS multiple HPLC interface method (called *MUX* by one vendor), the interface is set to sequence between multiple HPLC inputs (two or four), thus allowing simultaneous parallel HPLC separations and quantitation using just one mass spectrometer.[8] In the staggered input method, injections into multiple HPLCs are staggered in time and the effluent is switched to the MS just in time for the analyte peak to elute and be quantitated.[9]

Matrix suppression is one difficulty of LC/MS/MS analysis of plasma samples. Sample preparation often does not remove all of the plasma components. When some of these components (e.g., phospholipids) elute from the HPLC, they can suppress ionization of the analyte compound that co-elutes from the HPLC. This results in reduction and variability in the signal from the test compound. Various approaches have been developed to successfully deal with this problem.[10]

The concentration–time data are analyzed using software (e.g., WinNonLin). These rapidly fit the data to mathematical PK models and calculate the standard PK parameters.

37.4 Example Pharmacokinetic Data

An example of hypothetical pharmacokinetic data from an experiment is given in Table 37.2 and plotted in Figure 37.1. In this experiment, the compound was individually dosed PO at 10 mg/kg and IV at 1 mg/kg. Samples were collected over 24 hours, starting at 2 minutes. PK parameters were calculated using WinNonLin software. The values of C_0, C_{max}, and t_{max} are illustrated in Figure 37.2. The PK parameters and the equations used to calculate clearance (Cl), V_d, and %F are given in Table 37.3.

TABLE 37.2 ▶ Hypothetical Data from a Single-Compound PK Experiment

Time (h)	IV (ng/mL)	PO (ng/mL)
0.033	970	Not sampled
0.25	420	550
0.5	250	600
1	110	310
2	30	100
4	1.9	20
6	0	6.3
8	0	1.8
24	0	0

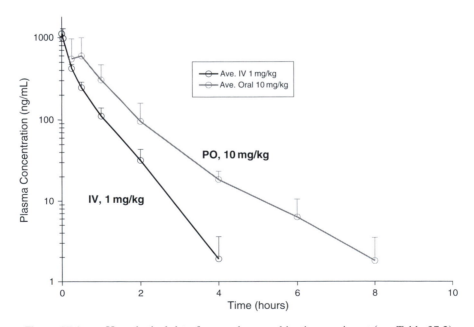

Figure 37.1 ▶ Hypothetical data from a pharmacokinetic experiment (see Table 37.2).

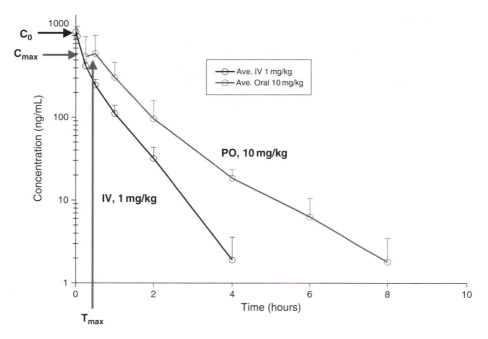

Figure 37.2 ▶ Determination of C_0, C_{max}, and t_{max} from PK data.

TABLE 37.3 ▶ PK Data Processed Using WinNonLin Software to Calculate C_0 and AUC, from Which Other PK Parameters are Calculated

	IV (1 mg/kg)	PO (10 mg/kg)
C_0 (ng/mL)	1000	
$t_{1/2}$ (h)	0.5	1
AUC (h*ng/mL)	420	840
t_{max} (h)		0.5
C_{max} (ng/mL)		600
CL (ml/min/kg)	40	
V_d (L/kg)	1	
% Bioavailability		20%

$CL = Dose/AUC = (1\,mg/kg)/(420\,h*ng/ml) * 10^6\,ng/mg * 1\,h/60\,min \sim 40\,mL/min/kg$

$V_d = Dose/C_0 = (1\,mg/kg)/(1000\,ng/mL) * 10^6\,ng/mg * 10^{-3}\,L/mL = 1L/kg$

$\%\ Bioavailability = AUC_{PO}/AUC_{IV} * Dose_{IV}/Dose_{PO} * 100\% = 840/420*1/10*100\% = 20\%$

37.5 Tissue Uptake

Tissue uptake is necessary in order for the compound to reach most therapeutic targets. Blood–organ barriers limit penetration into some tissues. For example, CNS drugs must penetrate into brain tissue through the blood–brain barrier (see Chapter 10). The penetration of cancer drugs into tumors may be hindered, compared to other tissues, by reduced blood flow and tumor morphology.

Performing these studies involves sampling of the tissue at selected time points. Blood should be purged from the tissue prior to analysis so that compound levels really derive from the tissue and not from the blood. Tissue samples are first homogenized and then analyzed in a manner similar to plasma. The data typically are analyzed by calculating the ratio of (a) tissue concentration to plasma concentration at a particular time point, (b) AUC_{tissue} to AUC_{plasma}, or (c) $C_{max,tissue}$ to $C_{max,plasma}$.

Problems

(Answers can be found in Appendix I at the end of the book.)

1. How does cassette dosing assist discovery PK studies?

2. What is the criticism of cassette dosing? How has this potential problem been addressed?

3. How does an automated sampling instrument help PK studies?

4. How does the CARRS approach differ from cassette dosing?

5. What is matrix suppression?

6. Why would a project team want tissue uptake data?

7. Calculate the PK parameters in the open cells below using the data provided:

cpd	Dose [IV, PO] (mg/kg)	AUC_{PO} (ng $\bullet$ h/mL)	C_0 (ng/mL)	AUC_{IV} (ng $\bullet$ h/mL)	Cl (mL/min/kg)	V_d (L/kg)	%F
1	1, 10	2,000	1,000	4,000			
2	1, 10	2,000	2,000	1,000			
3	5, 10	200	1,000	8,000			
4	1, 10	500	1,000	200			
5	5, 30	305	2,000	1,900			

References

1. Cox, K. A., White, R. E., & Korfmacher, W. A. (2002). Rapid determination of pharmacokinetic properties of new chemical entities: in vivo approaches. *Combinatorial Chemistry and High Throughput Screening, 5,* 29–37.

2. Olah, T. V., McLoughlin, D. A., & Gilbert, J. D. (1997). The simultaneous determination of mixtures of drug candidates by liquid chromatography/atmospheric pressure chemical ionization mass spectrometry as an in vivo drug screening procedure. *Rapid Communications in Mass Spectrometry, 11,* 17–23.

3. Ackermann, B. L. (2004). Results from a bench marking survey on cassette dosing practices in the pharmaceutical industry. *Journal of the American Society for Mass Spectrometry, 15,* 1374–1377.

4. White, R. E., & Manitpisitkul, P. (2001). Pharmacokinetic theory of cassette dosing in drug discovery screening. *Drug Metabolism and Disposition, 29,* 957–966.

5. Korfmacher, W. A., Cox, K. A., Ng, K. J., Veals, J., Hsieh, Y., Wainhaus, S., et al. (2001). Cassette-accelerated rapid rat screen: a systematic procedure for the dosing and liquid chromatography/atmospheric pressure ionization tandem mass spectrometric analysis of new chemical entities as part of new drug discovery. *Rapid Communications in Mass Spectrometry, 15,* 335–340.

6. Dunn-Meynell, K. W., Wainhaus, S., & Korfmacher, W. A. (2005). Optimizing an ultrafast generic high-performance liquid chromatography/tandem mass spectrometry method for faster discovery pharmacokinetic sample throughput. *Rapid Communications in Mass Spectrometry, 19,* 2905–2910.

7. Wainhaus, S., Nardo, C., Anstatt, R., Wang, S., Dunn-Meynell, K. W., & Korfmacher, W. (2007). Ultra fast liquid chromatography-MS/MS for pharmacokinetic profiling within drug discovery. *American Drug Discovery*, January 6–12.

8. Deng, Y., Wu, J.-T., Lloyd, T. L., Chi, C. L., Olah, T. V., & Unger, S. E. (2002). High-speed gradient parallel liquid chromatography/tandem mass spectrometry with fully automated sample preparation for bioanalysis: 30 seconds per sample from plasma. *Rapid Communications in Mass Spectrometry*, *16*, 1116–1123.

9. King, R. C., Miller-Stein, C., Magiera, D. J., & Brann, J. (2001). Description and validation of a staggered parallel high performance liquid chromatography system for good laboratory-practice-level quantitative analysis by liquid chromatography/tandem mass spectrometry. *Rapid Communications in Mass Spectrometry*, *16*, 43–52.

10. Bonfiglio, R., King, R. C., Olah, T. V., & Merkle, K. (1999). The effects of sample preparation methods on the variability of the electrospray ionization response for model drug compounds. *Rapid Communications in Mass Spectrometry*, *13*, 1175–1185.

Part 5
Specific Topics

Chapter 38

Diagnosing and Improving Pharmacokinetic Performance

Overview

▶ *When pharmacokinetic (PK) parameters are poor, in vitro property data can be used to diagnose the limitations.*

▶ *High clearance results from metabolism (phases I and II), biliary excretion, transporters, renal extraction, and plasma hydrolysis.*

▶ *Low bioavailability results from first-pass metabolism, low solubility, low permeability, high efflux, and intestinal decomposition.*

▶ *The identified property limitation guides structure modifications, and the new analog can be checked in vitro for improvement.*

In vivo animal studies are performed during discovery for compounds having a wide range of pharmacokinetic (PK) properties, including those having PK parameters that differ greatly from the majority of commercial drugs (i.e., not drug-like). To make up for these limitations, doses can be increased or administered more frequently, less preferable dosing routes can be used (e.g., IV), or an unusual dosage form can be administered. This may be necessary for early pharmacology proof-of-concept studies. However, there are inevitable tradeoffs if such compounds move into development. Non–drug-like compounds may require IV dosing instead of the preferred PO route, complicated or expensive formulations for insoluble compounds, sustained release of rapidly cleared compounds, or prodrugs for insoluble or impermeable compounds. Often a discovery compound is found to not achieve sufficient PK performance in vivo to produce efficacy or meet PK advancement criteria.

In this common scenario, it is helpful to use PK parameters as a guide to diagnosing the underlying physicochemical, biochemical, and structural property limitations. This provides insights for informed decisions on specific structural modifications that can be made to improve the limiting property. In vitro assays are very helpful in determining if the modified structure has been improved for the limiting property. The improved compounds then can be tested in vivo to determine if the PK parameters have been improved.[1,2] A scheme for this strategy is shown in Figure 38.1.

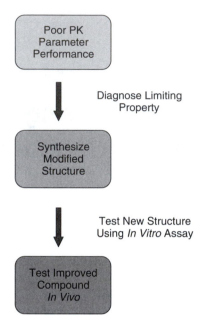

Figure 38.1 ▶ Scheme for diagnosis and improvement of a property that limits PK performance.

38.1 Diagnosing Underlying Property Limitations from PK Performance

The following sections list common PK issues that discovery project teams try to improve. The major possible causes of each of these limitations are provided as bullet points, and the in vitro assay/study that can be used to check for this limitation as a cause is indicated in brackets. The limiting causes are listed in terms of suggested priority of investigation in tracking down the cause. Strategies for modifying the structure to improve the property are provided in chapters on the individual properties.

38.1.1 High Clearance After IV Injection

▶ Liver metabolism (hepatic extraction)

  ▶ Phase I metabolic stability [microsomal + NADPH stability; use PK species]

  ▶ Phase II metabolic stability [microsomal + uridine diphosphate glucuronic acid stability; hepatocyte stability; use PK species]

▶ Liver biliary excretion (hepatic extraction)

  ▶ Permeation into bile [drug and metabolite concentration in bile from bile duct cannulated PK animal]

  ▶ Active transport into bile [P-glycoprotein ((Pgp) and other transporters in PK species]

▶ Renal extraction [drug and metabolite concentration in urine of PK animal; transporters involved in active secretion]

▶ Enzymatic hydrolysis in blood [plasma stability]

38.1.2 Low Oral Bioavailability

▶ High first-pass metabolism by liver and intestine; biliary extraction [phase I and II metabolic stability]

▶ Low intestinal solubility [solubility in simulated gastric fluid (SGF), simulated intestinal fluid (SIF), simulated intestinal bile salts–lecithin mixture (SIBLM), and pH 1–8 buffers;]

▶ Low intestinal permeability

    ▶ Low passive diffusion [parallel artificial membrane permeability assay (PAMPA); Caco-2 A>B and B>A; MDCK;]

    ▶ High Pgp efflux [MDR1-MDCKII Pgp assay; Caco-2 efflux ratio]

▶ Enzymatic or pH hydrolysis in intestine [stability in SGF, SIF, pH 1–8]

38.2 Case Studies on Interpreting Unusual PK Performance

Both transporter-mediated absorption and capacity-limited metabolism can lead to nonlinear PK profiles as illustrated by the following two case studies.

38.2.1 PK of CCR5 Antagonist UK-427,857

The compound shown in Figure 38.2 was found to have much higher C_{max} and area under the curve (AUC) (Table 38.1) when dosed in humans at 4.3 mg/kg than when dosed at 0.43 mg/kg after dose normalization.[3] Such behavior is indicative of saturation of an efflux transporter. With this diagnosis, Pgp efflux was studied using in vitro Caco-2 assay. P_{app} (A>B) was measured as $<1 \times 10^{-6}$ cm/s and P_{app} (B>A) was 12×10^{-6} cm/s, which was an efflux ratio ($P_{app, B>A}/P_{app, A>B}$) >10, indicating efflux. In follow-up studies, the Pgp inhibitor verapamil was found to reduce the efflux ratio, suggesting the compound was a Pgp substrate. The Pgp binding affinity was $K_m = 37\,\mu M$, and $V_{max} = 55$ nmol/mg/min. These findings were confirmed by in vivo studies using Pgp double knockout mice and wild type. Both C_{max} and AUC increased significantly in knockout mice compared to wild type (Table 38.2). This suggested the higher C_{max} and AUC at higher doses was caused by Pgp efflux in the intestine, which limited absorption at the lower dose, but was saturated at the higher dose, thus allowing higher absorption at the higher dose.

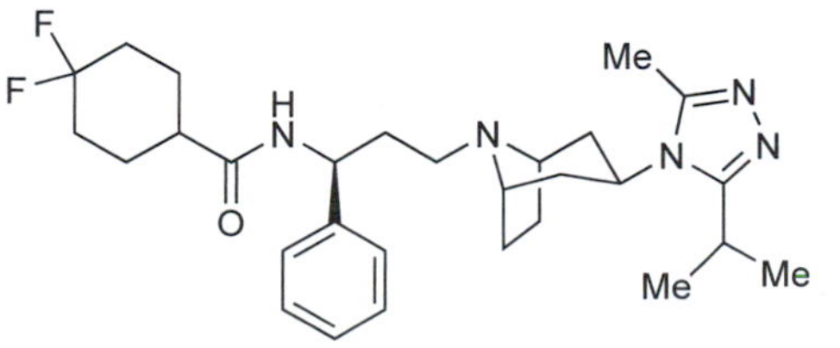

Log $D_{7.4} = 2.1$ H-bond Donor = 1
pKa = 7.3   H-Bond Acceptor = 6
MW = 514   cLogP = 3.11
Good solubility   Low permeability

Figure 38.2 ▶ Structure and physicochemical properties of CCR5 antagonist UK-427,857.

TABLE 38.1 ▶ **PK Parameters of CCR5 Antagonist UK-427, 857**

Parameter	Human		Comments
Oral Dose (mg/kg)	0.43 (30 mg)	4.3 (300 mg)	
Elimination half-life (h)	8.9	10.6	
C_{max} (ng/mL), dose normalized	36	144	increased
AUC (ng.h/mL), dose normalized	272	537	increased
T_{max} (h)	2.9	1.6	decreased

TABLE 38.2 ▶ **PK Parameters of CCR5 Antagonist UK-427, 857 in Pgp Knockout Mice**

PO 16 mg/kg	C_{max} ng/mL	AUC ng.h/mL	Elimin. $T_{1/2}$ (h)
wild-type fvb mice	536	440	0.7
mdr1a/1b knockout	1119	1247	1
% Increase	108%	183%	

38.2.2 PK of Triazole Antifungal Voriconazole

Voriconazole (Figure 38.3) has good solubility and excellent oral absorption, with <7% eliminated unchanged through feces. It is mostly eliminated by hepatic clearance. Oral bioavailability of the compound was greater than 70% in human. Voriconazole produced an unusual nonlinear PK profile (Figure 38.4) following PO or IV administration in rat, termed the *hockey-stick profile*.[4] The PK characteristics are gender dependent. The analog compound shown in Figure 38.5 does not have the nonlinear PK characteristics because of low Log D (0.5), which results in elimination mostly by the kidney. Table 38.3 lists the gender-dependent PK parameters of voriconazole. PO AUC (normalized) at 30 mg/kg was higher than IV AUC at 10 mpk, resulting in >100% oral bioavailability (%F = 159%) in male rat. This suggested capacity-limited elimination due to saturation of metabolizing enzymes, which is facilitated by the good absorption resulting in high exposure in systemic circulation. For both IV and PO administration, AUC for multiple dosing was lower than for single dosing. This is because voriconazole induces metabolizing cytochrome P450 (CYP450)

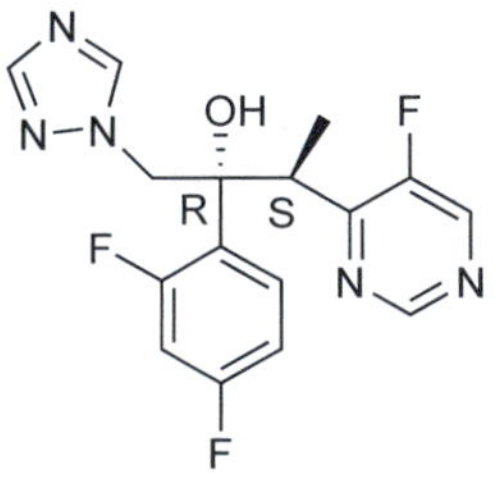

Aq. Solubility = 0.7 mg/mL, Log $D_{7.4}$ = 1.8
Excellent absorption, <7% in feces unchanged
Oral Bioavailability >70%

Figure 38.3 ▶ Structure and physicochemical properties of triazole antifungal voriconazole.

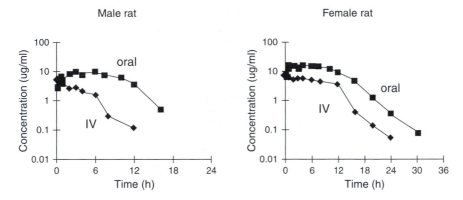

Figure 38.4 ▶ Nonlinear PK of voriconazole in rat. (Reprinted with permission from [14].)

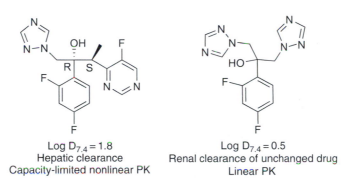

Log $D_{7.4}$ = 1.8
Hepatic clearance
Capacity-limited nonlinear PK

Log $D_{7.4}$ = 0.5
Renal clearance of unchanged drug
Linear PK

Figure 38.5 ▶ Effects of Log D on clearance and PK.

TABLE 38.3 ▶ Voriconazole: PK Data Interpretation in Rat

Sex	Male	Female	Comments
Plasma protein binding (%)	66	66	
IV			
Dose (mg/kg)	10	10	
Single dose AUC_t (ug .h/mL)	18.6	81.6	Gender dependent
Multiple dose AUC_t (ug .h/mL)	6.7	13.9	< S.D. CYP450 auto-induction
Oral			
Dose (mg/kg)	30	30	
Single dose C_{max} (ug/mL)	9.5	16.7	Gender dependent
Single dose T_{max} (h)	6	1	Gender dependent
Single dose AUC_t (ug .h/mL)	90	215.6	> IV, capacity-limited elimination
Multiple dose AUC_t (ug .h/mL)	32.3	57.4	< S.D. CYP450 auto-induction
Apparent bioavailability F (%)	159	88	Capacity-limited elimin. Good absorption

S.D., Single dose.

enzymes as indicated by the increase in liver weight and CYP450 enzymes with escalated doses (Table 38.4). As animals were exposed to voriconazole, more CYP450 enzymes were produced to metabolize the compound, resulting in quicker elimination. Hence, multiple dosing generated lower AUC than did single dosing.

TABLE 38.4 ▶ Voriconazole: Auto-Induction of CYP450s.

Dose (mg/kg)	Hepatic Microsomal Cytochrome P450 (nmol P450/mg protein)		Relative Liver weight		Voriconazole C_{max} (ug/mL)	
	Male	Female	Male	Female	Male	Female
Control	0.88	0.51	3.71	3.7	None	None
3	0.85	0.65	3.86	4.04	0.61	1.32
10	1.21	0.68	4.17	4.26	3.64	6.14
30	1.77	0.79	4.38	5.04	9.69	14.6
80	2.08	0.92	5.57	6.26	28.4	30.4

Problems

(Answers can be found in Appendix I at the end of the book.)

1. What dosing approaches can be tried to administer compounds that have poor absorption, short PK half-life, or low bioavailability after oral dosing?

2. What approach is preferable for enhancing absorption, PK half-life or bioavailability?

3. What physicochemical or biochemical properties, if they are low, will lead to poor bioavailability?

4. Which of the following properties can be a significant contributor to an observed high clearance in a PK study using an IV dose?: (a) low metabolic stability (liver), (b) low CYP inhibition, (c) high biliary excretion, (d) high plasma protein binding, (e) low renal extraction, (f) low plasma stability, (g) high RBC binding, (h) neutral pK_a, (i) hERG binding, (j) low stability at pH 4, (k) high phase I metabolism in intestinal epithelium, (l) high blood–brain barrier (BBB) brain to plasma ratio (B/P), (m) high renal extraction, (n) high metabolic stability (liver), (o) Pgp efflux.

5. Which of the following properties can be a significant contributor to an observed low oral bioavailability in a PK study using a PO dose?: (a) low metabolic stability (liver), (b) low CYP inhibition, (c) high biliary excretion, (d) high plasma protein binding, (e) low renal extraction, (f) low plasma stability, (g) high RBC binding, (h) neutral pK_a, (i) hERG binding, (j) low stability at pH 4, (k) high phase I metabolism in intestinal epithelium, (l) high BBB B/P, (m) high renal extraction, (n) high metabolic stability (liver), (o) low permeability, (p) low solubility, (q) Pgp efflux.

6. What effect may be observed if a compound is highly effluxed by Pgp in the intestine?: (a) high clearance, (b) oral dose-dependent C_{max}, (c) low V_d, (d) higher AUC at higher oral dose, (e) high AUC.

References

1. Gan, L.-S.L., & Thakker, D. R. (1997). Applications of the Caco-2 model in the design and development of orally active drugs: elucidation of biochemical and physical barriers posed by the intestinal epithelium. *Advanced Drug Delivery Reviews, 23,* 77–98.

2. Di, L., & Kerns, E. H. (2005). Application of pharmaceutical profiling assays for optimization of drug-like properties. *Current Opinion in Drug Discovery & Development, 8,* 495–504.

3. Walker, D. K., Abel, S., Comby, P., Muirhead, G. J., Nedderman, A. N. R., & Smith, D. A. (2005). Species differences in the disposition of the CCR5 antagonist, UK-427,857, a new potential treatment for HIV. *Drug Metabolism and Disposition*, *33*, 587–595.

4. Roffey, S. J., Cole, S., Comby, P., Gibson, D., Jezequel, S. G., Nedderman, A. N. R., et al. (2003). The disposition of voriconazole in mouse, rat, rabbit, guinea pig, dog and human. *Drug Metabolism and Disposition*, *31*, 731–741.

Chapter 39

Prodrugs

Overview

► *Prodrugs have a structure that improves solubility, permeability, stability, or targeting to a tissue in order to improve pharmacokinetics.*

► *The pro-moiety is cleaved in vivo to release the active structure.*

► *Prodrugs can improve properties when no other structural modification is sufficient.*

► *The prodrug strategy is successful only a portion of the times it is used.*

Of all drugs worldwide, 5% are prodrugs. About half of the prodrugs are activated by hydrolysis, suggesting most of them are esters. Twenty-three percent of prodrugs are activated by biotransformation, meaning there is no pro-moiety. Prodrugs typically are developed to overcome pharmaceutical, pharmacokinetic (PK), and pharmacodynamic (PD) barriers. Many successful prodrugs are discovered by accident rather than by design.

There are many benefits to using the prodrug approach. It can be applied to improve solubility and passive permeability for absorption improvement. Prodrugs can be prepared to enhance transporter-mediated absorption and to improve metabolic stability. Certain prodrugs have been developed to reduce side effects. For example, most nonsteroidal antiinflammatory drugs (NSAIDs) are carboxylic acids and can cause GI irritation. The ester prodrugs of NSAIDs can overcome this side effect. Table 39.1 shows drug development barriers that have been overcome by various prodrug strategies.[1] Figure 39.1 shows examples of blockbuster prodrugs and their indications.[2] Prodrugs can be quite successful.

The prodrug approach has many challenges. Development programs for prodrugs are complex. Prodrugs tend to show interspecies and intraspecies variability due to differences in enzyme activity for prodrug activation. Enzymes that activate prodrugs might have genetic polymorphisms, which can cause variability from subject to subject. If two drugs are competing for the same enzyme, drug–drug interaction can result. Certain pro-moieties can have toxicity. Prodrug strategies are generally considered as the last resort to achieve pharmaceutical/PK properties that are incompatible with a given pharmacophore.

TABLE 39.1 ▶ Drug Development Barriers and Issues that Can be Overcome by Prodrug Strategies[1]

Barriers	Issues
Permeability	Not absorbed from GI tract because of polarity
	Low brain permeation
	Poor skin penetration
Solubility	Poor absorption and low oral bioavailability
	IV formulation cannot be developed
Metabolism	Vulnerable drug metabolized at absorption site
	Half-life is too short
	Sustained release is desired
Stability	Chemically unstable
	Better shelf life is needed
Transporter	Lack of specificity
	Selective delivery is desired
Safety	Intolerance/irritation
Pharmaceutics	Poor patient/doctor/nurse acceptance
	Bad taste or odor problems
	Painful injection
	Incompatibility (tablet desired but liquid is active)

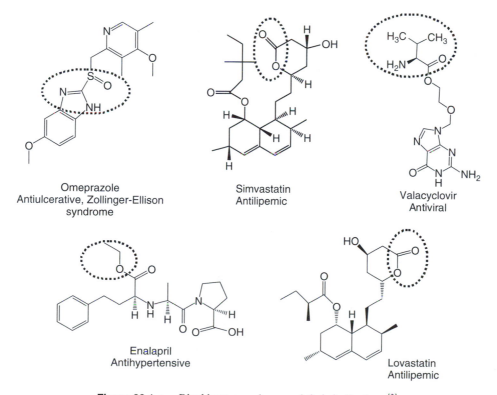

Omeprazole
Antiulcerative, Zollinger-Ellison syndrome

Simvastatin
Antilipemic

Valacyclovir
Antiviral

Enalapril
Antihypertensive

Lovastatin
Antilipemic

Figure 39.1 ▶ Blockbuster prodrugs and their indications.[2]

39.1 Using Prodrugs to Improve Solubility

Prodrug strategies can be applied to improve solubility. Table 39.2 shows commercially available prodrugs with improved solubility.[3] The structures of some prodrugs are illustrated in Figure 39.2. Prodrugs with a non-ionizable pro-moiety (e.g., glycol, polyethylene glycol [PEG], sugars) typically can improve solubility by two- to three-fold. Prodrugs with ionizable pro-moieties (e.g., phosphate) can increase solubility by orders of magnitude. There are three types of ionizable pro-moieties. Succinate-like derivatives were used early as prodrugs but

TABLE 39.2 ▶ Commercially Available Prodrugs with Improved Solubility[3]

Name	Solubility in water (mg/mL)
Clindamycin	0.2
Clindamycin-2-PO_4	150
Chloramphenicol	2.5
Succinate sodium	500
Metronidazole	10
N,N-dimethylglycinate	200
Phenytoin	0.02
Phosphate	142
Paclitaxel I	0.025
PEG-paclitaxel I	666
Celexicoxib	0.05
Parecoxib sodium	15

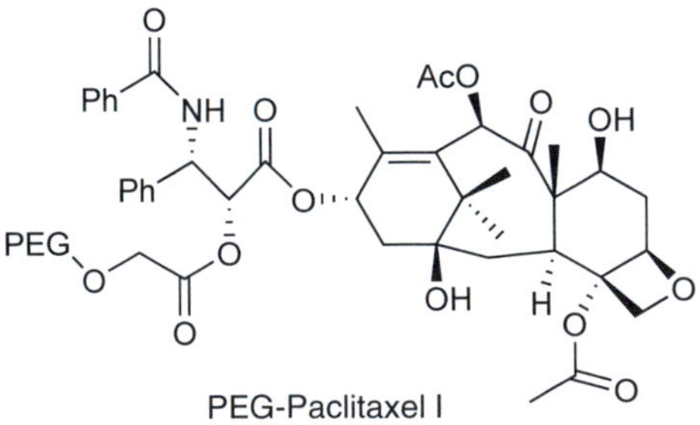

PEG-Paclitaxel I

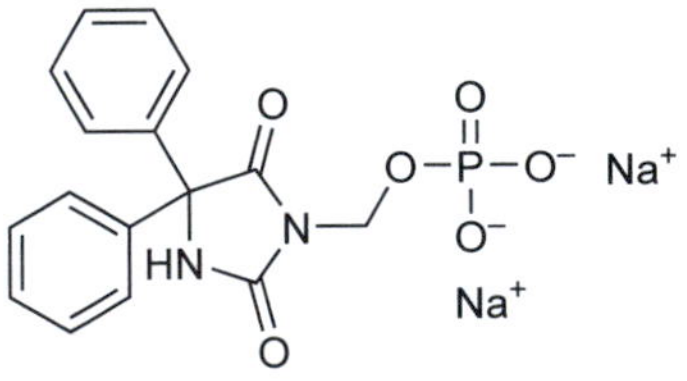

Fosphenytoin

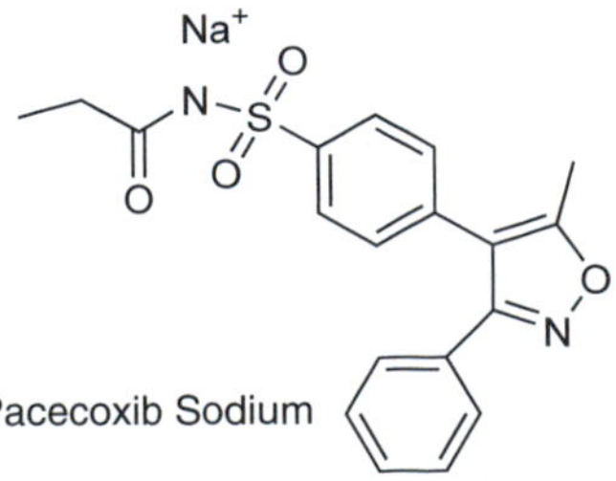

Pacecoxib Sodium

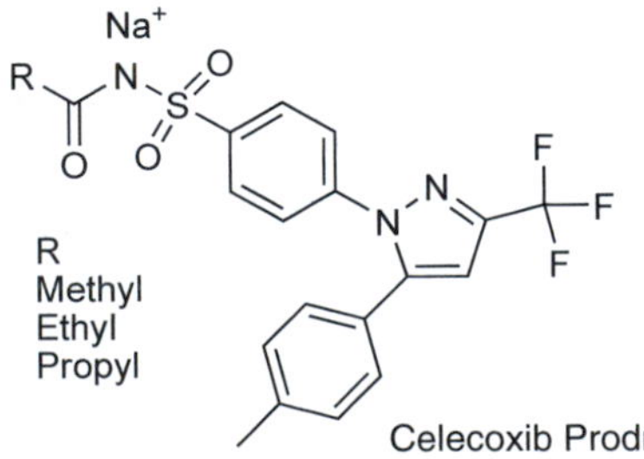

Celecoxib Prodrugs

Figure 39.2 ▶ Structures of prodrugs with improved solubility.

are chemically unstable. Amino acid type, attached to hydroxyls (e.g., glucocorticoids) and phosphate type, attached to hydroxyls or amines (e.g., fosphenytoin) are common approaches used in the industry to increase solubility (Figure 39.3).

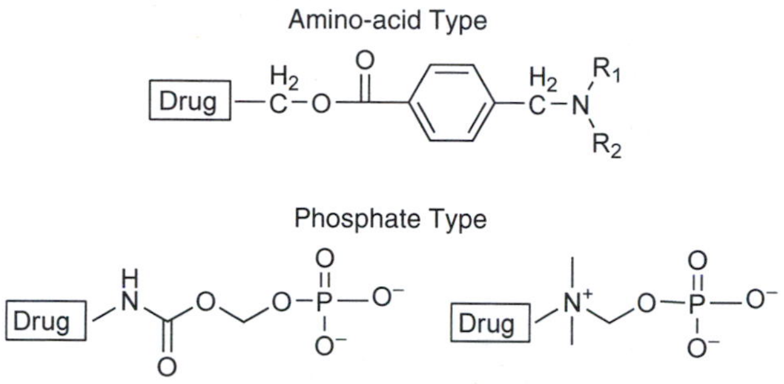

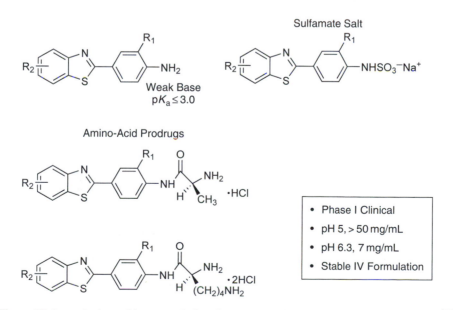

Figure 39.3 ► Amino acid–type and phosphate-type prodrugs used to increase solubility.

The antitumor agent shown in Figure 39.4 is a weak base with p$K_a \leq 3.0$. The low solubility and weak basicity of the compound limited options for parenteral formulations. The novel sulfamate salt prodrug was prepared. Although the prodrug was more soluble, it was unstable and converted back to the parent under acidic conditions. Subsequently, amino acid–type prodrugs were synthesized with good solubility and stability. The IV formulation of the dihydrochloride salt was used in phase I clinical trials.

Figure 39.4 ► Amino acid–type and phosphate-type prodrugs used to increase solubility.[15]

The phosphate-type prodrug gained a lot of popularity when fosphenytoin (see Figure 7.25) was marketed.[4] The mechanism of this phosphate prodrug is illustrated in Figure 39.5. Its activation enhances absorption.[5] Phosphate prodrugs of amines are made to increase solubility, due to the presence of the highly ionized species in the GI tract. The prodrug is hydrolyzed in the GI lumen by alkaline phosphatase, yielding hydroxymethyl amine intermediate and inorganic phosphate. The intermediate is highly unstable in physiological fluids and breaks down spontaneously to give the parent amine and one equivalent of formaldehyde. The parent amine can cross the GI membrane and be absorbed into systematic circulation. Figure 39.6 shows two examples of using phosphate prodrugs as potential approaches to enhance aqueous solubility of loxapine and cinnarizine. Both drugs (non-prodrugs) have low solubility with problematic formulation and erratic oral bioavailability.[5] One limitation of this approach is the formation of one equivalent of formaldehyde, which can have a toxic effect at a high dose or in chronic applications.

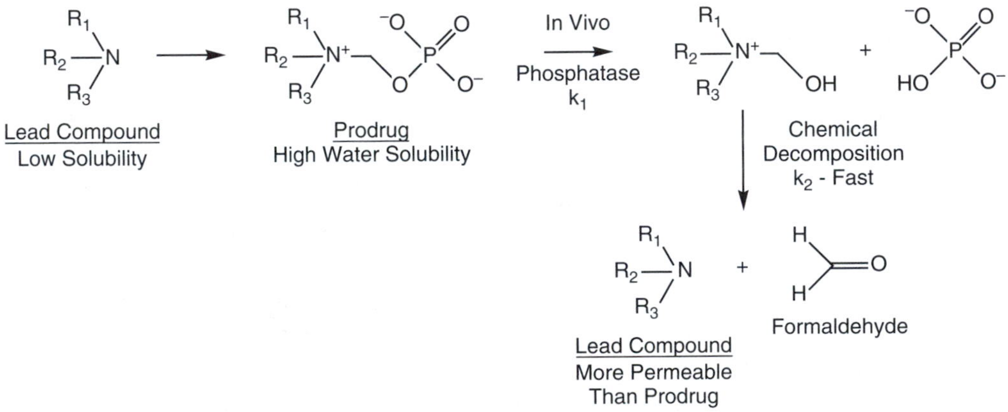

Figure 39.5 ▶ Mechanisms of phosphate prodrug used to increase aqueous solubility.[5]

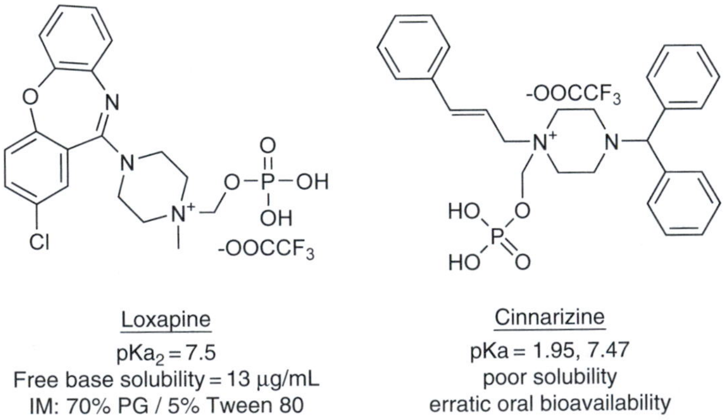

Figure 39.6 ▶ Examples of using phosphate prodrug approach to increase solubility. Properties of the active (non-prodrug) are listed.[5]

39.2 Prodrugs to Increase Passive Permeability

Prodrug strategies are most commonly used to increase permeability of compounds by masking the polar functional groups and hydrogen bonds with ester or amide linkers and increasing lipophilicity. Both permeability by passive diffusion and the transporter-mediated process (see Section 39.3) have been addressed with prodrug approaches.

Oral delivery of ester/amide prodrugs to the therapeutic target is confronted with many physiological, chemical, and biochemical barriers. In general, the highest oral bioavailability values that ester prodrugs can achieve clinically are 40% to 60%. This is due to incomplete membrane permeation, P-glycoprotein efflux, hydrolysis in the GI lumen and intestinal cells, nonesterase metabolism in the liver, biliary excretion, and metabolism of the parent.[6] Thus, a successful prodrug approach must consider the balance of all these issues. An ideal ester/amide prodrug should exhibit the following properties[6]:

▶ Weak or no activity against any pharmacological target

▶ Good chemical stability at physiological pHs

▶ Sufficient aqueous solubility

▶ High passive permeability

▶ Resistance to hydrolysis during absorption

▶ Hydrolyzed to parent rapidly and quantitatively after absorption

▶ The released pro-moiety has no toxicity or unwanted pharmacological effects

39.2.1 Ester Prodrugs for Carboxylic Acids

Simple alkyl esters are preferred for carboxylic acid prodrugs to increase passive diffusion permeability. Ethyl ester is the most common prodrug of this type. Other pro-moieties include aryl, double esters with diols, cyclic carbonates, and lactones. Examples of different types of prodrugs are shown in Figure 39.7.[2,6] Although simple esters are preferred, bioconversion of some simple alkyl or aryl esters is not mediated by esterases. This is nonideal for a prodrug approach because metabolism is nonproductive and leads to low systemic exposure.

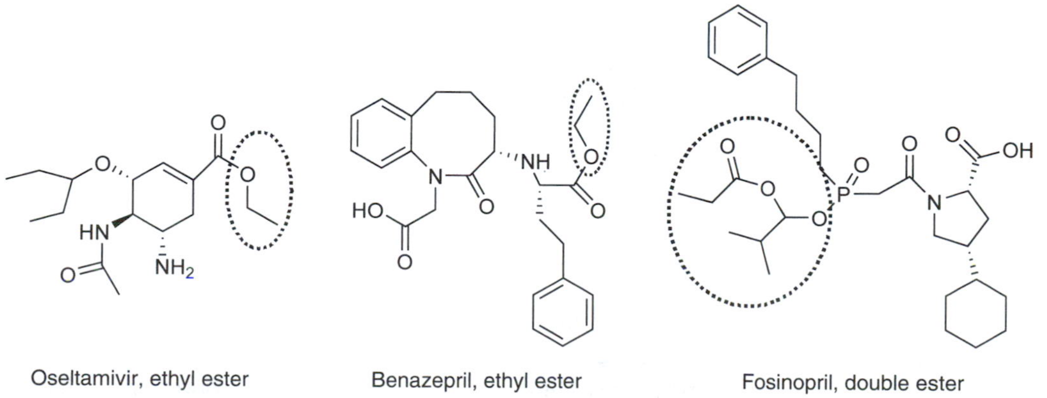

Oseltamivir, ethyl ester   Benazepril, ethyl ester   Fosinopril, double ester

Figure 39.7 ▶ Examples of ester prodrugs of acids used to enhance passive permeability. Pro-moieties are circled.

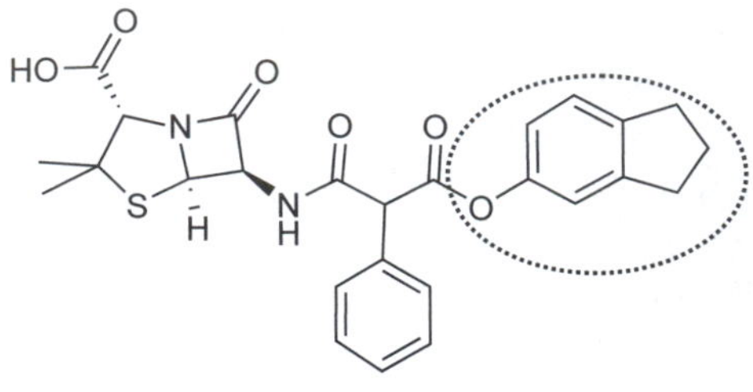

Carbinicillin indanyl ester, aryl ester

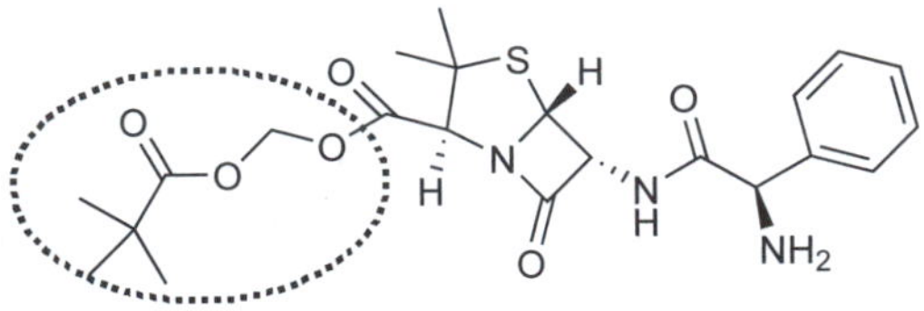

Pivampicillin, double ester

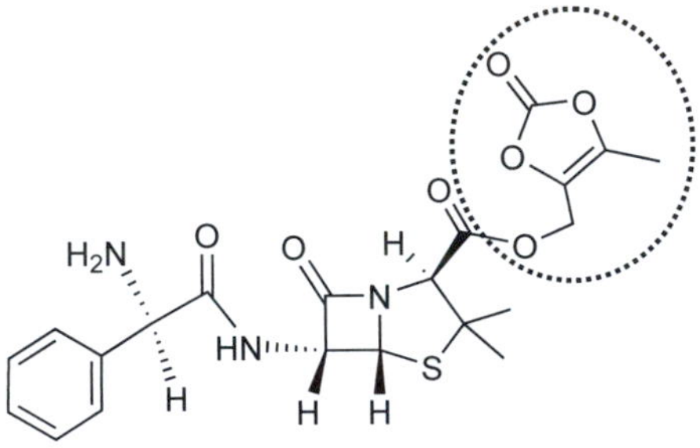

Lenampicillin, cyclic carbonate

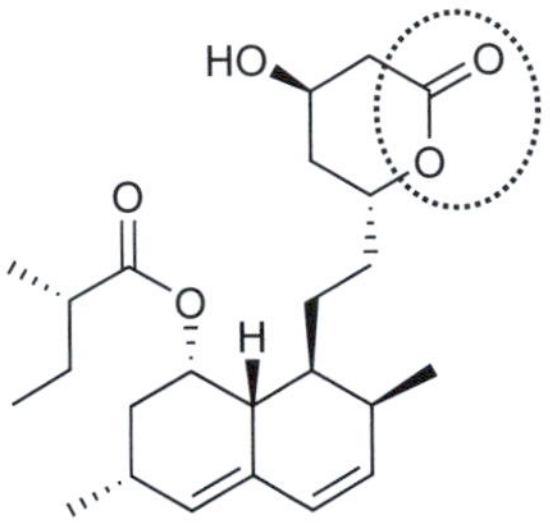

Lovastatin, lactone

Figure 39.7 ▶ *Continued.*

Double esters are prepared in order to increase the recognition by esterases through the second ester. However, chemical stability of double esters is low, and the liberated aldehyde fragment can have toxicity. Cyclic carbonate prodrugs (e.g., lenampicillin) are designed to be labile in plasma to avoid nonproductive metabolism by cellular esterases. Prodrugs that hydrolyze in blood or plasma by blood-borne enzymes are beneficial to increase oral bioavailability and systemic circulation of the active principle. Double esters and cyclic carbonate prodrugs are designed for this purpose. Lactone prodrugs are developed for specific targeting. Bioconversion of lovastatin lactone (a hydroxymethylglutaryl coenzyme A [HMG-CoA] reductase inhibitor) to the active acid occurs in the liver, which is the site of action. Although the oral bioavailability of the compound is only 30% due to first-pass liver extraction, the high local concentration at the target organ (liver) results in good efficacy.[6]

39.2.2 Ester Prodrugs for Alcohols and Phenols

An increase in the lipophilicity of alcohols and phenols can often be achieved by preparing ester prodrugs using carboxylic acids. Examples are shown in Table 39.3.[1] The prodrugs showed increased corneal permeability, brain penetration, and oral absorption. Enhanced oral bioavailability of timolol prodrug was due to increased permeability and solubility as a result of decreased crystal lattice energy compared to the parent.

TABLE 39.3 ► Examples of Ester Prodrugs for Alcohols and Phenols

Prodrugs	Limitations of parent	Benefits of prodrug
Dipivaloyl-epinephrine	Log P = −0.04 Low corneal penetration	Log P = 2.08 Four- to six-fold increase in corneal penetration
Dibenzoyl-Amino-Dihydroxy-tetrahydronaphthalene (ADTN)	No CNS penetration	Reaches CNS
Butyryl-Timolol	Low oral exposure	High oral exposure Enable IV formulation

39.2.3 Prodrugs Derived from Nitrogen-Containing Functional Group

Because of the slow hydrolysis rate of amides in vivo, prodrugs using amide approaches are generally not recommended, except for activated amides, such as *N*-benzoyl- or *N*-pivaloyl derivatives. Imines and enamines, stabilized through hydrogen bonds and small peptide derivatives, can be effective prodrugs for amines. Carbamates can be used as prodrugs for amidines. For compounds containing acidic NH functional groups, sulfonamides, carboxamides, and carbamates are effective prodrugs. Figure 39.8 shows examples of this type of prodrug.

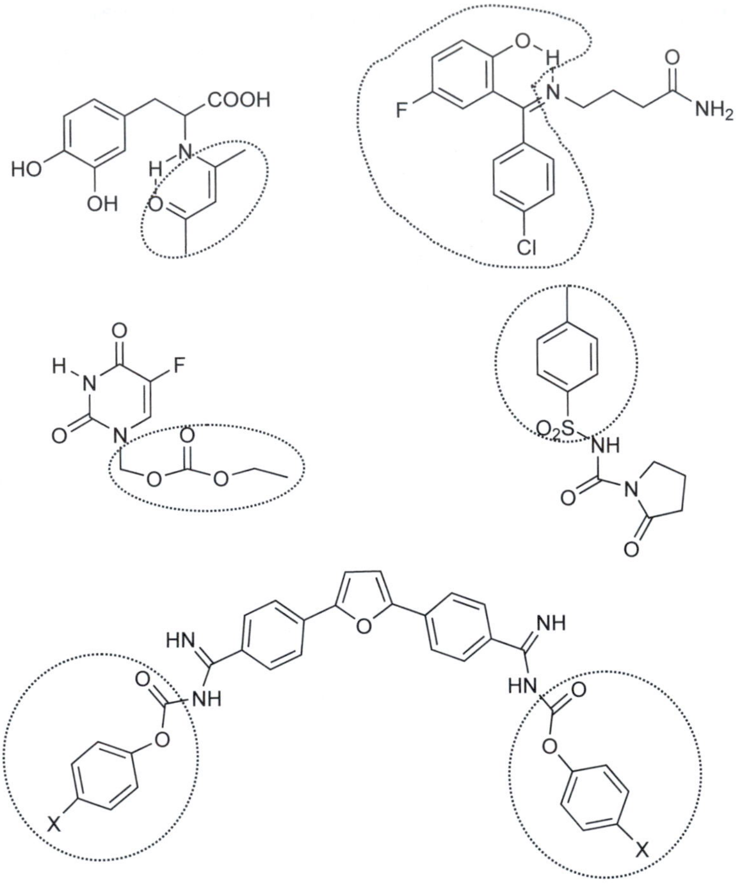

Figure 39.8 ▶ Examples of prodrugs for nitrogen-containing compounds.

39.3 Transporter-Mediated Prodrugs to Enhance Intestinal Absorption

Prodrugs can be designed to take advantage of the transporter-mediated process and enhance intestinal absorption. The transporters for which prodrugs have been made include peptide transporters, amino acid transporters, nucleoside transporters, bile acid transporters, and monocarboxylic acid transporters.[2,7,8] Table 39.4 lists examples of prodrugs that utilize transporter-mediated processes to enhance oral absorption.

Valacyclovir and valganciclovir are prodrugs of the natural amino acid valine.[9,10] They are substrates for the peptide transporters PEPT1 and PEPT2. The transporters increase oral absorption of the compounds.

Zidovudine is a synthetic nucleoside. It is converted by cellular kinases to the active metabolite zidovudine 5'-triphosphate. Zidovudine utilizes nucleoside transporters to increase oral absorption and cellular uptake.[11]

Enalapril is an angiotensin-converting enzyme inhibitor and a monoacid ester prodrug.[6] The oral bioavailability of the active principle diacid is only 3%, but the oral bioavailability of the monoacid is about 40%. This is because (1) an ethyl ester increases lipophilicity and results in increased transcellular absorption, and (2) PEPT1 transporter assists uptake of the compound.

TABLE 39.4 ►   Transported-Mediated Prodrugs for Oral Absorption

Prodrugs	Transporters	Benefits of prodrug
Valacyclovir (Valtrex)	PEPT1 and PEPT2[10]	Oral bioavailability Three- to five-fold higher than acyclovir
Valganciclovir	PEPT1 and PEPT2[9]	Oral bioavailability Ten-fold higher than ganciclovir
Zidovudine (AZT, Retrovir)	Nucleoside transporter[11]	Oral bioavailability 64%[14]
Enalapril	PEPT1[6]	Oral bioavailability is 36%–44% due to increase in lipophilicity and transporter-mediated absorption. Oral bioavailability of diacid parent is 3%.

Most transporter-mediated prodrugs were discovered by accident. The specificity and capacity of the transporters determine the success of this approach. One should be aware that transporters can be saturated at high concentrations, and there is a potential for drug–drug interaction if two drugs compete for the same transporter.

39.4 Prodrugs to Reduce Metabolism

A prodrug approach can be used to prolong the half-life of the parent drug by masking the labile functional groups, such as phenolic alcohols, which are susceptible to phase II metabolism. They are essentially slow-release drugs. Figure 39.9 shows examples of prodrugs with increased metabolic stability. Bambuterol is a dicarbamate prodrug of terbutaline. The phenolic alcohols are protected from phase II metabolism, and the carbamates are slowly hydrolyzed by nonspecific cholinesterases to release the parent terbutaline. The slow metabolism results in a longer half-life. Bambuterol is dosed once per day versus three times per day for terbutaline.

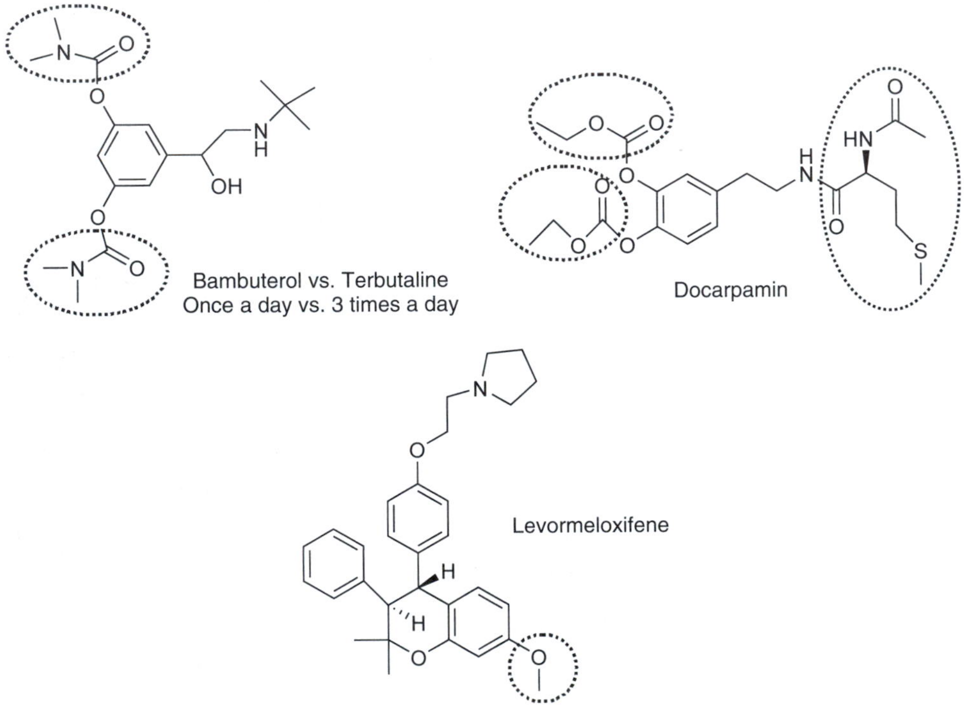

Figure 39.9 ►   Examples of prodrugs used to reduce metabolism.[2]

Dopamine is not orally available because of rapid metabolism. It is extensively metabolized by *O*-sulfation, *O*-glucuronidation, and deamination in the intestine and liver. Docarpamine is an orally active dopamine prodrug. The bisethylcarbonates are hydrolyzed in the intestine, and the amide is converted in the liver.[12]

Levormeloxifene is an *O*-methylated prodrug of a selective estrogen receptor modulator. The compound is activated in vivo by oxidative demethylation. The prodrug enhanced oral bioavailability by protecting the metabolically labile site (OH).[2]

39.5 Prodrugs to Target Specific Tissues

Selective tissue delivery can increase therapeutic activity and reduce side effects. For example, PEG-conjugated anticancer prodrugs (e.g. PEG-paclitaxel) are found to selectively accumulate in tumor cells, prolong half-life, and improve efficacy. Prodrugs can also be used to target brain, bone, colon, and other specific tissues. Organ- or tissue-specific delivery is also known as the *magic bullet*.

Capecitabine is an orally active prodrug of 5-fluorouracil (5-FU).[13] Bioactivation of capecitabine is shown in Figure 39.10. It is first hydrolyzed in the liver by carboxylesterase and decarboxylated to 5′-deoxy-5-fluorocytidine, which is further converted to 5′-deoxy-5-fluorouridine by cytidine deaminase. Transformation of 5′-deoxy-5-fluorouridine to 5-FU occurs selectively in tumor cells by thymidine phosphorylase. Distribution of 5-FU to tumor is impressive: six times higher than GI and 15 times higher than blood after oral administration of capecitabine.

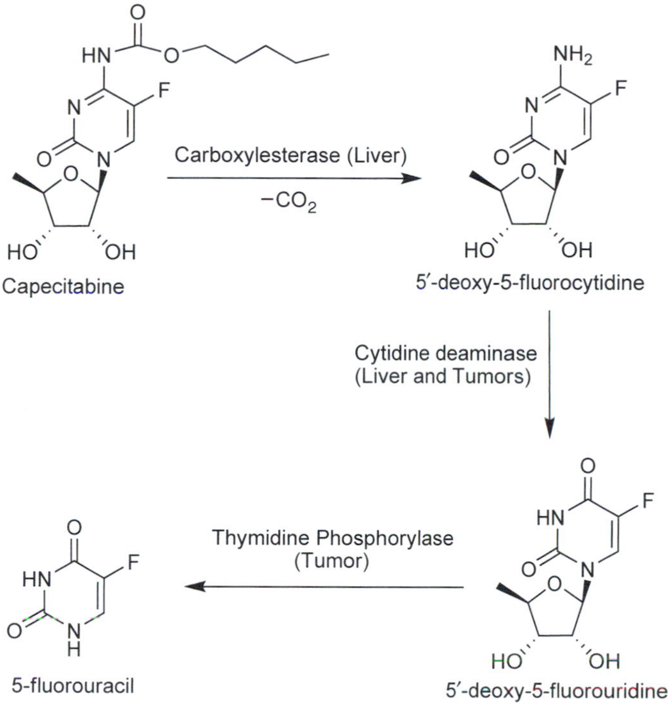

Figure 39.10 ▶ Activation of the tumor-specific prodrug capecitabine to 5-fluorouracil.[13]

39.6 Soft Drugs

Soft drugs are discussed in Chapter 12.

Problems

(Answers can be found in Appendix I at the end of the book.)

1. Which of these properties can be improved using prodrugs?: (a) toxicity, (b) permeability, (c) uptake transport, (d) hERG binding, (e) metabolic stability, (f) plasma protein binding, (g) solubility, (h) CYP inhibition?

2. How can the following structures be modified with pro-moieties to improve solubility?:

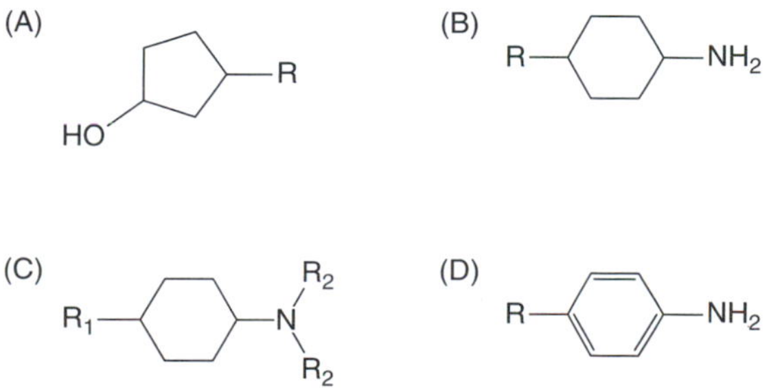

3. How can the following structure be modified with a pro-moiety to improve permeability?:

4. What hydrolyzes phosphate prodrugs in the intestine?

5. How is the active carboxylic acid shown in Problem 3 released from the prodrug after absorption (permeation)?

☐ References

1. Wermuth, C. G., & Corneille, P. (2003). Designing prodrugs and bioprecursors. In *Practice of medicinal chemistry* 2nd ed, London, UK: Elsevier, pp. 561–585.

2. Ettmayer, P., Amidon, G. L., Clement, B., & Testa, B. (2004). Lessons learned from marketed and investigational prodrugs. *Journal of Medicinal Chemistry, 47*, 2393–2404.

3. Garad, S. D. (2004). How to improve the bioavailability of poorly soluble drugs. *American Pharmaceutical Review, 7*, 80–93.

4. Stella, V. J. (1996). A case for prodrugs: fosphenytoin. *Advanced Drug Delivery Reviews, 19*, 311–330.

5. Krise, J. P., Zygmunt, J., Georg, G. I., & Stella, V. J. (1999). Novel prodrug approach for tertiary amines: synthesis and preliminary evaluation of N-phosphonooxymethyl prodrugs. *Journal of Medicinal Chemistry, 42*, 3094–3100.

6. Beaumont, K., Webster, R., Gardner, I., & Dack, K. (2003). Design of ester prodrugs to enhance oral absorption of poorly permeable compounds: challenges to the discovery scientist. *Current Drug Metabolism, 4*, 461–485.

7. Cho, A. (2006). Recent advances in oral prodrug discovery. *Annual Reports in Medicinal Chemistry, 41*, 395–407.

8. Majumdar, S., Duvvuri, S., & Mitra, A. K. (2004). Membrane transporter/receptor-targeted prodrug design: strategies for human and veterinary drug development. *Advanced Drug Delivery Reviews Veterinary Drug Delivery: Part VI, 56*, 1437–1452.

9. Sugawara, M., Huang, W., Fei, Y.-J., Leibach, F. H., Ganapathy, V., & Ganapathy, M. E. (2000). Transport of valganciclovir, a ganciclovir prodrug, via peptide transporters PEPT1 and PEPT2. *Journal of Pharmaceutical Sciences, 89*, 781–789.

10. Ganapathy, M. E., Huang, W., Wang, H., Ganapathy, V., & Leibach, F. H. (1998). Valacyclovir: a substrate for the intestinal and renal peptide transporters PEPT1 and PEPT2. *Biochemical and Biophysical Research Communications, 246*, 470–475.

11. Parang, K., Wiebe, L. I., & Knaus, E. E. (2000). Novel approaches for designing 5′-O-ester prodrugs of 3′-azido-2′,3′-dideoxythymidine (AZT). *Current Medicinal Chemistry, 7*, 995–1039.

12. Yoshikawa, M., Nishiyama, S., & Takaiti, O. (1995). Metabolism of dopamine prodrug, docarpamine. *Hypertension Research, 18*, S211–S213.

13. Testa, B. (2004). Prodrug research: futile or fertile? *Biochemical Pharmacology, 68*, 2097–2106.

14. *PDR electronic library*. Stamford, CT: Thomson Micromedex.

15. Hutchinson, I., Jennings, S. A., Vishnuvajjala, B. R., Westwell, A. D., & Stevens, M. F. G. (2002). Antitumor benzothiazoles. 16.1. Synthesis and pharmaceutical properties of antitumor 2-(4-aminophenyl)benzothiazole amino acid prodrugs. *Journal of Medicinal Chemistry, 45*, 744–747.

Chapter 40

Effects of Properties on Biological Assays

Overview

▶ *Low-solubility compounds may precipitate in bioassays and produce erroneous data.*

▶ *Bioassay development should include optimization for low-solubility compounds.*

▶ *Serial dilutions should be performed in dimethylsulfoxide (DMSO), followed by dilution into assay buffer.*

▶ *Permeability can be used to interpret cell-based assay data.*

▶ *Solution stability checks compound stability under assay conditions.*

▶ *Some compounds are insoluble in DMSO stocks or can become less soluble with freeze–thaw cycles.*

▶ *Discussion of these issues among the project team leaders ensures optimum data.*

The measurement of drug-like properties accelerated during the 1990s for the purpose of reducing the attrition of clinical compounds during development, owing to poor biopharmaceutical properties.[1] In vitro property measurement provided a cost-effective and successful strategy for improving absorption, distribution, metabolism, excretion, and toxicity (ADME/Tox) properties and led to improved human pharmacokinetics (PK) and bioavailability.

In addition to PK, property measurement has benefited another major drug discovery area in an unexpected manner. The availability of property data led to the recognition that physicochemical and biochemical properties are also related to the performance of compounds in biological assays. For example, if compounds are insoluble in the bioassay matrix, then IC_{50} values will be wrong. If a compound has poor passive diffusion permeability, it will not penetrate the cell's membrane to interact with an intracellular target protein. If a compound is chemically unstable in the bioassay matrix, the data will be erroneous.

The logic of the intimate involvement of compound properties with biological testing is apparent by examination of the discovery biological testing process (Figure 40.1). Drug-like property activities began with living systems (animals to humans), by improving the delivery of compounds to the therapeutic target through improvement of PK and reduction of toxicity. The linkage of efficacy and PK is a central concept of drug discovery. However, efficacy in living systems is just one stage of the biological testing process. If earlier steps of biological testing are considered (cellular assays, enzyme assays, high-throughput screening [HTS]), it is apparent that drug-like properties also are linked to efficacy with in vitro biological

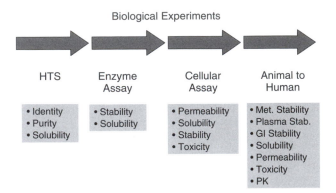

Figure 40.1 ▶ All stages of biological testing in drug discovery include property barriers.

tests. In each of these assays, compound properties affect exposure of the compound to the target.

The results of each assay are used to prioritize compounds in activity and selectivity. Fateful decisions on compound selection are made on the basis of biological data. Structure–activity relationships (SARs) are developed from biological data and are used to guide the activity optimization of leads. It is assumed that SAR is built on interaction with the therapeutic target alone. If the SAR is affected by solubility, permeability, or chemical stability, it will be misguided. It is crucial for the project team that SAR be founded only on activity. It would be unfortunate to overlook an important pharmacophore because the biological assay data were affected by properties. It is better to properly assess all compounds, even if they initially have poor properties, and then to improve properties of the active leads by structural modification. An active series is precious, and modern discovery scientists cannot afford to miss an opportunity because of inadequate experimental design.

Potential property barriers at each stage of biological testing are listed in Figure 40.1. HTS assay results can be affected by compound solubility, identity, and purity. Bench-top assays using enzymes and receptors also are affected by these properties. In addition, chemical instability of compounds in the bioassay medium can affect IC_{50}. Cell-based assays are additionally affected by permeability if the target is inside the cells. In vivo assays are subject to many of the property barriers discussed in Chapter 3.

This shift in perception of biological assays motivated considerable thinking about how to enhance discovery activities for the potentially major effect of properties. This has an impact on the following:

▶ Compound workflow: Store and handle compounds and solutions appropriately

▶ Experiment design: Optimize biological assays for properties

▶ Data interpretation: Recognize property effects; interpret results accordingly

Among the property issues, solubility in aqueous buffers and dimethylsulfoxide (DMSO) is the leading concern.[2,3] Biological assays universally rely on compound solubility in DMSO stock solutions and aqueous buffers. In recent years it has been recognized that many discovery compounds have low solubility. Low solubility in assay protocols results in falsely high IC_{50} values, low screening hit rates, poor SAR correlations, data inconsistency, differences in rank ordering between enzyme and cell-based assays, and poor in vitro ADME/Tox assay results. Recent studies indicate that discovery workflow could be improved by designing screening libraries with criteria for aqueous and DMSO solubility, improving the storage and handling of DMSO stock solutions, assaying hits and leads for solubility early in discovery, improving assay protocols for compound dilution, and developing assays that are proven to work properly with insoluble compounds.

An example of solubility issues illustrates the concerns: low solubility of compounds in screening libraries can cause low HTS hit rates. It has been shown that screening libraries can contain a high percentage of low aqueous solubility compounds.[4] In one study, libraries of soluble compounds had a much higher hit rate (32%) in screens than for libraries containing a high percentage of insoluble compounds (4% hit rate). Low solubility causes lower concentrations in screening assays, so the activity is not adequately assessed. In such libraries, impurities can be more soluble than the main component. Thus, the impurity may be the cause of activity or a measured property, resulting in erroneous SAR or property conclusions.

The following sections discuss the effects of individual properties on biological assays. Successful actions that can be taken in discovery to deal with these problems are described and are summarized in Table 40.1.

TABLE 40.1 ► Approaches to Dealing with Solubility Limitations in Biological Assays

Biological assay aspect	Approach
Assay development	• Develop and validate assays to work with low-solubility compounds
Assay protocol	• Perform serial dilution in DMSO (not buffer) and transfer to assay media
	• Mix DMSO stock directly with assay media; avoid dilution in pure water
	• Screen at lower concentrations
	• In-well sonication to redissolve
	• Reduce or eliminate freeze–thaw cycles
	• Retest HTS hits using 0.1% Triton X-100 to break aggregates
	• Correct activity values with concentrations in assay media
Assay conditions	• Assess assay tolerance for media modifiers that enhance solubility; use maximum amounts
Sample handling	• Store DMSO plates at room temperature and use them for a minimum of time
	• Dissolve salts in 1:1 DMSO/water
	• Store stocks in 9:1 DMSO/water at 4°C
	• Store compounds in solid arrays

40.1 Effects of Insolubility in DMSO

It is often assumed that compounds are universally soluble in DMSO. However, this is not the case. DMSO solubility can be limited.[5,6] Compounds that have a strong molecular lattice for crystal packing can have low DMSO solubility. Compounds in this class have lower molecular weight (MW) and are rigid and hydrophilic, such as organic salts. A second compound class, which has DMSO solubility limitations, is not well solvated by DMSO because of high MW, high Log P, large number of rotatable bonds, or high solvent-accessible surface area. Low solubility in DMSO can result in compound precipitation. This will cause concentrations in the bioassay that are lower than expected and a measured IC_{50} that is higher than the actual value.

Another problem is decreasing concentration over time. Biologists often observe that a compound is more active when it is freshly prepared than after it has been stored for a while. A primary cause of this phenomenon is precipitation of compound from the DMSO stock solution.

Common procedures can exacerbate this precipitation. Standard biological assay protocols dissolve compounds at a concentration of 10 to 30 mM in DMSO and then store the solutions at a cool temperature to reduce decomposition. Unfortunately, as many as 10% to 20% of compounds in libraries have DMSO solubility that is below this concentration,[7,8] and solubility drops further at reduced temperatures. The reduced concentrations of these DMSO stocks result in lower than expected concentrations in the biological assays using these stocks. Even if the compound has good intrinsic activity, it will appear to have low activity.

In addition, precipitate in the DMSO stock can have different effects when the stock is diluted for the assay (Figure 40.2). The IC_{50} dilution curve concentrations will be lower than planned when no precipitate is carried from the DMSO stock to the highest concentration aqueous solution, or when precipitate is carried over and does not dissolve. In these cases, the activity of the compound appears to be lower than it actually is. Conversely, the IC_{50} dilution curve concentrations will be higher than planned when precipitate is carried from the DMSO stock to the highest concentration aqueous solution and dissolves. In this case, the activity of the compound appears to be better than it actually is. Overall, precipitation of the DMSO stock will cause variable and unknown concentrations in the assay solutions that can make the compound appear more or less active than it actually is.

Another DMSO stock solution problem is "freeze–thaw cycles." When compound plates are reused, they are stored in the refrigerator. This is widely thought to reduce chemical decomposition. Unfortunately, the low temperature reduces the solubilities of compounds in solution.[9] The reduced solubility is favorable for crystal formation. These crystals usually have a lower solubility and dissolution rate than the amorphous material from which the

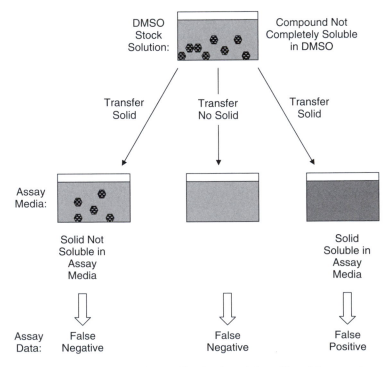

Figure 40.2 ▶ When a DMSO solution has undissolved particles, either false-negative or false-positive assay data can result, depending on whether solid is transferred and whether the solid is soluble in the assay media.

original DMSO solution was prepared.[5] Cooling also condenses water from the air into the DMSO (up to 10% w/w).[5,7,10–13] Unfortunately, many compounds have lower solubility in DMSO containing water than in DMSO alone.

40.2 Dealing with Insolubility in DMSO

One approach to the problem of low HTS hit rates caused by insoluble library compounds has been selecting only soluble compounds for screening libraries. Compounds are prescreened for solubility, and only those exceeding minimum DMSO or aqueous solubility criteria are placed in the screening library.[14]

Alternatively, insoluble salts can be dissolved in 1:1 DMSO/aqueous buffer, which increases the solubility of salts. Other water miscible organic solvents have been substituted for DMSO (e.g., methanol, ethanol, acetonitrile, tetrahydrofuran (THF), pyridine, dimethyl-formamide (DMF)).[15] With any solvent, it is important to determine the assay's tolerance to the organic solvent.

Sample storage approaches have been developed to reduce precipitation of DMSO stocks. One simple approach for reducing precipitation induced by freeze–thaw cycles is to use each DMSO solution only once. Individual-use tubes can be stored in automated systems, retrieved as needed, and then discarded after one use.[7] Another approach reduces freeze–thaw cycles by limiting the length of time that an individual DMSO solution is used (e.g., 2 weeks) and storing them at ambient temperature. This reduces the precipitation induced by cooling. Samples are not kept long enough for significant decomposition to occur at room temperature.[5]

Recognizing that water is difficult to keep out of DMSO solutions, some discovery organizations store compounds in 10% water/90% DMSO.[16] At a storage temperature of 4°C, the solutions remain liquid. Variability is reduced because there is no increase in volume from water absorption and no variable precipitation.

Alternatively, solutions can be prepared at lower concentrations (2 to 5 mM) to reduce DMSO precipitation.[4,16] A tradeoff of this approach is that the upper concentration of the assay is limited. For example, in a cell assay for which a maximum of 1% DMSO is tolerated,[17] only 1 µL of DMSO stock can be added to 99 µL of assay buffer, allowing an upper concentration of only 20 µM when a 2 mM DMSO stock is used. This can limit the determination of IC_{50}.

When compounds precipitate from DMSO or aqueous buffers, a low-energy sonicator has been successfully used to redissolve solids[7] or drive the solution to supersaturation. The energy is low enough to avoid compound decomposition.

New technologies for sample storage have begun to appear. NanoCarrier stores compounds in 1,536-well plates after evaporation of the DMSO to avoid limited DMSO solubility.[18] Compounds are stored as dry films in the DotFoil technology and rapidly dissolved for experiments.[19] Library compounds are spotted onto cards, dried, and sealed in lightproof pouches and inert atmosphere in the ChemCards product.[9]

It is useful for chemists to obtain an early estimate of the DMSO solubility of their project's compounds by using software.[6,8,20–22] Use of such tools can alert discovery scientists to compounds that may have low solubility in DMSO.

40.3 Effects of Insolubility in Aqueous Buffers

Biological assays typically test a compound's activity at various concentrations and determine IC_{50}. Typically, a high-concentration dilution (e.g., 100 µM) is made from DMSO stock into aqueous buffer, followed by serial dilution to the lower concentrations (Figure 40.3, *A*). Compounds with low aqueous solubility may not be fully soluble at the concentration of the highest aqueous solution. Serial dilution of this solution will produce a dilution curve that is

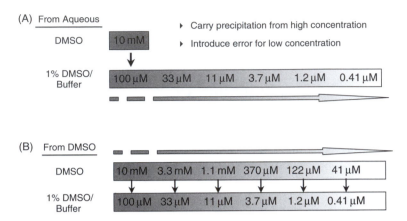

Figure 40.3 ► Schemes of serial dilution. **A,** Serial dilution in aqueous solutions can cause precipitation at the highest concentrations, which is carried to subsequent dilutions. **B,** Serial dilution in DMSO keeps compounds in solution, followed by transfer of a small aliquot of DMSO into the aqueous buffer. This also maintains the same DMSO concentration in each of the concentration solutions. (Reprinted with permission from [28].)

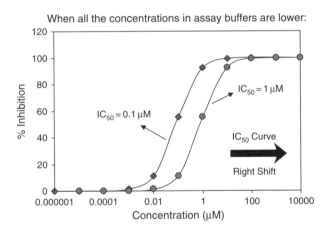

Figure 40.4 ► If the test compound is insoluble at the initial highest concentration of the dilution curve and serial dilutions are made by aqueous dilution (Figure 40.3, *A*), the IC_{50} curve is likely to be "right shifted," causing compounds to appear less active.

shifted to lower concentration than planned (Figure 40.4). The apparent IC_{50} will be higher than the compound's intrinsic IC_{50}, making the compound appear to be less active than it is.

Studies have indicated that about 30% of discovery compounds have an aqueous solubility <10 μM. This is the concentration that is commonly used for HTS.[23] Thus, it is likely that, for a portion of the library, HTS is providing IC_{50} values that are higher than the actual values.

An interesting result of limited aqueous solubility is unexplained discrepancies between different bioassays. Differences can be caused by the varying compositions of assay buffers. The composition differences occur because assays differ in their tolerance for DMSO and other components, but they also have an effect on the solubilities of compounds.[17] Some enzyme assays can tolerate up to 10% DMSO, but others can tolerate only up to 0.2% DMSO before the enzyme is inhibited. Significant differences in compound concentration can occur and result in differences in the assay results. An example of this is enzyme assays, which can tolerate additives, and cell-based assays, which have low tolerance. A low-solubility

Compounds	Solubility in Receptor Binding Assay Buffer (µM)	Solubility in Cell-based Assay Buffer (µM)
1	11	2.4
2	10	4.8
3	10	1.4
Buffer	5% BSA, 2.5% DMSO	0.1% DMSO

* Target assay concentration 10 µM

Figure 40.5 ▶ Cell-based assays customarily use fewer media modifiers, which can cause lower solubility and unexplained discrepancies between receptor/enzyme assays and cell-based assays.

compound's concentration in an enzyme assay containing 2.5% DMSO and 5% bovine serum albumin (BSA) can be 10-fold higher than in a cell-based assay with 0.1 % DMSO and 0% BSA (Figure 40.5).

In HTS, false ubiquitous hitters can be caused by compound aggregation at concentrations >10 µM.[24–26] The enzymes adsorb to the aggregates and appear to be competitively inhibited. False hits from aggregation will never be optimized by medicinal chemists.[27] Aggregates of 30- to 400-nm size pass through 0.2-µM filters (200 nm), so they are not removed by filtration. Aggregation may be triggered by supersaturation as concentrated DMSO solutions are added to aqueous buffers.[5] In one library of 1,030 compounds, the false hit rate was 19% at 30 µM but dropped to 1.4% at 5 µM, at which aggregation is less common.[27]

Not only activity and selectivity testing can be affected by low solubility. In vitro ADME/Tox assays are similarly affected. Erroneous data from metabolic stability, CYP inhibition, and hERG blocking assays can be generated because these assays are based on concentration. Low solubility can make these issues appear to be less of a problem.

40.4 Dealing with Insolubility in Aqueous Buffers

Various approaches are used to deal with insolubility of compounds in bioassays. Awareness of the potential problem is the first step. Once the discovery team realizes that biological assays may not perform properly with insoluble compounds, they can take steps to mitigate the situation. The most successful actions are as follows:

▶ Modify the dilution protocol to keep compounds in solution

▶ Assess compound solubility and concentrations

▶ Optimize assays for low-solubility compounds

In general, these are practical applications of physicochemical principles and best analytical chemistry practices.

40.4.1 Modify the Dilution Protocol to Keep Compounds in Solution

Serious problems can occur during the serial dilution protocol.[28] This is the first place to look to ensure bioassay reliability. It is important to perform the serial dilution in DMSO. Then a small volume of each DMSO dilution can be added to the aqueous assay matrix. This is illustrated in Figure 40.3, *B*. If the compound precipitates in the most concentrated solutions, this will not affect the actual compound concentrations in the lower concentration solutions. If the serial dilution is performed from one aqueous solution to the next, then the errors in the highest concentration solution are propagated to the subsequent dilutions. Accurate preparation of the dilution curve is the most important action to take to ensure quality biological assay results.

Many biologists have observed that assay results are more reproducible if they mix the highest concentration solution up and down with the pipetter before transferring an aliquot to the next solution in the serial dilution. This is because the precipitated particles are broken up into smaller particulates and form more homogeneous suspensions. The problem with this approach is that particulates may not be transferred evenly. It is much better to use assay conditions that completely solubilize the compounds and entirely avoid particles. Also, if precipitation is occurring in the first aqueous solution, it may be useful to reduce the concentration of the highest concentration in the dose–response curve. This will reduce precipitation and make the ensuing dilutions more reliable.

Some biological protocols prepare a high-concentration aqueous stock solution from DMSO stock and then use the aqueous stock (instead of DMSO stock) to make the dilution curve. Unfortunately, this procedure increases the amount of time that low-solubility compounds remain at high concentration. It is ideal to directly add DMSO dilutions into assay media, that contains components, such as proteins, cellular material, lipid membranes, and microsomes, which help to solubilize low-solubility compounds. Addition of DMSO solution to the assay media may cause supersaturation, which is a solution with a higher concentration than would be reached at equilibrium. Supersaturation reduces the rate of precipitation, allowing time for the compound molecules to interact with the target. Usually the DMSO concentration must be kept low, so only a small volume of DMSO should be added to the media. Many workers mistakenly believe that if their manual pipetter has volume markings down to 0.5 µL, then they can accurately and precisely deliver this volume. Pipetting error ranges are much higher at the low-volume limit of the pipetter.

Another mistake is diluting DMSO solutions into pure water. It is much better to dilute into buffer. Pure water has no buffering capacity, so ionizable compounds tend to convert to the neutral state where they have a much lower solubility. Buffer will maintain the pH at the prepared value. Ionized molecules have much higher solubility than neutral molecules.

40.4.2 Assess Compound Solubility and Concentrations

It is useful to assess the solubilities of compounds early. Several commercial software programs predict aqueous solubility from the molecular structure.[21,29] Although these calculations may not be absolutely accurate (usually within 10-fold of the actual solubility) and provide equilibrium solubility predictions, they can provide chemists with an early warning of potential solubility issues. Furthermore, they typically provide a profile of solubility versus pH, so chemists can see the trends and preferable pHs.[30,31]

In addition to software, most discovery organizations are alerted to potential solubility limitations of compounds using high-throughput solubility assays. These assays are more reliable than software models. Most solubility assays use generic aqueous buffers and conditions (see Chapter 25).[32–34] Kinetic solubility methods, which use initial dissolution in DMSO and addition of this solution to aqueous buffers, mirror the biological assay protocols. This is more appropriate than thermodynamic (equilibrium) solubility methods, which use aqueous buffer added to solid compound. Some groups use high-performance liquid chromatography (HPLC) with a chemiluminescent nitrogen detector (CLND) for increased accuracy in quantitation of the actual compound concentration. This detector has a molar response to compound concentration and does not require an analytical standard for quantitation. This distinguishes apparent concentration of added sample from actual concentration, which may be lower owing to impurities in the sample.[4,35−38]

For more thorough examination of solubility, it is good practice to use the conditions and protocol of the bioassay. This is because solubility is very dependent on the conditions of the assay media and the exact protocol. A customized solubility assay, using the bioassay conditions and protocol, is very useful.

Some organizations take an "adjustment" approach. The actual compound concentration in each of dilution curve solution is measured and the corrected IC_{50} calculated.[4] Although this approach is appealing, there are some drawbacks. If the measurement is done in a generic buffer, the concentrations will differ from those in assay media and may lead to an inaccurate IC_{50}. Also, the measurements are time and resource consuming, so they may not be efficient. This approach may be very appropriate if a key compound requires this level of accuracy, but it is inefficient if tens or hundreds of compounds are to be studied. Also, this approach could result in selection of several lead compounds with low solubility as the primary leads of a project. If the project team is never able to improve the solubility of these compounds, the team may be unnecessarily burdened with low-solubility leads. The most efficient approach is to optimize the assay protocol during assay development so that it works for low-solubility compounds and produces accurate activity without the need for burdensome analyses and IC_{50} correction to every tested compound.

40.4.3 Optimize Assays for Low Solubility Compounds

Solubility is strongly affected by assay media components (e.g., buffer, organic solvents, counter-ions, protein), dilution protocol, and incubation conditions (e.g., time, temperature). It is good practice to optimize the assay media components, percentages, dilution, and protocol during assay development to improve the solubilizing capabilities of the bioassay. Co-solvents (e.g., DMSO, methanol, ethanol, acetonitrile, DMF, dioxane) and excipients (e.g., cyclodextrin) have been used to improve solubility in bioassay media.[9,39,40] In many cases, the assay developer uses standard conditions or media components without testing for tolerance. The assay may tolerate a higher concentration of DMSO or other components, and this should be determined during assay development.

The ability of an assay to perform accurately for low-solubility compounds can be determined during assay development. A set of low-solubility compounds can be processed using the assay conditions and protocol. Their concentrations can be analyzed to determine if the assay is maintaining them in solution.

If compounds precipitate in aqueous buffer, in-well sonication can redissolve them.[7] A high-power sonicator with 96 individual probes can be used to simultaneously sonicate each well in a plate. This technique can salvage precipitated solutions, or it can ensure that all the wells are completely solubilized as part of the protocol. Precipitation is not always obvious by eye.

Running biological assays at a lower concentration can reduce precipitation. More compounds are soluble at a concentration of 3 µM than at 10 µM.

If the project team suspects that some of their hits may be due to aggregation, several approaches can help to identify the aggregators. Hits can be rescreened in the presence of 0.1% Triton X-100.[27,41] This detergent breaks up the aggregates. Another approach is to analyze the solutions using a dynamic light scattering plate reader, which detects aggregates. If the transition of the IC_{50} curve is sharper than a normal curve aggregation could be responsible because of the dependence of aggregation on concentration.[26] Screening can be performed at a lower concentration, at which aggregation occurs less frequently.

Finally, a computational model can help to identify structural features characteristic of aggregators.[27,42,43]

40.4.4 Effects of Permeability in Cell-Based Assays

Membrane permeability can greatly affect the observed biological activity of some compounds for intracellular targets. These compounds may have been very active in cell-free enzyme or receptor assays. However, their cellular membrane permeability may be low and reduce or eliminate their activity.

This effect usually is caused by low permeability by passive diffusion. Another well-known example is multidrug resistance cancer cells, which have high levels of P-glycoprotein and greatly reduce intracellular concentration by efflux. Most cell lines used in drug discovery have low levels of efflux transporters.

Compounds may have limited permeability in cell-based property assays. For example, measurement of metabolic stability using hepatocytes certainly is affected by the rate of enzymatic metabolism reactions, but it also can be affected by the hepatocyte membrane permeability of the compounds.

40.4.5 Dealing with Permeability in Cell-Based Assays

The simplest approach for checking passive diffusion permeability potential is structural rules. If polar surface area (PSA) is >140 Å^2, H-bond acceptor >10, H-bond donor >5, MW >500, or if the compound has a strong acid, it is possible that cell membrane permeability is limited.

High-throughput permeability assay data are useful. For example, if parallel artificial membrane permeability assay (PAMPA) permeability is low, then cell membrane permeability is likely to be low. Some chemists use Caco-2 data for estimating cell membrane permeability. It can be overkill to use the expensive Caco-2 assay for this purpose when less expensive assays could be used. Caco-2 is also more complex because it exhibits several permeability mechanisms that may be confusing to interpret for intracellular compound exposure.

Intracellular concentrations have been determined for this purpose (see Section 27.2.2). This is more common with radiolabeled compounds, but the availability of sensitive and rapid liquid chromatography/mass spectrometry/mass spectrometry (LC/MS/MS) methods make this more practical.

If reduced activity in cell-based assays is observed, the medicinal chemist can identify the structural cause of the low permeability. If the functional group is not crucial for the activity, then structural modification to reduce the polarity, hydrogen bonding, or size may greatly improve the permeability and cell-based activity. The permeability effect may be obvious by examining the activity differential between cell-free and cell-based assays to determine if there is a greater differential for the analogs that are permeability limited.

With limited resource investment, the permeability limitations of a series can be estimated, which can be useful for better understanding the behavior of the compounds.

40.4.6 Effects of Chemical Instability in Bioassays

Compounds may be chemically unstable in biological or property assays. They may react with an assay media component, they may be hydrolyzed, or a media condition (e.g., pH, temperature, light) can accelerate decomposition. Another complicating factor is that decomposition products can be either more or less active than the tested compound, thus confusing the SAR. Moreover, compounds in the same series could have differences in chemical decomposition, which could be mistaken as SAR. Several decomposition reactions can occur, including hydrolysis, hydration, oxidation, and isomerization (see Chapter 13).

40.4.7 Dealing with Chemical Instability in Bioassays

When structural features indicate the potential for decomposition or when inconsistent data are generated that may be caused by chemical instability, it is prudent to test the stability in vitro (see Chapter 31). The test should be conducted under conditions as close as possible to the biological assay conditions (e.g., media, time, temperature). HPLC is useful for measuring the compound concentration over time. LC/MS is useful for rapidly identifying the reaction products, from which the decomposition mechanism can be inferred. The decomposition time course will provide a means for estimating the level of the effect of the decomposition on the bioassay activity.

Problems

(Answers can be found in Appendix I at the end of the book.)

1. What property barriers does a compound encounter in an in vitro cell-based assay for an intracellular therapeutic target?

2. Why might it be counterproductive to eliminate from further consideration all compounds that do not have good activity in an in vitro assay, regardless of whether the low activity is due to poor target binding, solubility, permeability, or stability?

3. What are the characteristics of two classes of compound with low DMSO solubility?

4. How might low solubility in DMSO cause either higher or lower aqueous assay concentrations than intended?

5. What are two negative effects of freeze–thaw cycles?

6. What approaches can help to better solubilize compounds in an organic stock solution?

7. Would low compound solubility in the aqueous assay media cause higher or lower IC_{50}?

8. What two general approaches can improve the characteristics of a biological assay to better assess compound activity?

9. A serial dilution should be performed by which of the following protocols?: (a) prepare a high-concentration DMSO stock solution and pipette as small of a volume as possible into the aqueous buffer, (b) make the initial dilution from DMSO as a high-concentration aqueous solution and dilute with aqueous buffer to subsequently lower concentrations, (c) make the initial dilution from DMSO as a high-concentration DMSO solution and dilute with DMSO to subsequently lower concentrations that are each diluted into aqueous buffer.

10. To measure the solubility of compounds in the biological assay, it is best to: (a) use a generic aqueous buffer, (b) use water, (c) use the solution used in the bioassay, (d) follow the protocol of the assay.

11. What components can be included in bioassay media to maximize the concentration of low-solubility compounds?

12. What tools can discovery scientists use to estimate if permeability is a potential problem for a lead series in a cell-based assay?

13. Define IC_{50} right shift. What is the cause?

14. What solubility-related problem can cause erroneous activity in cell-based assays that are not consistent with enzyme/receptor assays in which the compounds were previously tested? What can be done during assay development to improve this situation?

References

1. Kennedy, T. (1997). Managing the drug discovery/development interface. *Drug Discovery Today*, *2*, 436–444.

2. Di, L., & Kerns, E. H. (2006). Biological assay challenges from compound solubility: strategies for bioassay optimization. *Drug Discovery Today*, *11*, 446–451.

3. Di, L., & Kerns, E. H. (2007). Solubility issues in early discovery and HTS. In P. Augustijns & M. Brewster (Eds.), *Solvent systems and their selection in pharmaceutics and biopharmaceutics* (pp. 111–136). New York: Springer.

4. Popa-Burke, I. G., Issakova, O., Arroway, J. D., Bernasconi, P., Chen, M., Coudurier, L., et al. (2004). Streamlined system for purifying and quantifying a diverse library of compounds and the effect of compound concentration measurements on the accurate interpretation of biological assay results. *Analytical Chemistry*, *76*, 7278–7287.

5. Lipinski, C. A. (2004). Solubility in water and DMSO: issues and potential solutions. *Biotechnology: Pharmaceutical Aspects*, *1*, 93–125.

6. Balakin, K. V., Ivanenkov, Y. A., Skorenko, A. V., Nikolsky, Y. V., Savchuk, N. P., & Ivashchenko, A. A. (2004). In silico estimation of DMSO solubility of organic compounds for bioscreening. *Journal of Biomolecular Screening*, *9*, 22–31.

7. Oldenburg, K., Pooler, D., Scudder, K., Lipinski, C., & Kelly, M. (2005). High throughput sonication: evaluation for compound solubilization. *Combinatorial Chemistry and High Throughput Screening*, *8*, 499–512.

8. Balakin, K. V. (2003). DMSO Solubility and bioscreening. *Current Drug Discovery*, 27–30.

9. Hoever, M., & Zbinden, P. (2004). The evolution of microarrayed compound screening. *Drug Discovery Today*, *9*, 358–365.

10. Cheng, X., Hochlowski, J., Tang, H., Hepp, D., Beckner, C., Kantor, S., et al. (2003). Studies on repository compound stability in DMSO under various conditions. *Journal of Biomolecular Screening*, *8*, 292–304.

11. Kozikowski, B. A., Burt, T. M., Tirey, D. A., Williams, L. E., Kuzmak, B. R., Stanton, D. T., et al. (2003). The effect of freeze/thaw cycles on the stability of compounds in DMSO. *Journal of Biomolecular Screening*, *8*, 210–215.

12. Lipinski, C. (2004). Solubility in the design of combinatorial libraries. *Chemical Analysis*, *163*, 407–434.

13. Lipinski, C. A., & Hoffer, E. (2003). Compound properties and drug quality. *Practice of Medicinal Chemistry*, *2*, 341–349.

14. Walters, W. P., & Namchuk, M. (2003). Designing screens: how to make your hits a hit. *Nature Reviews Drug Discovery*, *2*, 259–266.

15. Buchli, R., VanGundy, R. S., Hickman-Miller, H. D., Giberson, C. F., Bardet, W., & Hildebrand, W. H. (2005). Development and validation of a fluorescence polarization-based competitive peptide-binding assay for HLA-A*0201-a new tool for epitope discovery. *Biochemistry*, *44*, 12491–12507.

16. Schopfer, U., Engeloch, C., Stanek, J., Girod, M., Schuffenhauer, A., Jacoby, E., et al. (2005). The Novartis compound archive: from concept to reality. *Combinatorial Chemistry & High Throughput Screening*, *8*, 513–519.

17. Johnston, P. A., & Johnston, P. A. (2002). Cellular platforms for HTS: three case studies. *Drug Discovery Today*, *7*, 353–363.

18. Benson, N., Boyd, H. F., Everett, J. R., Fries, J., Gribbon, P., Haque, N., et al. (2005). NanoStore: A Concept for logistical improvements of compound handling in high-throughput screening. *Journal of Biomolecular Screening, 10*, 573–580.

19. Topp, A., Zbinden, P., Wehner, H. U., & Regenass, U. (2005). A novel storage and retrieval concept for compound collections on dry film. *Journal of the Association for Laboratory Automation, 10*, 88–97.

20. Lu, J., & Bakken, G. A. (2004). *Building classification models for DMSO solubility: comparison of five methods.* In Abstracts of Papers, 228th ACS National Meeting, Philadelphia, PA, United States, August 22–26, 2004, CINF-045.

21. Delaney, J. S. (2005). Predicting aqueous solubility from structure. *Drug Discovery Today, 10*, 289–295.

22. Japertas, P., Verheij, H., & Petrauskas, A. (2004). *DMSO solubility prediction.* In LogP 2004, Zurich, Switzerland.

23. Lipinski, C. A. (2001). Avoiding investment in doomed drugs. *Current Drug Discovery, 2001*, 17–19.

24. McGovern, S. L., Caselli, E., Grigorieff, N., & Shoichet, B. K. (2002). A common mechanism underlying promiscuous inhibitors from virtual and high-throughput screening. *Journal of Medicinal Chemistry, 45*, 1712–1722.

25. McGovern, S. L., Helfand, B. T., Feng, B., & Shoichet, B. K. (2003). A specific mechanism of nonspecific inhibition. *Journal of Medicinal Chemistry, 46*, 4265–4272.

26. McGovern, S. L., & Shoichet, B. K. (2003). Kinase inhibitors: not just for kinases anymore. *Journal of Medicinal Chemistry, 46*, 1478–1483.

27. Feng, B. Y., Shelat, A., Doman, T. N., Guy, R. K., & Shoichet, B. K. (2005). High-throughput assays for promiscuous inhibitors. *Nature Chemical Biology, 1*, 146–148.

28. Di, L., & Kerns, E. H. (2005). Application of pharmaceutical profiling assays for optimization of drug-like properties. *Current Opinion in Drug Discovery & Development, 8*, 495–504.

29. Jorgensen, W. L., & Duffy, E. M. (2002). Prediction of drug solubility from structure. *Advanced Drug Delivery Reviews, 54*, 355–366.

30. Oprea, T. I., Bologa, C. G., Edwards, B. S., Prossnitz, E. R., & Sklar, L. A. (2005). Post-high-throughput screening analysis: an empirical compound prioritization scheme. *Journal of Biomolecular Screening, 10*, 419–426.

31. van de Waterbeemd, H., & Gifford, E. (2003). ADMET in silico modelling: towards prediction paradise? *Nature Reviews Drug Discovery, 2*, 192–204.

32. Kerns, E. H., & Di, L. (2005). Automation in pharmaceutical profiling. *Journal of the Association for Laboratory Automation, 10*, 114–123.

33. Kerns, E. H. (2001). High throughput physicochemical profiling for drug discovery. *Journal of Pharmaceutical Sciences, 90*, 1838–1858.

34. Kerns, E. H., & Di, L. (2004). Physicochemical profiling: overview of the screens. *Drug Discovery Today: Technologies, 1*, 343–348.

35. Yurek, D. A., Branch, D. L., & Kuo, M.-S. (2002). Development of a system to evaluate compound identity, purity, and concentration in a single experiment and its application in quality assessment of combinatorial libraries and screening hits. *Journal of Combinatorial Chemistry, 4*, 138–148.

36. Kerns, E. H., Di, L., Bourassa, J., Gross, J., Huang, N., Liu, H., et al. (2005). Integrity profiling of high throughput screening hits using LC-MS and related techniques. *Combinatorial Chemistry & High Throughput Screening, 8*, 459–466.

37. Yan, B., Collins, N., Wheatley, J., Irving, M., Leopold, K., Chan, C., et al. (2004). High-throughput purification of combinatorial libraries I: a high-throughput purification system using an accelerated retention window approach. *Journal of Combinatorial Chemistry, 6*, 255–261.

38. Yan, B., Fang, L., Irving, M., Zhang, S., Boldi, A. M., Woolard, F., et al. (2003). Quality control in combinatorial chemistry: determination of the quantity, purity, and quantitative purity of compounds in combinatorial libraries. *Journal of Combinatorial Chemistry, 5*, 547–559.

39. Dean, K. E.S., Klein, G., Renaudet, O., & Reymond, J.-L. (2003). A green fluorescent chemosensor for amino acids provides a versatile high-throughput screening (HTS) assay for proteases. *Bioorganic & Medicinal Chemistry Letters, 13*, 1653–1656.

40. Schmidt, M., & Bornscheuer, U. T. (2005). High-throughput assays for lipases and esterases. *Biomolecular Engineering, 22,* 51–56.

41. Ryan, A. J., Gray, N. M., Lowe, P. N., & Chung, C.-W. (2003). Effect of detergent on "promiscuous" inhibitors. *Journal of Medicinal Chemistry, 46,* 3448–3451.

42. Roche, O., Schneider, P., Zuegge, J., Guba, W., Kansy, M., Alanine, A., et al. (2002). Development of a virtual screening method for identification of "frequent hitters" in compound libraries. *Journal of Medicinal Chemistry, 45,* 137–142.

43. Seidler, J., McGovern, S. L., Doman, T. N., & Shoichet, B. K. (2003). Identification and prediction of promiscuous aggregating inhibitors among known drugs. *Journal of Medicinal Chemistry, 46,* 4477–4486.

Chapter 41

Formulation

Overview

> ▶ *Formulation increases compound solubility to enhance absorption in vivo.*

> ▶ *Increased in vivo exposure provides more informative data for efficacy, pharmacokinetics, and toxicity during discovery.*

Formulation has numerous benefits in drug discovery and development. It enables efficacy, toxicity, and pharmacokinetic (PK) studies. Formulation can improve oral bioavailability, shorten onset of a therapeutic effect, enhance stability of drugs, and reduce dosing frequency. More consistent dosing can be achieved by reducing food effect through formulation. Formulation can reduce side effects (i.e., decreasing tissue irritation and improving taste). Tissue-(e.g., tumor) specific formulation can enhance efficacy and reduce toxicity. Novel formulations may be patentable.

Formulation is most commonly used to increase solubility and sometimes to improve stability (Figure 41.1). It is rarely used to enhance permeability because of toxicity.[1] Application of permeability enhancers is still an active research area. Enhancers work by opening the tight junctions between the cells to allow drugs to pass through by paracellular diffusion. However, this also allows toxic substances to go through and cause toxicity. Structure modification is required to increase permeability, and there are no formulation rescues.

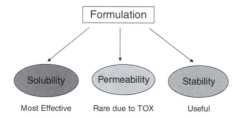

Figure 41.1 ▶ Applications of formulation.

Formulation in drug discovery faces many challenges.[2] Drug discovery typically has limited material. Only milligram quantities are prepared. This limits the options to screen for optimal formulations because insufficient material is available. Formulation in drug discovery requires short time lines. Usually dosing vehicles need to be developed in a few days for animal studies. There is not enough time to develop ideal formulation vehicles. At early stages of drug discovery, the compounds do not always have good potency. High doses typically are used in order to demonstrate efficacy and proof of concept. These factors make it more challenging for formulation to achieve high loading. At this stage, multiple animal species and multiple routes of administration are explored to evaluate the drug effects. This requires development of various dosing vehicles and dosage forms.

41.1 Routes of Administration

The different routes of administration for commercial drugs are summarized in Table 41.1. Figure 41.2 shows the distribution of pharmaceutical sales for each delivery route.[3] The majority of the drugs (70%) are delivery by oral administration and 16% by injection.

TABLE 41.1 ▶ Comparison of Different Routes of Administration

Routes	Dosage form	Tonicity	Ideal pH	Bioavailability
Oral	Solid Suspension Solution	Not required	Not required	Incomplete absorption First pass in gut and liver
IP	Suspension Solution	Isotonic preferred	5–8	No first pass in gut Has first pass in liver
IV bolus	Solution Emulsion	Isotonic	5–8	Complete
SC	Solution Emulsion Suspension	Isotonic preferred	3–8	Incomplete absorption No first pass in gut and liver
IM	Solution Emulsion Suspension	Isotonic	3–8	Incomplete absorption No first pass in gut and liver

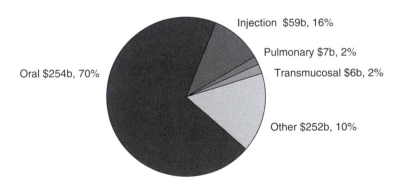

Figure 41.2 ▶ Distribution of pharmaceutical sales based on delivery route.[3]

Oral (PO)

Oral (PO) dosing is the most convenient, economic, safe, noninvasive route of administration. However, it requires patient compliance. PO tends to have limited and variable absorption for some drugs that have poor physiochemical properties, such as low solubility and/or low permeability. Drugs dosed PO are subject to first-pass metabolism in the gut and liver. They can have limited bioavailability due to absorption and metabolism barriers.

Gavage delivery is quite common for animal PO dosing. Drugs are delivered as a liquid into the stomach via a feeding needle.

Intravenous (IV)

Intravenous (IV) dosing has rapid onset and complete bioavailability. It precisely delivers the complete dose. IV requires that the dose be soluble and the solution be miscible with serum without precipitation. Oil/water emulsions and liposomal and nanoparticulate systems can be injected IV as long as the particle sizes are much smaller than erythrocytes. Certain formulations may cause hemolysis or precipitation after injection. An in vitro precipitation model can be used to determine whether compounds are likely to precipitate upon dilution and IV administration.[4,5]

IV delivery is commonly performed by needle injection into an easily accessible vein (e.g., tail, leg).

Intraperitoneal (IP)

Intraperitoneal (IP) injection is particularly useful in discovery laboratories for small animal studies, where it often is preferred over IV because of its ease of administration. The IP route also is used in clinical situations, particularly in intensive care units and during chemotherapy, where high concentration can be achieved locally while minimizing systemic side effects. IP bypasses first-pass metabolism from the gut but still is subject to first-pass metabolism by the liver[6] because absorption occurs via the portal system. Compounds that are very lipophilic will be quickly absorbed systemically by the IP route but not by the IM or SC route.

IP administration is performed by needle injection into the abdominal cavity. Both solution and suspension can be used for IP injection.

Subcutaneous (SC)

Drugs are injected in a subcutaneous (SC) site, which is beneath the skin, an area that is rich in fat and blood vessels. Solutions, suspensions, or implantation forms all can be used for SC delivery. Solutions are preferably isotonic. SC bypasses first-pass metabolism. Thus, SC is a useful route in proof-of-concept studies if a compound is highly metabolized. Injection volume is very small (0.5–2 mL) for SC delivery, so it typically is used for high-potency drug candidates.

Intramuscular (IM)

An intramuscular (IM) medication is given by needle into the muscle. It can be an isotonic solution, oil, or suspension. Drugs in aqueous solution are rapidly absorbed. However, very slow constant absorption can be obtained if the drug is administered in oil or suspended in other repository vehicles. IM administration of certain drugs can be painful. The drug might precipitate at the site of administration.

41.2 Potency Drives Delivery Opportunities

Routes of delivery are dependent on the potency of drug candidates. Figure 41.3 illustrates that the more potent the compound is, the more options are available for delivery. For very active compounds with a dose <10 mg (or 1 mg/kg), all the routes can be applied to deliver the drug. However, if the compound has an average potency with dose >100 mg (or 10 mg/kg), the route of administration is limited to only PO or IV. Therefore, potent compounds have more opportunities for delivery routes, whereas low-potency compounds have limited options.

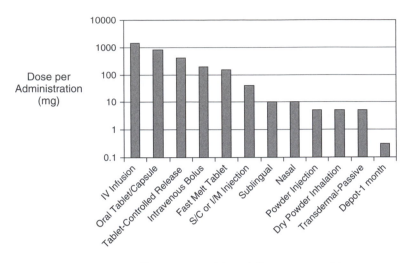

Figure 41.3 ▶ Drug potency drives delivery opportunities.

 ## 41.3 Formulation Strategies

Many strategies have been developed to formulate insoluble compounds, including pH adjustments, use of co-solvents, surfactants, and complexation agents, lipid-based formulation, and particle size reduction.

Adjust pH of Dosing Solution

For compounds containing ionizable groups, adjusting the pH of the dosing solution can favor the ionized form and increase solubility. The pH of the solution should be two units beyond the pK_a of the compounds to ensure complete ionization. For acids, basic buffers should be used; conversely, for bases, acidic buffers should be used. Table 41.2 summarizes common buffers used for pH adjustment to enhance solubility of ionizable compounds.[7] Dosing a basic amine in a citric acid solution is more likely to dissolve the compound and enhance in vivo exposure than dissolving the compound in water. Both in situ solution (salt formed in solution with counter-ion and pH adjustment) and salt form approaches can be used to increase solubility through ionization of acids and bases.

TABLE 41.2 ▶ Common Buffers for Formulation[7]

Buffering agents	pK_a(s)	Suitable for pH range	Commercial products
Maleic acid	1.9, 6.2	2–3	Teniposide
Tartaric acid	2.9, 4.2	2.5–4	Tolazoline HCl
Lactic acid	3.8	3–4.5	Ciprofloxacin
Citric acid	3.1, 4.8, 6.4	3–7	Labetalol HCl, Nicardipine HCl
Acetic acid	4.75	4–6	Mitoxantrone HCl, Ritodrine HCl
Sodium bicarbonate	6.3, 10.3	4–9	Cefotetan, Cyclophosphamide
Sodium phosphate	2.2, 7.2, 12.4	6–8	Warfarin, Vecuronium Br

Use Co-solvent

Co-solvents can help dissolve insoluble compounds by increasing solubility. Commonly used co-solvents and their toxicities are listed in Table 41.3. Examples of parenteral products (administered IV in the clinic) containing co-solvents and surfactants are listed in Table 41.4. Co-solvent and pH adjustment can be used in combination to further enhance the solubility of insoluble compounds. Precautions should be taken when using co-solvent in animal models for efficacy studies, to minimize potential interference of pharmacological effects by the co-solvent.

TABLE 41.3 ▶ Commonly Used Co-solvents and Their Toxicities

| Co-solvents | Mouse LD$_{50}$ (g/kg) | | | Rat LD$_{50}$ (g/kg) | | | Percent of commercial products containing |
	PO	IV	IP	PO	IV	IP	
DMSO	8	2	1	7	1	4	
Glycerin	24	8	10	20	7	7	
Dimethylacetamide	4	6	9	126	6	9	< 3
Ethanol	29	9	10	–	7	10	< 10
Propylene glycol	5	3	3	5	3	3	~ 40
PEG 400	17	6	3	18	5	8	~ 50

TABLE 41.4 ▶ Examples of Parenteral Products Containing Co-solvents and Surfactants

Trade name	Generic name	Manufacturer	Cosolvents/surfactant	Routes
Sandimmune	Cyclosporin	Novartis	Cremophor EL 50% Ethanol 27.8%	IV infusion
Lanoxin	Digoxin	GSK	Propylene glycol 40% Ethanol 10%	IV, IM
Ativan	Lorazepam	Wyeth	PEG 400 18% Propylene glycol 80%	IM, IV
Taxol	Paclitaxel	BMS	Cremophor EL 50% Ethanol 50%	IV infusion

Utilize Surfactants

Surfactants can provide many benefits for formulations[8]: (1) increase solubility of drugs through micellization, (2) prevent precipitation due to surface properties, especially after dilution, (3) improve stability of drugs in solution by incorporating them into micelles, and (4) prevent aggregation in protein formulation due to interfacial properties. Commonly used surfactants are summarized in Table 41.5.[9] Table 41.6 lists examples of marketed drugs containing surfactants.

Suspension formulation (undissolved particles in liquid) using surfactant is most common in drug discovery (1) when solution formulation is not feasible due to the limit of solubility, and (2) for toxicity and chronic studies. Suspensions can be dosed through PO/gavage administration, IP, SC, or IM delivery. A typical suspension formulation includes a surfactant (e.g., Tween 80) to wet the surface of particles and a bulking agent (e.g., Methocel) to suspend the solid particles. For suspension formulation, it is critical to reduce the particle size of the solid material to enhance the surface area and dissolution rate in order to maximize exposure. The commonly used suspension formulations in drug discovery for PK and toxicokinetics (TK) studies is Tween 80 (0.1%–2%)/Methocel (0.5%–1%).

TABLE 41.5 ▶ Solubilizing Excipients Used in Commercially Available Solubilized Oral and Injectable Formulations[9]

Water-soluble co-solvent	Water-insoluble co-solvent	Surfactants
DMA	Beeswax	Cremophor EL
DMSO	Oleic acid	Cremophor RH 40
Ethanol	Soy fatty acids	Cremophor RH 60
Glycerin	Vitamin E	Tween 20
NMP	Castor oil	Tween 80
PEG 300	Corn oil	TPGS
PEG 400	Cottonseed oil	Solutol HS-15
Poloxamer 407	Olive oil	Span 20
Propylene glycol	Peanut oil	Softigen 767
HPβCD	Safflower oil	Labrasol
SBEβCD	Sesame oil	Labrafil M-1944CS
αCD	Peppermint oil	Labrafil M-2125CS
Phospholipids (HSPC, DSPG, DMPC, DMPG)	Soybean oil	PEG 400 monostearate
		PEG 1750 monostearate

TABLE 41.6 ▶ Marketed Parenteral Products Containing Surfactants[8]

Trade name	Generic name	Manufacturer	Surfactants	Routes
Cordarone X IV	Amiodarone HCl	Sanofi-Aventis	Tween 80 10% Dilute 1:50	IV infusion
Neupogen	Filgrastim	Amgen	Tween 80 0.004%	IV
Proleukin	Aldesleukin	Chiron	SDS 0.18 mg/mL Dilute 1:42	IV infusion
Calcijex	Calcitriol	Abbott	Tween 80 0.4%	IV

Lipid-based Formulation

There are four different types of lipid-based formulations depending on the percentage of oil, water-soluble and water-insoluble surfactants, and co-solvents.[10,11] Lipid-based formulation can increase the solubility of lipophilic compounds. The simplest lipid delivery system is to dissolve a drug into pure oil (e.g., vegetable oil). For example, oil solution is the standard method for administration of lipid-soluble vitamins, such as vitamins A and D. For more complex systems, compounds are dissolved in oil or lipids and then dispersed into aqueous buffers in the presence of surfactants, with or without co-solvents. They can form emulsions, micelles, or liposomes. Their structures are illustrated in Figure 41.4. Examples of marketed products using lipid-based formulations are listed in Table 41.7.

An emulsion is a dispersion of two immiscible liquids (e.g., oil and water) with a surfactant or emulsifier. Emulsifier coats the droplets to stabilize the emulsion by creating repulsion between the droplets. There are two common types of emulsions. Water-in-oil (W/O) emulsions typically are used in sustained release of steroids and vaccines by IM delivery. Oil-in-water (O/W) emulsions are one of the most commonly used lipid formulations for various routes of administration, such as SC, IM, and IV. Excipients for emulsion formulation are various oils of polar triglycerides (e.g., soybean oil, sesame oil, corn oil, safflower oil), an aqueous phase (saline, D5W and buffers), and an emulsifier (egg lecithin and Tween 80). Drugs are incorporated into the oil droplets, and most of the drop sizes are around 0.2 to 0.6 μM. Parenteral lipid emulsions normally are ready-to-use formulations and conveniently stored at room temperature. Drug emulsions normally are formulated to

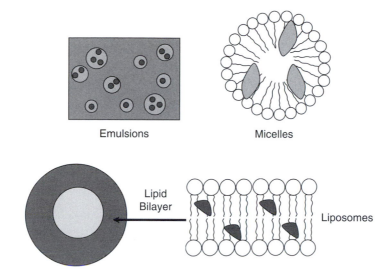

Figure 41.4 ▶ Structures of lipid-based formulations.

TABLE 41.7 ▶ Example of Commercially Available Injectable Lipid-based Formulation[8,18]

Solubilizer system	Trade name	Drug	Manufacturer
Lipid emulsion	Diazemuls	Diazepam	Dumex
Lipid emulsion	Diprivan	Propofol	AstraZeneca
Mixed micelles	Valium MM	Diazepam	Roche
Mixed micelles	Konakion/120	Vitamin K	Roche
Liposome	AmBisome	Amphotericin B	Gilead
Liposome	Doxil	Doxorubicin	Alza

isotonic concentration at pH 7 to 8, which reduces pain upon injection compared to solvent-based or solubilized formulations.[12] Loading of drugs for emulsion systems can be quite high and vary from 1 to 100 mg/mL, depending on formulation and process conditions.

Micelles are aggregates that self-associate to form a hydrophobic core and hydrophilic outer sphere. Lipophilic drugs can be incorporated into the core and, thus, dissolved in aqueous media. The sizes of micelles are approximately 5 to 50 nm.[13] Micelles typically consist of low–molecular-weight amphiphilic molecules, such as bile salts and phospholipids. The limitations of small molecular micelles are (1) low capacity for drug loading, (2) possible toxicity due to disruption of lipid bilayers, and (3) possible precipitation after injection due to breakdown of the micelles. Block copolymer micelles offer many advantages over traditional micellar formulations, with less toxicity, better stability, high loading, controlled-release properties, and targeted delivery.[13–17]

Liposomes are microscopic hollow spheres, typically made of bilayers of natural or semisynthetic phospholipids and/or cholesterol. Insoluble compounds can be solubilized in the hydrophobic space of liposomes. They vary in size (30 nm to 30 μM), bilayer rigidity, geometry, and charge. Beside the hydrophobic bilayer, the encapsulated aqueous compartment and the polar interface can be used to capture sparingly soluble compounds.[18] The versatility makes liposome delivery systems amenable to formulate a wide range of drug classes. Liposome technology currently is focused on applications in oncology, but it also can be used to formulate drugs to treat fungal, bacterial, and viral infections, alleviate pain, reduce inflammation, treat blood disorders, and for medical imaging and vaccines.[18]

Liposomes can be destabilized if overloaded with drug molecules. Drug loading of lipophilic drugs normally is substantially lower for liposomes than for emulsions. It typically is limited to high-potency compounds. Liposome formulations normally are lyophilized because of stability issues and require reconstitution before use.

Drug Complexation

Cyclodextrins are cyclic oligosaccharides consisting of (α-1,4)-linked α-D-glucopyranose units (Figure 41.5), with a hydrophilic outer surface and a lipophilic central cavity.[19] Cyclodextrins are able to form water-soluble inclusion complexes with many lipophilic poorly soluble compounds. An example of the guest–host complex of aspirin with β-cyclodextrin (βCD) is shown on Figure 41.6. Cyclodextrins are relatively large molecules with molecular weight between 1,000 and 2,000. The most abundant natural cyclodextrins are α-cyclodextrin (αCD), βCD, and γ-cyclodextrin (γCD), which contain six, seven, and eight glucopyranose units, respectively. Of these three cyclodextrins, βCD appears to be

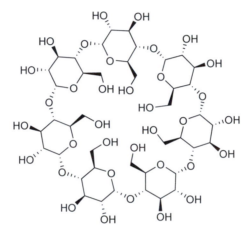

Figure 41.5 ▶ Structure β-cyclodextrin.

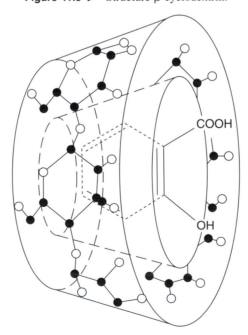

Figure 41.6 ▶ Structure of aspirin and β-cyclodextrin complex. (Reprinted with permission from [20].)

TABLE 41.8 ► **Examples of Cyclodextrin-containing Pharmaceutical Products**[19]

Drug/cyclodextrin	Trade name	Formulation
Alprostadil (PGE$_1$)/αCD	Prostavastin, Rigidur	IV solution
Itraconazole/HPβCD	Sporanox	Oral and IV solutions
Mitomycin/HPβCD	Mitozytrex	IV infusion
Voriconazole/SBEβCD	Vfend	IV solution
Ziprasidone mesylate/SBEβCD	Geodon, Zeldox	IM solution

αCD, α-Cyclodextrin; HPβCD, 2-hydroxypropyl-β-cyclodextrin; SBECD, sulfobutylether β-cyclodextrin.

the most useful pharmaceutical complexing agent because of its complexing abilities, low cost, and other properties.[20] Table 41.8 lists examples of marketed cyclodextrin-containing pharmaceutical products.[19] The major limitation of cyclodextrin complexes is toxicity, especially at high concentrations, which limits the dose level. Formation of cyclodextrin complexes requires specific molecular properties, and this approach may not work for certain compounds. Cyclodextrin derivatives with improved properties tend to be expensive.

Solid Dispersions

Solid dispersion systems can increase dissolution rate and bioavailability of water-insoluble drugs.[21–23] In solid dispersion systems, a drug may exist as an amorphous form in inert, hydrophilic polymeric carriers to form solid solutions (Figure 41.7). When they are exposed to aqueous media, the carriers dissolve, and the drug is released as very fine colloidal particles. This greatly reduces particle size and increases surface area, which results in improved dissolution rates and PO absorption. Furthermore, no energy is required to break up the crystal lattice of a drug (normally present in a crystalline solid dosage form) during the dissolution process. Drug solubility and wettability may be increased by surrounding hydrophilic carriers.

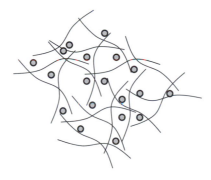

Figure 41.7 ► Amorphous solid solution.[22,24]

The methods used to prepare solid dispersions include the melting method, the solvent method, and the solvent wetting method.[23–25] Although solid dispersion is an area of active research, very few products relying on this technology have made it to the market (Table 41.9). The main reason is that solid dispersion is a high-energy metastable form. Phase separation, crystal growth, or conversion from the amorphous to the crystalline form during storage decrease solubility and dissolution rate and result in variable oral bioavailability.

Particle Size Reduction

If oral bioavailability is dissolution rate limited (not solubility limited), particle size reduction can increase the performance of the drug. The effect of particle size on the oral

TABLE 41.9 ▶ Examples of Marketed Products with Solid Dispersion Formulation[23]

Drug carrier	Trade name
Griseofulvin-poly(ethylene glycol)	Gris-PEG (Novartis)
Nabilone-povidone	Cesamet (Lilly)

bioavailability of a water-insoluble discovery compound is illustrated in Figure 41.8.[26] The smaller the particle size, the higher the in vivo exposure after PO dosing due to enhancement of dissolution rate. Microparticulate and nanoparticulate systems have particle sizes in the low micrometer to nanometer range. They can be delivered using all common routes of administration, that is, PO, injectable (IP, SC, and IM), and topical.

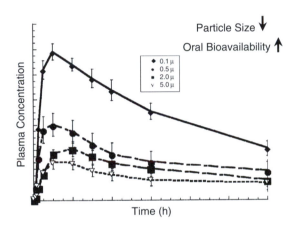

Figure 41.8 ▶ Effect of particle sizes on oral bioavailability of a discovery compound. (Reprinted with permission from [26].)

Many milling technologies have been developed for particle size reduction, such as the ball mill, fluid energy mill, cutter mill, hammer mill, pin mill, vibration mill, and media mill. Because only a small amount of material is available in drug discovery laboratories, industrial scale milling instruments may not be cost effective. A simple grinding apparatus, such as mortar and pestle or coffee mill, can be useful for reducing particle size to a narrow size distribution in order to increase oral bioavailability and reduce in vivo experimental variability due to wide particle size distribution. When particle size is small enough that dissolution is no longer a rate-limiting factor, variability due to food effect can be significantly reduced. A great advantage of nanoparticle technology is that drugs can be dosed at a significantly higher level than traditional approaches using co-solvents, for which dose is limited due to toxicity of the excipients. Figure 41.9 shows that a nanoparticle formulation of Taxol can be dosed at three times a higher level than the highest dose in the current commercial formulation of Cremophor EL/ethanol, which translates to greater efficacy.[26]

41.4 Practical Guide for Formulation in Drug Discovery

Although many advances have been made in formulating clinical dosages for humans, few reports have addressed formulation issues in preclinical studies.[2,7,27−29] The objective of preclinical in vivo studies affects formulation strategies.[29] For example, if the goal is to obtain an idea of oral efficacy, an optimal solution formulation is most effective in producing

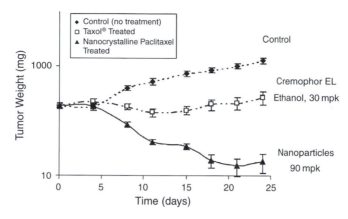

Figure 41.9 ► Effect of nanoparticles on efficacy of human lung tumor. (Reprinted with permission from [26].)

maximal exposure and reducing the variables in the studies compared to suspension or solid dosage form. On the other hand, if the purpose of an in vivo study is to explore the feasibility of developing the drug candidate into a commercial product, a more complex design is necessary using both solution and suspension formulation to determine both high and low solubility and dissolution effects on exposure. Different strategies are applied for PK, toxicokinetic, and pharmacodynamic studies.

41.4.1 Formulation for PK Studies

For early discovery PK screening, solution formulation is preferred to eliminate the effects of solubility and solid-state properties, such as crystal forms and particle size. Insufficient in vivo exposure will lead to a search for new and improved compounds through structural modification and the identification of reasons for low exposure (permeability, metabolism, etc.). Formulations should be tested for potential precipitation in simulated gastrointestinal fluids (simulated gastric fluid, simulated intestinal fluid, fasted state simulated intestinal fluid, fed state simulated intestinal fluid[30]). The same formulation should be used for the entire series of compounds within a discovery project for unbiased comparison and to minimize the potential impact of vehicle on the PK profiles. Although solution formulation is not always possible for all the compounds within a series, efforts should be made to formulate the first few compounds in the series to develop a robust vehicle for all the compounds in the projects. Table 41.10 lists possible vehicles for in vivo PK studies.[27]

TABLE 41.10 ► **Possible Formulation Approaches for Discovery in Vivo PK Screening**[27]

pH Adjustment and co-solvent	Surfactant solution	Lipid-based
pH buffers	Surfactants	Pure oil solution
Co-solvents:	Labrasol	Oil/buffer/surfactant
Polyethylene glycol	Tween 80	Emulsions
Propylene glycol	Cremophor RH 40	Micelles
Glycofurol 75	Lecithin	Liposomes
Glycerine		
Ethanol		
Transcutol		
pH buffers + co-solvents	Surfactant + co-solvents	

If compounds have excellent exposure using a solution formulation, PK studies with a suspension formulation should be initiated to (1) evaluate feasibility for solid dosage form, (2) predict exposure in toxicity studies, which is often performed in suspension formulation, and (3) measure exposure at maximum tolerated dose. If compounds have excellent exposure in a solution formulation but poor exposure with a suspension, physical properties can be modified to increase bioavailability by particle size reduction, such as micronization or nanoparticle technologies. Compounds with acceptable exposure in both solution and suspension formulations are considered for further development advancement.[27]

41.4.2 Formulation for Toxicity Studies

Suspension formulation with simple and safe excipients typically is used for acute and chronic toxicity studies. Vehicles often consist of a surfactant for wetting of particles (e.g., 2% Tween 80) and a bulking agent to increase viscosity and reduce sedimentation (e.g., 0.5% methylcellulose or hydroxyethyl cellulose). For hard-to-formulate or low systemic exposure compounds, due to low permeability or high metabolism, more sophisticated formulations (e.g., lipid-based or nanoparticle delivery systems) or a different route of administration (IV, IP, SC) can be used for toxicity studies.

41.4.3 Formulation for Pharmacological Activity Studies

Animal models are specific for the different diseases and therapeutic areas. Efficacy studies tend to be longer in duration and require more expensive animals (e.g., transgenic animals) compared to PK models. An optimal formulation is essential in order to demonstrate activity and proof of concept. Table 41.11 gives possible formulations for cardiovascular studies.[27]

TABLE 41.11 ▶ Possible Formulations for Efficacy Studies in Pharmacological Models[27]

Oral suspension	Oral solution	IV Solution
0.5% Hydroxyethylcellulose, wet milled	20%–50% PEG 400	20% PEG 400 in saline, 0.2 mL/kg (inject slowly)
0.5% Hydroxyethylcellulose/ 0.1% Tween 80, wet milled	3:1 (20%–50% glycofurol 75/Cremophor) to saline or buffer	50% PEG 400 in saline, 0.1 mL/kg (10-min infusion)
0.5% Hydroxyethylcellulose/ 50% Lipofundin, wet milled	Phosal 50 PG: mixture of 50% propylene glycol and 50% soybean lecithin	20%–50% glycofurol 75 in saline
N20 (10% soy bean oil in mixture)	Tween 80 (up to 10% in water)	20%–50% PEG/glycofurol/poloxamer 188 (39/10/1), in saline or buffer
Nanocrystals	1:1 Labrasol/Gelucire: diluted with 50%–90% water or buffer (emulsion)	3:1 (20%–50% glycofurol 75/Cremophor) to saline or buffer
	95:5 Miglyol/lecithin: to be homogenized with 50%–90% water (emulsion)	9:1 (20%–50% glycofurol 75/Solutol HS-15) to saline or buffer

Problems

(Answers can be found in Appendix I at the end of the book.)

1. What is the most common route of administration? What compound properties limit this route for some compounds?

2. What are advantages of IV administration?

3. Which of the following are advantages of formulation?: (a) reduce food effect, (b) reduce CYP inhibition, (c) reduce bioavailability, (d) increase stability, (e) achieve higher and earlier blood concentrations, (f) reduce phase II metabolism, (g) increase absorption.

4. What is an in situ salt, and how would you make one for a basic compound with $pK_a = 9.5$?

5. What three co-solvents are most commonly used in commercial drug products? What do they do?

6. Which of the following are effects of surfactants in formulations?: (a) incorporates insoluble compounds into micelles, (b) inhibits metabolism in the intestine, (c) surface effects reduce precipitation, (d) stabilizes particulate suspensions, (e) opens tight junctions for improved paracellular permeability.

7. What is a commonly used discovery formulation for in vivo dosing? What is the function of the components?

8. What different forms can lipid formulations take? Briefly describe each.

9. How do cyclodextrins work?

10. What aspects of a solid dispersion enhance intestinal solubility?

11. Why is it useful to reduce particle size?

12. Why is a solution formulation preferred for dosing in early discovery PK studies?

13. Why is it worthwhile to optimize a dosing formulation for efficacy studies?

14. What routes of administration can bypass first-pass metabolism?: (a) PO, (b) IP, (c) IM, (d) SC, (e) IV.

References

1. Ward, P. D., Tippin, T. K., & Thakker, D. R. (2000). Enhancing paracellular permeability by modulating epithelial tight junctions. *Pharmaceutical Science & Technology Today, 3*, 346–358.

2. Neervannan, S. (2006). Preclinical formulations for discovery and toxicology: physicochemical challenges. *Expert Opinion on Drug Metabolism & Toxicology, 2*, 715–731.

3. Devillers, G. Exploring a pharmaceutical market niche & trends: nasal spray drug delivery. In *Drug delivery technology*. Retrieved from http://www.drugdeliverytech.com/cgi-bin/articles.cgi?idArticle = 128.

4. Simamora, P., Pinsuwan, S., Alvarez, J. M., Myrdal, P. B., & Yalkowsky, S. H. (1995). Effect of pH on injection phlebitis. *Journal of Pharmaceutical Sciences, 84*, 520–522.

5. Johnson, J. L..H., He, Y., & Yalkowsky, S. H. (2003). Prediction of precipitation-induced phlebitis: a statistical validation of an in vitro model. *Journal of Pharmaceutical Sciences, 92*, 1574–1581.

6. Lukas, G., Brindle, S. D., & Greengard, P. (1971). Route of absorption of intraperitoneally administered compounds. *Journal of Pharmacology and Experimental Therapeutics, 178*, 562–566.

7. Lee, Y.-C., Zocharski, P. D., & Samas, B. (2003). An intravenous formulation decision tree for discovery compound formulation development. *International Journal of Pharmaceutics*, *253*, 111–119.

8. Sweetana, S., & Akers, M. J. (1996). Solubility principles and practices for parenteral drug dosage form development. *PDA Journal of Pharmaceutical Science and Technology*, *50*, 330–342.

9. Strickley, R. G. (2004). Solubilizing excipients in oral and injectable formulations. *Pharmaceutical Research V21*, 201–230.

10. Pouton, C. W. (2000). Lipid formulations for oral administration of drugs: non-emulsifying, self-emulsifying and "self-microemulsifying" drug delivery systems. *European Journal of Pharmaceutical Sciences*, *11*, S93-S98.

11. Pouton, C. W. (2006). Formulation of poorly water-soluble drugs for oral administration: physicochemical and physiological issues and the lipid formulation classification system. *European Journal of Pharmaceutical Sciences*, *29*, 278–287.

12. Collins-Gold, L., Feichtinger, N., & Warnheim, T. (2000). Are lipid emulsions the drug delivery solutions? *Modern Drug Discovery*, *3*, 44–46, 48.

13. Wang, J., Mongayt, D., & Torchilin, V. P. (2005). Polymeric micelles for delivery of poorly soluble drugs: preparation and anticancer activity in vitro of paclitaxel incorporated into mixed micelles based on poly(ethylene glycol)-lipid conjugate and positively charged lipids. *Journal of Drug Targeting*, *13*, 73–80.

14. Lavasanifar, A., Samuel, J., & Kwon, G. S. (2002). Poly(ethylene oxide)-block-poly(-amino acid) micelles for drug delivery. *Advanced Drug Delivery Reviews*, *54*, 169–190.

15. Liggins, R. T., & Burt, H. M. (2002). Polyether-polyester diblock copolymers for the preparation of paclitaxel loaded polymeric micelle formulations. *Advanced Drug Delivery Reviews*, *54*, 191–202.

16. Kwon, G. S. (2002). Block copolymer micelles as drug delivery systems. *Advanced Drug Delivery Reviews*, *54*, 167.

17. Aliabadi, H. M., & Lavasanifar, A. (2006). Polymeric micelles for drug delivery. *Expert Opinion on Drug Delivery*, *3*, 139–162.

18. Reimer, D., Eastman, S., Flowers, C., Boey, A., Redelmeier, T., & Ouyang, C. (2005). Liposome formulations for sparingly soluble compounds: liposome technology offers many advantages for formulation of sparingly soluble compounds. In *Pharmaceutical Formulation and Quality*. Retrieved from http://www.pharmaquality.com/mag/08092005/pfq_08092005_FO2.html.

19. Loftsson, T., & Duchene, D. (2007). Cyclodextrins and their pharmaceutical applications. *International Journal of Pharmaceutics*, *329*, 1–11.

20. Loftsson, T., & Masson, M. (2001). Cyclodextrins in topical drug formulations: theory and practice. *International Journal of Pharmaceutics*, *225*, 15–30.

21. Sethia, S., & Squillante, E. (2003). Solid dispersions: revival with greater possibilities and applications in oral drug delivery. *Critical Reviews in Therapeutic Drug Carrier Systems*, *20*, 215–247.

22. Kreuter, J. (1999). Feste dispersionen. In J. Kreuter & C.-D. Herzfeldt (Eds.), *Grundlagen der arzneiformenlehre galenik*, *2* (pp. 262–274). Frankfurt am Main: Springer.

23. Serajuddin, A. T. M. (1999). Solid dispersion of poorly water-soluble drugs: early promises, subsequent problems, and recent breakthroughs. *Journal of Pharmaceutical Sciences*, *88*, 1058–1066.

24. Leuner, C., & Dressman, J. (2000). Improving drug solubility for oral delivery using solid dispersions. *European Journal of Pharmaceutics and Biopharmaceutics*, *50*, 47–60.

25. Breitenbach, J. (2002). Melt extrusion: from process to drug delivery technology. *European Journal of Pharmaceutics and Biopharmaceutics*, *54*, 107–117.

26. Merisko-Liversidge, E., Liversidge, G. G., & Cooper, E. R. (2003). Nanosizing: a formulation approach for poorly-water-soluble compounds. *European Journal of Pharmaceutical Sciences*, *18*, 113–120.

27. Maas, J., Kamm, W., & Hauck, G. (2007). An integrated early formulation strategy: from hit evaluation to preclinical candidate profiling. *European Journal of Pharmaceutics and Biopharmaceutics*, *61*, 1–10.

28. Chaubal, M. V. (2004). Application of formulation technologies in lead candidate selection and optimization. *Drug Discovery Today*, *9*, 603–609.

29. Chen, X.-Q., Antman Melissa, D., Gesenberg, C., & Gudmundsson Olafur, S. (2006). Discovery pharmaceutics: challenges and opportunities. *The AAPS Journal*, *8*, E402–408.

30. Galia, E., Nicolaides, E., Horter, D., Lobenberg, R., Reppas, C., & Dressman, J. B. (1998). Evaluation of various dissolution media for predicting in vivo performance of class I and II drugs. *Pharmaceutical Research*, *15*, 698–705.

Appendix I

Answers to Chapter Problems

Chapter 1: Introduction

1. Having properties that produce acceptable ADME/Tox.

2. Pharmacology (efficacy, selectivity) and ADME/Tox (physicochemical, metabolic, and toxicity).

3. Biologists can optimize biological assays, dosing vehicles, and routes of administration to insure quality biological data.

4. a, b, c, d

Chapter 2: Advantages of Good Drug-like Properties

1. Structure modification.

2. Physicochemical properties: physical environment (e.g., pH, co-solutes); Biochemical properties: proteins (e.g., enzymes, transporters).

3. Exploration: use chemical libraries of compounds having drug-like properties. Lead selection: select leads that have acceptable properties. Lead optimization: optimize properties by structure modification. Development selection: advance compounds to development that meet or exceed established drug-like property criteria.

4. In vitro property assays and in vivo PK measurement.

5. Development: higher attrition, slower and more expensive development; Clinical: increased patient burden and lower compliance; Product lifetime: reduced patent life.

6. Precipitation or instability make compounds appear less active; low permeability limits cell penetration for intracellular targets; poor PK properties limit therapeutic target exposure in vivo and make compounds less active; poor penetration into the CNS.

7. Structure–property relationships. How different structures affect drug-like property values.

8. a, b, c, e

Chapter 3: Barriers to Drug Exposure in Living Systems

1. Inherent activity (target binding), exposure (drug-like properties that perform well at in vivo barriers).

2. Low dose, once per day oral tablet.

3. Property limitations for exposure include low solubility, low permeability, decomposition in the GI tract and plasma, low metabolic stability, high efflux transport, high plasma protein binding.

4. Increasing solubility results in increasing concentration at the surface of the membrane, which is favorable for higher absorption.

5. Increasing permeability results in increasing transfer of compound molecules through the membrane, resulting in higher absorption.

6. Stomach has lower surface area, shorter transit time, lower blood flow, and lower pH that can cause chemical instability.

7. pH is lower in the fasted state.

8. b, b

9. Bile, enhances solubility of lipophilic drugs. Pancreatic fluid, can catalyze hydrolysis.

10. b and c

11. b

12. Enzymatic hydrolysis, plasma protein binding, red blood cell binding.

13. e, f

14. Metabolism, biliary excretion.

15. Blood–brain barrier.

16. Metabolites are more polar and more readily extracted in the nephron.

17. d

18. a, b, c, d, f, g

19. a, c

20. all

Chapter 4: Rules for Rapid Property Profiling from Structure

1. a, d, e, h, j

2. H-bonds with water must be broken to permeate through the membrane.

3. High lipophilicity reduces solubility.

4. b

5. (a) HBD$=0$, HBA$=2$; (b) HBD$=1$, HBA$=2$

6.

Structure	#HBD	#HBA	MW	cLogP	PSA	Problem
1 Buspirone	0	7	385	1.7	70	none
2	5	10	418	−3.3	143	PSA
3 Paclitaxel	4	15	852	4.5	209	HBA, MW, PSA
4 Cephalexin	4	7	347	0.5	138	none
5 Cefuroxime	4	12	424	−1.5	199	HBA, PSA
6 Olsalazine	4	8	302	3.2	141	PSA

7. a, c, e, f

8. b, d

9. a, c

Chapter 5: Lipophilicity

1. For Log P, all drug molecules are neutral in solution, whereas for Log D, anywhere from none to all molecules are ionized, depending on the compound pK_a and aqueous pH.

2. Molecular volume, dipolarity, hydrogen bonding acidity, hydrogen bond basicity.

3. $1 < \text{Log D}_{7.4} < 3$

4. Low Log P: low passive diffusion permeability. High Log P: low solubility.

5. b

6. a, c, e

7. b, c, d

Chapter 6: pK_a

1. b, d, e

2. a, c, f

3. a

4.

Location	pH	$[HA]/[A^-] = 10(pK_a - pH)$	Ionization
Stomach	1.5	$10^{(4.2-1.5)} = 10^{(2.7)}$	Neutral
Duodenum	$10^{(4.2-5.5)} = 10^{(-1.3)}$	$(-)$ Ionized $\sim$95%	
Blood	7.4		$10^{(4.2-7.4)} = 10^{(-3.2)}$
$(-)$ Ionized $\sim$100%			

Location	pH	$[BH+]/[B] = 10^{(pK_a - pH)}$	Ionization
Stomach	1.5	$10^{(9.8-1.5)} = 10^{(8.3)}$	$(+)$ Ionized
Duodenum	$10^{(9.8-5.5)} = 10^{(4.3)}$	$(+)$ Ionized	
Blood	7.4	$10^{(9.8-7.4)} = 10^{(2.4)}$	$(+)$ Ionized

5. b, d

Chapter 7: Solubility

1. The same. Both the free base and HCl salt will reach the same concentration in the buffered solution, thus, the IC_{50} will be the same.

2. No. In pure water the pH will change when the compound is added, and the sodium salt will be more soluble. Sodium salt. Yes.

3. Increase solubility: add an ionizable group, reduce Log P, add H-bonding, add polar group, out-of-plane substitution, reduce MW, prodrug, and formulation. Best: introduce ionizable groups. Increase dissolution rate: reduce particle size, formulate with surfactant, and salt form.

4. Solubility-limited absorption.

5. 2,000 µg/mL

6. In order to improve target binding, lipophilic groups are often added to the template, which reduces aqueous solubility.

7. pH, counter-ions, protein, lipid, surfactants, salts, co-solvents (types and concentrations), buffer, temperature, and incubation time.

8. Lipophilicity, molecular size, pK_a, charge, crystal lattice energy.

9. Solubility is the highest sustainable concentration; dissolution rate is how much of the compound dissolves per unit time.

10. Discovery compounds usually are amorphous solids whose thermodynamic solubility can change from batch to batch. Nearly all discovery experiments first dissolve compounds in DMSO solution.

11. (a) 10 µg/mL, (b) 100 µg/mL, (c) 520 µg/mL

12. Permeability by passive diffusion.

13. Add ionizable group.

14. b

15.

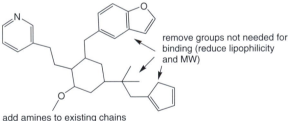

16. a, d

17. all

18. a, d

19. all

Chapter 8: Permeability

1. passive diffusion.

2. MW <180; polar.

3. (a) increase, (b) decrease.

4. Uptake into cells in cell-based assays, intestinal absorption, blood–brain and other organ barriers, therapeutic target cell penetration, entry and elimination from hepatocytes (liver), kidney nephron.

5. a, d, e

6.

7. a, c, e, f

8. d, b, c, a (H-bonding and polarity reduce permeability).

9. a, c, b (the lead compound already has high MW and lipophilicity, so adding more molecular mass is likely to reduce permeability).

10. a, c, b (the lead compound has low MW and lipophilicity, so adding more lipophilicity is likely to increase permeability).

Chapter 9: Transporters

1. a, b, c, e, f, h, i

2. b

3. all

4. Pgp. It affects oral absorption, brain penetration, drug excretion, multidrug resistance tumor cells, and resistance to antibiotics. Pgp has broad substrate specificity and impacts PK profiles of many drug candidates.

5. B, D

6. Decrease H-bond acceptors, steric hindrance adjacent to H-bond acceptor, reduce lipophilicity, add groups that can disrupt Pgp binding.

7. Pgp knockout mice; Pgp inhibitor co-dosing ("chemical knockout").

8. Efflux: b, e; Uptake: a, c, d, f.

9. c

10. all

11. a, b, c, d

12. a, b

13. C, A, B

Chapter 10: Blood–Brain Barrier

1. c

2. Drug concentration is much lower in the blood than in the GI lumen. Pgp can be saturated in the GI but not at the BBB.

3. e

4. B (MW, acid, total HB, PSA); C (MW); D (MW, acid, total HB, PSA).

5.

Remove unnecessary hydrogen bonders to get below 8 total (esp., donors)
Remove unnecessary groups to get MW below 400
Remove unnecessary polar and ionizable groups to increase cLogP

6. b, d, e, f

7. a, c, d, e, g

8. b, c, e, f

9. a, c, d, e, g

Chapter 11: Metabolic Stability

1. Intestinal decomposition (pH and enzymatic), intestinal metabolism, liver metabolism, plasma decomposition.

2. c

3. a, d

4. Change in metabolic reactions if one pathway is blocked.

5. Aliphatic and aromatic hydroxylation, *N*- and *O*-dealkylation, *N*-oxidation, *N*-hydroxylation, dehydrogenation, (others shown in Figure 11.5).

6. Glucuronidation, sulfation, *N*-acetylation, glutathione conjugation (others shown in Figure 11.6).

7. c, e

8. Possible structure modifications to investigate:

X = F, Cl, CN, CH$_3$, (CH$_3$)$_2$, =CH$_2$, CF$_3$, =O
Change NH to CH$_2$ if not necessary for activity

To block M1, M2

X = F, Cl, CN, CH$_3$, (CH$_3$)$_2$, =CH$_2$, CF$_3$, =O
Change ethyl to cyclopropyl
Remove entire ether if not necessary for activity
Change ether oxygen to CH$_2$ if not necessary

To block M3

Y = CH$_2$F, CHF$_2$, CF$_3$, Cl, CN, cyclopropyl

To block M4

Z = F, Cl, CN, CH$_3$, CH$_2$F, CHF$_2$, CF$_3$
Move N around ring or add second N

To block M5

9. Potential sites of phase II metabolic reactions:

Glucuronidation
Glycination

Glucuronidation
Sulfation

Acetylation
Glucuronidation

10. a, b, c, e

11. a, b, d, e, f

Chapter 12: Plasma Stability

1. B, D, G, I, J, K

2. Antedrugs, prodrugs.

3. The following are among the possible structural modifications:

Change ester to amide

Change ester to amide, Steric hindrance

Steric hindrance

Eliminate hydrolyzable group

4. No. The hydrolysis enzymes in microsomes and plasma are different and should be assessed separately.

5. c, d, f, g

Chapter 13: Solution Stability

1. All

2. No. The stability might be improved without reducing activity by structural modification, thus allowing a valuable pharmacophore/series to continue optimization toward a quality clinical candidate.

Chapter 14: Plasma Protein Binding

1. (a) metabolic clearance decrease, (b) tissue concentration decrease, (c) tissue distribution decrease, (d) blood concentration increase, (e) renal clearance decrease, (f) PK half-life increase, (g) pharmacological effect decrease, (h) brain penetration decrease.

2. Albumin, α_1-acid glycoprotein, lipoprotein.

3. Compound binds to albumin and is restricted from interacting with the target protein.

4.

A

Reduce acidity
Reduce lipophilicity (cLogP = 4.4)
Increase PSA (44)
Increase HBD (+1)

Reduce acidity
Reduce lipophilicity (cLogP = 3.9)
Increase PSA (44)
Increase HBD (+1)

Reduce acidity
Reduce lipophilicity (cLogP = 1.2)
Increase PSA (52)
Increase H-bonding (+1 HBD, +1 HBA)

B

Reduce lipophilicity (cLogP = 3.5)
Increase PSA (22)
Increase H-bonding (+1 HBA)

Reduce lipophilicity (cLogP = 1.7)
Increase PSA (39)
Increase H-bonding (+2 HBA)

Increase basicity
Reduce lipophilicity (cLogP = 0.5)
Increase PSA (51)
Increase H-bonding (+1 HBD, +3 HBA)

(For Example 4B, increasing basicity will likely increase binding to AGP.)

Chapter 15: Cytochrome P450 Inhibition

1. $10\,\mu M$

2. Greater than 10 times C_{max}. When concentration is greater than K_i.

3. Seldane (terfenadine) was removed from the market because its metabolism at CYP3A4 was inhibited by 3A4 inhibitors, such as erythromycin. The resulting higher concentrations of Seldane led to hERG blocking and TdP arrhythmia in some patients.

4. In reversible inhibition, the inhibitor binds and releases from the enzyme. In mechanism-based inhibition, the inhibitor binds to the enzyme by a covalent reaction or strong complexation interaction. Dialyzing the inhibitor away from the enzyme can reduce reversible inhibition and inhibition does not increase with incubation time. Mechanism-based inhibition is not reduced by dialysis and inhibition increases with incubation time.

5. Among the possible structural modifications are the following:

Decrease lipophilicity (cLogP = 3.0)

Decrease lipophilicity (cLogP = 3.5)
Add steric hindrance

Decrease lipophilicity (cLogP = 3.8)
Reduce the pK_a of the nitrogen

6. c

7. No. CYP inhibition involves competitive binding to a specific CYP isozyme when two compounds are present. A metabolic reaction does not need to occur to the inhibitor; it only needs to bind. In CYP inhibition, the reaction occurs to the substrate (not necessarily to the inhibitor). Metabolic stability involves one compound that (1) binds to any of the CYP isozymes and (2) reacts to form a metabolite.

Chapter 16: hERG Blocking

1. Potassium ion channel in heart muscle.

2. Outflow of K^+ ions from the cell is part of the action potential and reestablishes the internal negative potential of the cells.

3. Lengthened QT interval on electrocardiogram (ECG).

4. Torsades de pointes arrhythmia, which can be triggered by LQT.

5. 1 in 10^5 to 10^6 patients for antihistamines, 1 in 5×10^4 patients for terfenadine.

6. hERG $IC_{50}/C_{max, unbound} > 30$, or < 5 seconds lengthening of QT interval.

7. c

8. c, d

9. Among possible structural modifications are the following:

Reduce pK_a of amine
Reduce lipophilicity of substructure (cLogP = 3.2)

Reduce lipophilicity of substructures (cLogP = 2.1)

10. c

Chapter 17: Toxicity

1. Ratio (or range) of compound concentrations at which it is toxic versus efficacious. Large.

2. Reactive metabolites undergo covalent reactions with protein, DNA. They can be detected using glutathione or another trapping reagent.

3. a, b, e

4. Redox cycling by some structures (e.g., quinines) deplete the reducing capacity of the cell (e.g., glutathione), allowing free radicals and peroxides to increase.

5. Higher levels of metabolizing enzymes are induced, resulting in greater chance of forming reactive metabolites or increasing the clearance of a co-administered drug.

6. Antagonism or agonism of another biochemical target can cause side effects, which can be toxic.

Chapter 18: Integrity and Purity

1. Activity could be due to the impurity; measured property could be inaccurate due to interference by impurity; impurity may cause a toxic effect; erroneous SAR conclusions from wrong structure.

2. Mishandling, mislabeling, decomposition as a solid or in solution, inaccurate original structural assignment.

3. All

4. b

Chapter 19: Pharmacokinetics

1.

PK parameter	Definition choices
CL	b. rate a compound is removed from systemic circulation
C_{max}	f. highest concentration reached in the blood
V_d	e. apparent volume into which the compound is dissolved
$t_{1/2}$	c. time for the compound's concentration in systemic circulation to decrease by half
C_0	g. initial concentration after IV dosing
AUC	d. compound's exposure as determined by blood concentration over time
%F	a. percentage of the oral dose that reaches systemic circulation unchanged

2. IV introduces the entire dose of a compound directly into the bloodstream. PO involves a delay for stomach residence, dissolution (if a solid), intestinal absorption over a couple of hours, and first-pass metabolism, which reduces the amount of compound that reaches the bloodstream.

3. (a) high nonspecific tissue binding, higher lipophilicity, lower PPB; (b) higher plasma protein binding, lower lipophilicity, higher hydrophilicity, lower nonspecific tissue binding.

4. b

5. Liver, kidney

6. 0.1: highly bloodstream restricted; 1: evenly distributed throughout the body; 100: highly tissue bound.

7. b

8. a

9. b

10.

IV dose (mg/kg)	PO dose (mg/kg)	AUC_{PO} (ng•h/mL)	AUC_{IV} (ng•h/mL)	Bioavailability
1	10	500	500	10%
2	10	1000	500	40%
5	10	300	200	75%

11. b

Chapter 20: Lead-like Compounds

1. Optimization typically synthesizes new compounds by adding moieties that increase the MW, hydrogen bonds, and lipophilicity. Such new compounds can exceed the rule of 5 and be at greater risk for poor solubility, permeability, and absorption.

2. (a) not lead-like: MW, PSA, HBD, and HBA; (b) not lead-like: PSA, HBD, HBA; (c) just over the guidelines for lead-like: MW, HBA; (d) not lead-like: MW, PSA, HBD, HBA.

3. a, c

Chapter 21: Strategies for Integrating Drug-like Properties into Drug Discovery

1. Performing target-binding optimization first can lock the project into substructures that reduce drug-like properties. There is great reluctance to eliminate structural features and reduce binding to achieve good properties. When planning structural modifications to improve target binding, an alternative is to consider the effects of the changes on the drug-like properties.

2. Faster decision-making. More new structures can be planned and synthesized to increase the chances of finding a successful structure modification that improves a property.

3. With multiple property assays, medicinal chemists do not receive specific guidance on how to modify the structure to improve the property. For example, Caco-2 permeability may have permeability components from passive diffusion, paracellular, active uptake, and efflux mechanisms. PAMPA permeability only has passive diffusion, and medicinal chemists can readily identify the structural modifications that will increase passive diffusion (e.g., increase lipophilicity, decrease hydrogen bonds, decrease polarity, change pK_a).

Chapter 22: Methods for Profiling Drug-like Properties: General Concepts

1. No. The pH 7.4 conditions of the generic solubility assay are very different than the low pH conditions of the intestine.

2. No. The Caco-2 assay is too expensive at this level. Most of the HTS hits can be deprioritized using other, less expensive data or calculated properties. After less expensive triage, a smaller number of HTS hits can be run on Caco-2 for the final decisions on leads.

3. No. Clinical candidate nomination should have a more rigorous data package that includes Caco-2 for permeability assessment.

4. No. Hydrolysis can occur in the plasma and intestine and be catalyzed by numerous enzymes, which have different substrate specificity than microsomal hydrolysis enzymes.

5. Yes. Metabolic stability varies among species. Therefore, it is useful to have data for metabolic stability in rat (often used as the toxicology species), the project's pharmacology/efficacy species (e.g., stability in transgenic mice can help with decisions on in vivo dosing and data interpretation if the pharmacology/efficacy species is transgenic), and human (assists predictions for clinical studies).

Chapter 23: Lipophilicity Methods

1. Drug discovery compounds may contain substructures not covered adequately in the training set for the software. Also, they generally have fewer drug-like properties than the commercial drugs on which the in silico models were built and behave more poorly.

2. Yes. Log P has been studied for many more years, so the underlying mechanisms are better understood. High-quality measurements of Log P, which are used to build the in silico models, are more available than for other properties.

3. a, b, d, e, f, g

4. Standards, whose Log D values were previously measured using an in-depth method, are run on the same system. Their retention times are plotted versus their Log D values.

5. A factor of 10 or 1 Log unit.

6. a, d, e

Chapter 24: pK_a Methods

1. As the ionizable group is titrated from neutral to charged (or vice versa), the following can be detected: (1) change in the UV absorbance of a group near the ionizable group, (2) change in pH of the solution, or (3) change in electrophoretic mobility.

2. Throughput: in silico > SGA > CE > pH metric.

3. Features: high throughput; pK_a is calculated for different tautomers; the software labels each ionization center with its pK_a on the structure.

4. Yes. As with *all* assays, low solubility can affect the measurement. Either no data should be reported for very-low-solubility compounds, or methods should be modified and validated to work accurately for low solubility compounds.

Chapter 25: Solubility Methods

1. Log S = 0.8 − Log P − 0.01 (MP-25)

Compound	Melting point (°C)	Log P	Estimated solubility (mol/L)
1	125	0.8	0.1
2	125	1.8	0.01
3	225	1.8	0.001
4	225	3.8	0.00001

2. $S_{tot} = S_{HA} (1 + 10^{(pH - pK_a)})$

Compound	Intrinsic solubility (g/mL)	pK_a	pH	Total solubility (g/mL)
1	0.001	4.4	7.4	1
2	0.001	4.4	4.4	0.002
3	0.001	4.4	8.4	10
4	0.00001	4.4	7.4	0.01

3. Nephelometric: light scattering by precipitated particles. Direct UV: dissolved compound in aqueous solution.

4. Customized solubility methods use the same solution conditions as the experiment in a discovery team experiment (e.g., bioassay buffer) in order to better estimate the solubility of a compound under the conditions of interest.

5. Compound form: kinetic solubility uses a DMSO solution, whereas equilibrium solubility uses solid. Compound addition: kinetic solubility adds a small volume of DMSO solution

to an aqueous buffer, whereas equilibrium solubility adds buffer to solid compound. Time: kinetic solubility uses 1 to 18 hours of incubation, whereas equilibrium solubility uses 24 to 72 hours of incubation.

6. a, d, e

Chapter 26: Permeability Methods

1.

Method	Passive diffusion	Active uptake	Efflux	Paracellular
IAM	X			
PAMPA	X			
Caco-2	X	X	X	X

2. IAM uses HPLC instrumentation, which is widely available in drug discovery laboratories.

3. (a) 21 days of cell culture prior to use; (b) LC/MS quantitation; (c) Caco-2 is often run in both A>B and B>A directions; (d) more expensive supplies; (e) more scientists' time.

4. Active uptake transport, efflux transport, paracellular.

5. IAM uses phospholipids bonded to the solid support, whereas reversed phases commonly use bonded octadecane.

6. Bases appear to be effluxed. Acids appear to be actively taken up.

7.

Compound's permeability mechanism(s)	PAMPA relatively higher than Caco-2	Caco-2 relatively higher than PAMPA	PAMPA and Caco-2 relatively the same
Passive diffusion only			X
Passive diffusion and active uptake		X	
Passive diffusion and efflux	X		
Passive diffusion and paracellular		X	

Chapter 27: Transporter Methods

1. $ER = P_{B>A}/P_{A>B}$. $ER > 2$ indicates significant compound efflux.

2. (a) Measure $P_{A>B}$ with and without an efflux transporter-specific inhibitor. If $P_{A>B}$ is greater in the presence of the inhibitor, the compound is an efflux substrate. (b) Measure $P_{A>B}$ with a cell line (e.g., MDCKII) transfected with the gene of one specific efflux transporter (e.g., MDR1) and with the wild-type cell line. If $P_{A>B}$ is greater in the wild type, the compound is likely to be an efflux substrate.

3. Uptake is useful when (a) the cell line does not form a confluent monolayer with tight junctions that allow for a transwell experiment, (b) the discovery project team is interested in a specific cell line.

4. Binding of a test compound to an ABC transporter (e.g., Pgp). The ATP hydrolyzes to form ADP and Pi. The Pi is detected colorimetrically.

5. Inhibition of Pgp, which limits efflux of a known Pgp substrate (calcein-AM).

6. In vivo Pgp assay (Pgp knockout mouse) demonstrates proof-of-concept that Pgp affects the ADME characteristics of the compound.

7. Genetic knockout is a permanent condition in which no Pgp gene is present and none is expressed, thus no Pgp efflux can occur. The chemical knockout experiment involves the co-administration of a Pgp inhibitor with the test compound or saturation with test compound to reduce efflux. Chemical knockout can be used in the efficacy/pharmacology species model to test biological effects of Pgp substrates.

Chapter 28: Blood–Brain Barrier Methods

1. BBB permeation is the velocity of compound transfer across the BBB into the brain; brain distribution is the partitioning of compound between blood and brain tissue.

2. Hydrogen bonding, PSA, lipophilicity, MW, basicity/acidity.

3. Classification of passive BBB permeability (CNS +, CNS −)

4. Plasma and buffer; brain homogenate and buffer. It is a brain distribution method. The method indicates free drug concentration in the brain, free drug concentration in plasma, and predicted B/P.

5. Hydrogen bonding capacity.

6. BBB permeability by passive diffusion.

7. The isolation process is difficult and time consuming. Cells do not form tight junctions. Transporters may be overexpressed or underexpressed.

8. MDR1-MDCKII is primarily used for Pgp efflux and, occasionally, as a model of total BBB permeability. MDCK (without MDR1) is used for passive diffusion.

9. B/P limitations are it is resource intensive, B/P is a distribution ratio that is heavily affected by plasma protein and brain tissue binding but not necessarily by free drug.

10. BBB permeability (unaffected by brain tissue binding, plasma binding, or metabolism).

Chapter 29: Metabolic Stability Methods

1. S9 and hepatocytes contain more metabolizing enzymes than microsomes; thus, they survey a broader number of potential metabolic reactions. Hepatocytes also involve cell membrane permeability.

2. When the compound has a phenol, carboxylic acid, or other hydroxyl that might be susceptible to glucuronidation. In other compounds, glucuronidation may be observed as a minor metabolite but is not likely to be the rate-determining reaction for metabolic stability. Sulfation is rapid but is saturable and has limited capacity.

3. CYP-containing materials: rhCYP, microsomes, S9, hepatocytes, and liver slices contain CYPs. S9, hepatocytes, and liver slices contain sulfotransferases.

4. NADPH.

5. CYP enzymes can be saturated at higher concentration and give a falsely high metabolic stability value.

6. Metabolic stability usually varies among species, which may be important to the project.

7. The relationship between log (% remaining) and time is linear, so two points define the line. Additional time points add precision and accuracy but require additional resources.

8. Activity varies with each batch. If a project team is rank ordering compounds that were measured for metabolic stability using different microsome batches, it is important that the activity be consistent or the rank ordering, or the structure–metabolism relationships, will be erroneous.

9. Uridine diphosphate glucuronic acid (UDPGA), 3′-phosphoadenoside-5′-phosphosulfate (PAPS).

10. None. Hepatocytes generate their own co-factors.

11. (a) guiding structure modification for improvement of metabolic stability, (b) anticipating potential DDI.

12. Determine specifically where the metabolism occurred on the structure, to guide structural modification of a compound to block metabolism.

13. LC/MS/MS provides a profile of major metabolites and unambiguously identifies some metabolites (e.g., dealkylations). NMR provides regiospecific identifications of sites of hydroxylation and other metabolic reactions (e.g., position of the hydroxyl on a phenyl ring).

Chapter 30: Plasma Stability Methods

1. Hydrolytic enzymes of widely different binding specificity, e.g., esterases, lipases, and phosphatases.

2. Check the activity for comparison to previous batches; adjust the method conditions for consistent activity on QC compounds. Activity of different plasma batches can vary significantly.

3. (a) false (as long as you stay within the ranges discussed in the chapter); (b) true; (c) true; (d) true; (e) false.

Chapter 31: Solution Stability Methods

1. Bioassay media components, pH buffers, intestinal fluid, gastric fluid, enzymes, light, oxygen, temperature.

2. Quenching the reaction.

3. Use a programmable HPLC autosampler that can add reagents, mix, inject at predetermined time points, and perform these functions for multiple samples.

4. Use the same conditions and protocol that are relevant to the project's experiment in question.

5. Reaction kinetics (used to predict long-term stability); structures of decomposition products (used to modify structures for improved stability).

6. Diagnose unexplained results from in vivo or in vitro experiments; rank ordering compounds for stability; apply kinetics for planning other experiments or clinical studies; predict how much compound remains at various times; determine which moiety is unstable; guide structural modifications to improve stability.

Chapter 32: CYP Inhibition Methods

1. Recombinant human CYP isozymes (rhCYP); human liver microsomes.

2. High substrate turnover; high test compound metabolism; protein binding.

3. Fluorogenic; drug compounds; luminogenic; radioactive.

4. b

5. Fluorescent test compound or metabolites will interfere with the results.

6. Multiple CYP isozymes are simultaneously assayed, thus increasing throughput.

7. High substrate and test compound metabolism (turnover), for some isozymes, and protein binding increase the apparent IC_{50} of CYP inhibition, making the test compound appear less inhibitory. Conditions are not optimal for enzyme kinetics of most of the isoforms.

8. The concentrations of each isozyme and substrate can be set up independently from the other isozymes, thus allowing optimum enzymatic conditions for accurate initial enzyme kinetic rates.

Chapter 33: Plasma Protein Binding Methods

1. Equilibrium dialysis.

2. The equilibrium concentration of the chamber containing buffer without plasma protein.

3. The compound will appear to have higher clearance because it is not detected in the plasma.

4. The membrane filter prevents plasma proteins from passing, thus allowing the unbound compound concentration to be independently measured.

Chapter 34: hERG Methods

1. Alerting the project team early to potential hERG blocking problems. This allows the team to investigate further before major investment in synthesis of analogs for a lead series that may later fail.

2. Patch clamp.

3. Human ECG studies.

4. a, c

5. High IC_{50} values compared to patch clamp.

6. b, c

7. K$^+$; Rubidium (Rb) is the same size and charge as K$^+$ and can permeate through the K$^+$ channel. Rb can be sensitively detected using radioactivity with ^{86}Rb or atomic absorption spectroscopy with stable Rb.

8. c, d

9. The current required to hold the transmembrane potential at a constant voltage.

10. The standard patch clamp method is manual and time consuming, and requires highly skilled scientists to perform accurately.

11. a, c, d

Chapter 35: Toxicity Methods

1. Structural rules decided by expert committees and QSAR models.

2. An indication or alert to potential toxicity problems that can be followed up experimentally.

3. CYP3A4 induction.

4. Test the activity of each isozyme before and after incubation with the test compound. If the activity increases with test compound treatment, the compound induced the enzyme.

5. c, d

6.

Assay	DNA fragments move faster in gel electrophoresis than normal DNA	Abnormally divided DNA is observed by microscope	Reversion mutations allow colonies to grow without histidine	Unusually shaped chromosomes	Normal mammalian cells mutate so TMP does not kill them
Micronucleus		X			
Chromosomal aberration				X	
Comet	X				
Ames			X		
TK mouse lymphoma					X

7. c

8. Microscopic examination of embryos from rodents or zebrafish.

9. Glutathione

10. Animal dosing studies to determine maximum tolerable dose, such as acute dose, chronic dose (e.g., 2 weeks), safety pharmacology with full microscopic histology. In vitro genotoxicity/mutagenicity studies, such as Ames, micronucleus, and chromosome aberration.

11. Toxicometabonomics (measures changes in normal endogenous biochemical intermediates); toxicoproteomics (measures changes in protein expression); toxicogenomics (measures changes in mRNA expression).

Chapter 36: Integrity and Purity Methods

1. To increase throughput and deliver data to teams sooner. For example, 1,000 samples require 42 days to complete at 30 minutes per analysis versus 3.5 days at 5 minutes per analysis.

2. NMR spectral interpretation makes the time per sample too long. Also, impurities can be observed in NMR, but the number of impurities and their relative amounts are hard to determine because there is no chromatographic separation.

3. The undissolved material may be a different compound. If it is the putative compound, impurities will appear to be a higher relative concentration.

4. The small-particle columns enhance resolution and allow short analysis times (1–1.5 minutes per sample).

5. No HPLC detector responds on a molar basis for all compounds. Response per number of molecules varies with compound. Therefore, purity is a relative response versus the more desirable relative number of moles. Furthermore, no detector produces a response for all compounds. Some compounds do not respond to the detector and remain undetected.

6. No MS interfaces produces ions for all compounds. Nor do they produce the same types of ions for all compounds. For example, the commonly used electrospray interface makes positive $(M+1)^+$ ions for basic compounds, negative $(M-1)^-$ ions for acidic compounds, adduct ions (e.g., $M+NH_4)^+$ ions for neutral compounds, and no ions for some compounds. A trained scientist should examine the data for consistency with expected ionization.

Chapter 37: Pharmacokinetic Methods

1. Co-dosing multiple compounds reduces the number of animals used in PK experimentation. It also shortens the experimental time, so the project team receives the data faster. It is useful for initial PK screening of many compounds and can be later followed up with more detailed studies of selected compounds.

2. The criticism of cassette dosing is that compounds may interact and affect each other's PK parameters, especially with regard to metabolizing enzymes and transporters. This has been addressed by reducing the dosing level (mg/kg) to reduce interaction. Also, a compound with known PK parameters is added to the cocktail, and, if its PK parameters are affected, the study is repeated with individual compounds.

3. Samples are obtained unattended around the clock, thus allowing fewer scientist hours and improving scheduling.

4. In CARRS, the compounds are individually dosed. The samples from two animals, dosed with the same compound, are pooled to reduce the analyses by half.

5. Matrix suppression is the co-elution of plasma matrix material from the HPLC column into the mass spectrometer along with the test compound. This material suppresses the ion signal from the test compound and causes a falsely low or variable signal.

6. The team wants to know the concentration of a compound in the disease target tissue, in order to correlate in vivo concentration to pharmacological effects or to determine if the compound is penetrating into the tissue (e.g., brain, tumor) to a sufficient concentration.

7.

cpd	Dose [IV, PO] (mg/kg)	AUC_{PO} (ng•h/mL)	C_0 (ng/mL)	AUC_{IV} (ng•h/mL)	CL (mL/min/kg)	V_d (L/kg)	%F
1	1, 10	2,000	1,000	4,000	4.2	1	5%
2	1, 10	2,000	2,000	1,000	17	0.5	20%
3	5, 10	200	1,000	8,000	10	5	1.3%
4	1, 10	500	1,000	200	83	1	25%
5	5, 30	305	2,000	1,900	44	2.5	2.7%

Chapter 38: Diagnosing and Improving Pharmacokinetic Performance

1. Alternate formulation; salt forms; prodrugs; different dosing route (e.g., IV, IP, SC, IM).

2. Structure modifications that improve the limiting property.

3. Solubility, permeability, first-pass metabolism (GI and liver); intestinal solution stability.

4. a, c, f, g, m

5. a, c, j, k, o, p, q

6. b, d

Chapter 39: Prodrugs

1. b, c, e, g

2. Among the possible prodrug analogs are the following:

A.

B.

C.

D.

3. Ester prodrug:

4. Alkaline phosphatase.

5. Hydrolyzed by esterase in blood

Chapter 40: Effects of Properties on Biological Assays

1. Solubility in cell assay media, stability in cell assay media, permeability through the cell membrane, and toxicity to the cell.

2. If the property limitation can be diagnosed, structural modifications might improve the property without losing a valuable pharmacophore. If the assay characteristics are improved (e.g., to better solubilize or maintain the stability of a series), more accurate SAR may be derived. When multiple variables (i.e., properties, target binding) contribute to the final activity, discovery project teams are left without clear guidance for decision making and planning structural modifications to improve activity or properties.

3. Class 1: lower MW, rigid, hydrophilic (e.g., organic salts); Class 2: high MW, high Log P, many rotatable bonds.

4. Higher concentrations: if precipitate is carried from the DMSO stock solution into the aqueous solution and dissolves. Lower concentrations: if only the solution portion of the DMSO is carried into the aqueous solution; if both solution portion and particulate is carried into the aqueous solution and the precipitate does not dissolve.

5. Cooling of the solution can enhance crystallization and crystals are harder to redissolve. Condensation of water in the DMSO stock solution can lower the solubility of the compound.

6. (1) Class 1 may better dissolve in 1:1 aqueous/DMSO solution. (2) Another solvent can be substituted for DMSO. (3) Use the DMSO solution for a minimum time or a single time. (4) Store DMSO solutions at room temperature for a short term use (e.g., 2 weeks). (5) Make DMSO solution at a lower concentration. (6) Sonication.

7. Higher IC_{50}.

8. (1) Modify the dilution protocol. (2) Optimize the solutions to better solubilize compounds.

9. c

10. c, d

11. DMSO, other organic co-solvents, protein, excipients.

12. Molecular properties (H-bond donors, H-bond acceptors, MW, PSA, acidity), PAMPA, Caco-2.

13. IC_{50} right shift is caused when the concentrations of each point in the dilution curve are actually lower than intended. This is caused when the concentration of the initial (highest) concentration point is low because of compound precipitation or another error.

14. Often lower DMSO concentration and other co-solutes are used in cell-based assays, resulting in precipitation of low-solubility compounds. During method development, the tolerance of the assay for higher DMSO or other co-solvents can be tested and the DMSO concentration can be maximized for the final assay. The assay may also tolerate higher concentrations of other co-solvents, which can be used instead of DMSO.

Chapter 41: Formulation

1. Oral (PO). This route is limited by low solubility, low permeability, or high first-pass metabolism.

2. Rapid onset, bioavailability of the entire dose.

3. a, d, e, g

4. An in situ salt is a solution of the compound in a buffer that fully ionizes the compound and provides a counter-ion to keep it solubilized. For a basic compound of pK_a 9.5, HCl could be added to adjust the pH to <7.5 and provide the counter-ion.

5. Co-solvents: ethanol, propylene glycol, PEG400. Function: increase solubility of the compound.

6. a, c, d

7. Tween 80/Methocel. Tween 80 is a surfactant that wets the surface of the compound particles. Methocel helps keep the particles in suspension so that they do not settle.

8. Pure oil (100% oil dissolves a highly lipophilic compound), emulsion (a compound is dissolved in oil and dispersed as droplets into aqueous buffer with the aid of surfactant or emulsifier), micelle (spherical monolayer with hydrophilic shell and lipophilic core with which the lipophilic compound associates), liposome (spherical bilayer in which hydrophilic compounds are held in the central core or lipophilic compounds are associated with the lipophilic portion of the bilayer).

9. The hydrophilic outer shell interacts with water and the lipophilic compound is held in the central core, which is more lipophilic.

10. Small compound particle size, amorphous particles are more soluble than crystals, the particles are released directly into an aqueous environment.

11. Increases surface area for faster dissolution.

12. Eliminates the effects of solubility and solid-state properties.

13. Provides maximal exposure to study the pharmacological effects rather than being confused by delivery effects.

14. c, d, e

Appendix II

General References

1. Avdeef, A. (2003). *Absorption and drug development: solubility, permeability, and charge state*. New York: John Wiley & Sons.

2. Berg, J. M., Tymoczko, J. L., & Stryer, L. (2002). *Biochemistry*. New York: WH Freeman and Company, New York.

3. Birkett, D. J. (2002). *Pharmacokinetics made easy*. North Ryde, Australia: McGraw-Hill Australia.

4. Borchardt, R. T., Kerns, E. H., Lipinski, C. A., Thakker, D. R., & Wang, B. (Eds.). (2004). *Pharmaceutical profiling in drug discovery for lead selection*. Arlington, Virginia: AAPS Press.

5. Borchardt, R. T., Kerns, E. H., Hageman, M. J., Thakker, H. R., & Stevens, J. L. (Eds.). (2006). *Optimizing the "drug-like" properties of leads in drug discovery*. Arlington, Virginia: AAPS Press and Springer.

6. Kwon, Y. (2001). *Handbook of essential pharmacokinetics, pharmacodynamics and drug metabolism for industrial scientists*. New York: Springer.

7. Martin, A. (1993) *Physical pharmacy*. Philadelphia: Lea & Febiger.

8. Rodrigues, A. D. (Ed). (2002). *Drug-drug interactions*. New York: Marcel Dekker.

9. Silverman, R. B. (2004) *The organic chemistry of drug design and drug action* (2nd ed.). San Diego: Elsevier Academic Press.

10. Smith, D. A., Van de Waterbeemd, H., & Walker, D. K. (2001). *Pharmacokinetics and metabolism in drug design*. Weinheim, Germany: Wiley-VCH.

11. Testa, B., van de Waterbeemd, H., Folkers, G., & Guy, R. (Eds.). (2002). *Pharmacokinetic optimization in drug research, biological, physicochemical and computational strategies*. Postfach, Switzerland: Verlag Helvetica Chimica Acta.

12. Testa, B., Krämer, S. D., Wunderli-Allenspach, H., & Folkers, G. (Eds.). (2006). *Pharmacokinetic profiling in drug research*. Zurich, Switzerland: Verlag Helvetica Chimica Acta.

13. Testa, B., & van de Waterbeemd, H. (Eds.). (2007). *Comprehensive medicinal chemistry II, vol. 5, ADME-Tox approaches*. Amsterdam: Elsevier.

14. Van de Waterbeemd, H., Lnnernäs, H., & Artursson, P. (Eds.). (2003). *Drug bioavailability*. Weinheim, Germany: Wiley-VCH.

15. Wermuth, C. G. (Ed.). (2003). *The practice of medicinal chemistry*. San Diego: Elsevier-Academic Press.

16. Yan, Z., Caldwell, & Gary W. (2004). *Optimization in drug discovery; in vitro methods*. Totowa: Humana Press.

Appendix III

Glossary

Term	Acronym	Definition
absorptive direction		having the tendency to absorb; moving in the direction of absorption or uptake (e.g., from intestinal lumen into bloodstream, or from apical to basolateral faces of cells)
absorption		process of absorbing
acidity		level of hydrogen ion concentration in solution as measured on the negative logarithmic pH scale (1 is highly acidic [10^{-1} M] and 14 [10^{-14} M] is highly basic or alkaline); the more readily a compound gives up a proton, the more acidic it is
action potential		change in voltage across a cell membrane with the opening and closing of ion channels (e.g., in a muscle or nerve cell)
active transport		facilitated movement of a molecule across a cellular membrane from a point of lower concentration to a point of higher concentration; requires energy
activity		capacity to produce physiological or chemical effects by the binding of the molecule to a biological macromolecule
acute		occurring with rapid onset, in a short time, after a single dose, or lasting a short time
ADME	ADME	absorption, distribution, metabolism, and excretion
ADME/Tox	ADME/Tox	absorption, distribution, metabolism, excretion, and toxicity
administer		dose a compound
advance		select a compound to move forward in the discovery and development pipeline
albumin (serum)		major plasma protein (60% of total); involved in regulating osmotic pressure and transporting large organic anions (e.g., fatty acids, bilirubin, many drugs, hormones)

Term	Acronym	Definition
Ames assay		procedure for testing a compound's ability to cause DNA mutation, which may cause cancer
amorphous solid		solid in which atoms or ions are not arranged in a definite crystal structure
amphipathic		molecules that have both hydrophilic and hydrophobic parts
amphiphilic		molecule that combines hydrophilic and lipophilic properties
amylase		digestive enzyme that catalyzes the hydrolysis of starch (carbohydrate)
analog		structural derivative of a parent compound whose chemical and biological properties may be quite different
anion		negatively charged ion (e.g., chloride)
antedrug		locally active synthetic derivative of a drug that undergoes biotransformation to a readily excretable inactive form upon entry into the systemic circulation, thus minimizing systemic side effects; also called a *soft drug*
apical	A	initial face of a cell encountered by compound moving in the absorptive direction (e.g., top side of Caco-2 cells)
area under the curve	AUC	pharmacokinetic term for the integrated area under a plot of compound plasma concentration (y-axis, ordinate) vs time (x-axis, abscissa) from a dosed test animal; used to evaluate total exposure to the compound over time
barrier		obstacle to the access of molecules to the therapeutic target which results in reduced compound concentration at the target or delayed access (e.g., membrane, metabolism, pH)
basicity		level of hydrogen ion concentration in solution as measured on the negative logarithmic pH scale (1 is highly acidic [10^{-1} M] and 14 [10^{-14} M] is highly basic or alkaline); the more readily a compound accepts a proton, the more basic it is
basolateral	B	exiting face of a cell for molecules moving in the absorptive direction (e.g., bottom side of Caco-2 cells)
bilayer membrane		dual layer of phospholipids having the nopolar side chains oriented inward and polar head groups toward the aqueous solutions
bile canaliculus		capillaries between hepatocytes for collection of bile and solutes (e.g., bile salts, drugs, metabolites), which merge to form bile ductules, which merge into the bile duct
bile salts		steroid acids that form micelles in the intestinal lumen and assist solubilization of lipophilic compounds (e.g., lipids, fatty acids, lipophilic drugs) by emulsification

Term	Acronym	Definition
biliary excretion		elimination of compound via bile
binding		energetic association of compound with a macromolecule (e.g., therapeutic target protein)
bioactivity		degree of response produced when a compound is administered to a living system, living tissue, or biochemical assay
bioassay		procedure for testing if a compound produces a biochemical or biological response compared to a standard
bioavailability	F	fraction of administered compound that is detected in systemic circulation after administration; losses result from lack of absorption into the systemic circulation and/or metabolic clearance; oral bioavailability is associated with oral administration
bioequivalence	BE	degree to which one compound formulation acts in the body with the same strength and bioavailability as a standard formulation
bioequivalent		formulations are bioequivalent if the nature and extent of therapeutic and toxic effects are equal following the administration of equal doses
bioisosteric		compound resulting from exchange of a moiety with another in order to have similar biological properties to the parent compound, but modified properties
biomarker		endogenous biochemical compound whose concentration indicates the progress of a disease or the effects of treatment
Biopharmaceutics Classification System	BCS	method of evaluating compound solubility, permeability, and dose for the purpose of granting waivers by regulatory agencies for bioequivalence and bioavailability studies
biotransformation		chemical alteration of a molecule by enzymes
blockbuster		drug product with sales greater than $1 billion (U.S.) per year
blocking (ion channel)		partial or complete obstruction of an ion channel so that ions cannot pass at a normal rate
blood flow to organ	Q	flow rate of blood to an organ (e.g., liver)
blood–brain barrier	BBB	endothelial cell layer of the brain microvessels between blood and brain; restricts passage of some compounds into brain tissue, depending on the compound structure
blood–cerebrospinal fluid barrier	BCSFB	barrier located at the tight junctions that surround and connect the cuboidal epithelial cells on the surface of the choroid plexus
bound drug	C_{bound}, C_B	compound molecules bound to a protein or lipid
brain microvessel endothelial cells	BMEC	endothelial cells removed from microcapillary blood vessels of brain and cultured for BBB permeability studies

Term	Acronym	Definition
brain/plasma ratio	B/P	ratio of compound in brain to plasma, which is variously calculated using AUC, C_{max}, or concentration at a time point
breast cancer resistance protein	BCRP	membrane efflux transporter family
brush border		specialization of the free surface of a cell to have microvilli that increase the surface area (e.g., on epithelial cells lining the intestine)
buccal		inner lining of cheeks or lips
buffer		solution of ionic compound(s) that resists changes in pH
Caco-2	Caco-2	human colon carcinoma cell line for which one use is intestinal permeability studies
candidate		compound considered for approval of entry into drug development; compound undergoing studies from preclinical through phase I–IV clinical trials until NDA approval
cannulation		insertion of a tube into a body cavity, duct, or vessel for the drainage of fluid or administration of medication
capillary electrophoresis	CE	separation technique using a capillary tube filled with buffer and forming a circuit across a high voltage
carcinogenic		compound that causes cancer
cassette dosing		co-administration of multiple compounds for simultaneous assessment of pharmacokinetic parameters using a single test animal
cation		positively charged ion
central nervous system	CNS	brain and spinal cord
cerebrospinal fluid	CSF	fluid that is continuously produced by the brain's choroid plexus, flows in the ventricles and around the surface of the brain and spinal cord, and is absorbed into the venous system; it absorbs shock and maintains constant pressure
chiral		having asymmetric centers that are not superimposable
chiral center		tetrahedral carbon atom having four different attached groups
chromosome		assembly in the cell nucleus consisting of a single long thread of DNA (containing many genes) and associated proteins, which tightens up into a defined structure for cell division; occurs in pairs with one from the father and one from the mother
chronic		occurring over a long duration, frequently reoccurring, being long lasting, or continual regular dosing or exposure
classification		categorization of things into classes or categories of the same type
clearance	Cl	volume of blood from which the drug is completely removed per unit time by the various elimination processes; amount eliminated is proportional to the concentration of the drug in the blood
cLogP	cLogP	Log P calculated using structure

Term	Acronym	Definition
CNS−	CNS−	compound does not penetrate appreciably into brain tissue after dosing, as determined by absence of a measurable concentration or lack of the expected pharmacological response
CNS+	CNS+	compound penetrates appreciably into brain tissue after dosing as determined by a measurable concentration or positive pharmacological response
co-administer		dose at the same time
cocktail		mixture of compounds used for in vivo PK cassette dosing or for in vitro assays that simultaneously assess multiple compounds or properties (e.g., cocktail CYP450 inhibition assay of multiple isozymes)
cost/benefit ratio		resources required to obtain a certain measurement vs the benefit that data have for a project team
counter-ion		ion with an opposite charge to that of another ion in the solution or salt
crystal form		geometric configuration in which a compound forms a crystal (often two or more polymorphs of a compound have different crystal energies)
cytochrome	CYP	enzyme family containing a heme porphyrin to which an iron atom is attached; important in cell respiration as catalysts of oxidation–reduction reactions
cytochrome P450	CYP450	family of cytochrome isozymes that absorb light at 450 nm and oxidize compounds in many tissues; found in high abundance in the liver
cytotoxicity		degree of a compound's ability to damage or kill cells
Dalton	Da	unit of mass used for molecules, equal to one twelfth of the atomic mass of ^{12}C
degradation product		chemical product of an undesired reaction of a compound owing to its environment (e.g., oxidation of a compound caused by air O_2)
delta Log P	ΔLog P	difference in Log P values for partitioning between aqueous and organic solvents [e.g., Log $P_{(octanol-water)}$ minus Log $P_{(cyclohexane-water)}$]
development		studies following discovery that takes a compound with desired biological effects in animal models and prepares and tests it as a drug product that can be used in humans; includes formulation, stability, chemical process, human pharmacokinetics, toxicity, and clinical efficacy

Term	Acronym	Definition
dialysis		separation of smaller molecules (e.g., drugs) from larger molecules (e.g., proteins) in solution by selective diffusion through a semipermeable membrane
discovery		research that finds compounds with desirable biological effects in animal models, which have potential to become new drugs in humans
disposition		what happens to a compound after it is administered to an organism; fate
dissolution		dissolving in a solution
distribution	D	movement of compound molecules into tissues of an organism
DMSO	DMSO	dimethylsulfoxide
dosage form		physical combination of a compound (e.g., drug) with additives (e.g., excipients, encapsulation) for administration to the test animal or human (e.g., tablet, lotion, solution)
dosing regimen		systematic plan of dosing (i.e., dosage, route, frequency)
dosing solution		solution of compound for administration to a test animal which may include additives to solubilize the compound
drug product		dosage form that is dispensed to patients
drug substance		pharmacologically active component of a drug product
drug–drug interaction	DDI	interference of molecules of one drug with the normal disposition of molecules of another drug when they are co-administered
drug-like		having properties that are consistent with most commercial drugs and lead to acceptable pharmacokinetics and toxicity in humans
DTT	DTT	dithiothreitol
duodenum		first segment of the small intestine following the stomach
EC_{50}	EC_{50}	median effective concentration (concentration that induces a 50% effect in a functional in vitro assay)
ED_{50}	ED_{50}	median efficacious dose (dose that produces the desired effect in 50% of population in an in vivo assay)
efficacy		ability to produce the desired pharmacological effect (control or cure) on the test animal or human
efflux		transport of a molecule out of a cell by a transporter with the expenditure of energy
efflux ratio	ER	permeability in the secretory direction (basolateral to apical) divided by permeability in the absorptive direction (apical to basolateral) using an in vitro cell layer assay (e.g., Caco-2)
electrocardiogram	ECG	graphic record from measurement of voltage changes on the surface of the heart resulting from conductance of action potentials across the muscle

Term	Acronym	Definition
elimination		disappearance of a dosed compound from a living organism, usually by metabolic biotransformation and/or excretion by intestine, kidneys, lungs, skin, or any other bodily fluid
elimination rate constant	k	first-order kinetics indicating the rate of compound elimination; used in pharmacokinetics studies
endogenous		compound that naturally occurs within the organism
endothelial		layer of epithelial cells lining heart cavities, blood vessels, and serum cavities of the body
enterohepatic circulation		recurring movement of molecules from intestine to bloodstream (via absorption), then to the liver, intestine (via biliary excretion), then back into the bloodstream (via absorption)
enzyme		protein that catalyzes a specific biochemical reaction
epithelial		layer of cells that line the inner and outer surfaces of organs, vessels, and cavities
equilibrium dialysis		performing dialysis until equilibrium is established across the membrane
equilibrium solubility		solubility assay in which buffer is added to solid compound and the solution is stirred for an extended time (e.g., 72 hours) until equilibrium is established between the solution and the excess solid
esophagus		tube that leads from the throat to the stomach
excipients		substances added to drug to produce a dosing vehicle or drug product to enhance solubility, dissolution, stability, taste, consistency, or other properties
excretion	E	removal of dosed compound molecules or metabolites from the body, usually via the urine or feces
exposure		concentration and/or duration of compound molecules in the body, tissue, or in vitro assay that can interact with the therapeutic target
extracellular fluid	ECF	body fluid excluding that in cells; includes plasma and fluid between cells (interstitial fluid)
extraction ratio	E	fraction of compound in blood that is removed with each pass through the organ of elimination (i.e., kidney, liver)
fasted state		condition (e.g., pH, bile salts) of fluids in gastrointestinal tract when no ingestion of food has occurred for several hours
First-pass metabolism (or effect)		metabolism occurring to compound molecules, mostly in intestine and liver, prior to reaching systemic circulation
FLIPR	FLIPR	fluorometric imaging plate reader method; uses fluorescent reagents for assay end point/quantitation
fluorogenic		production of a fluorescent product from a nonfluorescent starting substance for quantitation in an assay
flux		rate of flow of molecules across a membrane

Term	Acronym	Definition
Food and Drug Administration	FDA	U.S. Food and Drug Administration, regulatory agency that approves and monitors commercial drugs (and foods) in the United States
formulation		mixture of excipients (vehicle) and compound solid to make a dosage form
fragment		molecule with lower MW, fewer H-bonds, and more moderate Log P than typical drugs; for use in screening for ligands to a therapeutic target using NMR or x-ray crystallography
free drug	$C_{unbound}$, C_U	concentration of compound in blood that is unbound to plasma proteins (e.g., albumin) or lipids, or in tissues that is unbound to proteins or lipids
gallbladder		organ in which bile is stored prior to release into duodenum during stomach emptying
gastric		associated with the stomach
gastrointestinal	GI	system consisting of stomach, small intestine, and large intestine
genomics		study of the complement of genes and other genetic material present in the cell, animal, or human
genotoxicity		degree of a compound's ability to cause DNA or chromosomal damage
glomerular filtration rate	GFR	flow rate of fluid passing by filtration from the glomerulus into the Bowman's capsule of all the nephrons in the kidney
glomerulus		cluster of blood capillaries surrounded by the Bowman's capsule in the nephron of the kidney
glucuronide		product of reaction of glucuronic acid with a hydroxyl (e.g., phenol, carboxylic acid, alcohol) or amine, catalyzed by UDP-glucuronosyltransferase
glutathione		endogenous compound that reacts with reactive metabolites of compounds for detoxification
G-protein–coupled receptor	GPCR	transmembrane protein family that binds a signaling compound on the extracellular side, which induces a change on the intracellular side, initiating biochemical reactions within the cell
half-life	$t_{1/2}$	time for half the quantity of a compound in a living organism to be metabolized or eliminated by normal biological processes
heme group		protoporphyrin ring with a central iron atom
hepatic		pertaining to the liver
hepatic portal vein	HPV	blood vessel that carries blood to the liver from the stomach and intestine
hepatocytes		liver cells in which a large portion of metabolism of xenobiotic compounds occurs; many other endogenous biochemical reactions also occur here

Term	Acronym	Definition
high-performance liquid chromatography	HPLC	technique for mixture separation, compound identification or quantitation in which compounds partition between stationary and mobile phases within a tube
hERG	hERG	human ether-a-go-go-related gene that encodes a subunit of the potassium channel, which contributes to cardiac repolarization
high throughput	HT	techniques that allow for fast testing of a large number of compounds
high-throughput chemistry		synthesizing compounds at a high rate compared to conventional one-at-a-time compound synthesis; sometimes called *combichem*
high-throughput screening	HTS	performing assays at a high rate using large compound libraries
hit		compound that is active in HTS or in initial screens following virtual screening
hit selection		process of choosing the most favorable hits for further study during the hit-to-lead stage of drug discovery
hit-to-lead		time period in which a large number of hits are studied and structurally modified to select a few leads for the optimization phase
HLM	HLM	human liver microsomes
human serum albumin		a plasma protein; important in maintaining fluid balance in blood, maintaining blood pressure, regulating fatty acids, and transporting hormones
hydrogen bond		bond that exists between an electronegative atom (e.g., O, N, F) and a hydrogen atom bonded to another electronegative atom
hydrogen-bond acceptor	HBA	electronegative atom (e.g., O, N, F) that may accept a hydrogen bond
hydrogen-bond donor	HBD	hydrogen atom attached to a relatively electronegative atom that may form a hydrogen bond
hydrolysis		chemical reaction in which a compound reacts with water
hydrophilicity		tendency of a molecule to be solvated by water
IC_{50}	IC_{50}	median inhibition concentration (concentration that reduces the effect by 50%)
idiosyncratic toxicity		unpredicted toxicity observed occasionally in large populations and is theorized to be triggered by reaction of a drug metabolite with a protein
ileum		final section of the small intestine
immobilized artificial membrane	IAM	HPLC stationary phase in which phospholipid is bonded to the stationary phase and used to predict membrane permeation
immunotoxicity		toxicity caused by a mechanism of the immune system following compound administration (e.g., immune reaction resulting from reaction of a drug metabolite with a protein)
in silico		performed using a computer and specially developed software
in situ		performed in the natural position (e.g., in the living organism)

Term	Acronym	Definition
in vitro		performed in a laboratory vessel (e.g., test tube, well of a titer plate) outside the living system
in vivo		performed in a living organism
induction		increase in enzyme concentration, caused by dosing a compound that triggers a nuclear receptor, leading to production of mRNA and synthesis of more copies of the enzyme
influx		transport of a molecule into a cell by a transporter with the expenditure of energy (i.e., uptake)
inhibitor		compound that binds to an enzyme and prevents the normal enzyme-substrate binding and subsequent catalytic reaction
initial concentration	C_0	initial concentration in blood following intravenous compound administration
insoluble		having a solubility that is very low
integrity		condition in which the analytical data for a compound are consistent with the putative structure
intestinal epithelium		monolayer of cells that form the inner surface of the gastrointestinal lumen
intramuscular	IM	administered by needle into the muscle
intraperitoneal	IP	administered within the peritoneum (lining of intestinal cavity)
intravenous	IV	administered directly into the bloodstream via a vein as a bolus injection or infusion
intrinsic solubility		solubility of the neutral form of the compound
inverted vesicle		vesicle that is turned inside out
investigational new drug application	IND	application to FDA to begin phase I trials
ion channel		transmembrane protein complex that serves as a gate for the entrance or exit of ions from the cell
ionic strength		measure of the average electrostatic interactions among ions in an electrolyte
irreversible inhibition		covalent or coordination binding of an inhibitor to a protein rendering it permanently inactive
irreversible inhibitor		compound that irreversibly modifies and deactivates an enzyme (covalent or coordination); often contains reactive functional groups (e.g., nitrogen mustards, aldehydes, haloalkanes, or alkenes)
isosteric		similar electronic arrangements in chemical compounds
isozyme		member of a family of enzymes that catalyze the same type of reaction but differ from each other in primary structure and/or electrophoretic mobility
jejunum		longest part of the small intestine extending from the duodenum to the ileum
K_i	K_i	dissociation constant of an inhibitor (enzyme kinetics)
kinetic solubility		solubility assay in which a small volume of organic solvent (e.g., DMSO) containing concentrated test compound is added to aqueous buffer, followed by measurement of the concentration after an established incubation period

Term	Acronym	Definition
knockout		when an protein's activity (e.g., efflux transport) is eliminated by deleting the gene that codes for it (genetic knockout) or co-administering an inhibitor (chemical knockout)
knowledge-based expert system		experts decide on rules for classifying a substructure as conferring a likelihood of having certain property behavior and for constructing software that evaluates new compounds based on these rules (e.g., nitroso confers likelihood of mutagenicity)
labeling		required information included with prescription drugs that describes vital information for patients and physicians, such as indications, precautions, warnings, and side effects
LD_{50}	LD_{50}	median lethal dose (i.e., lethal to 50% of the population of test animals in a prescribed time)
lead		compound that is currently most favorable in a discovery project and serves as a template for the design of analogs during lead optimization
lead optimization		time period in which a few leads are structurally modified and assayed to study the structure–activity and structure–property relationships in order to obtain the optimum structure as a clinical candidate
lead-like		having properties that are consistent with leads that will undergo structural augmentation during lead optimization to result in a clinical candidate that stays within drug-like property space
library		assembly of compounds, often analogs made by parallel synthesis (compound library) or assembled from multiple sources for HTS (screening library)
ligand		molecule that binds to another chemical entity to form a larger complex (e.g., substrate to enzyme)
lipase		member of a family of enzymes that catalyze the hydrolysis of fats (e.g., monoglycerides, triglycerides) to fatty acids and glycerol
Lipinski rules	Rule of 5	set of structural characteristic guidelines for drug-like structures
lipophilicity		affinity of a molecule or a moiety for a lipid (nonpolar) environment
loading dose		initial high dose of a compound, given to bring the compound in the body to the steady-state amount
Log BB	Log BB	$\log_{10}$ value of brain to plasma concentration ratio; a blood–brain partition coefficient
Log D	Log D	$\log_{10}$ of the distribution coefficient of the equilibrium concentrations of all species (unionized and ionized) of a molecule in octanol to the same species in the water phase under given solution conditions; differs from Log P in that ionized species are considered as well as the neutral form of the molecule

Term	Acronym	Definition
Log P	Log P	$\log_{10}$ of the partition coefficient; measure of differential solubility of a compound in two solvents; most commonly partitioning is between 1-octanol and water, a measure of the hydrophobicity or hydrophilicity of a substance
LQT	LQT	lengthening of the QT interval on an electrocardiogram
lumen		inner open space of a tubular organ (i.e., of a blood vessel or the intestine)
luminogenic		produces a luminescent product for quantitation from a nonluminescent starting substance
lyse		break the cell membrane, destroy, or disorganize cells using chemicals, enzymes, or viruses
maximum concentration	C_{max}	maximum plasma concentration of compound reached after administration in vivo
maximum tolerated dose	MTD	maximum daily (chronic) dose that an animal species can tolerate for a major portion of its lifetime without significant impairment or toxic effect other than carcinogenicity; can be determined by extrapolating a 90 day study
MDCK	MDCK	Madin Darby Canine Kidney cell line
MDR1-MDCKII	MDR1-MDCKII	Madin Darby Canine Kidney cell line transfected with human MDR1 gene, which codes for Pgp
mechanism-based inhibition	MBI	irreversible inhibition owing to formation of a covalent or quasi-irreversible bond between the inhibitor or inhibitor metabolite and the enzyme (e.g., CYP450) that inactivates the enzyme
metabolic phenotyping		determination of which metabolic enzymes and isozymes metabolize a particular compound
metabolic switching		if the primary scheme of metabolism (e.g., isozyme, site of metabolism) is blocked by DDI or chemical modification of the structure, other routes of metabolism can increase
metabolism		enzymatic structure modification of a compound in an organism
metabonomics		study of the identities and concentrations of endogenous small molecules present in the cell or organism (i.e., metabolite pool) during normal life and following a stimulus (e.g., compound dose)
metastable crystal		crystal form that is not the most thermodynamically stable
micelle		aggregate of amphipathic molecules in water with the nonpolar portions in the interior and the polar portions on the exterior exposed to water
microdialysis		technique in which a dialysis membrane capillary or probe is implanted into a tissue or fluid compartment, through which compounds in the extracellular fluid are collected for analysis
micronucleus		vesicle smaller than nucleus that contains chromosomal material that was cleaved from a chromosome

Term	Acronym	Definition
microsome		vesicles prepared from tissue (e.g., liver) by homogenization and differential centrifugation, which contain enzymes and ribosomes attached to the endoplasmic reticulum
microtiter plates		flat plate with multiple "wells" (4–3,456) used as small test tubes; has become a standard tool in biomedical research and clinical diagnostic testing laboratories; as called a *microplate*
microvessels		capillary blood vessels (e.g., in brain)
molecular weight	MW	sum of the atomic weights of all the atoms in a molecule
mRNA	mRNA	messenger RNA that is transcribed from DNA by RNA polymerase and translated into a protein sequence on the ribosome
multidrug resistance protein	MRP	family of efflux transporters whose existence was first indicated as a mechanism for resistance to drug therapy, such as in cancer
mutation		permanent change (i.e., structural alteration) in DNA or RNA
NADPH	NADPH	reduced form of nicotinamide adenine dinucleotide phosphate (NADP), a coenzyme involved in numerous enzymatic reactions (e.g., metabolism by CYP450), in which it serves as an electron carrier by being alternately oxidized ($NADP^+$) and reduced (NADPH)
nanoparticles		microscopic particle with at least one dimension <100 nm
natural products		compounds produced by a living organism (e.g., plants, microorganisms) that have pharmacological or biological activity and may be used as drugs or in drug design
nephelometry		technique used to measure the size and concentration of particles in a liquid by analysis of light scattered by the liquid (e.g., for solubility assays)
nephron		unit of the kidney that removes waste materials (e.g., drug, metabolites) from the blood and excretes them via the urine
new drug application	NDA	new drug application to FDA; to market a new drug
NMR	NMR	nuclear magnetic resonance spectroscopy; used for structural studies
no observable adverse effect level	NOAEL	highest dose or exposure that causes no detectable *adverse* effect (no adverse alterations compared to control organisms of morphology, functional capacity, growth, development, or life span) in test animals, in which higher doses or concentrations resulted in an adverse effect; any nonadverse affects are manageable

Term	Acronym	Definition
no observable effect level	NOEL	highest dose or exposure that causes no detectable effect (no alterations compared to control organisms of morphology, functional capacity, growth, development, or life span) in test animals, in which higher doses or concentrations resulted in an effect
nonspecific binding		binding of compound to lipid or protein in tissue other than at the active site
NSAID	NSAID	nonsteroidal antiinflammatory drug (e.g., aspirin, ibuprofen)
off-target		interaction of compound with a biochemical material (e.g., receptor) other than the intended therapeutic target
oocyte		developing female gamete before completion and fertilization
optimization		process of synthesizing chemical analogs of a lead compound with the goal of creating compounds with improved pharmacological properties
oral	PO	dosing by mouth
oxidative stress		increased oxidant production in animal cells characterized by the release of free radicals and peroxides resulting in cellular degeneration
PAMPA-BBB	PAMPA-BBB	variation of PAMPA permeability technique to predict passive diffusion through the blood–brain barrier
paracellular		movement of molecules across a cell membrane via pores between the epithelial cells
parallel artificial membrane permeability assay	PAMPA	in vitro permeability method using phospholipid dissolved in organic solvent and placed in the pores of a filter membrane held between two aqueous compartments
parallel synthesis		reaction of an intermediate with a number of different reagents to produce a library of analogs
passive diffusion		movement of molecules across a cell membrane from the region of higher concentration to the region of lower concentration without the expenditure of energy; different compounds have different rates of passive diffusion owing to the selective permeability of the membrane
patch clamp		in vitro assay in which an electrical circuit is established across a cell membrane to study the current changes produced by movement of ions through ion channels
perfusion		bathing a vessel (e.g., intestine, brain blood vessels) with a solution; often used to study the permeation of compound through the vessel wall
peripheral tissue		tissue situated away from the central tissue being considered; on the outer part of an organ or body
permeability	P_e, P_{app}	ability of a compound to penetrate or pass through a membrane; velocity of flow

Term	Acronym	Definition
permeability surface area coefficient	PS	Permeability times surface area of brain capillary endothelium, which is approximately 100 cm^2/g of brain (see Chapter 28, Tanaka and Mizojiri[50])
perpetrator		enzyme inhibitor or inducer that alters the pharmacokinetics of co-administered drugs
Pgp	Pgp	P-glycoprotein, an ABC family transporter of the MDR subfamily; extensively expressed in some normal cells (e.g., intestinal epithelium, liver, renal proximal tubule, BBB endothelial cells) also called ABCB1, MDR1, and PGY1
pH	pH	negative log$_{10}$ of the hydrogen ion concentration; measure of the acidity or alkalinity of a solution; neutral solutions are pH 7, alkaline solutions have pH >7, and acidic solutions have pH <7
pharmaceutics		science of preparing and dispensing drugs; includes formulation, stability, and salt form selection
pharmacodynamics	PD	study of the biochemical and physiological effects, duration, mechanisms of action, and concentration effects of a compound on an organism; what a drug does to the body
pharmacokinetics	PK	study of the concentration–time course fate of a compound and its metabolites in an organism, which is affected by absorption, distribution, metabolism, and excretion (ADME); what the body does to the drug
pharmacology		study of the action of compounds on organisms, including the biochemical interactions, biological effects, and applications in treating disease; includes drug composition and properties, interactions, toxicology, PK, PD, therapy, and medical applications
pharmacophore		structure (ensemble of steric and electronic features) that effectively binds to the specific therapeutic target to produce the desired effect by triggering or blocking its biological response; serves as a template for the synthesis of analogs during optimization
Phase 0 clinical trials		human clinical studies in which single doses (usually subtherapeutic) of an investigational drug are administered to a small number of human volunteers (10–15) to obtain initial data on the compound's PK, PD, and mechanism of action; assesses whether the compound behaves in human subjects as predicted by preclinical studies; reduces time and cost for decisions on continued development

Term	Acronym	Definition
Phase I development (clinical trials)		human clinical studies in which an investigational drug is administered to a small group of healthy volunteers (20–80) who do not have the subject medical condition; assesses the human safety, tolerability, and PK
Phase I metabolism		enzymatic modifications of the molecular structure of the compound (e.g., oxidation, dealkylation); polar groups are either introduced or unmasked, resulting in more polar metabolites; can lead either to activation or inactivation of the drug; produces sites to which polar molecules are more readily conjugated (see *phase II metabolism*)
Phase II development (clinical trials)		human clinical studies in which an investigational drug is administered to a larger group of human volunteers (20–300) who have the subject medical condition; assesses the human efficacy and continues the safety assessments in a larger group
Phase II metabolism		enzymatic addition (conjugation) of a polar moiety to the compound's structure (e.g., glucuronic acid, sulfate); the metabolite has increased polarity
Phase III development (clinical trials)		human clinical trials in which an investigational drug is administered to a large group of volunteer patients (300–3,000 or more) who have the subject medical condition as randomized controlled trials in multiple clinics; provides a definitive assessment of human efficacy in comparison with the current best drug treatment for the condition ("gold standard") that can be extrapolated to the general population; provides information for product labeling
phospholipid		lipid containing phosphorus, including those with glycerol or sphingosine backbones; the primary lipids in cell membranes
physicochemical		physical and chemical properties of a compound
pinocytosis		type of endocytosis in which molecules are taken up from outside the cell through invagination of the cell membrane to form vesicles
pK_a	pK_a	acid dissociation constant; equilibrium constant for the dissociation of a weak acid (or negative logarithm of the ionization constant K of an acid); pH of a solution in which half of the acid molecules are ionized
plasma		liquid component of blood in which blood cells are suspended
plasma clearance		clearance of compound from the blood (plasma)
plasma protein		proteins found in blood plasma; various proteins serve different functions, including circulatory transport (for lipids, hormones, vitamins, metals), enzymes, and complement components
plasma protein binding	PPB	degree to which a compound binds to the proteins in blood plasma

Term	Acronym	Definition
polar surface area	PSA	surface sum over all polar atoms (e.g., oxygen, nitrogen) and attached hydrogens
polymorph		solid that shares chemical composition with another material but is composed of a different crystal lattice or form
potentiometric titration	pH-metric	technique using two electrodes (neutral and standard reference); voltage across the compound in solution is measured as titrant is added; graph of voltage vs volume of added titrant indicates the end point of the reaction as halfway between the increase in voltage
preclinical		research conducted prior to clinical studies
predevelopment		development activities conducted prior to clinical studies, especially just before phase I
primary cells		cultured cells derived directly from living tissue; typically lack the ability to remain viable for many further passages; certain genes may be up-regulated or down-regulated compared to tissue from which they were obtained
prodrug		compound with low activity that, once administered, is metabolized in vivo into an active compound
proof of concept	POC	short experiment whose purpose is to verify a theory
property		physical, chemical, metabolic, or biochemical characteristic of a compound that affects compound pharmacokinetics, exposure to the therapeutic target, or toxicity
property-based design		structure design for the purpose of improving physicochemical, pharmacokinetic, and toxicity properties
protease		any enzyme that catalyzes proteolysis by hydrolysis of peptide bonds linking amino acids together in a polypeptide chain
proteomics		study of the identities, concentrations, and locations of endogenous proteins present in the cell or organism during normal life or following a stimulus (e.g., compound exposure)
QT prolongation		lengthening of the QT interval on an electrocardiogram (see *LQT*)
quantitative structure–activity relationships	QSAR	process by which chemical structure is quantitatively correlated with biological activity
quantitative structure–property relationships	QSPR	process by which chemical structure is quantitatively correlated with properties
racemate		mixture of equal amounts of left-handed and right-handed enantiomers of a chiral molecule

Term	Acronym	Definition
reactive metabolite		compound metabolite that reacts covalently with an endogenous macromolecule (e.g., protein, DNA) to form a stable conjugate; usually disrupts normal function
receptor		protein in the cell membrane, cytoplasm, or nucleus to which a specific ligand (e.g., neurotransmitter, hormone) binding initiates a cellular biochemical response
recombinant human cytochrome proteins	rhCYP	human CYP450 isozymes prepared by recombinant DNA technology
renal		pertaining to the kidney
repolarization		change in membrane potential that returns it to a negative value (after depolarization of an action potential changed it to a positive value) by movement of potassium ions out of the cell through ion channels
reversed-phase HPLC	RP HPLC	high-performance liquid chromatography in which the solid chromatographic particles are derivatized to form a nonpolar stationary phase (e.g., bonding of octadecylsilane on silica or polymer particle surface) and an aqueous/organic mixed mobile phase is used
reversible inhibition		binding to enzyme with noncovalent interaction
reversible inhibitor		compound that binds reversibly to an enzyme with noncovalent interactions (e.g., hydrogen bonds, hydrophobic interactions, ionic bonds) to produce multiple weak bonds between the inhibitor and the active site to provide strong and specific binding; inhibits normal ligand–enzyme binding
rotatable bond	RB	single non-ring bond, on a nonterminal heavy atom (not hydrogen)
S9	S9	material prepared from liver tissue by homogenization and differential centrifugation at $9,000g$; contains metabolizing enzymes attached to membranes from the endoplasmic reticulum (microsomes) and cytosol; used for some in vitro metabolism studies
safety		study of toxicity risk
safety window		range between the concentration that is predicted to cause human toxicity and the concentration that is predicted to produce human efficacy
salt form		material in which ionized molecules of a compound are paired with ions of the opposite polarity to enhance solubility or another physical property (e.g., buspirone hydrochloride)
scaffold		central structure (lead) that binds to the therapeutic target and is modified during lead optimization to improve activity, selectivity, and drug-like properties (see *template*)

Term	Acronym	Definition
scale-up		make a larger batch of compound for studies requiring large amounts
secretory		in the direction of elimination in intestine (e.g., from blood into intestinal lumen)
selectivity		degree to which a compound produces the desired result compared to adverse side effects; ratio between the IC_{50} at another target for which binding is possibly deleterious to the IC_{50} at the therapeutic target
selectivity screen		measurement of IC_{50} of a compound in a large number of biochemical assays to check for activity at other biological targets to avoid causing side effects
serum		aqueous fluid of blood, containing dissolved compounds and proteins, from which blood cells and clotting factors have been removed
shake flask		laboratory vessel in which partitioning (e.g., Log P) or equilibrium solubility experiments are performed (i.e., capped vials, sealable tube, separatory funnel)
simulated gastric fluid	SGF	artificial solution whose recipe is prescribed by the United States Pharmacopeia to mimic gastric fluid contents
simulated intestinal bile salts–lecithin mixture	SIBLM	artificial solution to mimic intestinal fluid contents
simulated intestinal fluid	SIF	artificial solution whose recipe is prescribed by the United States Pharmacopeia to mimic intestinal fluid contents
soft drug		see *antedrug*
solid dispersion		dispersing one or more active ingredients in an inert matrix in the solid state in order to achieve increased dissolution rate, sustained release of drugs, altered solid-state properties, enhanced release of drugs from ointment and suppository bases, or improved solubility and stability
solubility	S	measure of how much of a compound will dissolve in a specific liquid; usually measured in weight per unit volume
solubilizer		material added to enhance a compound's solubility
steric hindrance		diminished reactivity of a site in a molecule owing to the presence of adjacent moiety (s)
stock		concentrated chemical solution that is diluted before use
structure alert		substructure in compound has been reported as causing toxicity
structure elucidation		determination of the structure of an unknown compound using analytical techniques (e.g., spectroscopy)

Term	Acronym	Definition
structure–activity relationship	SAR	relationship linking chemical structure and pharmacological activity
structure-based design		structure design for the purpose of enhancing pharmacological activity
structure–property relationship	SPR	relationship linking chemical structure and an ADME/Tox properties
subcutaneous	SC	administered using a needle just under the skin into the subcutaneous tissues; after injection the compound moves into the small blood vessels and bloodstream
sublingual		administered beneath or on the underside of the tongue, where the tablet dissolves and the drug is absorbed through the sublingual gland
substrate		compound on which an enzyme acts
surfactant		material that in small quantity markedly affects the surface characteristics of a system; also called a *surface-active agent*
tandem mass spectrometry	MS/MS	instrument consisting of two mass spectrometers in series connected by a collision chamber; molecular ions are sorted in the first mass spectrometer, broken into fragments (product ions) in the collision chamber, and fragments are sorted in the second mass spectrometer; used for structure elucidation and quantitative analysis
target		endogenous enzyme, receptor, ion channel, or other protein in the body that may be affected by a compound to produce a therapeutic effect
template		core structure that serves as a pattern for the synthesis of other molecules; also called a *scaffold*
teratogenicity		degree of a compound's ability to damage, kill, or morphologically alter a fetus
therapeutic index		ratio between the toxic dose and the therapeutic dose of a drug, used as a measure of the relative safety of the drug for a particular treatment
therapeutic target		see *target*
time of maximum drug concentration	t_{max}	time after dosing to reach the peak concentration
tissue uptake		absorption of compound molecules into a tissue
torsades de pointes	TdP	sudden ventricular tachycardia in which the ECG shows a steady undulation in the QRS axis in runs of 5–20 beats and with progressive changes in direction
toxicity	TOX	extent, quality, or degree of being poisonous
transepithelial electrical resistance	TEER	electrical resistance across a membrane layer in an in vitro permeability experiment
transfect		introduction of DNA into a recipient eukaryote cell and its subsequent integration into the chromosomal DNA
transport		movement of a compound across a membrane barrier

Term	Acronym	Definition
transporter		type of protein that actively transports molecules across a cell membrane that would not otherwise allow such compounds across
turbidimetry		method for determining the concentration of a compound in solution by measuring the loss in intensity of a light beam through a solution that contains suspended particulate matter
turbidity		cloudiness or opacity in a liquid caused by suspended particulate matter
UDP-glucuronosyltransferase	UGT	class of metabolic enzymes that catalyze addition of glucuronic acid to drugs and metabolites, a phase II process that increases their solubility in water and enhances their excretion
ultrafiltration		technique for separation of low-MW compounds from high-MW proteins using centrifugation and a membrane that excludes proteins
unbound "free" drug	$C_{unbound}$, C_U	fraction of compound not bound to carrier proteins in blood or nonspecifically to protein or lipid in tissue
uptake		movement of compound into a cell or tissue
vehicle		additives and solvents used to solubilize a compound for dosing
venous sinusoid		capillary blood vessel (branching from the portal vein) from which drug molecules permeate into hepatocytes
victim drug		compound that is metabolized by enzymes whose activities are inhibited or induced by a co-administered perpetrator drug
virtual screening		selection of compounds that may bind to the target using a computational model; also called *in silico screening*
volume of distribution	V_d	apparent volume in which the compound is dissolved in the living organism; indicates how widely the compound is distributed in the body
withdraw		remove a commercial drug from the market
xenobiotic		compound that is not naturally found in the organism (e.g., drug); also called a *foreign* or *exogenous substance*
zwitterion		dipolar ion containing ionic groups of opposite charge

Index

	Lead	Analog	Desired Profile
MW	330	445	<450
clogP	1.9	5.19	<4.0
IC50 (µM)	4.2	>20	<1.0 µM
Binding to target (STD, FP, Trp-Fl.)	X-ray		Yes (NMR, FP)
MIC			
B. subtilus	>200 µM	50 µM	<200 µM
S. aureus MRSA	>200 µM	25 µM	<200 µM
S. aureus ATCC	>200 µM	200 µM	<200 µM
S. pneumo +	>200 µM	25 µM	<200 µM
Selectivity: *C. albicans* (MIC µg/mL)	>200	>200	>10 fold
Aqueous Solubility (µg/ml @ pH 7.4)	>100	26.5	>60
Permeability (10^{-6} m/s @ pH 7.4)	0	0.15	>1
CYP 3A4 (% inhibition @ 3 µM)	11	7	<15
CYP 2D6 (% inhibition @ 3 µM)	0	1	<15
CYP 2C9 (% inhibition @ 3 µM)	NT	23	<15
Microsome stability (% remaining @ 30 min)	NT	NT	>80
Definable Series	Yes	Yes	Yes
Definable SAR	Yes	Yes	Yes

Plate 3 ▶ Example of goals used by Wyeth Research exploratory medicinal chemists for hit selection, initial structural modification, and lead selection in an acyl carrier protein synthase (AcpS) inhibitor project. (see Figure 20.2 on p. 245)

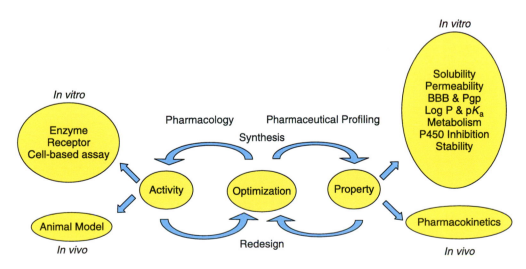

Plate 4 ▶ Iterative parallel optimization by simultaneous assessment of both activity and properties. (Reprinted with permission from [2].) (see Figure 21.3 on p. 251)

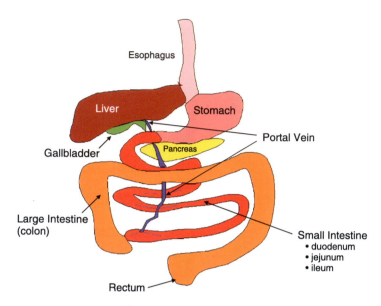

Plate 1 ▶ Diagram of the gastrointestinal tract. (see Figure 3.2 on p. 20)

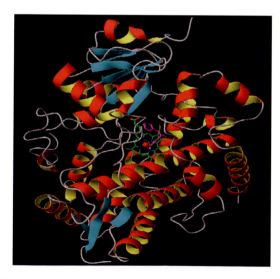

Plate 2 ▶ Structure of human cytochrome P450 3A4 with heme and inhibitor metyrapone. (Drawing courtesy Kristi Fan.) (see Figure 11.3 on p. 140)